DIGITAL TERRESTRIAL TELEVISION BROADCASTING

IEEE Press
445 Hoes Lane
Piscataway, NJ 08854

Technical Reviewers

Xinyi Liu, *Hong Kong Applied Science and Research Institute*
Pablo Angueira, *University of the Basque Country*

DIGITAL TERRESTRIAL TELEVISION BROADCASTING

Technology and System

Edited by

JIAN SONG
ZHIXING YANG
JUN WANG

The ComSoc Guides to Communications Technologies
Nim K. Cheung, ***Series Editor***

IEEE PRESS

WILEY

Published by John Wiley & Sons, Inc., Hoboken, New Jersey.
Published simultaneously in Canada

For general information on our other products and services or for technical support, please contact our Customer Care Department within the United States at (800) 762–2974, outside the United States at (317) 572–3993 or fax (317) 572–4002.

Wiley also publishes its books in a variety of electronic formats. Some content that appears in print may not be available in electronic formats. For more information about Wiley products, visit our web site at www.wiley.com.

Library of Congress Cataloging-in-Publication Data is available.

ISBN: 978-1-118-13053-7

Printed in the United States of America

10 9 8 7 6 5 4 3 2 1

CONTENTS

PREFACE

The goal of this book is to serve as a comprehensive reference book for readers in the field of electronic engineering with a background in digital signal processing and telecommunications (only fundamentals and not necessarily with advanced knowledge in this area). The target readers include researchers, engineers, service providers, market analyst, policy makers, and IT staff who work in the digital video broadcasting area. The book may serve as a textbook for undergraduate courses of one semester or short courses if the instructor only focuses on the fundamental concepts and as a graduate textbook if details need to be addressed. It can also serve as a continuing education textbook for those in the DTV industry who want to obtain the latest updates.

Chapter 1 introduces the basic concepts of the digital television (DTV) system, including a historical perspective and the constitution of the DTV system with emphasis on the terrestrial broadcasting system. Chapter 2 presents the characteristics of the harsh terrestrial transmission environment for DTV signals, including propagation loss, the "shadow effect" by terrain factors, the multipath effect, and the Doppler effect when transmitters and/or receivers are under mobility. Chapter 3 covers the fundamentals of channel coding, including interleavers that help convert burst errors into random errors for better error correction capability, especially those being adopted in the digital television terrestrial broadcasting (DTTB) system. Chapter 4 mainly introduces the modulation techniques in various DTTB systems. The basic concepts of coded modulation, which jointly optimizes channel coding and digital modulation to best control errors by nonideal effects for transmission are also addressed. Chapter 5 provides information on the frame structure, channel coding, modulation, and major parameters of the existing first-2 generation international standards, especially for the newly developed digital television/terrestrial multimedia broadcasting (DTMB) system. Chapter 6 gives general information on the second generation of the DTTB system, which can provide over 30% increase in data throughput by adopting higher constellation mapping and longer coding length for better error correction capability. Chapter 7 focuses on the design and implementation issues of the DTV receiver, including carrier synchronization, timing recovery, channel estimation, equalization, decoding, and de-interleaving, with concrete implementation examples of these algorithms. Chapter 8 addresses issues such as coverage and network planning of DTTB networks with a detailed introduction of the single-frequency network (SFN). A brief introduction of the characteristics and implementation of the diversity technology is also provided. Chapter 9 gives a general description of the system-level performance test of the DTTB systems, including the physical meaning of the test item, test methodologies/procedures, and for information purposes the test requirements using DTMB as an example. Chapter 10 describes the technical features in detail of four multimedia mobile broadcasting systems. Even though some systems are out of favor nowadays due mainly to the spectrum issue and the tough competition, the featured technologies of those systems have been adopted by other systems.

With 10 chapters and quite broad topics, the instructor may arrange the topics in different ways depending on the time length of the course. Chapter 2 is more physics related while Chapters 3 and 4 provide fundamentals of coding and modulation. These three chapters, together with Chapters 5, 6 and 10, can help readers better understand the major design issues and constraints of the DTTB systems. If the targeting audience is interested in knowing receiver design, Chapter 7 and some of the references can serve this purpose, and network planning is addressed for service providers. For engineers whose job function is testing, Chapter 9 provides a good topic for continuing education purposes.

The authors of this book have actively been involved with fundamental research on the core technologies of DTTB systems (i.e., time-domain synchronous OFDM, TDS-OFDM), hardware implementation and the performance validation of the DTTB receiver (more specifically, the DTMB receiver), and the international standardization process, and some technical context directly comes from their research and development work. This valuable experience has motivated the authors to write this book and share their research results and comprehensive understanding of the DTTB system with readers who work in this area.

The authors would like to express their sincere appreciation for the contributions of Dr. Nim Cheung. Without his kind recommendation, encouragement, and always on-time help, this book would not have been completed. The book is a joint effort of researchers working at Tsinghua University, China. The authors are indebted to Professor Jintao Wang, Professor Chao Zhang, Professor Changyong Pan, Professor Zhaocheng Wang, Professor Fang Yang, Professor Yonglin Xue, Professor Kewu Peng, Professor Yu Zhang, Professor Hui Yang, Dr. Qiuliang Xie, and other team members for their much valuable contributions.

The authors would also like to thank all the comments from the reviewers of this book proposal as well as this book. Their much valuable comments and suggestions allowed us to better choose and arrange all the context of the book. Finally, we also thank the great help and patience from Mary Hatcher and Brady Chin of John Wiley & Sons, Inc. Without their kind guidance and assistance, it would have taken much longer with more painful effort to finish the book.

1

BASIC CONCEPTS OF DIGITAL TERRESTRIAL TELEVISION TRANSMISSION SYSTEM

1.1 INTRODUCTION AND HISTORIC REVIEW

Television is a word of Latin and Greek origin meaning "far sight." In Greek, *tele* means "far" while *visio* is "sight" in Latin. A television (TV) system transmits both audio and video signals to millions of households through electromagnetic waves and is one of the most important means of entertainment as well as information access. With the never-ending technological breakthroughs and the continuously increasing demands of audio and video services, the TV system has evolved over generations with several important developmental periods in less than a century.

1.1.1 Birth and Development of Television Black-and-White TV Era

In the mid-1920s, the Scottish inventor John Logie Baird demonstrated the successful transmission of motion images produced by a scanning disk with the resolution of 30 lines, good enough to discern a human face. In 1928, the first TV signal transmission was carried out in Schenectady, New York, and the world's first TV station was established by the British Broadcasting Corporation in London eight years later. After World War II, the black-and-white TV era began. Detailed technical and implementation specifications of TV service, including photography, editing, production, broadcasting, transmission, reception, and networking, were gradually formulated. With the ever-growing popularity of TV viewers, the color TV with better watching experience was invented to simulate the real world.

Digital Terrestrial Television Broadcasting: Technology and System, First Edition. Edited by Jian Song, Zhixing Yang, and Jun Wang.

1.1.2 Analog Color TV Era

In 1940, Peter Carl Goldmark with CBS (Columbia Broadcasting System) Lab invented a color TV system known as the field-sequential system. This system occupied an analog bandwidth of 12 MHz and was carried by 343 lines (~100 lines less than that of the black-and-white TV) at different field scan rates, and hence was incompatible with black-and-white TV. The system started field trial broadcast in 1946, and this is the dawn of the color TV age.

In the 1950s, a color TV signal system called NTSC (National Television Standards Committee) was developed in the United States that was compatible with the black-and-white TV. This scheme uses a luminance–chrominance encoding scheme with red, green, and blue (RGB) primary signals encoded into one luminance signal (Y) and two quadrature-amplitude-modulated color (or chrominance) signals (U and V), and all are transmitted at the same time. An NTSC TV channel occupies 6 MHz bandwidth with the video signal transmitted between 0.5 and 5.45 MHz baseband. The video carrier is 1.25 MHz and the video carrier generates two sidebands, similar to most amplitude-modulated signal, one above the carrier and one below. The sidebands are each 4.2 (5.45–1.25) MHz wide. The entire upper sideband will be transmitted while only 1.25 MHz of the lower sideband (known as a vestigial sideband, VSB) is transmitted. The color subcarrier is 3.579545 MHz above the video carrier and quadrature amplitude modulated with the suppressed carrier while the audio signal is frequency modulated. The NTSC system was deployed in most of North America, parts of South America, Myanmar, South Korea, Taiwan, Japan, the Philippines, and some Pacific island nations and territories. This invention is considered as the landmark of the second stage of the development—the analog color TV era.

A group of French researchers started their work in parallel and this led to the invention of the Sequential Color with Memory (SECAM) system in 1956, and the system was successfully demonstrated in 1961. In the SECAM system, two color difference signals are transmitted alternately (line by line) and frequency modulated by the color subcarrier. This system was adopted by France, the Soviet Union, Eastern European countries (except for Romania and Albania), and Middle East countries and was the first color TV standard in Europe. In 1962, Walter Bruch, a German engineer at Telefunken, put forward the Phase Alternate Line (PAL) system based on the NTSC system in the Federal Republic of Germany. This system performs line-by-line phase inversion of the quadrature component of the chrominance signal in the NTSC system and can effectively offset the phase error and increase the tolerance for differential phase error from ±12° to ±40° in the NTSC system. This new system was adopted by more than 120 countries successively, and in 1972 China decided to adopt it as well.

In the first 70 years of the twentieth century, even though the development of TV had gone through two different phases (black and white and color), the fundamental characteristics of TV signal transmission was unchanged, that is, the TV signal was continuous, or analog, and hence why both black-and-white and color TVs were called analog. In analog TV signal transmission, the amplitude, frequency, phase, or a combination of these parameters of the carrier are changed in accordance with the contents to be transmitted. Linear modulation as well as transmission are therefore achieved in one step. Though simple and straightforward, the analog TV system has the following issues in practice [1]:

1. In terms of the quality, long-term storage, and dissemination of the video programs, the analog TV program source suffers from color–luminance interference, large-area flicker, and poor image definition, and it is difficult to replicate the content for too many times.
2. In terms of signal transmission efficiency, the analog TV network is largely restricted by the bandwidth available. For example, the PAL system can accommodate only one analog video signal and one analog audio signal in 8 MHz bandwidth, and the spectral efficiency is low. In addition, due to cochannel and adjacent-channel interference in neighboring areas, different analog channels have to be used to carry the same programs to different areas to avoid mutual interference. Therefore, the spectral efficiency is further decreased, and it is very difficult to introduce new programs by assigning additional channels in the same region due to the limited available spectrum.
3. In terms of the quality of the signal transmission, the analog TV signal may suffer from "ghosts" from terrestrial broadcasting due to its poor anti-multipath interference ability, which severely affects the viewer's experience. In addition, if the analog TV signal needs to be amplified for a longer transmission distance, the noise accumulation will make the signal quality very poor due to the deteriorating signal-to-noise ratio.
4. In terms of the circuitry, network equipment, and terminals of the analog TV system, the geometric distortion of images is inevitable due to the nonlinearity of the circuitry while the phase distortion of the amplifiers would cause color deviation, aggravating the Ghost phenomenon. In addition, the analog TV system suffers from poor stability, time-domain aliasing, low degree of integration, difficulty in calibration, automatic control, and monitoring.

1.1.3 Digital TV Era

People's demands for better audio and video quality of the TV signal has always been a tremendous driving force for the broadcasting industry, and this led to the invention of the digital television (DTV). Also, due to significant technical breakthroughs in the digital signal processing field (including signal acquisition, recording, compression, storage, distribution, transmission, and reception), the semiconductor industry, and other related industries in the past half century, the broadcasting industry is now embracing the third important stage in its history, i.e. the DTV era.

The visual information received by human eyes in daily life is always analog, and the mission of both the first and second generation of TV broadcasting systems (black and white or color) is to transmit these analog signals to the numerous TV sets with the highest possible quality. Although the definition or the structure of the different DTV systems may be slightly different, the core definition or the major functional blocks are the same. They must include the sampling, quantization, and encoding of analog TV programs to convert them into digital format before they are further processed, recorded, stored, and distributed. The sequence segmentation, scrambling, forward error correction coding, modulation, and up conversion are done in the baseband to form the DTV radio frequency (RF) signals after up conversion at the transmit side. At the receive side, after achieving the system synchronization and signal equalization based on accurate channel estimation, inverse operations on the received signal to that at the transmit side will be performed before the final program can be finally displayed on the TV screen. Digital broadcasting technologies not only provide better reception and display performances compared to its analog counterpart but also introduce new functions that are not available with the analog broadcasting technologies. Considering all the advantages digital technology can provide over its analog counterpart, it is obvious that a DTV system can offer high-quality audio visual experiences and more comprehensive services for consumers. Given all these featured services the DTV system can support, digitization is widely considered a fundamental change and new landmark in the TV broadcasting industry, after the introduction of the black-and-white TV and the color TV.

The advantages of DTV over the traditional analog TV can be summarized as follows:

1. *Better Anti-Interference Ability, No Noise Accumulation, and High-Quality Signals.* After digitization, the analog signal is changed into a binary (two-level) sequence. Unless the amplitude of the noise exceeds a certain level, noise introduced during processing or transmission can be effectively eliminated. Error-free transmission can also be achieved by means of forward error correction coding. During transmission of the DTV signal, the quality of the image and sound received by the users in the coverage is almost identical to that originally transmitted from the TV station. Thus the quality of programs in DTV would not be degraded if the system is well designed, whereas the processing or transmission of the analog TV signals may introduce additional noise which is difficult to remove, and the quality of the image and sound will thus be gradually degraded due to the noise accumulation.
2. *Higher Transmission Efficiency and More Flexibility in Multiplexing.* Digital TV broadcasting can utilize the precious spectrum resources more efficiently. Using terrestrial broadcasting as an example, DTV can use the so-called taboo channel, which is not allowed in analog TV systems, and adopt the single-frequency network (SFN) technology. When SFN is adopted, the same DTV channel can be used to carry the same TV programs with different transmitters to cover a very large area (even the countrywide SFN is possible). Depending on the video coding compression scheme used in a DTV system, one analog TV channel can at least contain one HDTV (high-definition TV) program, or ~10 SDTV (standard-definition TV) programs, or more than 20 DTV programs with VHS quality. Digital TV technology helps reduce the bandwidth requirement for each program, and the spectrum efficiency increases greatly. With the

spectrum saving from DTV broadcasting, broadcasters can use the saved spectrum to either provide more TV programs or offer new services.

3. *Easy to Encrypt and Support Interactive Services.* DTV systems can be extended from a point-to-multipoint broadcasting system to a point-to-point interactive system to support value-added services so that the user can either watch TV programs or search/exchange information based on personal preferences. Digitization in the whole process also facilitates the encryption, and existing encryption techniques can be easily used in the DTV system.
4. *Easy to Store, Process, and Distribute under Network Environment.* The advantage of a DTV signal over its analog counterpart is that it is easy to store, process, and exchange. This facilitates the integrated transmission of images, data, and voice as well as TV program sharing under the network environment.

In summary, the introduction of the DTV concept relies on the latest technical breakthroughs from video compression and information transmission/processing. The digital video compression coding technique is applied to the video source to minimize redundancy with high compression ratio at no (or almost no) loss in quality. The transmission data rate for any TV program is therefore reduced and the transmission efficiency of the whole system is improved. Using error correction coding technology which introduces certain redundancy to the compressed information sequence and the highly efficient digital modulation technologies, better transmission performance in the presence of noise, interference, and other nonperfect conditions can be achieved. Also, due to the latest development in drive and display technologies, DTV systems can surely offer better viewer experience, including sharper images, better color, and more exquisite sound quality, all with improved spectral efficiency.

Looking forward, ultrahigh-definition TV (UHDTV) with UHDTV-1, representing 4K of 3840×2160, and UHDTV-2, representing 8 K of 7680×4320, systems have been proposed by NHK Science & Technology Research Laboratories and are accepted by the International Telecommunication Union (ITU) [2]. Definitely UHDTV will be one development trend from the display point of view, while three-dimensional (3D) TV technologies following the recent popularity of 3D movies will be another clear trend for the display. From the users' experience point of view, intelligent TV systems will surely attract more and more people due its great simplicity and interactive capability. With the massive development of DTV networks and the increasing number of DTV users, various systems and applications for DTV have been and will continue to be introduced.

There are three types of TV broadcasting networks regardless of whether they are analog or digital: terrestrial (also known as over the air), cable, and satellite TV networks, as shown in Figure 1-1. Satellite TV broadcasting provides coverage of a large area, especially in rural areas with sparse population while cable TV broadcasting uses the coaxial cable to deliver information to the home, with an emphasis on serving densely populated areas. As the most commonly used method of TV broadcasting, the terrestrial system uses transmitting stations to send radio waves over the air to cover certain service areas and users can receive TV programs by all

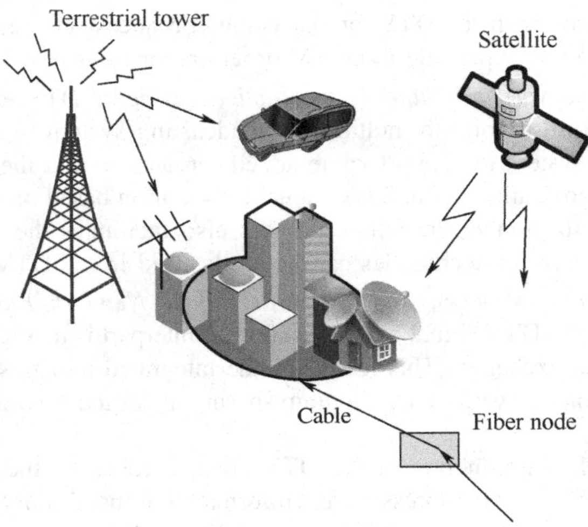

FIGURE 1-1 Classifications of DTV infrastructure.

kinds of receiving antennas and various terminals. This makes it the most direct and reliable approach to reach people nationwide in case of emergency. Statistics show that most people in the world still rely on terrestrial broadcasting networks to receive TV programs, with the percentage in China over 60%. This book mainly focuses on the core technologies and performance of the DTV terrestrial transmission system, which lay down the foundation for the various applications of DTV systems. The key video compression concepts will also be introduced in this chapter.

It is generally acknowledged that the transmission environment for a satellite or cable channel is very similar to the ideal additive white Gaussian noise (AWGN) channel, and adoption of both advanced channel coding and modulation can make the performance of both satellite or cable broadcasting approach the theoretical limit. Being the most commonly used DTV networks worldwide, the Digital Television Terrestrial Broadcasting (DTTB) networks support the largest number of users. The term digital terrestrial television (DTTV or DTT) is also used to refer to the DTTB system, and they are used interchangeably within this book. The terrestrial broadcasting channel, however, presents the harshest transmission conditions due to the high degree of interference, especially with rapid changes in both time delay and amplitude of the multipath interference. This channel is far more complicated compared to that of either satellite or cable networks. The transmission environment for a terrestrial DTV broadcasting channel is obviously not an AWGN channel, and this presents a great challenge for the DTTB system designer. Laboratory test results for DTTB system performance under an AWGN environment may be significantly different from that in the real world. In another words, the coding scheme with decent gain for an AWGN channel may not be applicable to the actual transmission environment. Therefore, system performance should be carefully evaluated when choosing the appropriate transmission scheme not only in an AWGN channel but also in a multipath channel. Another important issue that needs to be addressed is interference from the terrestrial broadcasting network itself. With the inevitable coexistence of both analog and digital terrestrial TV services during the transition period, the system must have strong capability to deal with both adjacent and cochannel interference from the analog transmission and also minimize its interference to the existing broadcasting systems (both analog and digital). This helps guarantee the overall reception performance for all end users.

1.2 MAJOR INTERNATIONAL AND REGIONAL DTV ORGANIZATIONS [3]

Almost all countries and regions have been or are now seriously considering the deployment of DTV broadcasting networks based on the advantages the DTV system can provide. Countries such as the United States, Canada, United Kingdom, Germany, Japan, Netherlands, Finland, Switzerland, South Korean, and Sweden have successfully completed their DTV transition (also known as the digital switch-over or analog switch-off) for their TV broadcast networks, while many countries in the world are still in the process of transitioning their TV broadcasting networks from analog to digital.

1.2.1 International DTV Broadcasting Standards

Even though the application scenarios for terrestrial DTV broadcasting are very similar, different international transmission standards for DTTB systems have been proposed, including ATSC (Advanced Television Systems Committee) by the United States, DVB-T (Digital Video Broadcasting-Terrestrial) by Digital Video Broadcasting organization, ISDB-T (Integrated Service Digital Broadcasting-Terrestrial) by Japan, and DTMB (Digital Terrestrial Television Multimedia Broadcasting) by China. All four DTTB standards have been accepted by the ITU, and they have already been commercialized in many countries and regions worldwide.

In the United States, the Federal Communications Commission (FCC) developed its own DTV broadcasting standard in 1987, which is required to be compatible with the existing NTSC TV standard. In 1992, the ATSC, consisting of members who passed its qualification and obtained authentication, was founded with the aim of creating advanced TV system standards. In the same year, ATSC put forward four candidate proposals, and eventually integrated them into a unified standard by Grand Alliance (GA) in 1995. This standard includes the AC-3 standard for multichannel audio source coding and the MPEG-2 standard for video source coding, system information, and multiplexing. The ATSC/8VSB describes a single-carrier system for terrestrial broadcasting with a throughput of 19.39 Mbps when the system bandwidth is 6 MHz. The ATSC/16VSB is a standard for digital cable TV systems with total throughput of 38.78 Mbps. The ATSC standard is generally believed to have a higher spectral efficiency and power efficiency but usually requires a better receiving environment. The FCC adopted ATSC as the DTV standard for the United States on December 24, 1996, and revised it in 2009. H.264/AVC video coding was introduced to the ATSC system in 2008. All terrestrial TV broadcasters are required to deliver over-the-air TV programs using the ATSC standard, and even cable operators are requested to carry ATSC signals from terrestrial broadcasters. By June 12, 2009, the United States had successfully replaced almost all analog NTSC TV system with ATSC. Canada and South Korea decided to use ATSC as well.

The European Launching Group (ELG) was founded in 1991 with the help of the German government. The ELG realized that mutual respect and trust had to be established between members and became the Digital Video Broadcasting (DVB) program in September 1993. Currently the DVB organization has more than 270 members from nearly 40 countries, who are dedicated to the establishment of a technical system for digital broadcasting systems. The DVB project provides a series of standard frameworks (DVB-C, DVB-S, and DVB-T) for digital video broadcasting systems using different transmission media (e.g., coaxial cable, satellite, and terrestrial) and has announced over 60 DTV broadcasting standards which have been accepted worldwide. The DVB-S is the transmission standard for satellite digital broadcasting in which one analog TV channel which previously delivered one PAL program can now support four DTV programs, and this greatly increases the efficiency of the satellite broadcasting system. The DVB-C is the transmission standard for DTV within the cable TV network in which one analog TV channel that previously delivered one PAL program can now provide four to six DTV programs. The DVB-T is the transmission standard for terrestrial digital broadcasting in which one analog TV channel that previously delivered one PAL program can now provide four to six DTV programs. DVB-T was first published in 1997, and the first broadcasting took place in the United Kingdom in 1998. These standards were all adopted by both the European Telecommunications Standards Institute (ETSI) and ITU. Like the ATSC, the DVB also initially selects MPEG-2 as the standard for audio and video source coding, system information, and multiplexing. Unlike ATSC, DVB-T is a multicarrier system which uses the coded orthogonal frequency division multiplexing (C-OFDM) technology for transmission. Compared to the ATSC, the DVB-T can effectively support both fixed and mobile reception under a complicated environment at very little expense of both spectral and power efficiencies and can support the single-frequency network application well. The extended application, the mobile TV standard DVB-H, has also been introduced. So far, over 60 countries have officially chosen DVB-T as the terrestrial DTV transmission scheme and more than 30 countries are now covered by DVB-T signals, among which some have finished the analog switch-off.

DVB decided to study options for an upgraded DVB-T standard in March 2006 and a formal study group named Technical Module on Next Generation DVB-T was established to develop an advanced modulation scheme as the second-generation digital terrestrial television standard in June 2006. In June 2008, DVB announced its second-generation DTTB standard, known as DVB-T2, and some countries and regions have shown strong interest in adopting it as it can provide more than 30% throughput than the first-generation DTTB standards. More details regarding DVB-T2 will be given in the following chapters.

The Japanese authority started the development of DTV broadcasting standards in 1994. They have also decided to use MPEG-2 as the standard for source coding and system information. The core standards of ISDB, the Japanese standard, are ISDB-S (satellite), ISDB-T (terrestrial), and ISDB-C (cable). Similar to DVB-T, the developers of the ISDB-T standard also chose OFDM as the modulation scheme, while using frequency segmentation to deliver both terrestrial and hand-held TV programs within the same 6-MHz frequency band. In other words, broadband and narrow-band information is transmitted using the same facility and within the same channel for the same coverage area, which greatly facilitates mobile reception by portable devices. This mixed transmission scheme turns out to be a big success to support mobile TV users. Japan finished its analog switch-off in 2012 and ISDB-T has also been adopted in several countries and regions.

The effort on developing the Chinese DTTB standard was officially started in 1999 through the call for proposals from the Chinese government. With several individual proposals being successfully merged in 2005 and an independent test by a third party, the Chinese national DTTB standard was approved by the Standardization Administration of the People's Republic of China and announced on August 18, 2006 [4]. The standard is called "Framing Structure, Channel Coding and Modulation for the Digital TV Broadcasting System," with an official label of GB20600-2006. The English translation is "Digital Television Terrestrial Multimedia Broadcasting (DTMB)." DTMB can satisfy various requirements of broadcasting services, such as HDTV, SDTV, and multimedia data broadcasting. It provides large-area coverage and supports both fixed and as mobile reception. DTMB adopts both single- and multi-carrier modulation with a unique frame structure called time-domain synchronous OFDM (TDS-OFDM) and uses the low-density parity code (LDPC). It can therefore provide fast system synchronization, better receiving sensitivity, and excellent system performance against the multipath effect plus the advantage of high spectrum efficiency and flexibility for future extension. The massive deployment of DTMB in China started in 2008 [5] and DTMB became the ITU standard in 2011.

1.2.2 Related International and Regional Organizations

To overcome the engineering problems that arose from the development and deployment of DTV networks and ensure smooth analog-to-digital migration, many international organizations have been working closely and developed a series of DTV-related frameworks and supporting standards. These standards cover all fields related to DTV broadcasting implementation, e.g., compression/decompression, coding/decoding, modulation, framing, frequency allocation, content encryption, conditional access, and signal distribution for the DTV signal. These organizations that have contributed significantly are as follows:

1. The Moving Image Experts Group (MPEG), a working group of the International Organization for Standardization (ISO) and the International Electrotechnical Commission (IEC), has the responsibility of developing the standards for compression, decompression, and processing on video, audio,

and a combination of both. The MPEG is a subsidiary organization of the ISO/IEC technical committees dedicated to standardizing the information technology related equipment.

2. The Multimedia and Hypermedia Information Coding Expert Group (MHEG) is another working group under the same subcommittee to which the MPEG belongs. MHEG is dedicated to the coding of both multimedia and hypermedia information by defining the encapsulation format for the multimedia documents such that communication can be performed by a special data format.
3. The Digital Audio Video Council (DAVIC) was founded in Switzerland in 1994 as an international, nonprofit organization with a membership of 220 corporations from 25 countries. The DAVIC is dedicated to providing end-to-end interoperation standards for both digital video and audio between different countries and different applications and delivers open interface and protocols for digital services and applications.
4. The European Broadcasting Union (EBU) is a nongovernmental and nonprofit organization. Any non-European broadcasting company can also become a member of EBU. It supports both DVB projects and the Digital Terrestrial Television Action Group (DigiTAG) as well as work by other standard groups, e.g., European Committee for Electrotechnical Standardization (CENELEC), ETSI, ITU, and IEC.
5. ITU, a subsidiary organization under the United Nations, is perhaps the most important international standardization organization in both the telecommunication and radio communication fields in the world. It is a major publisher for telecommunication technologies, rules, and standards and is dedicated to spectrum management. The ITU Radiocommunication Sector (ITU-R) formulates the DTV broadcasting standards.
6. ETSI and the American National Standards Institute (ANSI) have made a joint effort for the interconnection between video transmission circuits and telecommunication devices, and formulated two major standards: ETS 300 174 (equivalent to ITU-T Rec. J. 81) and ANSI TI. 802.01. These two standards distribute a video channel to each bit stream and describe coding, multiplexing, encryption, and network matching for the video channel so that devices are able to connect to the telecommunication devices directly. ETSI was founded in 1988 aiming to help establish the unified telecommunication market in Europe by formulating the related telecommunication standards. The ETSI technical committee formulated standards for interconnection between public networks and private networks. ETSI's multimedia Codec is used for the interconnection between the broadcasting networks and the telecommunication networks. ANSI's Codec is similar to ETSI's Codec except for the audio interface and the SMPTE control function and has good connection to the telecommunication networks in the United States with a transmission rate of 45 Mbps.
7. IEC is responsible for standardization of electrical equipment. ISO is a nongovernmental international alliance for standardization responsible for

formulating industrial standards. Both ISO and IEC are dedicated to the standardization of global personal and industrial equipment. They have established many joint technical committees in these fields.

8. The JTC 1 (Joint Technical Committee formed by the ISO and IEC) aims at formulating standards for information technology related equipment. JTC 1 establishes a subsidiary organization with the acronym of MPEG to formulate standards for digital video coding and audio compression equipment as described above.
9. The DigiTAG was founded in 1996 and is dedicated to creating a framework for digital terrestrial television applications in accordance with DVB-T specifications. The DigiTAG has around 40 members from 14 countries and is managed through EBU.
10. CENELEC was founded in 1973 and is a nonprofit European organization for electrotechnical standardization. CENELEC members are the national electrotechnical standardization bodies of most European countries. They are dedicated to solving the integration issues between the member states of the European Commission (EC). CENELEC cooperates with technical experts from 19 EC and European Free Trade Association (EFTA) member states to prepare voluntary standards which help facilitate trade between countries, create new markets, cut compliance costs, and support the development of a single European market. CENELEC also works closely with other technical committees in fields such as television and cable classification.

1.3 COMPOSITION OF DTV SYSTEM

1.3.1 Constitution of DTV System

A complete DTV broadcasting system consists of three key components; the transmitting head-end system, transmission system/distribution network, and user terminal system.

1.3.1.1 Transmitting Head-End System for DTV Broadcasting A transmitting head-end system for DTV broadcasting refers to the professional equipment for the TV station, and it mainly comprises the video cameras, video recorders, storage devices, special effect machines, editing machines, subtitling machines, audio and video encoders. Considering MPEG-2 has been used and is still used for video compression by most of DTV standards currently, MPEG-2 will be used as an example for the following discussion. The equipment is mainly used for source processing, information processing, storage, and play as well as other functionalities.

The source processing unit usually includes audio and video encoders, an adaptor, a data encapsulation device, a VOD (video on demand) system, and an editing processor. The MPEG encoder compresses and encodes the recorded audio and video signals into MPEG-2 format; the adaptor adaptively receives MPEG-2 signals from other networks such as synchronous digital hierarchy (SDH) and satellite and then sends them to the multiplexers for multiplexing purposes or to the program libraries for storage as well as further editing; the data encapsulation device helps packetize the Internet Protocol (IP) data and data in other formats used for data broadcasting as well as interactive services into signal format for DTV broadcasting and transmits these signals together with other signals to the users; the VOD system sends the programs and information requested by users; the editing processor edits and helps manage the stored digital programs.

The information processing unit usually comprises the program scheduling system, the user management system, the multiplexer, and the conditional access (CA) system. The program scheduling system is a platform for the service management and system applications. The user management system is responsible for handling users' account information. The multiplexer is the core part of the unit and is responsible for content scheduling, including reselection, allocation, multiplexing, and distribution of the contents gathered from different places to different channels with control of the program scheduling system. CA applies the encryption mechanism to the different program contents via a scrambler and multiplexer so that the program contents are encrypted according to different time periods and user groups according to service modes and user demands. As a new and attractive service of DTV broadcasting, an electric program guide (EPG) helps provide more program information to the end users by inserting the corresponding information into a real-time bit stream at the head end.

1.3.1.2 Transmission System/Distribution Network for DTV Broadcasting The typical networks for transmitting and distributing the DTV signals include terrestrial broadcasting, cable, and satellite.

Statistics show that terrestrial broadcasting is still the most important and popular TV broadcasting scheme. To accommodate the most complicated transmission environment for terrestrial broadcasting, the technologies and functional blocks in the DTTB system are different not only from that of the analog television but also possibly from that of satellite or cable DTV broadcasting. The terrestrial broadcasting transmission network mainly comprises the SFN adapters, exciters, and transmitters.

Cable TV is the major TV transmission method in densely populated regions such as metropolitan areas. Because the signal is sent through a coaxial cable, a very stable quality of signal transmission with a large number of programs can be supported. Cable DTV is also convenient to offer pay-per-view (PPV), VOD, and other bidirectional as well as value-added services.

Satellite TV provides large coverage and its signals can be received in urban, suburban, and rural areas if there exists a line-of-sight (LOS) path between the satellite and the receiving antenna. The equipment in a satellite TV transmission network mainly comprises the satellite modulators, RF power amplifiers, and satellite transponders.

1.3.1.3 User Terminal System for DTV Broadcasting In the DTV era, either a digital TV set or a set-top box (STB) matching the analog TV set is needed to watch DTV programs, and STB is very popular due to its low cost and user convenience

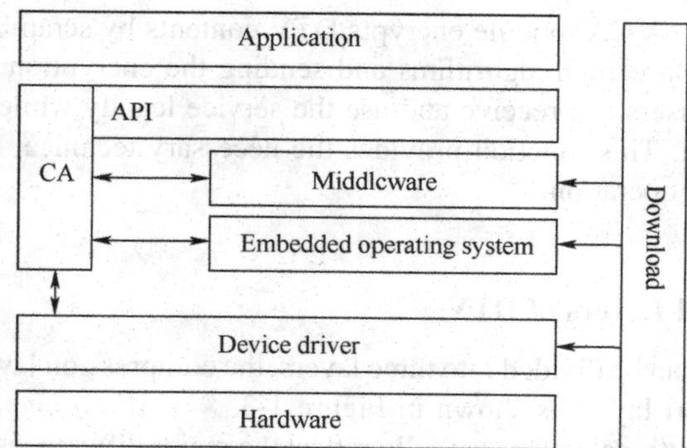

FIGURE 1-2 Layered structure of STB for DTV broadcasting.

during the switch-over period. Generally, each DTV STB consists of the hardware platform and the associated software. The structure of STB can be typically divided into four layers: hardware, device driver, middleware, and application software from the bottom to the top. In addition, there is also a CA module, as shown in Figure 1-2.

1. *Hardware Layer.* This layer provides the hardware platform for STB, which mainly comprises a receiving front end for the DTV broadcasting signal, MPEG-2 decoder, video/audio and graphic processing unit, CPU, memory, and various interface circuitries. The receiving front end for DTV broadcasting includes a tuner and a digital demodulator; the demodulator receives, demodulates, and decodes the RF signal to obtain the MPEG-2 transport streams. The MPEG-2 decoding part includes the demultiplexer, a descrambling engine, and the MPEG-2 decompression module and outputs audio and video data as well as data of other services. The video, audio, and graphic processing part provides digital and analog output of video/audio and graphic processing functions. The CPU and memory module are used to store and run the software and control all other modules. The interface circuitry supports various peripheral interfaces, including the universal serial interface (USB), Ethernet interface, RS232, and the video/audio interface.
2. *Device Driver.* The device driver provides the operating system (usually an embedded real-time operating system) kernel and various hardware drivers for STB.
3. *Middleware Layer.* The middleware layer separates the application software from underlying software which relies on the hardware, and this provides a unified functional interface for applications independent of specific hardware platforms. This layer is typically composed of various virtual machines.
4. *Application Software Layer.* The application software layer performs the functions required by the end user. It can be stored locally or downloaded through the broadcasting network.

5. *CA Module.* A CA module encrypts DTV contents by scrambling the service data based on certain algorithms and sending the encryption keys so that all authorized users can receive and use the service legally while those unauthorized cannot. This function provides the necessary technical means for DTV commercial operation.

1.3.2 Functional Layers of DTV

The DTV system can be divided into three layers, the compression layer, multiplexing layer, and transport layer, as shown in Figure 1-3.

These three functional layers can fully reflect the major differences between digital TV and analog TV broadcasting and can explain why DTV is superior to its analog counterpart from a technical point of view, as given in Table 1-1.

1.3.2.1 Compression Layer Source coding and decoding usually refer to video and audio compression and decompression. One of the most important tasks for the compression, especially for high-definition DTV, is to compress the video signals. Uncompressed SDTV (4:2:2) video has data throughput of 216 Mbps while for HDTV it is about 1.2 Gbps. Therefore, the DTV signal cannot be directly transmitted like the analog TV signal and compression (video encoding) is required to reduce the data rate.

The main function of video encoding technology is to compress the images to reduce the data rate from 1.2 Gbps down to ~20 Mbps for HDTV signals and from

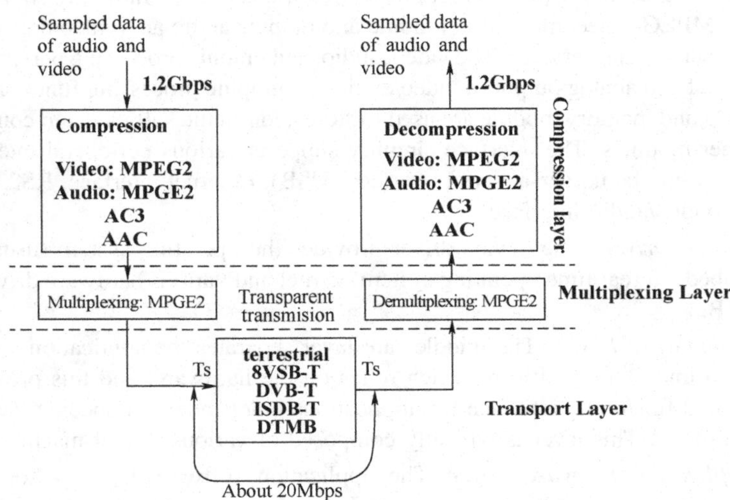

FIGURE 1-3 Functional layers of DTV broadcasting system.

TABLE 1-1 Major Technical Differences between Digital TV and Analog TV Broadcasting

	Digital TV Broadcasting	Analog TV Broadcasting
Source coding and decoding (compression layer)	The data transmission rate for DTV signals without compression is very high and a high-quality video/audio compression scheme must be applied.	Analog TV signal needs no compression.
Multiplexing (multiplexing layer)	The DTV system needs to packetize and multiplex encoded video/audio signals as well as auxiliary data into a single stream to ensure scalability, interactivity, and network interconnectivity.	Analog TV does not need multiplexing.
Channel coding and decoding and modulation and demodulation (transport layer)	The DTV signal no longer has vertical and horizontal flags after both compression and multiplexing. A DTV system uses error correction and equalization to improve the anti-interference capability and more DTV programs can be supported by one analog channel with the use of high-order constellation, and therefore, system transmission efficiency is highly improved.	Analog TV signals are arranged by line and field, and compensation is done with the help of both horizontal and vertical synchronization signals; pre- and postequalization pulse frequency or amplitude modulation is used as the modulation scheme.

216 Mbps to ~4 Mbps for SDTV signals. Video compression can be achieved mainly based on the following:

1. *Time Correlation between Consecutive Images.* Usually, the adjacent images of the video signal are highly correlated, and this helps reduce the information to be transmitted.
2. *Space Correlation in Image.* For example, it will be unnecessary to store all the pixels if the large portion of the image has a single color.
3. *The Visual Characteristics of Human Eye.* The degree of sensitivity of the human eye to the distortion in different portions of an original image is quite

different. For example, the human eye is usually insensitive to the distortions (even the total loss) on the insignificant information of the image. However, the distortion of information to which the human eye is quite sensitive should be minimized if not fully eliminated.

4. *Statistical Characteristics of Input.* The smaller the occurrence probability of the data pattern, the greater the entropy it will be, and this means that a longer code word is required. On the other hand, the larger the occurrence probability of the data pattern, the smaller the entropy, and this means a shorter code word should be assigned.

Similar to video coding and decoding, the main function of audio coding and decoding is to compress the sound information after digitization. Compression of the audio signals is mainly based on the following auditory features of the human ear:

1. *The Auditory Masking Effect.* As to the auditory sense of human beings, the presence of one sound masks the presence of another, and this masking effect is a relatively complex psychological and physiological phenomenon, including both the frequency-domain and time-domain masking effects of the human ear.
2. *Directional Characteristics of Human Ear to Sound.* The ear can barely figure out the direction of acoustic signals with frequency over 2 kHz, and therefore it is unnecessary to repeatedly store the high-frequency components of stereo broadcasting.

There are different international standards for digital image compression, and several examples of them are H.261, mainly used for TV conference; the JPEG standard, mainly used for still images; and the MPEG standard, mainly used for sequential images. As for the HDTV video compression coding and decoding standard, the MPEG-2 standard has been and is still widely used around the world. As for the audio coding, the MPEG-2 standard is used in Europe and Japan. The United States has adopted the Dolby AC-3 scheme with MPEG-2 as an alternate. With the progress in compression technologies, other excellent video compression standards, such as H.264, MPEG-4, AVS, and H.265, have been announced with higher compression ratios for the same image quality. Therefore, the bandwidth needed for video transmission can be further reduced.

1.3.2.2 Multiplexing Layer A multiplexing layer multiplexes several compressed information streams into one single stream, which makes it possible to transport all these data through one analog TV channel. For data flow from the transmitter side, the multiplexing layer packet processes and multiplexes all the output information streams from the encoders of video, audio, auxiliary data, etc., into one stream based on a certain rule and then sends this stream for the channel coding and modulation module before the frequency upconversion. The multiplexing layer is the basis to ensure the extensibility, scalability, and interactivity of the DTV system. There is no multiplexing needed in the analog TV system as video and audio signals are separately modulated and transmitted. Using the widely used MPEG-2 as an example for the DTV multiplex transport standards, the data format after the multiplexing is in transport stream (TS) format with a fixed data packet length of 188 bytes. The TS stream is convenient for the channel transmission, and various time tags used for indication and synchronization can be easily inserted into the TS streams.

Pay TV is very popular now and also widely believed to be an important feature for the future TV. The multiplexing layer helps to support this functionality by CA: The packetized program data are scrambled so that unauthorized receivers are unable to descramble the data and retrieve the original stream.

1.3.2.3 Transport Layer In the DTTB system, the transport layer mainly consists of the channel coding and signal modulation functional blocks, which are the major deterministic factors for DTV transmission system performance. Different DTTB systems adopt different forward error correction coding and modulation schemes with the general schematic diagram shown in Figure 1-4. The cascade correcting codes, including outer error correction code, time-domain interleaving, inner error correction code, and time- and/or frequency-domain interleaving, are basically applied to the error correction coding module. For the existing DTTB systems, there are two modulation technologies: single-carrier and multicarrier modulations. American ATSC [6] and the DTMB system (with parameter $C = 1$) [4] are examples of systems using single-carrier modulation, while the European DVB-T system (using coded-OFDM [7]), Japanese ISDB-T BST system (using segmented OFDM technology [8]), and the DTMB system (with $C = 3780$) (using TDS-OFDM) are examples of systems using multicarrier modulation.

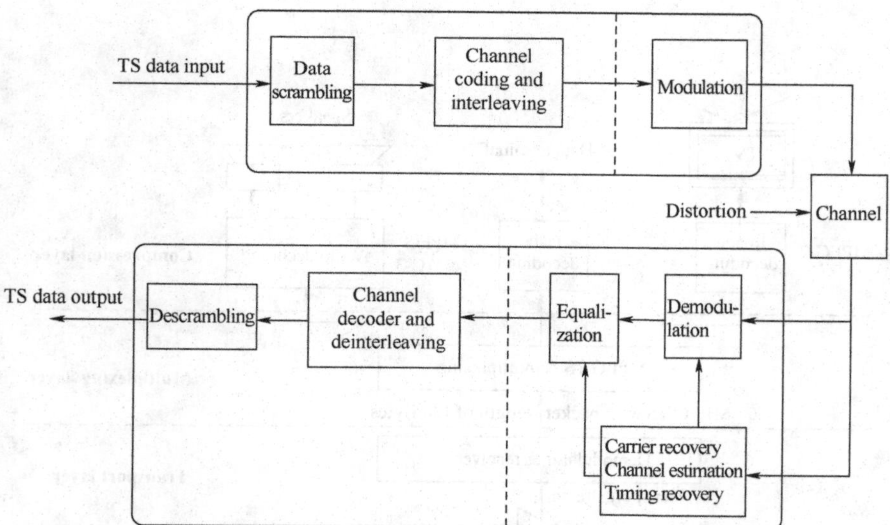

FIGURE 1-4 Transport layer of DTV.

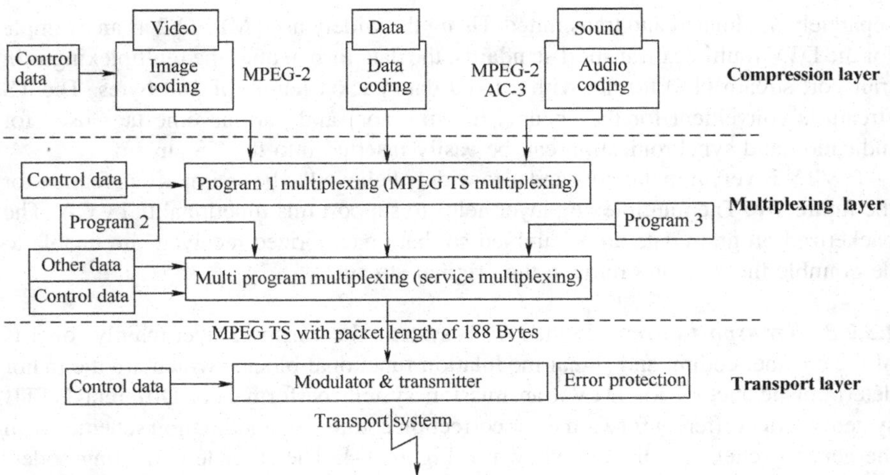

FIGURE 1-5 Layered structure of head-end transmitting system for DTV broadcasting.

This book focuses on the transport layer, with the transmission technologies involved in various DTTB standards described one by one in the following chapters. The compression and multiplexing layers are only introduced in this chapter to give readers an idea of the general principles so they have a complete understanding of the DTV system as well as the relationship between each of these two layers and the transport layer, as shown in Figures 1-5 and 1-6.

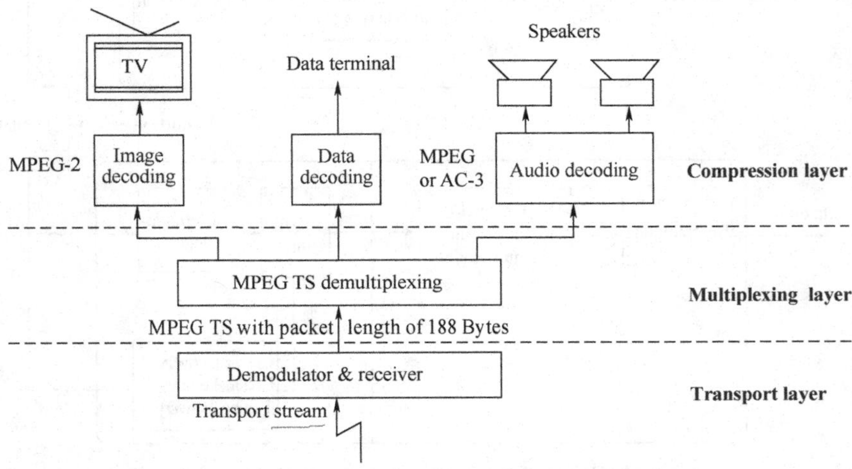

FIGURE 1-6 Layered structure of user terminal system for DTV broadcasting.

TABLE 1-2 Parameters of Video Formats of SDTV and HDTV

Category	Image Resolution	Scanning Mode	Display Mode (Aspect Ratio)
HDTV	1920 × 1080	P; I	16:9
	1920 × 1035; 1440 × 1152	I	16:9; 4:3
	1280 × 720	P	16:9
SDTV	576 or 480 × (720, 640, 544, 480, 352)	I; P	16:9; 4:3
	576 or 480 × (720, 640, 544, 480, 352)		

1.4 COMPRESSION LAYER AND MULTIPLEXING LAYER

1.4.1 Image Format

The major difference between HDTV and SDTV lies in the image quality (in terms of image resolution or definition), and therefore the bandwidth needed for the transmission is different. From the perspective of visual effects, the image quality of HDTV (over 1000 lines) reaches or comes close to the level of the 35-mm wide-screen film. The image quality of SDTV roughly corresponds to images with resolution of about 500 lines and is equivalent to that of DVD. If the MPEG-2 compression encoding standard is used for SDTV, the video code rate will be ~4 Mbps while that for HDTV will be around 20 Mbps. The parameters of SDTV and HDTV video formats are listed in Table 1-2, where I represents interlaced scanning and P represents progressive scanning.

Digital TV has various display modes (e.g., 4:3 and 16:9), among which the 16:9 wide-screen mode is the most common in HDTV. According to different display modes, there are different MPEG image formats. The basic formats are classified into four levels: low level, main level, high-level narrow screen (4:3), and high-level wide screen (16:9). For example, the parameters of the image format of the high-level wide screen are shown in Table 1-3.

1.4.2 Compression Modes for DTV Signal

The ITU BT.601 standard is believed to be the first formal step for the parameter standardization of DTV broadcasting systems. It specifies the basic parameters of signal coding for the TV studio using the 625-line (PAL) or 525-line (NTSC) system.

TABLE 1-3 Parameters of Image Format for Wide Screen (16:9)

Corresponding TV system	625 lines/50 fields/4:3
Wide-screen TV system	1250 lines/50 fields/16:9
Effective number of pixels	1920 × 1080
Sampling frequency	Luminance: 72 MHz; chrominance: 36 MHz
Line frequency	31250 Hz
Image bandwidth	27 MHz

The BT.601 standard specifies video formats for TV studios as well as the coding mode, the sampling rate, and the sampling structure for the color TV signal as follows:

1. Component coding should be used for color DTV signals, which means that composite color TV signals are first separated into luminance signal (denoted as Y) and chrominance signals (denoted as B–Y and R–Y, respectively) before they are quantized, encoded separately, and then combined into one DTV signal.
2. Using 4:2:2 coding as an example, the sampling frequency of the luminance signals and color difference signals are specified as 13.5 and 6.75 MHz, respectively, and in a quadrature structure. That is, the R–Y and B–Y are sampled on each line (and repeated by line, field, and frame) at the same sampling position as the Y signal having the odd index. With this arrangement, the sampling structure for the image is fixed, and the relative position of each sample point will be unchanged on the TV screen.
3. Linear PCM coding is to be performed on both luminance signal Y and two color difference signals (R–Y, B–Y), and the value of each sampling point is quantized by 8 bits. It also specifies that the digital coding not use the entire dynamic range of analog-to-digital (A/D) conversion; only 220 quantized levels are allocated to the luminance signal with the black level corresponding to the quantized level 16 and white level corresponding to the quantized level 235. There are 224 quantized levels allocated to each color difference signal with the zero level of the color difference signal corresponding to the quantized level 128.

In summary, the data throughput in component signal coding is very high. Taking the 4:2:2 coding standard as an example, the data rate of the bit stream is $13.5\times 8 + 6.75\times 8\times 2 = 216$ Mbps, and this is only the sampling frequency required for the SDTV definition, as shown in Table 1-3. The sampling frequency for HDTV is 72 MHz for the luminance signal and 36 MHz for the chrominance signal, and the corresponding data rate is around 1.2 Gbps. Therefore, the challenge for DTV is how to adopt an efficient compression method to eliminate the redundant information (redundancy) from the DTV signal so as to reduce the required transmission rate by orders of magnitude.

The 4:2:2 encoding mode also has a variant mode of 4:2:0, and this format is widely applied to DTV transmission systems when the bandwidth is limited. In this mode, the chrominance signal is sampled on every other line to drop the vertical sampling frequency to one-half of the horizontal sampling frequency, and it helps to decrease 25% of the data.

1.4.3 MPEG-2 for Video Compression

MPEG-2 is a frame-based video compression standard and supports both interlaced scanning video and different aspect ratios. MPEG-2 is a hierarchical compression scheme to adapt the different channel transmission conditions. This standard

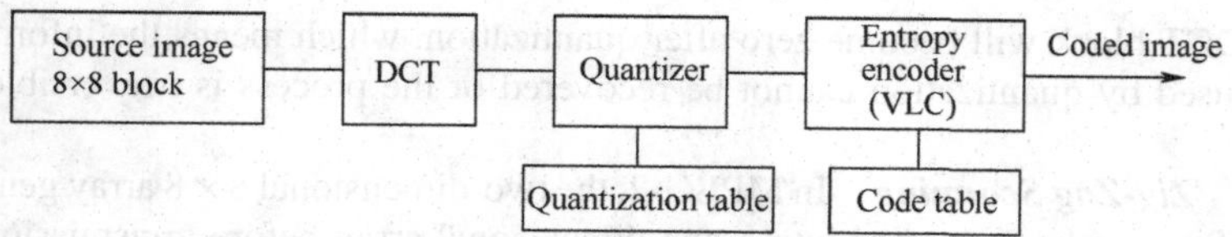

FIGURE 1-7 Intraframe coding scheme based on DCT.

defines the syntax of the data streams and the decoding process, not the encoding process.

The MPEG-2 compression scheme includes intraframe and interframe coding. During the compression, intraframe coding is performed by utilizing the spatial correlation within a frame while interframe coding uses time as a parameter and utilizes the adjacent image frames as a reference for the prediction and estimation to eliminate the temporal redundancy of video signals, and this can reduce the data rate significantly.

1.4.4 Intraframe Coding

Intraframe coding mainly comprises the discrete cosine transform (DCT) and quantizer and entropy coding, as shown in Figure 1-7.

1.4.4.1 Discrete Cosine Transform DCT is used for spatial transform in MPEG-2, which is performed in an 8×8 image block to generate one 8×8 DCT coefficient block. In general, the energy of the image will concentrate on the a few low-frequency DCT components after DCT, i.e., only values for a few lower-frequency components on the left upper corner of the 8×8 DCT coefficient block are greater, while the rest are very small. This makes it possible to transmit only those large values in the DCT coefficient block without significantly degrading the image quality.

DCT does not directly compress the images but helps concentrate the energy of the image, which is the foundation of video compression.

1.4.4.2 Quantizer In MPEG-2, quantization is performed on the DCT coefficients of the image by dividing the DCT coefficient by the quantization step (also known as quantization precision). The smaller the quantization step, the more accurate the quantization precision will be. As more detailed information of the image is preserved, the required transmission bandwidth is higher. According to the visual reaction principle, different DCT coefficients have different significance to the human eye; hence, an encoder will apply the different quantization precisions to the 64 DCT coefficients to ensure enough specific frequency information is contained in DCT while the quantization precision does not exceed the required limit. Among the DCT coefficients, the lower frequency coefficients have more significance to the human eye; therefore, a simple quantization step should be used. On the other hand, as high-frequency coefficients have much less significance to the human eye, a more complex quantization step is appropriate. In general, most high-frequency coefficients in the DCT block will become zero after quantization, which means the information loss caused by quantization cannot be recovered or the process is irreversible.

1.4.4.3 Zig-Zag Scanning In MPEG-2, the two-dimensional 8×8 array generated by DCT has to be converted into a one-dimensional array before transmission, and there are two conversion methods or so-called scanning modes: zig-zag scanning and interlaced scanning. Zig-zag scanning is commonly used. After quantization, most nonzero DCT coefficients will be within the upper left corner of the 8×8 matrix (i.e., low-frequency component area). These nonzero DCT coefficients will be at the beginning of the one-dimensional array after zig-zag scanning, followed by a long string of zeros (due to the quantization). This creates a favorable condition for run length coding.

1.4.4.4 Entropy Coding Coding on the bit stream generated by the quantization must be performed before transmission. A simple coding method uses a fixed-length code, i.e., every quantized coefficient value is expressed by a fixed number of bits, and this method has very low efficiency. Adoption of entropy coding can improve coding efficiency as it has the advantages of the statistical characteristics of the signals and therefore can reduce the averaged bit rate of the coded signal. Huffman coding is one of the most commonly used entropy coding and is adopted by MPEG-2 video compression systems. In Huffman coding, a code table is created after determining the probability of all the coded signals so that fewer bits are assigned to represent signals with high probability while more bits are assigned to represent signals with low probability, which makes the average length of the overall code streams as short as possible.

1.4.4.5 Channel Buffer The data rate of the bit streams generated by entropy coding generally varies with the statistical characteristics of the video images while the data throughput assigned for transmission is usually constant in practice. It is therefore required to introduce the channel buffer before the coded bit stream is sent to the transmitter. The bit stream is written into the channel buffer from the entropy encoder at the variable bit rate and read out at the nominal constant data throughput of the transmission system. The size of the buffer (or the so-called capacity) is preset but the instantaneous output bit rate of the encoder is often significantly above or below the data throughput of the transmission system, and this may cause overflow or underflow of the buffer. Therefore the control mechanism for the buffer is required to adjust the bit rate of the encoder by controlling the compression algorithm through feedback so that the written-in data rate and read-out data rate of the buffer eventually become balanced. Buffer control to the compression algorithm is achieved by changing the quantization step of the quantizer. When the instantaneous output rate of the encoder is too high and overflow of the buffer immediately occurs, the quantization step should be increased to lower the coded bit rate at the cost of the image quality. If the instantaneous output rate of the encoder is too low and underflow of the buffer is about to happen, the quantization step is decreased to increase the coded bit rate.

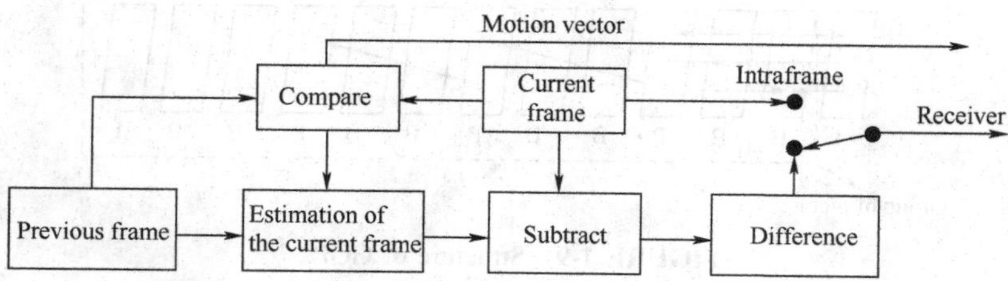

FIGURE 1-8 Motion estimation and compensation.

1.4.5 Interframe Coding Method

MPEG-2 is a frame-based compression scheme and is able to reduce interframe redundancy. In general, there is not too much difference between adjacent frames, and the MPEG-2 scheme tries to predict the current frame by the previous frame (as the reference frame), called interframe prediction. Popular methods of interframe coding are motion estimation and compensation, as shown in Figure 1-8.

When motion estimation is used for interframe coding, the compressed image is estimated by referring to the image of the reference frame. The accuracy of motion estimation is very important to the compression effects. For good motion estimation, only a very few bits are left after subtracting the estimated image from the current image for the compression. Motion estimation is performed on each macroblock to compute the position displacement at the corresponding positions between the current and reference images. While the basic coding unit used to eliminate the spatial redundancy in DCT an is 8×8 block, motion compensation is usually based on 16×16 blocks denoted as macroblocks. Position displacement is described by a motion vector characterizing this displacement along both horizontal and vertical directions. As mentioned before, the MPEG-2 standard only defines the decoding process and motion estimation can be realized in many ways, and hence the performance of different processing methods are also different. The most important parameter is the searching range for interframe motion estimation, and there are many modes to achieve motion estimation and prediction. Those modes include forward estimation, which predicts the macroblocks in the current frame by the corresponding macroblocks in the previous frame; backward estimation, which predicts the macroblocks in the current frame by the corresponding macroblocks in the future frame; and internal coding, which does not make the macroblock prediction. These prediction modes are chosen based on the picture type.

There are three common types of pictures in the MPEG-2 standard: I-, P-, and B-frames. The I-frame, the intraframe, should be coded without referring to any other pictures. Since its compression cannot eliminate the temporal redundancy, the compression ratio is quite limited. The P-frame is the predictive frame. The compression of the P-frame utilizes the previous I- or P-frame picture for the motion estimation and prediction on the current picture and uses the previous picture as the reference for the sequential prediction. This type of picture can eliminate both temporal and spatial redundancies at the same time, and the compression ratio is

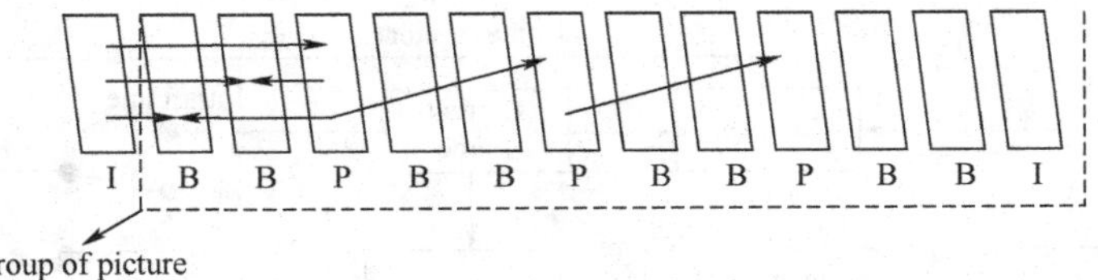

FIGURE 1-9 Structure of GOP.

higher than that of the I-frame picture. The B-frame is the bidirectional predicted picture, and its compression utilizes both previous and future I- or P-frame pictures at the same time to perform motion estimation. To support the backward prediction, which needs future frames, the encoder must buffer and readjust the input order of the image frames so that the B-frame can refer to the future frame picture for prediction. Continuous B-frame pictures would cause certain delay; however, this type of picture has the highest compression ratio.

A group of pictures (GOP) consists of all three types of pictures mentioned above and is shown in Figure 1-9. Two parameters are used for GOP description and they are N (the number of pictures in the group) and M (the spacing of the predictive pictures). Each GOP represents the integrity of a series of pictures for coding, and any editing and division of the coded bit stream must be performed between GOPs.

1.4.6 Audio Compression

The sampling frequency of audio signals is relatively low compared to that of video signals. Part 3 of the MPEG-1 standard is the digital audio coding standard, which is divided into three layers (1, 2, and 3 in ascending order) according to their performance and complexity; a higher layer is backward compatible with a lower layer. The three sampling frequencies are 32, 44.1, and 48 kHz and consist of a monochannel, a dual channel, and other sound modes. A cyclic redundancy check (CRC) code is applied to the code stream to improve the error correction ability at the receiver. An example of an MPEG audio signal encoder is shown in Figure 1-10. A digital audio signal is split into a number of subband signals by the analysis filter

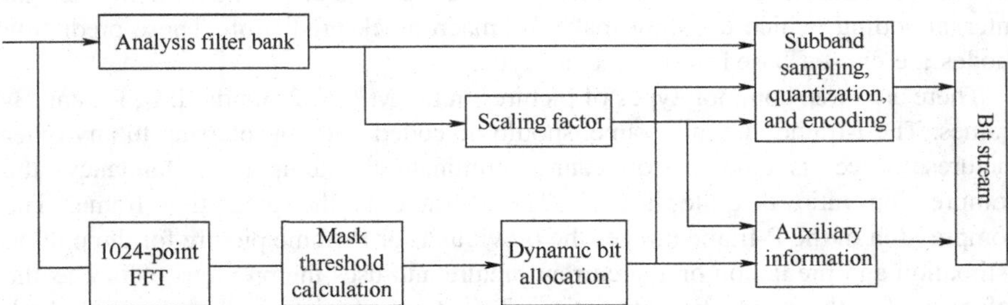

FIGURE 1-10 MPEG audio encoder.

bank, and each subband signal is then sampled, quantized, and coded separately. The quantization steps and the number of bits for quantization are dynamically allocated based on the psychoacoustic model. More accurate spectral analysis is performed on a group of sample points after fast Fourier transform (FFT) to compute the masking threshold. The allocation of the sampling points to each subband signal is different, and the quantization levels of each subband signal are also provided to the decoder as the auxiliary information.

Part 3 of the MPEG-2 standard introduces new digital audio coding methods. It is compatible with Part 3 of the MPEG-1 standard and supports multichannel coding (front center C, front left L, front right R, surround left Ls, surround right Rs, and low-frequency enhancement LFE). The MPEG-2 Advanced Audio Coding (AAC) as the high-quality digital audio coding standard was added to the Part 7 of the MPEG-2 standard.

1.4.7 MPEG-2 Coding

The MPEG-2 coding process is shown in Figure 1-11. The left is the compression layer and the right is the multiplexing layer. The compression layer encoding consists of both video and audio encodings in compliance with ISO/IEC13818-2 and ISO/IEC 13818-3, respectively, and both video and audio elementary streams (ESs) are generated after encoding. In the MPEG-2 multiplexing layer, each ES is packed into the packetized ES (PES) after a packet header is added. Each PES packet and finally video and audio PESs are multiplexed into one TS or program stream (PS). The multiplexing layer complies with the ISO/IEC 13818-1 standard. PES is a logical structure for the multiplexing process and is not used for either storage or transmission. PS has a flexible packet length that is suitable for video storage and editing and is applicable to DVD, interactive multimedia services, etc., with either fixed or variable code rates. TS is generally suitable for channel transmission such as DTV broadcasting.

Based on the above introduction and Figure 1-11, MPEG-2 encoding can be divided into three steps and each generates the ES, PES, and TS, respectively. Each

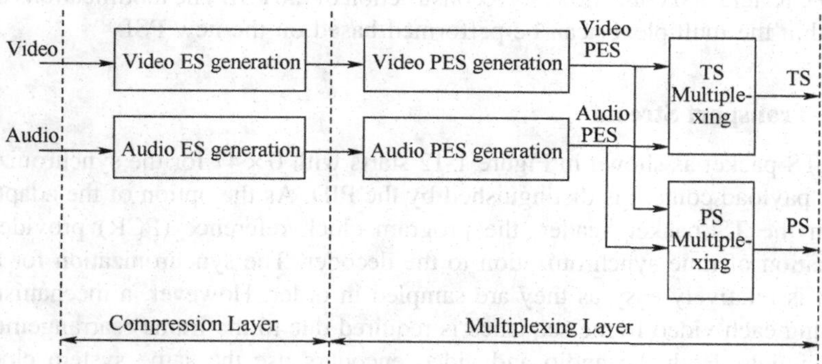

FIGURE 1-11 Generation process of MPEG-2 transport stream.

ES contains the compressed video, audio, and auxiliary data. A packet header is added to the whole or part of the ES to form the PES packet with fixed or variable packet length. A packet header is added to the whole or part of the PES packet to create the TS packet with fixed packet length of 188 bytes. A series of TS packets of audio, video, and auxiliary information are multiplexed together based on certain rules to generate the transport stream.

1.4.8 MPEG-2 Multiplexing

For DTV, MPEG-2 TS multiplexing includes two steps. In the first step, various transport streams of the audio, video, and data PES packets of the same program are multiplexed by a certain ratio to generate a single transport stream for the program [program-specific information (PSI) should be inserted during multiplexing for program identification purposes], so that one complete program transport stream is generated (this process is performed by the MPEG-2 encoder). In the second step, TSs of the different programs are multiplexed into one TS containing multiple programs, and this process is generally performed by a special multiplexer. Multiple DTV programs can be transported using one analog TV channel with bandwidth of 6, 7 or 8 MHz after video–audio compression and digital modulation. To fully utilize the most valuable spectrum resources, TSs of multiple programs are multiplexed again, and this is the second step of the multiplexing. The multiplexer is one of the core units in the DTV video compression system, which packetizes multiple DTV programs based on the protocols specified by MPEG-2 multiplexing layer and provides the interface for the channel transmission. In the TS, the program association table (PAT) and program map table (PMT) are necessary for the multiplexing, and all multiplexed programs are decoded based on the indications of the PAT and the packet identifier (PID) in the PMT.

The multiplexing layer multiplexes the synchronized audio and video information of multiple programs into a single serial bit stream, and synchronization of both the audio and video streams is crucial to the multiplexing layer. The key point of the first step is to introduce an adjustment field by the program clock reference (PCR) to restore the system clock. The key operations for the second step are the readjustment of the PCR field and collection and reconstruction of the PSI. The modifications on the TS within the multiplexer can be performed based on the new PSI.

1.4.9 Transport Stream

Every TS packet as shown in Figure 1-12 starts with 0×47 for the synchronization and its payload content is distinguished by the PID. As the option of the adaptation field in the TS packet header, the program clock reference (PCR) provides the information of time synchronization to the decoder. The synchronization for audio signals is relatively easy as they are sampled in order. However, a mechanism for retrieving each video frame sequence is required due to the frame rearrangement of video signals. Both the audio and video encoders use the same system clock of 27 MHz as the synchronization reference for the programs, and this system clock

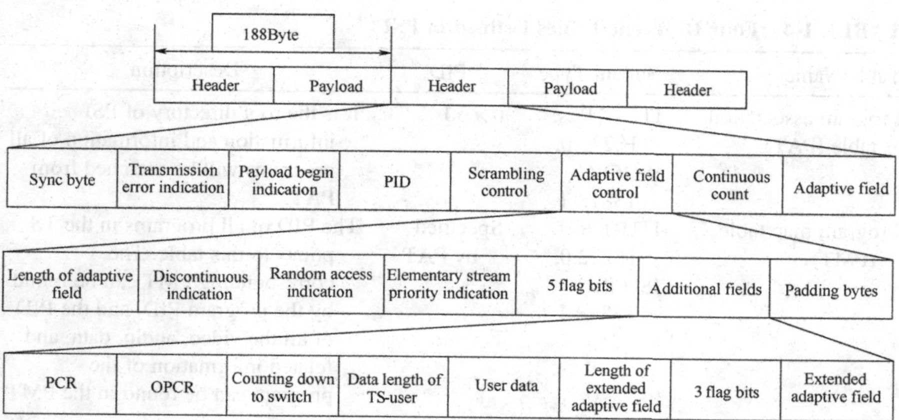

FIGURE 1-12 Syntax structure of transport layer for MPEG-2 TS.

reference information is carried by the PCR field within the TS (for the TS containing multiple programs, each program has its own system synchronization information). The fields indicating audio/video signal decoding and display time are called the decoding time stamp (DTS) and the presentation time stamp (PTS), respectively. At the transmitter side, the PCR of all programs are introduced to the TS while the PTS and DTS based on the PCR are introduced to the PES. At the receiver, the decoder restores the STC (system time clock) according to the PCR value of the program to be received from the TS and adjusts the time delay for the data in the buffer according to the PTS and DTS values to synchronize the time with the transmitter. After time synchronization is achieved, both audio and video data are decoded and displayed according to the DTS and PTS in the PES so as to realize the synchronization of the system clock, audio and video, and also the synchronization between audio and video signals. However, the "synchronization" used here does not necessarily mean the synchronization of the reception and reproduction with the pictures and the sound from the studio in real time. In fact, it is always necessary to have buffers at both transmitter and receiver and the synchronization for the broadcasting services is considered to be achieved if the system only has a constant delay between the transmitter and receiver.

To extract management and synchronization information from programs, the information related to decoding at the receiver, such as PSI and SI (service information), needs to be inserted into the TS to make sure that the receiver can identify the program to be received. PSI is equivalent to the PES packet from the syntax level, which belongs to the lower level TS packet structure. PSI is classified into four different tables as listed in Table 1-4, which become the payload in TS packets after segmentation. The PIDs of these TS packets are specified or given from the PAT. Each transport stream requires at least one complete and effective PAT, which provides the program number and the PID of the TS packet containing the program map table (PMT PID) to make sure that the program can be received well.

TABLE 1-4 Four Different Tables Defined in PSI

Table Name	Stream Type	PID	Description
Program association table (PAT)	ITU-T Res. H.222.0, ISO/IEC 13818-1	0×00	It is the root directory of PSI information and information of all programs will be searched from PAT.
Program map table (PMT)	ITU-T Res. H.222.0, ISO/IEC 13818-1	Specified by PAT	The PID of all programs in the TS points to this table. The corresponding PMT can be found by the program PID, and the PID of all the video, audio, data, and related information of the program can be found in the PMT table.
Network information table (NIT)	Private	Specified by PAT	Physical network parameters: FDM frequency, transmitter ID, etc.
Conditional access table (CAT)	ITU-T Res. H.222.0, ISO/IEC 13818-1	0×01	All the entitlement management messages (EMMs) for the encrypted/scrambled network are provided.

PMT provides the elementary PID of the TS packet belonging to the PES, and the corresponding TS packet can therefore be identified within the stream. PCR is generally included in the TS packet with the same PID within the video PES or is contained in a separate TS packet. The conditional access table (CAT) describes the ES encryption mode for a program and only the authorized decoder can receive the key from CAT for decoding the corresponding data stream. The content in the network information table (NIT) belongs to the service provider for private use, is not defined by the MPEG-2 standard, and is typically contained in the user-selected services. Since PSI is very important, and CRC is utilized.

1. The decoder first extracts the effective PAT, then finds the PID from the TS packet containing the PMT of the desired programs, and finally restores the PMT from the payload in the corresponding TS packet. After that, it extracts PCR and the corresponding audio and video PES of the desired program from the payload of the corresponding TS packets based on the PID in PMT. Eventually, the audio and video ESs are restored from the PES load.
2. The MPEG-2 standard specifies that all program information tables must be sent at a certain frequency of not less than 20 times per second so that the decoder can obtain the PSI in a timely fashion and perform the decoding correctly. However, if there is only PSI in MPEG-2, the receiving decoder still cannot provide the corresponding service information of the program. Therefore, the standard also defines the auxiliary SI to complement the PSI. SI mainly

provides the necessary information for the reception and decoding, such as program type, program time, program source, etc.

Additional tables such as the running status table (RST) and EPG are also multiplexed into the data stream to assist in successful reception. In this way, multiple programs can be carried by one physical channel using MPEG-2 transport stream multiplexing technology.

1.5 CURRENT DEPLOYMENT OF DTTB SYSTEMS

Currently, many countries are experiencing the migration from analog TV to the DTV [9]. Considering to the previous evolution from all-analog technologies to digital technologies in the telecommunication area, driven primarily by both customer demand and the progress of science and technology, DTV technology is more mature and advanced. While providing higher quality, more functional, and personalized audio and video programs, both DTV services and the method of reception have become much more diversified; the image resolution can be HDTV, SDTV, and mobile low-definition TV (LDTV). LDTV has horizontal scan lines (also known as pixels of vertical resolution) of no less than 250 with typical resolution of 340×255, and this corresponds to the resolution level of VCD. Users can receive DTV services over the air or by cable, satellite dish, Internet protocol television (IPTV), and mobile DTV receiver.

Direct-to-home (DTH) TV signals generally refer to direct broadcasting services provided by either telecommunication or broadcasting satellites, while DBS specifically refers to direct broadcasting services provided by broadcasting satellite. The DBS service is generally used for nationwide coverage, and the coverage area of the DTH service could easily be across several countries. In urban areas, DTV terrestrial broadcasting equipment can also be used (other than its own over-the-air DTV programs) to relay direct satellite broadcasting TV programs, which allows use of in-home equipment (fixed or mobile, all are capable of receiving terrestrial DTV programs) to receive high-quality direct broadcasting TV programs with a very simple antenna.

IPTV provides either service in one direction or interactive service, including the direct broadcast program, relaying program, VOD program, time-shifted program, and other services using the Internet as the transmission medium and the STB together with TV or PC as the terminal equipment. One important advantage that the Internet can provide is bidirectional channels to support interactive services, especially games. The unique interactive capability and various program sources of IPTV have made and will continue to have significant impact on cable DTV as well as other DTV networks. Moreover, with continuous advances in the technology, the TV industry is reshaping itself. More and more TVs are providing Internet connectivity while the computer monitor has been used to watch TV programs for a long time.

Another interesting observation on DTV deployment and application refers to two extremes for DTV reception: "big" and "small." Big usually means a large screen and high definition for not only in-home use but also big squares and large offices (screen wall). Small means the DTV receiver for mobile devices. Of course, the video resolution for a small TV is not as good as that of big TV, and much efforts has been made to minimize this difference. Considering the strong demand for mobile TV reception, many countries around the world are seriously considering creating or adopting the DTTB standard, which could potentially support portable mobile terminals. Portable mobile terminals include but are not limited to mobile phones which can receive the TV signal and other portable equipment such as DTV receivers on automobiles, laptop computers, PDAs, and MP4. The broadcast network can support music, text, image, data, and other multimedia services in addition to conventional DTV programs.

DTV has greatly shaped the industry and changed people's lives forever, not only the way people watch TV but also their daily life style.

1.5.1 Developments of ATSC, DVB-T, and ISDB-T

The United States is one of the countries that initiated the development of DTV and has successfully and smoothly completed the analog-to-digital switchover. As early as 1996, the U.S. Congress passed the Telecommunications Act of 1996, requiring that the transition from analog TV to DTV in terrestrial networks be completed by the end of 2006, which was a very ambitious plan. Due to problems in development, this deadline was postponed several times, and analog TV broadcast shut-down was eventually accomplished nationwide on June 12, 2009. Thereafter, there were no more full-power analog TV transmitters carrying analog TV signals. However, some low-power analog transmitters still provide analog TV service to people in rural areas, although both the areas and the number of households using analog TV services is very small [10]. The U.S. government has made it clear that all analog TV transmitters must shut down by September 1, 2015. Currently, ATSC has nearly achieved 100% coverage, and around 90% of households can receive at least eight ATSC programs. Over the whole country, more than 1600 TV stations broadcast the ATSC signals, which are mainly HDTV programs, and more than 600 TV stations are broadcasting SDTV programs.

During the transitional period, the U.S. government provided more than 30 million households with gift certificates to purchase the ATSC STB with a total cost of about $1.34 billion. This effort has greatly promoted DTV in the United States and ensured the postponed deadline will be met. In addition, the U.S. Congress also passed legislation in 2006 requiring that large-screen TV sets have built-in demodulators for the terrestrial DTV signal. Meanwhile the FCC established a timeline for all local TV broadcasters to finish the switchover and the copyright protection rules for the DTV programs, which paved the road for the DTV promotion in the country.

Several other countries, including Canada, Mexico, South Korea, and Honduras, have adopted the ATSC standard. As the ATSC M/H standard supporting mobile and portable equipment reception based on ATSC has been published, mobile and portable ATSC receivers are now sold in large quantities in many countries after more local broadcasters started carrying ATSC M/H signals.

ATSC 2.0 is a major new revision of the standard backward compatible with ATSC 1.0 announced in 2013 as a candidate standard and supports interactive and hybrid television technologies when the TV connects to the Internet [11]. The new features include the content delivered to the home either over the air or via the Internet; the ability to store and watch (pushed) video on demand in non–real time (NRT); broadcast of interactive content or Hypertext Markup Language (HTML) applications; advanced audio and video compression such as H.264 and high-efficiency AAC; and audience measurement, enhanced programming guides, the ability for the receiver to forward content to other devices; etc.

ATSC is now developing a new system/standard, ATSC 3.0, which is non–backward compatible with the existing ATSC 1.0. ATSC 3.0 will adopt new physical layer transmission technologies which fully reflect the latest technical breakthroughs to satisfy the ever-increasing demand of users for multimedia service from the broadcasting networks. A call for proposals for the ATSC 3.0 physical layer supporting video with a resolution of 3840×2160 at 60 fps was announced on March 26, 2013. The ATSC 3.0 Technology Standards Group (called TG3) has since then been established with the objective of developing standard(s) for a future broadcast television system, including, but not limited to, standard(s) for the physical, transport, and presentation layers. There are two subgroups in TG3: TG3-1, which aims to develop system requirements and implementation scenarios and conduct research on potential future scenarios and use cases, which are expected to generate further potential system requirements, and TG3-2, with the objective of evaluating the physical layer and conducting research on physical layers starting with relevant attributes, evaluation criteria, and candidate solutions.

Even though not all technical details of the proposals are available to the public and the deliberation process is still ongoing, TG3 is expected to establish requirements for high spectrum efficiency and support of multiple services. The following features and technologies for the physical layer transmission have been proposed and seriously considered: OFDM, bit-interleaved coding and modulation (BICM, with or without iterative decoding/demodulation) for capacity-approaching performance, including LDPC with code rate from very low to high, nonuniform quadrature amlitude modulation (QAM) or amplitude-phase shift keying (APSK), bit-to-cell word demultiplexing, etc.; multi-input–multioutput (MIMO) technology for either diversity or multiplexing gain; layered transmission called cloud transmission (Cloud Txn) [12]; enhanced SFN (eSFN) and SFN planning considering both large-area coverage and local services; support of nontraditional TV services such as non-real-time and emergency alert services, etc.

There is no clear deadline on when ATSC 3.0 will be finalized, but it is been widely expected to be done around 2016–2017.

Europe promoted DTV very quickly and successfully. By the middle of 2013, nearly 20 countries, including Germany, Spain, Norway, The Netherlands, Belgium, Finland, and the United Kingdom (UK), have completed their transition from analog to digital TV. Similar to the United States, these countries have also used legislation and government funding to ensure the smooth transition from analog to DTV [13].

The U.K. is among the countries to initiate efforts in promoting DTV. It has the most influential satellite pay-TV operator, British Sky Broadcasting Group PLC (BSkyB), and public TV broadcaster, British Broadcast Corporation (BBC). The pay mode for commercial operation, however, failed completely in the early stages of promotion. Later on, it promoted the free DTV broadcasting plan, Freeview, making most of the channels free while providing some pay services in 2002. The plan has successfully sped up the deployment process of DTV in the U.K. and made DTV service very popular. Thereafter, the number of DTV users increased steadily and switchover in the U.K. from analog to digital TV was completed by the end of 2012. Germany completed its switchover to DTV by regions and in stages. In April 2003, the transition was first completed in Berlin and then gradually to other areas in the country. By June 2009, the transition in the whole country was accomplished. The analog signal of both cable and satellite TVs was also completely shut down. The Netherlands was the first country in the world to transition from analog to digital TV with the switchover completed by the end of 2006.

Currently, the framework and standards for the second generation of digital video broadcasting (namely DVB-S2, DVB-T2, and DVB-C2) in Europe have been fully implemented. The major consideration for this framework and these standards was to improve overall system performance and reduce the research and development risk, cost, and time by sharing the newly developed technologies. For example, the forward error-correcting code in the second generation of DTV terrestrial transmission standards adopts the high-performance LDPC code which has already been used in the second generation of the satellite DTV transmission standard DVB-S2, and DVB-C2 not only chooses this LDPC code but also adopts the C-OFDM modulation technology used in terrestrial standard DVB-T2. The DVB T2-lite profile was added in June 2011 to the DVB-T2 standard to support mobile and portable TVs and to reduce implementation costs. The T2-Lite profile is mostly a subset of the DVB-T2 standard, and two additional code rates were added for improvement of mobile performance. T2-Lite is the first additional transmission profile type that makes use of the future extended frame (FEF) of DVB-T2. The FEF mechanism allows T2-Lite and T2 to be transmitted in one RF channel, even when the two profiles use different FFT sizes or guard intervals. The DVB next-generation broadcasting system to hand held (NGH) is a draft ETSI EN 303 105 standard [14,15]. This standard defines the next-generation transmission system for digital terrestrial and hybrid (combination of terrestrial with satellite transmissions) broadcasting to hand-held terminals. The NGH standard is based on the DVB-T2 standard and reuses or expands many concepts introduced in the DVB-T2 specification, such as nonuniform constellation, 4D rotated constellations, and the hierarchical modulation for local service insertion.

Japan has also progressed very rapidly in the research and development of DTV using its own home-grown ISDB-T standard [16,17]. In 2003, Japan started the DTV terrestrial broadcasting test in Tokyo, Osaka, Nagoya, and several other cities carrying digital HDTV programs. On April 21, 2006, Japan launched the so-called One-Seg mobile DTV service. By using one of the 13 segmented spectra for each analog TV channel, both terrestrial and mobile DTV services are carried by the same network. With this arrangement, the cellular phone allows viewers to watch a TV program for free other than for telecommunication purpose. One-Seg was originally launched by 29 TV stations and was popularized nationwide by December 2006. By the end of 2009, a system combining terrestrial and mobile DTV programs has been very successful in the commercial promotion, leading to the growth of mobile customers of One-Seg to over 30 million. At the same time, the broadcast operators (NHK and the private broadcast industry) still carried out the transmission of digital broadcasts. By the end of 2008, NHK had established 1444 relay stations, covering 96% of users. By January 2009, the number of DTTB receivers had grown to 46 million. Japan succeeded in the complete digitalization of terrestrial TV broadcasting by terminating analog broadcasting on July, 24, 2011 (except in some regions afflicted by the earthquake/tsunami damage). Several other countries in Asia and South America have decided to implement the ISDB-T standard [17].

1.5.2 Development and Deployment of DTMB System

China ranks number one in both the production and consumption of TVs worldwide: The total shipment is around 100 million sets (export and domestic sales account for about 50% each) in the past year and altogether there are around half billion TV sets in Chinese households. The research and development efforts on DTMB started in the late 1990s. After careful and thorough investigation, it was decided that the main design objective for the DTMB system should focus on high-spectrum efficiency and stable fixed as well as mobile reception capability. Specifically, the following technical features and requirements should be met:

1. The system should at least have the same reception performance as that of existing DTV systems. Users should easily receive stable DTV signals through existing facilities for analog TV, especially through the very simple, low-gain small antennas. The DTV signal should also be received successfully using directional, high-gain outdoor antennas at the edge of the service area or other places with weak signal strength. In other words, it should support stable operation whether it is under a strong static or a dynamic multipath reception environment.
2. The system transmission capacity for its typical working modes should be more than 20 Mbps. It is required that the DTTB system provide such services as HDTV programs or multiple SDTV programs plus multiple audio signals, auxiliary data, system service/control information, and program guides in one analog TV channel (i.e., 8 MHz bandwidth). This means the system should have a high enough payload bit rate to support these services simultaneously. System data capacity is one of the most important metrics to evaluate the design of the DTV system, especially for terrestrial broadcasting with total transmission capacity limited by a very harsh transmission environment. Improvement of other metrics of the DTTB system should not be at the expense of the data transmission capacity.
3. The system must support mobile reception. Portable (mobile) reception is one of the most fundamental requirements for DTTB service, and this helps provide

the satisfactory experience for the seamless reception at any time and place. This requires integration between the DTV receiver and mobile devices such as a mobile phone, PDA, digital camera, and portable computer. The integration also needs to address the design issues of the low power consumption and low implementation complexity of the system.

4. The system should support single-frequency network application for higher spectrum efficiency. The DTTB system should provide excellent anti-interference capability regardless of both cochannel and adjacent-channel interference, support the use of "taboo" channels of the analog TV system, and use either multifrequency or single-frequency networks to increase coverage and fill up the "shadow" area.

The DTMB system successfully integrates the unique contributions from different proposals and harmonizes them to form a uniform standard. It adopts the frame structure and multicarrier modulation technique of the TDS-OFDM proposed by the proposal from the consortium led by Tsinghua University, adopts the single-carrier modulation technique as well as the system information definition proposed by the consortium led by Shanghai Jiaotong University, and accepts the proposal of using LDPC code as the forward error correction code proposed by the Academy of Broadcasting Science of the State Administration of Radio, Film and Television (SARFT). One important feature of this standard is that it combines both single- and multicarrier modulation schemes under the same system structure with the same system frame structure, scrambling method, forward error correction code, system clock, signal bandwidth, and time interleaving so that both the transmitter and receiver can be automatically supported by the same hardware platform with very little increase of complexity.

The whole industry chain for DTMB is now pretty mature and the related standards for terrestrial DTV broadcasting have been issued. There are 17 standards on the transmit side as well as network planning that include the standards of implementation guidance, frequency, and service planning with the corresponding data interface, system assessment and measurement, equipment requirements and measurement, single-frequency network planning, and service information and monitoring. They are drafted and published by the Digital TV Terrestrial Technique Standard Group, which is led by SARFT. In addition, the DTV standard work group led by the Ministry of Industry and Information Technology has drafted 16 supporting standards, focusing on equipment and terminals, including transmitting equipment, receiving terminal equipment, data interface, display devices, system software, measuring instrument, and assessment methods for the video reception quality.

Hong Kong was the first city in the world to launch commercial DTMB broadcasting, after confirming that DTMB meets all the performance requirements for both fixed and mobile reception under different conditions such as LOS, non–line of sight (NLOS), single-frequency network, and tidal fading. In June 2007, the Office of the Telecommunications Authority (OFTA) in Hong Kong issued an official statement adopting the DTMB as the digital terrestrial TV broadcasting system for Hong Kong.

On December 31 of the same year, commercial broadcasting was formally launched in Hong Kong. In 2008, Macau also adopted the DTMB standard for broadcasting. By the end of 2010, a single-frequency network consisting of 7 transmitting stations had been successfully established in Hong Kong. The network can broadcast 14 programs, including HDTV and SDTV simultaneously, and both coverage and reception quality are far superior to that of analog broadcasting. The statistics data show that by the end of 2011, the DTMB system covered nearly 99% of the area in Hong Kong, with around 70% of the population using DTMB, and there are now more than 200 different models of DTMB receivers available in the Hong Kong market.

In Mainland China, the massive deployment of DTMB was planned and executed step by step. Announcements by SARFT indicated that, by 2009, DTMB signal broadcasting was accomplished in 37 cities, including the provincial capitals, municipalities directly under the central government, separate planning cities, and the cities hosting the 2008 Beijing Olympic Games. DTMB signal broadcasting over 300 large cities was also completed in 2010. Over 50% of the population in China is now covered by the DTMB signal. The next step is to launch DTMB signal broadcasting in close to 3000 county-level cities.

In the meantime, MIIT announced its plan to promote DTMB user terminals. MIIT required that all newly produced TV sets with a size of over 46 inches should have a built-in DTMB demodulator starting January 1, 2014, and all newly produced TV sets should have a built-in DTMB demodulator starting January 1, of 2015.

1.5.3 Network Convergence with DTTB Systems

Network convergence, recently a hot topic, is not the integration of different physical networks or infrastructures but the integration of different services. That is, the interactive multimedia services, including video/audio, data, and voice, should be supported at the same time within the same network, such as the triple play in telecommunications. The vision for network convergence is to fully utilize the resources of the existing network to generate the highest possible revenue and profit by providing comprehensive services to subscribers and users.

For the existing TV network infrastructure, which was originally designed for one-way broadcasting, especially for both terrestrial and satellite systems, network convergence presents a huge challenge to support bidirectional services based on the existing network architecture to tens of thousands of users within the coverage. If successful, it could potentially provide a huge return as it will help retain those subscribers seeking triple-play-like service. DVB-RCT (DVB-return channel terrestrial) and DVB-RCS (DVB-return channel by satellite) were proposed by the DVB organization to support the bidirectional functionality within the broadcasting network, yet both failed to make a case for the successful commercial application due to frequency requirement, costs, and other issues. Another way to support two-way services is the concept of the Integrated Communication and Broadcasting Network (ICBN). In this case, all services are carried not only by a single network but also by either broadcast or telecommunication networks depending on the nature of the requested service. For example, due to the capability of high-speed and broad

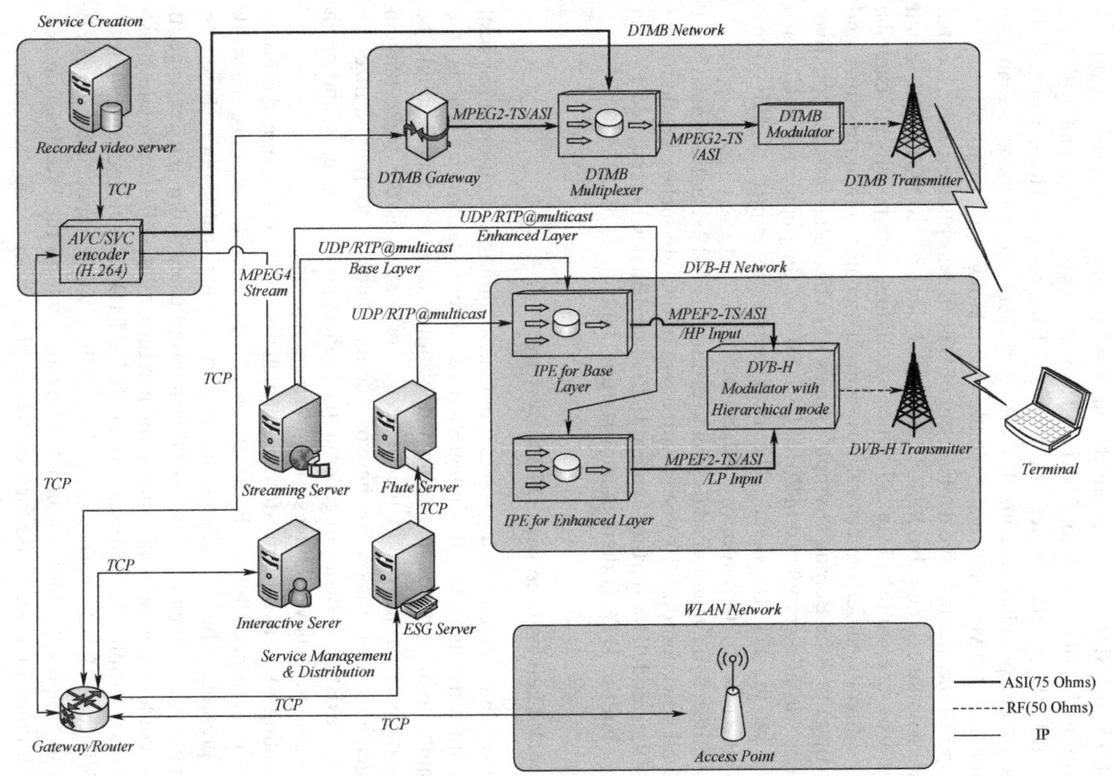

FIGURE 1-13 Schematic diagram of application scenario of MING-T project.

coverage (usually means more users to be supported) of the broadcasting network, commonly or repeatedly requested information or content-pushing services can be supported by the broadcasting network, which is good at delivering the information to lots of users simultaneously and efficiently. The individualized and interactive services, authentication, authorization, and accounting can be effectively supported by telecommunication or data networks. In this way, two-way services can be supported at very low cost without significantly modifying the existing network or requiring additional spectrum. One example of this ICBN concept is the MING-T (the acronym of Multistandard Integrated Network convergence for Global Mobile and Broadcast Technologies) project, a joint European and Chinese project funded by the European Commission within the framework of the FP6-IST program. The basic idea of the project is to use the broadcasting network as the downstream channel for bulk information which mainly provides multimedia information commonly required by a great many users and to use the telecommunication or data network (Internet) to offer the personalized interactive services (both downlink and uplink) or serve as the upstream service request channel. The schematic diagram for the application scenario of the MING-T project is shown in Figure 1-13. The research emphases include interoperability between different broadcasting systems with different standards and a combination of broadcasting and telecommunication systems, real-time handoff, and how to apply scalable video coding to the network SVC to ensure graceful degradation under different reception conditions and other core technologies. The project team has developed hardware and software architecture meeting application requirements for network integration and has carried out related laboratory demonstration and on-site tests [18,19] which could be referenced in the future as a potential way to support the implementation of the network convergence.

1.6 SUMMARY

This chapter has presented the history and evolution of analog and digital TV over the past century. The evolution of DTV has had significant impact on the broadcasting industry above and beyond the analog TV (including both black-and-white and color TV) age. DTV can efficiently use the spectrum and provide more capacity than analog TV broadcast (hence, more programs), better quality images, and lower operating costs.

However, the digital terrestrial TV broadcasting system uses over-the-air broadcast to users instead of a satellite dish or cable TV connection. These broadcasting systems would present the toughest propagation environment, which requires a very robust system design.

REFERENCES

1. Y. Wu, S. Hirakawa, U. H. Reimers, and J. Whitaker, "Overview of Digital Television Development," *Proceedings of the IEEE*, vol. 94, no. 1, pp. 8–21, Jan. 2006.
2. http://www.ultrahdtv.net/.

3. S. O'Leary, *Understanding Digital Terrestrial Broadcasting*, Boston and London: Artech House, 2000.
4. Chinese National Standard GB 20600-2006, *Framing Structure, Channel Coding and Modulation for Digital Television Terrestrial Broadcasting System*, Aug. 2006.
5. C.-Y. Ong, J. Song, C. Pan, and Y. Li, "Technology and standard of digital television terrestrial multimedia broadcasting," *IEEE Communications Magazine*, vol. 48, no. 5, pp. 119–127, May 2010.
6. Advanced Television System Committee. A/53, *ATSC Digital Television Standard*, Washington, DC: ATSC, 1995.
7. ETSI.300 744, *Digital Broadcasting Systems for Television, Sound and Data Services, Framing Structure, Channel Coding and Modulation for Digital Terrestrial Television*, Sophia-Antipolis, France: ETSI, 1999.
8. ARIB STD-B31, *Transmission System for Digital Terrestrial Television Broadcasting*, Tokyo: ARIB, Mar. 2014.
9. http://en.wikipedia.org/wiki/Digital_terrestrial_television.
10. http://en.wikipedia.org/wiki/DTV_transition_in_the_United_States.
11. Advanced Television System Committee, *A/103:2014, Non-Real-Time Delivery*, Washington, DC: ATSC, 2014.
12. Y. Wu, B. Rong, K. Salehian, and G. Gagnon, "Cloud transmission: A new spectrum-reuse friendly digital terrestrial broadcasting transmission system," *IEEE Transactions on Broadcasting*, vol. 58, no. 3, pp. 329–337, Sept. 2012.
13. http://en.wikipedia.org/wiki/Freeview_%28United_Kingdom%29.
14. https://www.dvb.org.
15. DVB BlueBook A160, *Next Generation broadcasting system to Handheld, physical layer specification (DVB-NGH)*, Nov. 2012.
16. http://en.wikipedia.org/wiki/Digital_television_in_Japan.
17. http://en.wikipedia.org/wiki/ISDB.
18. Z. Niu, L. Long, J. Song, and C. Pan, "A new paradigm for mobile multimedia broadcasting based on integrated communication and broadcast networks," *IEEE Communications Magazine*, vol. 46, no. 7, pp. 126–132, July 2008.
19. MING-T website: http://ming-t.informatik.uni-hamburg.de/.

2

CHANNEL CHARACTERISTICS OF DIGITAL TERRESTRIAL TELEVISION BROADCASTING SYSTEMS

2.1 INTRODUCTION

Compared with the coaxial cable used for cable systems, it is easy to imagine that the wireless propagation environment or channel is much worse, even more so than that of the satellite system. The radio signal carrying DTV programs experiences not only propagation loss but also the so-called shadow effect by terrains, including hills, trees, and buildings. The signal will arrive at the receiver from different paths by multiple reflections and scattering. For each path, the amplitude, phase, and arrival time of the radio signal are different from other's. The superposition of radio signals from different paths will generate fast amplitude variation (i.e., fast fading) as well as delay spread of the radio signal. If the receiver is under high-speed mobility, the Doppler effect and random frequency modulation occur. The harsh transmission environment has put very stringent requirements on physical layer transmission technologies to make the digital terrestrial television broadcasting (DTTB) system (with its block diagram shown in Figure 2-1) as reliable and robust as possible [1].

The propagation characteristic depends on the frequency as well as the transmission mode, and DTTB services usually occupy the RF VHF/UHF bands [2]. Unlike when it passes through the coaxial cable, the radio signal of DTTB systems will not only reach the receiver through the direct path but also experience reflections, inflections, and scattering. Terrain factors such as those between the transmitter

Digital Terrestrial Television Broadcasting: Technology and System, First Edition. Edited by Jian Song, Zhixing Yang, and Jun Wang.

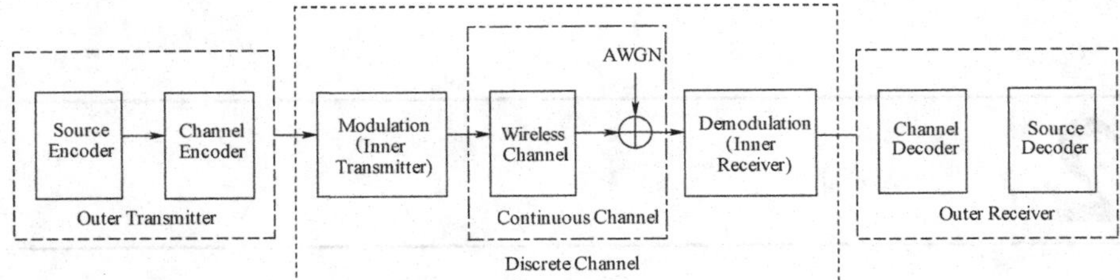

FIGURE 2-1 The functional blocks of the DTTB system.

and the receiver also play an important role. The effects mentioned in this paragraph are not always static but are time-varying. All these factors will add nonnegligible randomness to the received signal and make reliable transmission and robust reception challenging.

Statistically, one can classify the impact of the wireless channel on the received radio signals into two categories depending on the propagation distance: large-scale and small-scale effects. If the receiver is fixed at a certain location, the local mean power of the received radio signal power will be influenced by the large-scale effect, which includes LOS path loss and shadowing. Figure 2-2 shows the impact of large-scale effects, on the received signal power.Path loss (or path attenuation) is defined as the reduction in power density (i.e., attenuation) of an electromagnetic wave as it propagates through space. Study shows that for a wireless channel the path loss can sometimes be represented by the path loss exponent n, with its value normally between 2 and 4. For propagation in free space $n=2$, and for a relatively lossy environment $n=4$. For buildings, stadiums, and other indoor environments, n can be as high as 6. While the tunnel may act as a waveguide, n will be less than 2.

Shadowing refers to the phenomenon of the radio electromagnetic wave being blocked by hills, trees, buildings, etc., which causes significant variation in the

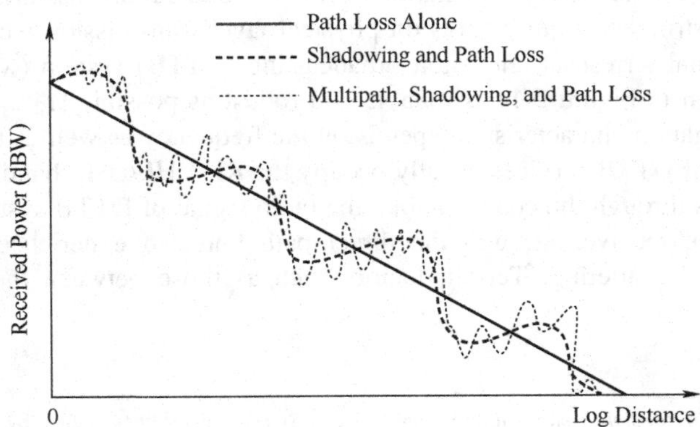

FIGURE 2-2 Large- and small-scale effects of wireless radio channel.

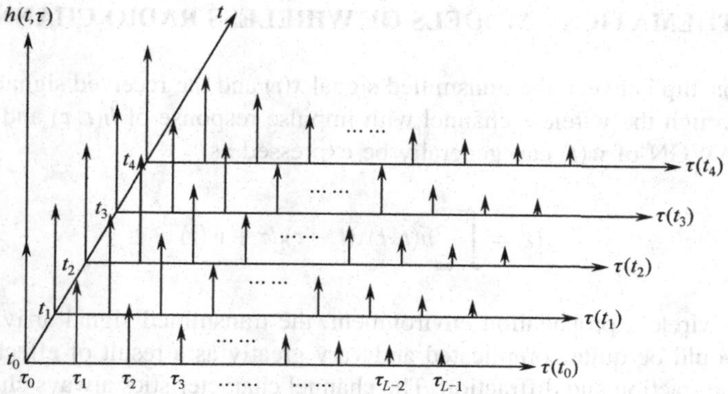

FIGURE 2-3 Channel impulse response $h(t, \tau)$, reflecting both multipath and time-varying effects.

average intensity of the electromagnetic field of the radio signal. Shadowing is also considered to have a large-scale effect with its statistical characteristic satisfying the lognormal distribution.

In summary, the large-scale effect is used to predict or measure the variation of the average radio signal power with a transmission distance of several hundreds of wavelengths and beyond. Path loss and shadowing jointly and statistically reflect the large-scale effect of the wireless radio channel.

The small-scale effect describes the rapid and strong variation of the received radio signal power within a short time (on the order of seconds) or short distance (within a few wavelengths). The major contribution to this variation is because the transmitted radio signal arrives at the receiver through different paths. For the small-scale effect, two wireless radio channel properties are of most importance: the Doppler shift (from the Doppler effect), produced by the relative movement between the transmitter and the receiver, i.e., the radio channel is time-variant, and the propagation delay characteristics of different paths in the multipath environment. Apparently different delays from the multipath channels cause time delay expansion on the received radio signal. The impact of these two characteristics is illustrated in Figure 2-3. The small-scale effect of the wireless radio channel is the main consideration of this book as it is crucial for the research and development in transmission technology selection as well as receiver design to ensure satisfactory performance of DTTB systems.

This chapter is organized as follows. After the brief introduction in Section 2.1, Section 2.2 introduces two mathematical modeling methods of the wireless radio channel: statistical and deterministic. Section 2.3 gives a detailed description of the properties of the wireless radio channel focusing on time and frequency selectivity and based on which the commonly used wireless radio channel model is derived in Section 2.4. Section 2.5 conceptually presents the multipath channel model for the DTTB system, especially under the SFN environment, and a summary is given in Section 2.6.

2.2 MATHEMATICAL MODELS OF WIRELESS RADIO CHANNEL

The relationship between the transmitted signal $x(t)$ and the received signal $y(t)$ after passing through the wireless channel with impulse response of $h(t, \tau)$ and contaminated by AWGN of $w(t)$ can generally be expressed as

$$y(t) = \int_{-\infty}^{+\infty} h(t, \tau)x(t - \tau)d\tau + w(t) \tag{2-1}$$

Under the wireless propagation environment, the transmitted signal arriving at the receiver would be quite complicated and vary greatly as a result of effects such as reflection, refraction and diffraction. The channel characteristics always change with time and space.

Therefore, the channel impulse response $h(t, \tau)$ can be considered as a stochastic process of both time and space. One can use the random process to describe the channel and this is considered stochastic or statistical channel modeling.

In reality, however, there is another channel modeling method which greatly simplifies the channel analysis. If one only considers the properties of the channel impulse response within a relatively short time period and a relatively narrow space, $h(t, \tau)$ will remain the same and the deterministic channel model can be used. The above discussion means there are two ways of wireless channel modeling depending on the time and space scale. If the time or space scale is large enough, one should use the statistical channel model, and the deterministic channel model can be adopted when both the time and space scales are small enough.

In most cases, the channel is taken as deterministic, linear, and time-invariant. Even when the statistical channel model is adopted, the parameters of this model must be deterministic for analytical purposes, e.g., delay spread and Doppler shift. One can conclude that both statistical and deterministic models can be used to describe the DTTB channel characteristics: The former describes the macrocharacteristics of the wireless radio channel while the latter presents the microcharacteristics of the channel suitable for engineering applications. A brief introduction of these two models follows.

2.2.1 Statistical Model of Channel Impulse Response

Stochastic process modeling usually requires knowledge of the joint probability density function determined by a set of random variables. Since it is quite difficult to obtain an accurate joint probability density function, people tend to use correlation functions of the channel impulse response to characterize the statistical property of the channel.

The time-variant channel response $h(t, \tau)$ can be expressed as a finite impulse response (FIR) filter:

$$h(t, \tau) = \sum_{l=0}^{L-1} h(t, l\Delta\tau)\delta(\tau - l\Delta\tau) \tag{2-2}$$

where L is the number of taps determined by the delay spread of the channel and the sampling interval of $\Delta\tau$.

The output signal $y(t)$ can then be expressed as

$$y(t) = x(t) \otimes h(t,\tau) = \sum_{l=0}^{L-1} h(t, l\Delta\tau)x(t - l\Delta\tau) \tag{2-3}$$

Because the channel is random, the autocorrelation function can be used to describe the channel impulse response, i.e.,

$$R_h(t, t'; \tau, \tau') = E\{h^*(t,\tau)h(t',\tau')\} \tag{2-4}$$

The autocorrelation function presented by equation 2-4, however, is still difficult to use as it contains four variables. One can use the wide-sense stationary, uncorrelated scatter (WSS-US) model to simplify future analyses [3].

Mathematically, if the autocorrelation function is dependent not on either t or t' but on $\Delta t = t' - t$, then the channel is considered WSS, that is,

$$R_h(t, t'; \tau, \tau') = R_h(\Delta t; \tau, \tau') \tag{2-5}$$

Strictly speaking, stability requires that the second- and higher-order statistics do not change over time, and here one can only request second-order statistics independent of time for study purposes. Mathematically speaking, the stability must be satisfied within an infinite time, but in practice, the stability is defined such that the second-order statistics do not change within 10 times that of the channel coherence time (i.e., reciprocal of the Doppler frequency). WSS does not mean that the channel impulse response is time-invariant. Using a flat-fading WSS channel as an example, the signal amplitude over time satisfies the Rayleigh distribution with its variance being time invariant but the signal envelope changing with time.

Scattering defines the physical process where the radio signal is forced to deviate from a straight line by localized nonuniformities in the medium through which it passes. Uncorrelated scattering (US) assumes that the scatters from different delays are uncorrelated, that is,

$$R_h(t, t'; \tau, \tau') = R_h(t, t', \tau)\delta(\tau - \tau') \tag{2-6}$$

The WSS-US model is commonly used for multipath fading channels, which is a combination of WSS and US. The WSS-US channel assumes that the channel correlation function is time-invariant while scatterings with different path delays are uncorrelated. This assumption is realistic to describe the short-term variations in the radio channel. Combining 2-5 and 2-6, it can be described as

$$R_h(t, t'; \tau, \tau') = R_h(\Delta t, \tau)\,\delta(\tau - \tau') \tag{2-7}$$

A time-variant channel impulse response consisting of L recognizable paths can be expressed as

$$h(t,\tau) = \sum_{l=0}^{L-1} \sqrt{P_l} g_l(t)\, \delta(\tau - \tau_l) \tag{2-8}$$

where τ_l stands for the time delay of the lth path, P_l stands for the power attenuation factor of the lth path time delay, and g_l stands for the time-variant, complex Gaussian process of the lth path with its power spectrum equal to the Doppler spectrum of the lth recognizable path. Again, the power attenuation factor P_l and the delay τ_l of all recognizable paths jointly determine the multipath characteristic of the frequency-selective fading channel while g_l controls the fading characteristics of each recognizable path.

2.2.2 Channel Impulse Response with Deterministic Parameters

When the WSS-US channel model is applied, the received signal can be expressed as the summation of the input signal with different delay multiplied by a coefficient independently determined by the time-variant, zero-mean complex Gaussian process. For different τ, $h(t,\tau)$ are uncorrelated to each other, while for a particular τ, $h(t,\tau)$ is modeled by a time-variant, complex Gaussian process with flat fading property:

$$h(t,\tau) = \sum_{l=0}^{L-1} \rho_l e^{j2\pi f_d t}\, \delta(\tau - \tau_l) = \sum_{l=0}^{L-1} h_l(t)\, \delta(\tau - \tau_l) \tag{2-9}$$

where ρ_l, τ_l, and f_d stand for the complex attenuation factor, time delay, and Doppler frequency of the lth path, respectively, and $h_l(t)$ stands for the time-variant, complex fading of the lth path. Without losing generality, the channel power attenuation is normalized and satisfies

$$\sum_{l=0}^{L-1} E\{\rho_l \rho_l\} = 1 \tag{2-10}$$

Let $H(t,f)$ be the corresponding channel transfer function of $h(t,\tau)$ known as the channel frequency response, expressed as

$$H(t,f) = \mathrm{FFT}\{h(t,\tau)\} = \sum_{l=0}^{L-1} \rho_l e^{j2\pi f_d t} e^{-j2\pi f \tau_l} = \sum_{l=0}^{L-1} h_l(t) e^{-j2\pi f \tau_l} \tag{2-11}$$

Since the channel is WSS-US and the power attenuation has been normalized, one can obtain the correlation function of the frequency response as

$$R(\Delta t,\Delta f) = E\{H^*(t,f)H(t+\Delta t, f+\Delta f)\} = E\left\{e^{j2\pi f_d \Delta t} e^{-j2\pi \Delta f \tau_l}\right\} = R(\Delta t)R(\Delta f) \tag{2-12}$$

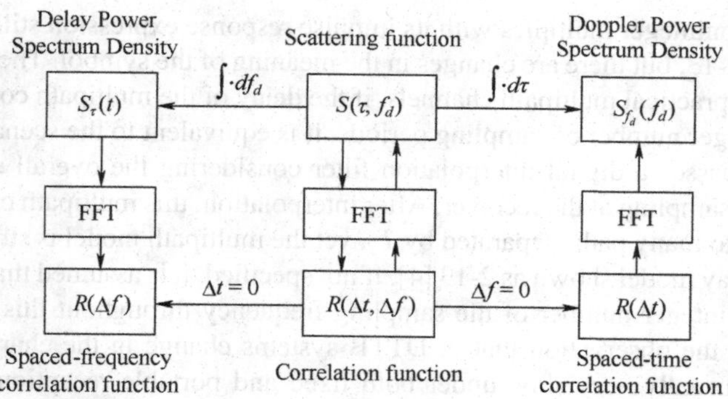

FIGURE 2-4 Relationship between correlation and scattering functions of WSS-US channel.

where $R(\Delta t) = E\{e^{j2\pi f_d \Delta t}\}$ is the spaced-time correlation function; the Fourier transform $S_{f_d}(f_d)$ is the Doppler power spectral density, which can also be obtained by the integral of the scattering function $S(\tau, f_d)$ over τ; and $R(\Delta f) = E\{e^{j2\pi \Delta f \tau_l}\}$ is the space-frequency correlation function with its inverse Fourier transform $S_\tau(\tau)$ equal to the power spectral density of the channel, which can also be obtained by the integral of scattering function $S(\tau, f_d)$ over f_d [6]. The relationship between the correlation function and the scattering function of a WSS-US channel is shown in Figure 2-4.

In DTTB systems, the baseband digital signal will become analog after modulation and pulse shape filtering. It is then sent to the receiver over the air from the transmit antenna after being up-converted into the radio frequency. At the receiver, the front end performs automatic gain control to compensate for the signal power loss due to the propagation. The receiver then converts the analog signals back to baseband digital signals after frequency down conversion and sampling. The match filter is also required before the demodulation.

Based on the above description, the channel impulse response $h(t, \tau)$ of DTTB systems should not only reflect the wireless radio channel itself but also include all the impacts of the pulse shape filter, the actual wireless radio channel (including the frequency conversion, transmission, and receiving antennas), the automatic gain control module, and the matched filter. One can use the equivalent baseband digital filter to characterize the relationship between input and output signals. When considering AWGN noted as $z(n)$, this relationship can be expressed as

$$y(n) = \sum_{l=0}^{L-1} h(n; n_l)x(n - n_l) + z(n) \tag{2-13}$$

where $h(n; n_l)$ is the equivalent baseband channel impulse response and can be taken as an FIR filter, $n_l = \tau_l / T_s$, with T_s the sampling interval or sampling period. It should be noted that if and only if n_l is an integer, $h(n; n_l)$ gives the channel impulse response exactly at each sampling point. When n_l is not an integer, that is, the time delay τ_l of this path is not an integer number of sampling intervals, $h(n; n_l)$ becomes the sampling channel of noninteger multiples with its impulse response expression still in the form of equation 2-13, but there are changes in the meaning of the symbol. The reason is as follows: For practical multipath channels, if the delay of the multipath components τ_l is not an integer number of sampling periods, it is equivalent to the scenario that this component passes a digital interpolation filter considering the overall effect of the filtering and sampling at the receiver. After interpolation, this multipath component is converted into many paths separated by T_s, yet the multipath model is still equivalent to the tap delay model shown as 2-13 [4]. If not specified, it is assumed that the system works at an integer number of the sampling frequency throughout this book.

Based on the observation that in DTTB systems change in the channel impulse response is usually very slow under both fixed and portable reception conditions, one can safely assume that the channel is time-invariant within the time interval the received DTTB signal is processed. In this case, the variable n in $h(n; n_l)$ can be omitted and 2-13 can be simplified as

$$y(n) = \sum_{l=0}^{L-1} h(n_l)x(n - n_l) + z(n) \tag{2-14}$$

The above equation indicates that the channel impulse response can be characterized as a linear, time-invariant FIR filter as follows:

$$h(n) = \sum_{l=0}^{L-1} h(n_l)\delta(n - n_l) \tag{2-15}$$

2.3 PROPERTY OF WIRELESS FADING CHANNEL PARAMETERS

This section investigates fading parameter properties based on the well-established mathematical model of the wireless radio channel given in Section 2.2.

Due mainly to the multipath effect of the wireless radio channel, the relative movement between the transmitter and the receiver, and the scattering environment around the mobile devices, the channel impulse response exhibits the dispersive property in both the time and frequency domains. These effects are called frequency- and time-selective fading, respectively, and the signal will suffer from delay spread and Doppler spread when passing through the wireless channel.

To characterize these two coexisting yet independent effects, two sets of parameters are needed: coherent bandwidth and coherent time. These two effects will be discussed separately in the following.

2.3.1 Multipath Delay Spread and Frequency-Selective Fading

Multipath delay spread and coherence bandwidth are two parameters used to describe the dispersive property of wireless radio channel time as shown in Figure 2-5, and one

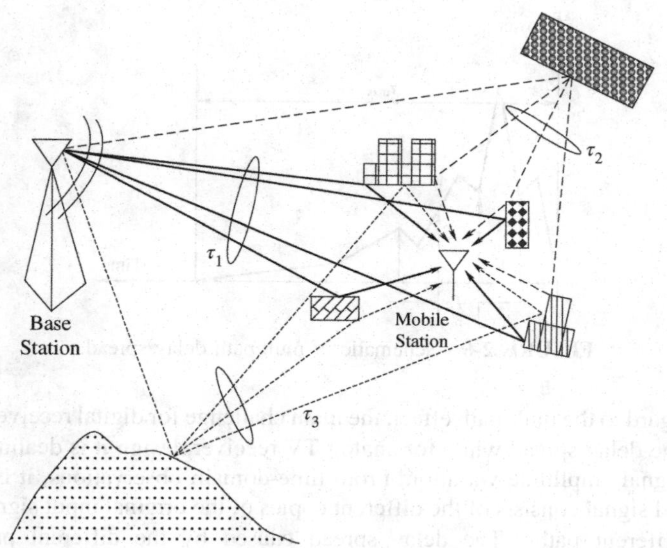

FIGURE 2-5 Schematic of multipath propagation over the air.

can see that the transmitted signal arrives at the receiver through different propagation mechanisms such as scattering, reflection, etc. Different propagation paths introduce different arrival times and arrival phases of the transmitted signal at the receiver, resulting in the so-called delay expansion phenomenon or delay spread of the received signal, which can also be seen in 2-9. The superposition of the transmitted signal with different amplitudes, phases, and time delays after passing through different paths will lead to the dramatic amplitude variation of the received signal, i.e., fading. The fading introduced by the multipath propagation effect is commonly known as the multipath fading.

Assume an impulse signal $x(t) = a_0\delta(t)$ is transmitted over a wireless radio channel. Then the received signal $y(t)$ will contain a series of impulses with different delays and channel attenuation coefficients due to the multipath fading. Since the channel characteristic of each path is an independent random process, one can have

$$y(t) = \sum_{l=0}^{L-1} a_0 h_l(t)\, \delta(\tau - \tau_l) \tag{2-16}$$

It should be noted that the multipath channel changes over time, that is, $h_l(t)$ and τ_l are functions of time. The phenomenon of pulse width expansion of the received signal caused by the multipath effect is known as delay spread. Assume $\tau_0 < \tau_1 < \cdots < \tau_{L-1}$. The maximal delay spread can be characterized by $\tau_{\max} = \tau_{L-1} - \tau_0$, i.e., the time difference between the first and last noticeable arrival signals from the same transmitted signal. Figure 2-6 illustrates how $\tau_{\max}$ is measured from the given channel impulse response.

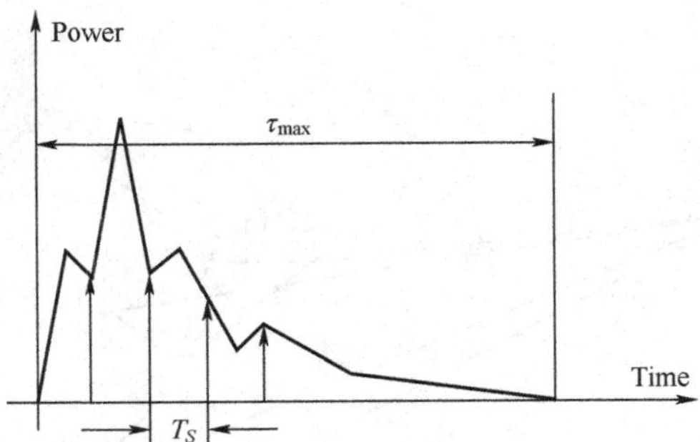

FIGURE 2-6 Schematic of multipath delay spread.

With regard to the multipath effect, the main challenge for digital receiver design is handling the delay spread while for analog TV receiver design it is dealing with the received signal amplitude variation. From time-domain observations, it is seen that the received signal consists of the different copies of the original input signal passing through different paths. The delay spread caused by the different propagation distances introduces severe waveform overlapping, called intersymbol interference (ISI). Therefore, to ensure satisfactory system performance, it is required that the width of symbol duration be much larger than τ_{max}, that is, the symbol rate should be much lower than $1/\tau_{max}$.

Assume $P(\tau)$ is the normalized delay power distribution of the multipath effect shown in Figure 2-7.

The average delay $\overline{\tau}$ of the multiple paths, which is the first-order momentum of $P(\tau)$, can be obtained as

$$\overline{\tau} = \int_0^{\infty} \tau P(\tau)\, d\tau \tag{2-17}$$

The delay spread Δ is defined as the root-mean-square (RMS) of $P(\tau)$ and can be obtained as

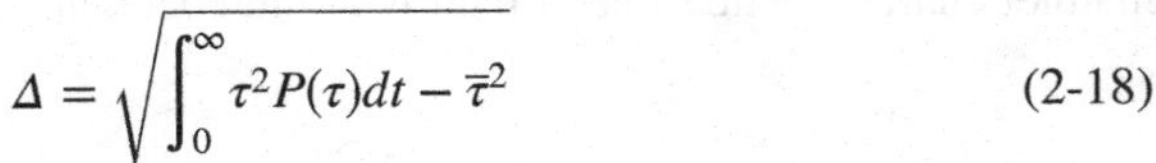

$$\Delta = \sqrt{\int_0^{\infty} \tau^2 P(\tau) dt - \overline{\tau}^2} \tag{2-18}$$

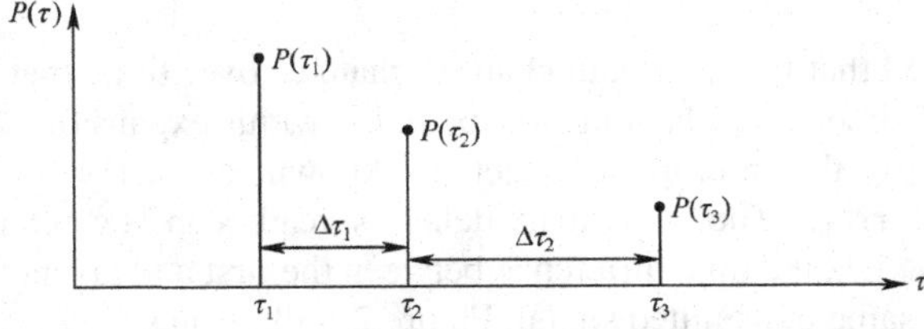

FIGURE 2-7 Power distribution of multipath channel.

where Δ is used to quantitatively characterize the influence of the channel expansion; a larger Δ means a much more severe delay spread effect.

From time-domain observations, the multipath effect introduces the delay spread of the input signal waveform when arriving at the receiver. This waveform overlap, or ISI, causes dramatic signal amplitude variation and hence degrades the system performance. If the observation is done in the frequency domain, one can clearly see that the superposition of multiple input signals with different delays causes the nonflat channel frequency response. This is called frequency-selective fading, that is, the wireless radio channel will have nonuniform attenuation coefficients for the different frequency components. In this case, the distortionless transmission criterion (which requires the constant attenuation and group delay) will not be satisfied and waveform degradation at the receiver side is inevitable due to the multipath channel propagation.

Similar to the time-domain parameters of the multipath delay spread Δ, coherence bandwidth is an important parameter characterizing the frequency-domain multipath channel, which means that the wireless radio channel is not frequency selective or its frequency response will remain flat within the coherent bandwidth. In words, the channel will generate the same (or almost the same) attenuation coefficients for the different frequency components within its coherent bandwidth. That is, a band-limited input signal with all its frequency components falling into the coherent bandwidth will no longer suffer from any (or noticeable) distortion after passing through the wireless radio channel. However, if the input signal bandwidth exceeds this coherence bandwidth, frequency selectivity will surely occur. The nonflat channel frequency response will even have zeros, introducing significant degradation of the signal spectrum, and hence ISI. In practice, the corresponding relationship between the coherence bandwidth B_c and Δ is as follows [5]:

$$B_c = \frac{1}{2\pi\Delta} \tag{2-19}$$

For a high-speed signal with bandwidth much wider than B_c as shown in Figure 2-8, all signal frequency components will experience different changes when passing

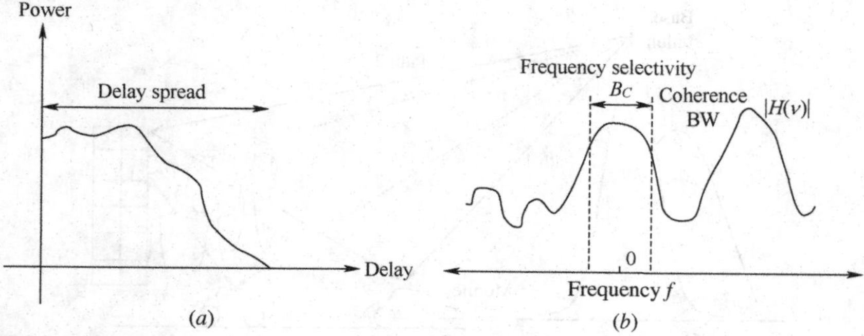

FIGURE 2-8 Multipath frequency-selective fading channel: (*a*) time-domain power distribution profile versus channel delay; (*b*) frequency-domain response of channel.

through the wireless radio channel, causing waveform distortion due to frequency-selective fading. In contrast, when the signal bandwidth is narrower than B_c, the channel frequency response can be considered approximately the same, i.e., each frequency component suffers from the same attenuation. This is called flat fading and no waveform distortion occurs. From the above description, it is seen that B_c is an important parameter characterizing the channel frequency selectivity in the frequency domain. Usually we take delay spread Δ and channel coherence bandwidth B_c as the time dispersion parameters of the channel. These two parameters together with the signal symbol rate decide which kind of fading the transmitted signal experiences, frequency selective or flat fading.

Delay spread Δ and channel coherence bandwidth B_c can be used to determine whether a system is narrowband or broadband from the time and frequency domains, respectively. If the system bandwidth B_s is much less than the coherence bandwidth B_c, then it is a narrowband system. When tB_s is much greater than the coherence bandwidth B_c, it is a broadband system. It is clear that one cannot simply claim a system is narrowband or broadband solely based on the bandwidth it occupies. The appropriate criterion is the ratio of B_s to B_c instead of just B_s. For example, a satellite system usually occupies several tens and even more than hundreds of megahertz of bandwidth, but in most cases, the multipath spread of the satellite channel is quite small, i.e., the reciprocal of the bandwidth is much greater than Δ and is therefore generally treated as a narrowband system.

2.3.2 Doppler Shift and Time-Selective Fading

The time diffusion parameters discussed above could not characterize the time-variant characteristics of the channel. This property refers to the channel characteristics varying over time, that is, when sending the same signal over the channel at different times, the received signal will not be the same, as shown in Figure 2-9. One can clearly see that due to the movement of the receiver inside the car, the channel is no longer identical at two different time instants: t and $t+\alpha$.

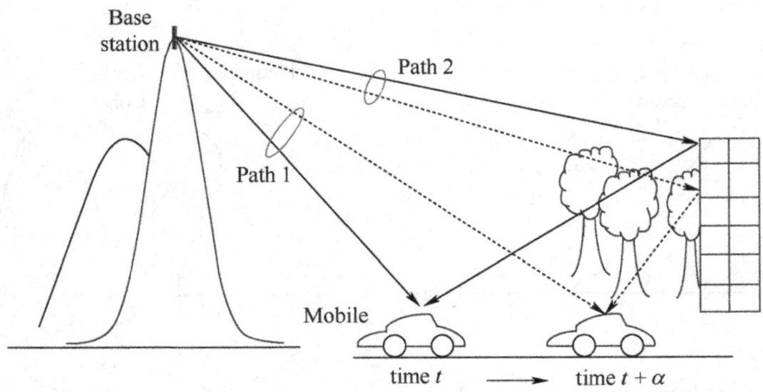

FIGURE 2-9 Time-variant characteristics of wireless radio channel.

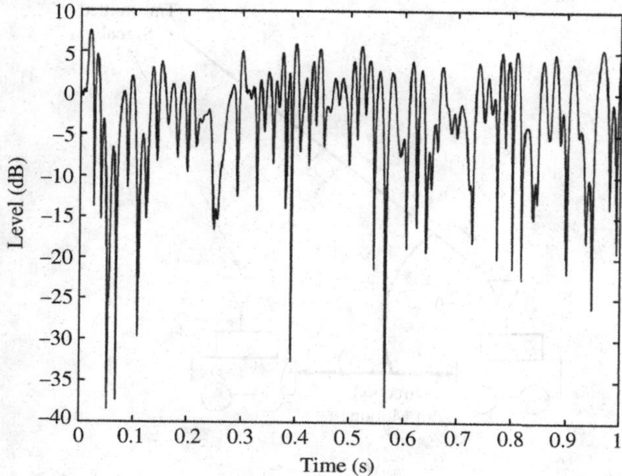

FIGURE 2-10 Channel attenuation versus time after time-variant channel.

Figure 2-10 illustrates how the channel attenuation changes over the time under the time-variant channel. This time-variant characteristic may be caused by the relative movement between the receiver and transmitter or the movement of objects along the propagation path of the signal [4]. The relative movement (displacement) between the receiver and transmitter also leads to changes in the central frequency of the received signal, resulting in additional frequency shift. That is, after a single-tone signal passes through the time-variant fading channel, the received signal is no longer single tone and becomes a signal with a certain bandwidth (having a certain frequency profile), which is called the spectrum expansion of the received signal. This additional frequency shift due to the relative movement shown in Figure 2-10 is also known as the Doppler frequency shift.

As shown in Figure 2-11, if the transmitter is fixed, the mobile receiver moves at speed v, and the angle between its moving direction and the incident radio wave is θ, the Doppler frequency shift is given as

$$f_d = \frac{\nu}{\lambda_c} \cos\ \theta = f_m \cos\ \theta \tag{2-20}$$

where c is the speed of light in a vacuum, λ_c is the wavelength of the signal central frequency, and f_m is the maximum Doppler shift values assuming $\theta = 0$.

After passing through the channel with Doppler frequency shift, the single-tone transmitted signal with frequency f_0 is no longer single tone but within the range $(\pm f_0 - f_m, \pm f_0 + f_m)$, as shown in Figure 2-12. The maximum Doppler frequency shift f_m is proportional to the velocity of the receiver and signal central frequency. For example, when the receiver moves at a speed of 100 km/h with the signal central frequency set at 770 MHz (at which the DTTB signal is transmitted), f_m will be 71.3 Hz. If the receiver moves at a walking speed of 6 km/h, f_m will be 4.3 Hz.

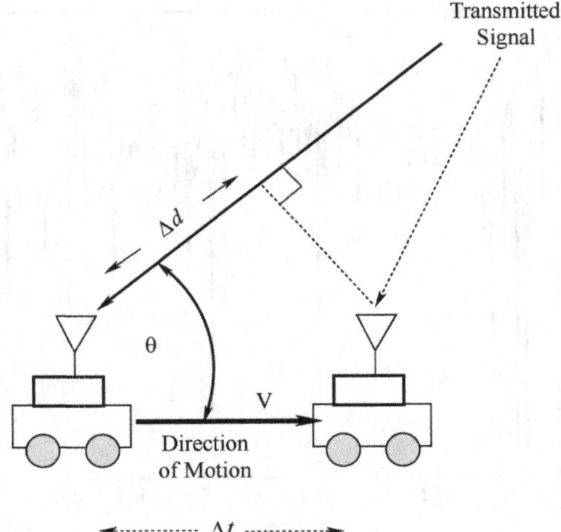

FIGURE 2-11 How relative movement between receiver and transmitter causes Doppler frequency shift.

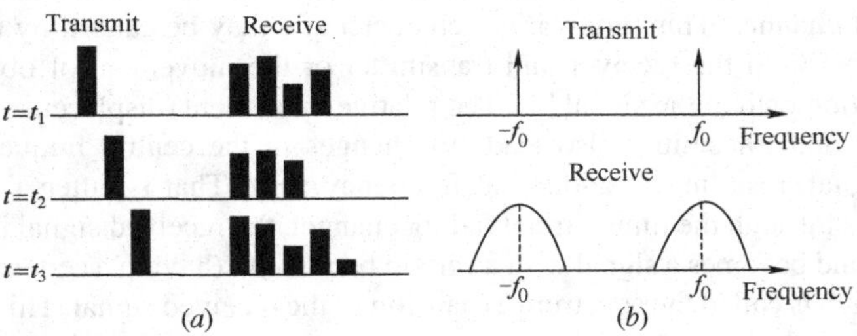

FIGURE 2-12 Time-selective fading of channel: (*a*) channel time-variant impact to received signal; (*b*) impact of Doppler frequency shift to received signal.

In reality, the signal can arrive at the receiver from any incident angle. If an omnidirectional antenna is used at the transmitter and the incident angle from the radio signal to the receiving antenna is a uniformly distributed random variable within ([0, 2 π]), then for a single-tone transmitted signal, the power spectral density function of the received signal is

$$P_s(f) = \frac{2E_s}{\omega_m \sqrt{1 - (f/f_m)^2}} \tag{2-21}$$

where E_s is the average power of the received signal, $\omega_m = 2\pi f_d$.The power spectral density function obtained here is the well-established Jakes model. This spectrum can fully reflect the power spectral density distribution of the frequency-dispersive channel out of the measurement results and is a widely accepted Doppler power spectrum for the wireless radio channel. The channel correlation function in the time domain can be obtained by performing the inverse Fourier transform on $P_s(f)$.

The relative movement between the receiver and the transmitter as well as the changes from the signal propagation environment will lead to the time-variant effect of the wireless radio channel, called time-selective fading. Time-selective fading can also introduce signal distortion as the radio channel has experienced certain changes during the signal propagation. Assuming the symbol duration is quite short and the channel characteristics barely change within this duration, the time-selective fading impact will be negligible and the channel is a slow-fading channel. In contrast, if the channel has undergone significant changes within the data symbol duration and the time-selective fading impact is nonnegligible, the channel is a fast-fading channel. Similar to multipath delay spread, Doppler spread B_D is defined as the standard deviation of the Doppler power spectral density $P_s(f)$, that is,

$$B_d = \sqrt{\frac{\int (f-\overline{B})^2 P_s(f)df}{\int P_s(f)df}} \tag{2-22}$$

where $\overline{B}$ is the mean value of the power spectral density.

Doppler spread B_d is proposed to measure the time-variant characteristics of the wireless radio channel. When the signal bandwidth B_s is much larger than the Doppler spread B_d, that is, $B_s \gg B_d$, the impact of time-selective fading caused by Doppler frequency shift to the signal can be ignored.

Doppler spread B_d is an important parameter characterizing the time-variant property of the channel in the frequency domain. If analyzing the time-selective fading channel in the time domain, the Doppler frequency shift effect can be represented by the channel coherence time T_c. Coherence time is the statistical average time interval within which the channel impulse response remains unchanged. Coherent time is inversely proportional to the Doppler frequency shift and is used to describe the time-variant characteristics of the frequency-dispersive channel. If the symbol duration of the signal (or the reciprocal of the channel bandwidth) is greater than the channel coherence time, the transmitted signal passing through this channel will suffer from the distortion when arriving at the receiver.

Since coherence time is the statistical average of the time interval within which the channel impulse response remains the same, the two or multiple signals arriving at the receiver within this time interval will have strong correlation in amplitude. Usually, the relationship of the coherence time T_c and the maximum Doppler frequency shift f_d can be approximated by [6]

$$T_c \approx \frac{6}{16\pi f_d} \tag{2-23}$$

Doppler spread B_d and coherence time T_c are the two important parameters representing the time-selective fading shown in Figure 2-13, generally known as parameters

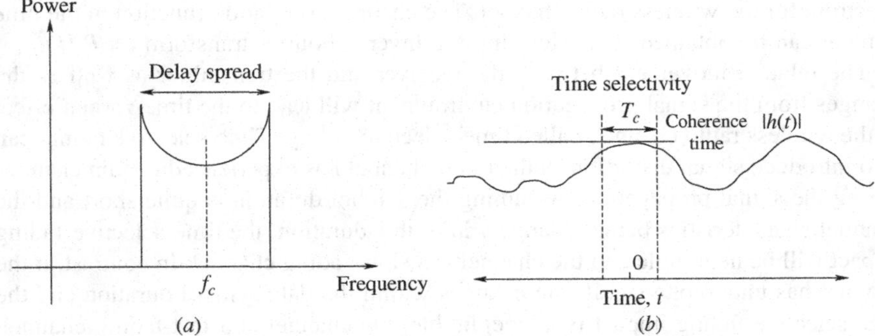

FIGURE 2-13 Time-selective fading of wireless radio channel with (*a*) Doppler spread in frequency domain and (*b*) coherence time in time domain.

characterizing the channel frequency dispersion. These two parameters together with the signal data rate (therefore, system bandwidth B_s) determine whether the transmitted signal experiences a fast or slow fading.

2.3.3 Time- and Frequency-Selective Fading of Wireless Radio Channel

From the above discussion, the transmitted signal may practically incur the multipath delay spread (time dispersion) in the time domain and/or the Doppler frequency shift (frequency dispersion) in the frequency domain. Both channel parameters and signal parameters determine four different types of fading that the transmitted signal may possible experience [7], as illustrated in Figure 2-14. According to the strength of the time dispersion effect, channels can be divided into flat-fading and frequency-selective fading channels. According to the strength of frequency dispersion impact (also known as time-selective fading) caused by the Doppler frequency shift, the channel can be divided into fast- and slow-fading channels. For fast-fading channels, the channel's coherence time is shorter than the symbol duration of the signal to be transmitted.

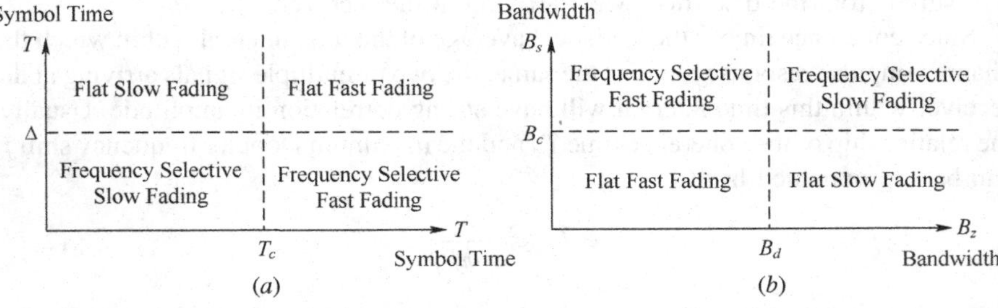

FIGURE 2-14 Relationship between signal parameters and channel fading type: (*a*) between fading and data symbol duration; (*b*) between fading and signal bandwidth.

2.4 COMMONLY USED STATISTICAL MODELS FOR FADING CHANNEL

2.4.1 Rayleigh Fading Model

Rayleigh distribution is often used to statistically model the flat-fading characteristics of wireless radio channels or statistically describe the time-variant characteristics of the envelope at the receiving side due to the independently distributed multipath components.

Flat-fading channel refers to a wireless radio channel with a maximum delay spread that is much smaller than the symbol duration of the signal. In this case, many multipath signals become nondistinguishable at the receiver and the received signal will be the summation transmitted signal passing through the different propagation paths.

Suppose the transmission signal is given as

$$x(t) = \exp\,[j(\omega_0 t + \varphi_0)] \tag{2-24}$$

where ω_0 is the angular central frequency and φ_0 is the initial phase for the central frequency. Assume $y_l(t)$ is the signal from the lth path to reach the receiver with amplitude a_l, phase φ_l, and incident angle θ_l; $y_l(t)$ can be expressed as

$$y_l(t) = a_l \exp\,[j(f_m \cos\ \theta_l + \varphi_l)]\ \exp\,[j(\omega_0 t + \varphi_0)] \tag{2-25}$$

Suppose the distributions of the signal amplitude and the antenna azimuth angle of each path are random and statistically independent. Then the received signal $y(t)$ is

$$y(t) = \sum_{l=0}^{L-1} y_l(t) = [\mathrm{Re}\,(y) + j\,\mathrm{Im}(y)]\ \exp\left[j(\omega_0 t + \varphi_0)\right] \tag{2-26}$$

where

$$\mathrm{Re}\,(y) = \sum_{l=0}^{L-1} a_l \cos\,(f_m \cos\ \theta_l + \varphi_l) \quad \mathrm{Im}(y) = \sum_{l=0}^{L-1} a_l \sin\,(f_m \cos\ \theta_l + \varphi_l) \tag{2-27}$$

According to the central limit theorem of probability, when L approaches infinity, the distribution of the random variables x and y tends to be the Gaussian distribution, and the probability density function can be expressed as

$$p(x) = \frac{1}{\sigma\sqrt{2\pi}} e^{-(x-x_0)^2/(2\sigma_x^2)} \quad p(y) = \frac{1}{\sigma\sqrt{2\pi}} e^{-(y-y_0)^2/(2\sigma_y^2)} \tag{2-28}$$

where σ_x^2 and σ_y^2 are the standard deviations of x and y, respectively, with x_0 and y_0 the mean of x and y, respectively.

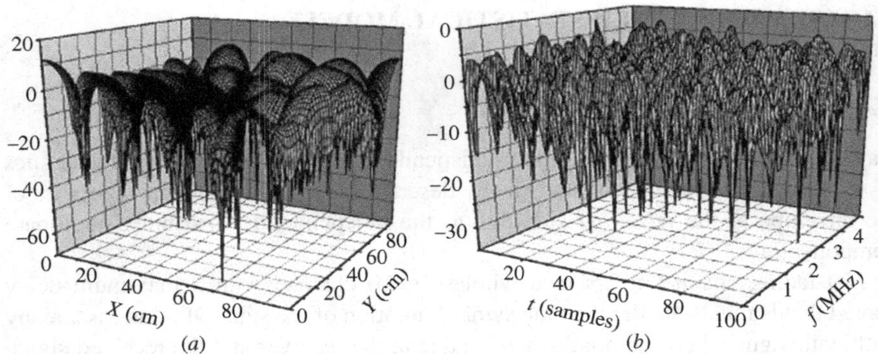

FIGURE 2-15 Rayleigh fading channel characteristics of (*a*) spatial distribution and (*b*) frequency response.

The envelope of the received signal is $v = (x^2 + y^2)^{1/2}$. Let $\sigma_x^2 = \sigma_y^2 = \sigma^2$ and $x_0 = y_0 = 0$. One can see that the signal constructed by two orthogonal Gaussian-distributed signals will have its envelope $v$satisfy the following Rayleigh distribution:

$$p(v) = \frac{v}{\sigma^2} e^{-v^2/(2\sigma^2)} \qquad (v \geq 0) \tag{2-29}$$

where σ is the RMS value and σ^2 is the average power of the received signal before envelope detection. Figure 2-15 illustrates the spatial distribution and frequency response characteristics of the Rayleigh fading channel.

One can have the following observations on the Rayleigh fading signal from 2-29:

1. Average value

$$\bar{v} = \int_0^\infty v p(v) dv = \sqrt{\frac{\pi\sigma}{2}} \approx 1.253\sigma \tag{2-30}$$

2. Mean-square value

$$\overline{v^2} = \int_0^\infty v^2 p(v) dv = 2\sigma^2 \tag{2-31}$$

2.4.2 Ricean Fading Model

Among all signal components arriving at the receiver under the fading environment, if there is one static component with highest intensity such as the direct path of the line-of-sight (LOS) transmission case, the envelope distribution of the small-scale fading will follow a Ricean distribution. In this case, other multipath components with much smaller amplitudes and random arrival angles will superimpose on this direct-path component. Since satellite communication usually requires the LOS transmission condition, its channel can be characterized by the Ricean fading model.

If the amplitude of this direct-path component becomes much smaller, i.e., comparable with others, the Ricean distribution degenerates to a Rayleigh distribution. The probability density function of the Ricean distribution is

$$p(\nu) = \begin{cases} \frac{\nu}{\sigma^2} e^{-(\nu^2+A^2)/(2\sigma^2)} I_0\left(\frac{A\nu}{\sigma^2}\right) & (A \geq 0, \nu \geq 0) \\ 0 & (\nu < 0) \end{cases} \tag{2-32}$$

where A is the peak value of the amplitude of the direct path signal component and $I_0(x)$ is the zero-order modified Bessel function. The parameter commonly used in Ricean distribution is the Ricean factor K, defined as the power ratio between the direct-path component and scattered-multipath components, that is,

$$K = \frac{A^2}{2\sigma^2} \tag{2-33}$$

The Ricean factor completely determines the property of the Ricean distribution, as shown in Figure 2-16. When K drops to zero, that is, the amplitude of the direct-path signal gradually decreases to zero, i.e., there is no direct path, the Ricean distribution becomes a Rayleigh distribution.

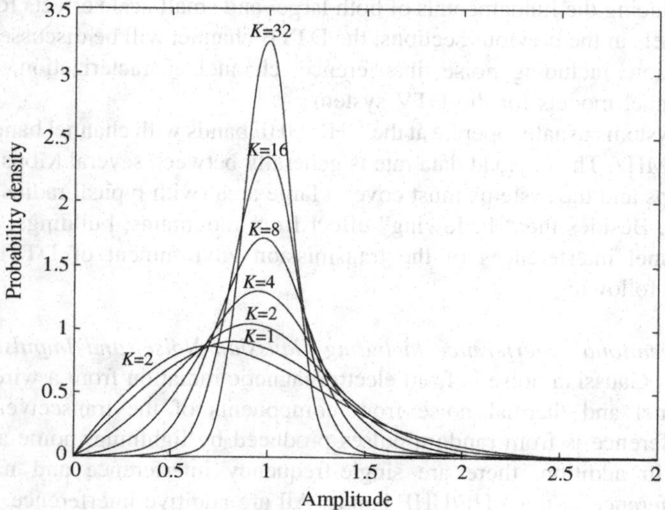

FIGURE 2-16 Ricean distribution with different Ricean factors K ($K=0$, Rayleigh distribution).

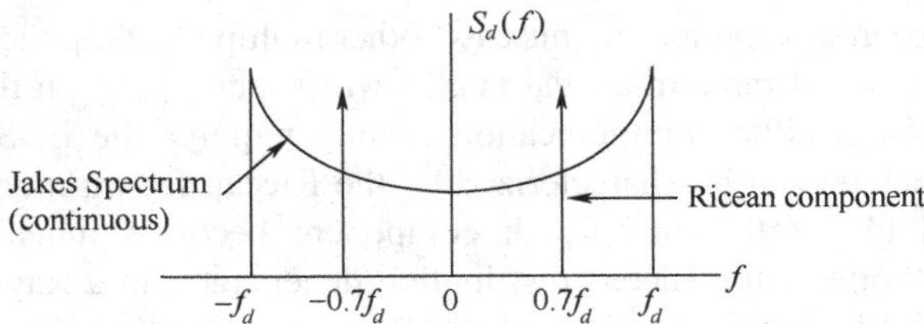

FIGURE 2-17 Doppler power spectrum of Ricean channel.

In characterizing the wireless radio channel, Rayleigh distribution is commonly used to describe the signal envelope variation after passing through the flat-fading channel, and Ricean distribution can be considered as a special case of the Rayleigh distribution with a direct path having much stronger amplitude. As the different incident angle of the arriving signal leads to the different Doppler frequency shift, the receiving signal consisting of all the multipath components will result in a continuous Doppler spectrum, depending on the different emissions and scattering environments. The two most common Doppler power spectra are the Jakes model and the Gaussian model. Figure 2-17 gives the power spectrum for the dynamic Ricean channel.

2.5 DTTB CHANNEL MODEL

2.5.1 Typical DTTB Channel Model

After introducing the fundamentals of both large- and small-scale effects for wireless radio channels in the previous sections, the DTTB channel will be discussed in detail in this section, including noise, interference, channel characterization, and some typical channel models for the DTV system.

DTTB systems usually operate at the VHF/UHF bands with channel bandwidths of 6, 7, and 8 MHz. The payload data rate is generally between several Mbps to several tens of Mbps and the systems must cover a large area (with typical radius of tens of kilometers). Besides the "shadowing" effect from mountains, buildings, etc., other major channel interferences of the transmission environment of DTTB systems include the following:

1. *Conventional Interference Including Gaussian Noise, and Impulse Interference.* Gaussian noise is from electromagnetic radiation from a wireless radio channel and thermal noise from components of the transceiver. Impulse interference is from random pulses produced by lightning, home appliances, etc. In addition, there are single-frequency interference and narrowband interference in the VHF/UHF bands. All are additive interference.
2. *Interference from Broadcasting Systems.* During the transition from analog TV to DTV broadcasting service, frequency planning usually allocates the

newly launched DTV programs to the "taboo" channel (referring to those channels not allowed to transmit as intermodulation effects and other impairments caused by signals from taboo channels will lead to degradation of the current analog TV service) to save the spectrum. Therefore, DTV services are likely to be interfered by the cochannel analog TV service from the neighboring area and from analog TV services operated at adjacent channels within the same service area (nonlinearity of the high-power transmitter strengthens the interference), and these interferences are also additive interference. Even though DTV signals will also introduce interference on analog TV signals, its interference is relatively small as the power of the DTV signal is generally lower than that of the analog TV signal.

3. *Multipath Effect.* Due to reflections from mountains, buildings, and moving objects, the phase of the RF signal reaching the receiver through different paths will be different, which results in signal fading. The impact of multipath interference has been discussed in the previous section and the impact of the multipath on the analog TV signal is called "ghosting." Ghosting is defined as a replica of the transmitted image offset in position and superimposed on top of each other for the analog broadcast. For the DTV services, ISI is generated due to the multipath effect and may cause the reception to fail. As reception failure is not caused by the noise or interference from other systems/signals, it will not be able to be solved by simply increasing the transmitter power. By choosing the appropriate diversity schemes, equalization methods, interleaving length, forward error correction coding approaches, and other anti-fading techniques, there is a good chance that this issue can be resolved effectively.
4. *Doppler Effect.* As mentioned earlier, the relative movement between a transmitter and receiver as well as the movement of objects between the transmitter and receiver causes Doppler frequency shift, which can be calculated by using 2-20. To ensure that the DTTB signal is successfully received by the terminals under mobility, the Doppler effect needs to be taken into consideration. Taking TV channel 28 as an example, with the central frequency of 634 MHz and the corresponding wavelength of 0.47 m, the maximum Doppler frequency shift can be up to 93.93 Hz if the receiver is moving at 160 km/h. This means the system design should be very robust to handle this issue.

In 2000, field tests on multiple sites were conducted in Brazil to obtain the typical channels for DTTB systems [8]. The results listed in Table 2-1 clearly show that the multipath effect is inevitable.

The interference listed in Table 2.1 makes the DTTB channel change with time, location, and frequency therefore, it should be considered carefully in the design of DTTB systems. In general, the maximum delay spread for a DTTB channel will be several tens of microseconds. The channel coherence bandwidth is on the order of 10 kHz according to (2-19), much smaller than the DTTB channel bandwidth (in MHz). Therefore, the DTTB channel is frequency selective and its multipath

TABLE 2-1 Existence Probability of Interfering Effects of Channel at Test Sites in Brazil

Interference Effect	Probability
Multipath effect	100%
Pulse clutter	23%
Doppler effect	2%
Clutter fluctuations	2%
Low level (30–51 dB μV/m)	15%

components are nonnegligible. The channel impulse response can be expressed by (2-9), and typical DTTB channel models can be classified into the following two types:

1. *Static Multipath Channel Model.* As the channel is time invariant, the impulse response $h(t, \tau)$ becomes $h(\tau)$, and (2-9) can be simplified as

$$h(\tau) = \sum_{l=0}^{L-1} h_l \delta(\tau - \tau_l) \tag{2-34}$$

For the outdoor, fixed-antenna reception scenario, the statistical characteristics of the channel change slowly over the time, and the channel can be approximated as time invariant within a short period of time.

For testing purposes, ATSC proposes a static channel model while DVB-T provides channel models of both fixed (F1) and portable (P1) reception, respectively [9]. These models are all time-invariant and can be expressed as

$$h(\tau) = \frac{1}{\sqrt{\sum_{l=0}^{L} \rho_l^2}} \sum_{l=0}^{L} \rho_l e^{-j2\pi\theta_l} \delta(n - \tau_l) \tag{2-35}$$

where L again represents the total number of echoes again; θ_l, ρ_l, and τ_l represent the phase shift, amplitude, and relative delay of the lth path, respectively; and $l = 0$ represents the direct path.

For the channel model of F1, $L = 20$, where F1 is a Ricean channel with factor K given as

$$K = \frac{1}{\rho_0^2} \sum_{l=1}^{L} \rho_l^2 \tag{2-36}$$

If K is set to 10, one can obtain

$$\rho_0 = \sqrt{10 \sum_{l=1}^{L} \rho_l^2} \tag{2-37}$$

TABLE 2-2 Value of θ, ρ, and τ in DVB-T P1 Channel Model

Echo Index	τ_l (μs)	ρ_l	θ_l (rad)
1	1.003019	0.057662	4.855121
2	5.422091	0.176809	3.419109
3	0.518650	0.407163	5.864470
4	2.751772	0.303585	2.215894
5	0.602895	0.258782	3.758058
6	1.016585	0.061831	5.430202
7	0.143556	0.150340	3.952093
8	0.153832	0.051534	1.093586
9	3.324866	0.185074	5.775198
10	1.935570	0.400967	0.154459
11	0.429948	0.295723	5.928383
12	3.228872	0.350825	3.053023
13	0.848831	0.262909	0.628578
14	0.073883	0.225894	2.128544
15	0.203952	0.170996	1.099463
16	0.194207	0.149723	3.462951
17	0.924450	0.240140	3.664773
18	1.381320	0.116587	2.833799
19	0.640512	0.221155	3.334290
20	1.368671	0.259730	0.393889

P1 is a Rayleigh fading channel, that is, there is no direct path ($\rho_0 = 0$). The typical values of τ, ρ, and θ are given in Table 2-2. The frequency response of the DVB-T P1 channel is shown in Figure 2-18 with channel bandwidth of 8 MHz.

For testing on DTTB systems, two other channel models are commonly used: the Brazil channel model, including Brazil channels A–E [8], and the SARFT of China, including SARFT channels 1–8. Both can objectively reflect the characteristics of the DTTB channel objectively. Table 2-3 gives the channel impulse response of some typical channel models from these two sets, all of which are with six paths. In these models, the coefficient of each path is a WSS, narrowband complex Gaussian randomly process with its power spectrum being a Jakes model. The coefficients of each path are independent and have been normalized in power.

2. *Dynamic Multipath Model.* This model includes the time-variant characteristics of the channel and can be constructed by the following two methods [10]:

1. *Phase Modulation Fading Simulator (PMFS).* The channel impulse response is given by

$$h(\tau) = \sum_{l=0}^{L-1} h_l e^{j2\pi f_l t} \delta(\tau - \tau_l) \tag{2-38}$$

where f_l is the Doppler frequency shift of the lth path. In this model, every path is assumed to be noncorrelated. The PMFS provides a simple method for

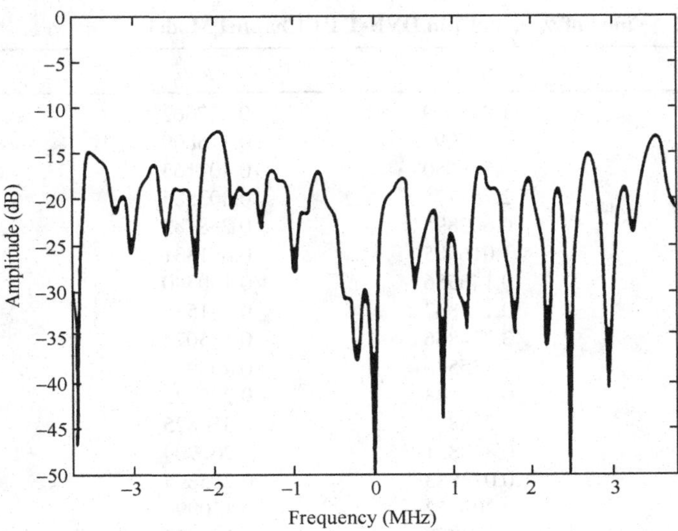

FIGURE 2-18 Frequency response of DVB-T P1 channel model.

TABLE 2-3 Channel Impulse Response of Several Typical DTTB Channel Models

Multipath Model	Multipath Index	Amplitude (dB)	Delay (μs)
SARFT 1	0	0	0
	1	−20	−1.8
	2	−20	0.15
	3	−10	1.8
	4	−14	5.7
	5	−18	18
SARFT 6	0	0	0
	1	−10	−18
	2	−20	−1.8
	3	−20	0.15
	4	−10	1.8
	5	−14	5.7
Brazil A	0	0	0
	1	−13.8	0.15
	2	−16.2	2.22
	3	−14.9	3.05
	4	−13.6	5.86
	5	−16.4	5.93
Brazil B	0	0	0
	1	−12	0.3
	2	−4	3.5
	3	−7	4.4
	4	−15	9.5
	5	−22	12.7

the construction of dynamic multipath while it has significant drawbacks. Despite the fact that $\{h_l, \tau_l, f_l\}$ is randomly generated, once selected, their values will not change over the entire simulation process. Obviously, this is different from actual situation.

2. *Quadrature Modulation Fading Simulator (QMFS).* The channel impulse response is given by (2-8), and the coefficient of each path is a WSS, narrowband complex Gaussian random process. Here, the mean value and the power spectrum of each path are solely determined by the transmission environment, and the multipath components are noncorrelated. In reality, the signal can arrive at the receiver from any incident angle. Assume the incident angle is uniformly distributed between 0 and 2π when an omnidirectional antenna is used, and the Doppler power spectrum of the received signal will be the Jakes model as described earlier.

2.5.2 Single-Frequency Network of Channel Model for DTTB Systems

The SFN is a very important feature of DTV systems which differentiates it from analog TV systems. As mentioned earlier, if one channel is used in one service area, it won't be allowed to be used in the service areas next to the current area. This is the main reason the taboo channel was introduced, and the so-called multifrequency network (MFN) is used in analog TV systems. However, the SFN requires that a number of transmitters transmit the same signal exactly at the same time and the same frequency to achieve the reliable coverage within a certain service area [11]. Figure 2-19 shows the difference between MFN (with reuse factor of 7, meaning each transmitter uses a different frequency indicated by different filling patterns), and SFN (with reuse factor of 1, meaning all seven transmitters use the same frequency). Compared to the MFN, the SFN has the following advantages: The SFN can obviously achieve better spectrum utilization and can provide more evenly distributed signal strength within the desired coverage area. The diversity gain from the different transmitters within the same service area can enhance the reception performance

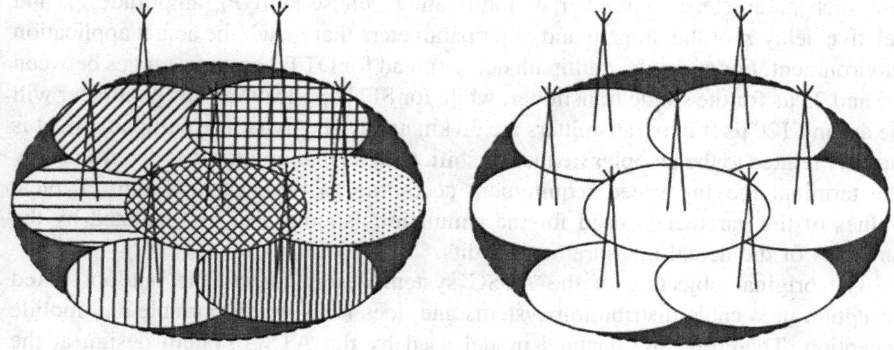

FIGURE 2-19 Comparison of MFN and SFN structures.

greatly and/or help reduce the overall transmit power without sacrificing coverage. By adjusting the density of the transmitters, the height as well as the location of towers, and the transmit power of each transmitter, better coverage and spectrum efficiency can be achieved with blind spots being effectively reduced or even completely removed [12].

Applying the SFN concept to DTTB systems attracts lots of attention due to the advantages of high spectrum efficiency. There are two basic schemes for SFN construction. One uses several high-power transmitters to achieve the coverage for a large area, even for a country. The other is to realize the SFN by using a high-power transmitter and a number of "gap fillers" or "on-channel repeaters" with much lower power. Gap fillers are mainly used to provide additional coverage on those blind spots not covered by the high-power transmitter.

Since the SFN requires all transmitters within the service area to transmit the same signal exactly at the same time and frequency, the receiver in the service area will receive a lot of replicas of the same original signal from different transmitters through different paths. The use of the SFN could generate much more severe multipath effects potentially, that is, in addition to the multipath from reflections and scatterings, signals sent by other transmitters at the same frequency are likely to produce a large amount of artificial multipath, and some of them have long delay and high signal strength. Artificial delay depends on the distance of the transmitters. For example, when the distance of two transmitters is 100 km, the delay could reach 330 μs. For SFN design, the distance of the two transmitters must be carefully chosen to make sure that the artificial delay plus the maximum multipath spread is still less than the length of the guard interval. The typical normalized received signal power versus delay of the single transmitter and two-transmitter SFN are shown in Figure 2-20. It is seen that those long-delay and strong-intensity multipath components will greatly increase the frequency-selective fading of the radio broadcasting channel and therefore put much stringent requirements on receiver design.

The SFN channel model has been incorporated into both the Brazil and SARFT channel models as the Brazil channel E and SARFT channel 8, respectively. The specific channel parameters are provided in Table 2-4.

The channel model should be chosen as close to the actual application situation as possible and be validated by the testing results if applicable. All channel models must use such parameters as number of multipath L, phase shift P_l, amplitude g_l, and relative delay τ_l of the lth path and other parameters that match the actual application environment. For example, multipath delay spread for DTTB systems ranges between 10 and 25 μs for the single transmitter, while for SFN applications this parameter will be around 120 μs if two transmitters are 36 km apart. The dynamic multipath effect is heavily related to the Doppler frequency shift, with its maximum value determined by the terminal moving speed requirement according to the system design. Typical values of the parameters used for the simulations should also be validated by the statistics of the actual measurement results.

The original objective of the ATSC system design targets the outdoor, fixed reception uses cable distribution systems and does not support portable and mobile reception. Therefore, the channel model used by the ATSC system design at the beginning is a static one with a direct path (Ricean channel). Similarly, the original

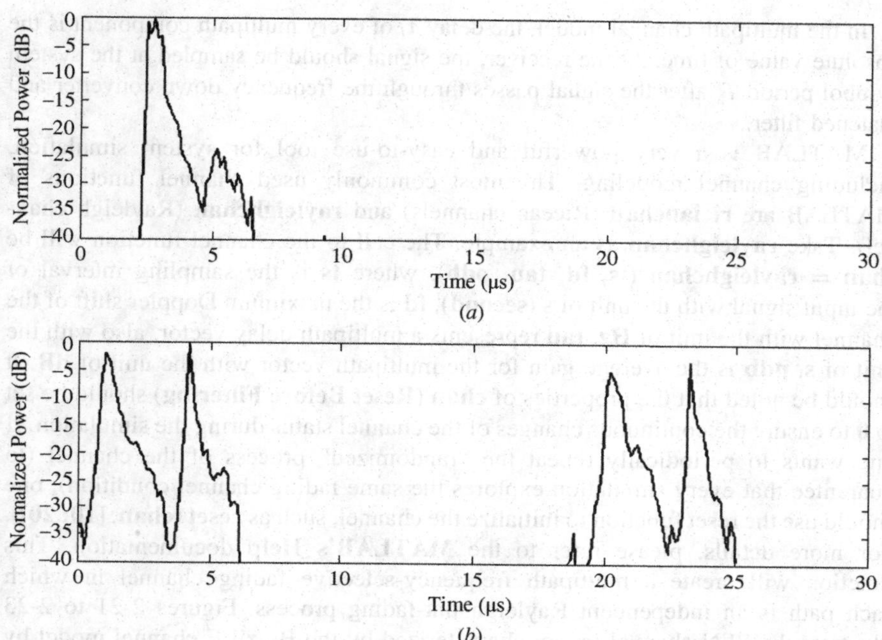

FIGURE 2-20 Normalized received signal power versus delay for (*a*) single-transmitter and (*b*) two-transmitter SFN.

goal of the DVB-T system design is for both indoor and outdoor fixed reception, portable instead of mobile reception. Therefore, the model is a static Rayleigh channel. With introduction of the mobile reception of DVB-T, the dynamic channel model is also required to test mobile reception performance. This also applies to both ISDB-T and DTMB systems.

TABLE 2-4 Parameters of SFN Channel Models

Multipath Model	Multipath Index	Amplitude (dB)	Delay (μs)
SARFT 8	0	0	0
	1	−18	−1.8
	2	−20	0.15
	3	−20	1.8
	4	−10	5.7
	5	0	30
Brazil E	0	0	0
	1	0	1.00
	2	0	2.00

In the multipath channel model, the delay τ_l of every multipath component is the absolute value of time. At the receiver, the signal should be sampled at the system symbol period T_s after the signal passes through the frequency down converter and matched filter.

MATLAB is a very powerful and easy-to-use tool for system simulation, including channel modeling. The most commonly used channel functions of MATLAB are **ricianchan** (Ricean channels) and **rayleighchan** (Rayleigh channel). Take **rayleighchan** as an example. The call to the channel function will be **chan = rayleighchan (ts, fd, tau, pdb)**, where **ts** is the sampling interval of the input signal with the unit of **s (second), fd** is the maximum Doppler shift of the channel with the unit of **Hz, tau** represents a multipath delay vector, also with the unit of **s, pdb** is the average gain for the multipath vector with the unit of **dB**. It should be noted that the properties of **chan** (**Reset Before Filtering**) should be set to **0** to ensure the continuous changes of the channel status during the simulation. If one wants to periodically repeat the "randomized" process of the channel (to guarantee that every simulation explores the same fading channel condition), one should use the reset function to initialize the channel, such as **reset (chan, [10; 20]**). For more details, please refer to the **MATLAB's Help** documentation. This function will create a multipath frequency-selective fading channel in which each path is an independent Rayleigh flat-fading process. Figures 2-21 to 2-23 illustrate the SFN channel model characterized by the Brazil E channel model by using the **rayleighchan** function from MATLAB with the parameters listed in Table 2-4. Figure 2-21 is the impulse response of the Rayleigh channel, Figure 2-22 gives the channel frequency response, and Figure 2-23 shows the changes in the channel impulse response over certain period of time.

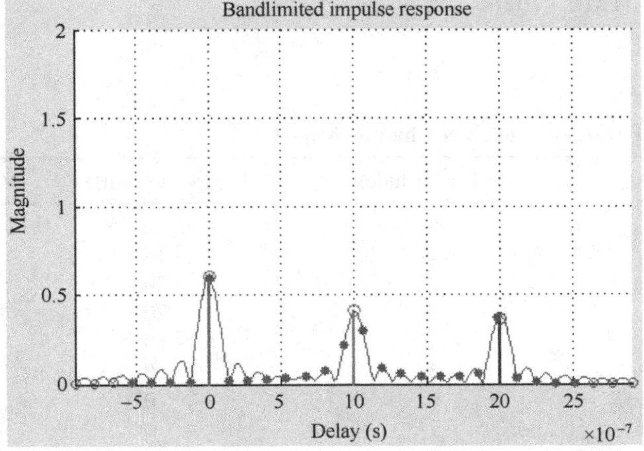

FIGURE 2-21 Impulse response of SFN channel modeled by Brazil E channel model.

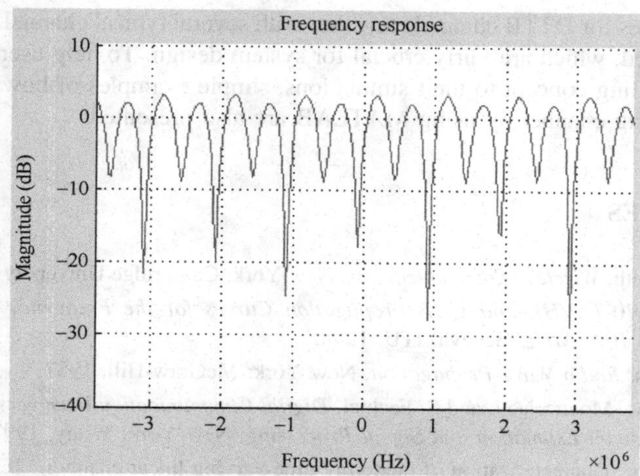

FIGURE 2-22 Frequency response of SFN channel modeled by Brazil E channel model.

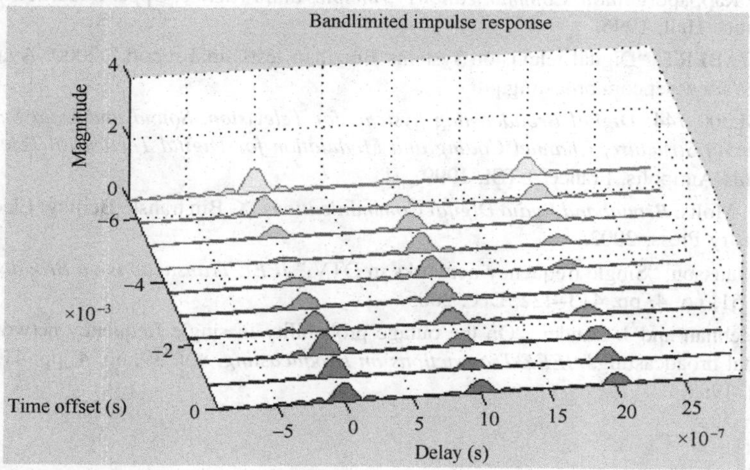

FIGURE 2-23 Impulse response of SFN channel modeled by Brazil E channel model over certain period of time.

2.6 SUMMARY

This chapter starts with a brief introduction of the basic characteristics of the wireless radio channel with two fundamental mathematical models. Based on that, a detailed analysis of channel characteristics such as frequency-selective fading, time-selective fading, delay spread of the multipath, Doppler spread, and other important concepts are presented. The commonly used fading channel models are then derived and the modeling issues for DTTB channels together with several typical channel models are then introduced, which are fairly crucial for system design. To help users apply the channel modeling concept to their simulations, simple examples of how to create a Rayleigh fading channel by using MATLAB are also presented.

REFERENCES

1. A. Goldsmith, *Wireless Communications*, New York: Cambridge University Press, 2005.
2. ITU-R P.370-7, *VHF and UHF Propagation Curves for the Frequency Range from 30 MHz to 1000 MHz*, Geneva: ITU, 1995.
3. L. Boithias, *Radio Wave Propagation*, New York: McGraw-Hill, 1987.
4. H. Meyr, M. Moeneclaey, and S. Fechtel, *Digital Communication Receivers: Synchronization, Channel Estimation and Signal Processing*, New York: Wiley, 1997.
5. P. A. Bello, "Characterization of randomly time-varying linear channel," *IEEE Transactions on Communication*, vol. 11, no. 4, pp. 360–393, Dec. 1963.
6. J. G. Proakis, *Digital Communications*, 5th ed., Columbus, OH: McGraw-Hill, 2007.
7. T. S. Rappaport, *Radio Communications Principles and Practice*, Upper Saddle River, NJ: Prentice Hall, 1996.
8. SET/ABERT, "Digital television systems Brazilian tests final report," 2000, Available: http://www.set.com.br/testing.pdf.
9. ETSI.300 744, *Digital Broadcasting Systems for Television, Sound and Data Services, Framing Structure, Channel Coding and Modulation for Digital Terrestrial Television*, Sophia-Antipolis, France: ETSI, 1999.
10. A. F. Molis, *Broadband Radio Digital Communications*, X. Bin transl., Beijing: Electronic Industry Press, 2002.
11. A. Mattsson, "Single frequency networks in DTV," *IEEE Transactions on Broadcasting*, vol. 51, no. 4, pp. 413–422, Dec. 2005.
12. R. Rebhan and J. Zander, "On the outage probability in single frequency networks for digital broadcasting," *IEEE Transactions on Broadcasting*, vol. 39, no. 4, pp. 413–422, Dec. 1993.

3

CHANNEL CODING FOR DTTB SYSTEM

3.1 CHANNEL CAPACITY AND SHANNON'S CHANNEL CODING THEOREM

The DTTB system is aimed at delivering digital video programs reliably over the air to user terminals. As mentioned in Chapter 2, the band-limited, wireless RF channel for DTTB systems is a very harsh propagation environment, which means that the information transmission capability of this physical channel is quite limited. According to information theory, the information transmission capability is measured by the uncertainty of the information, which can be characterized by its "entropy." The highest possible symbol rate that can be supported is solely determined by the transmission bandwidth. If the symbol to be transmitted is X with entropy $H(X)$ and the symbol received is Y with entropy $H(Y)$, the information quantity which could be delivered over the channel is the mutual information of X and Y and is defined as

$$I(X, Y) = H(X) - H(X|Y) = H(Y) - H(Y|X) \tag{3-1}$$

where $H(X|Y)$ and $H(Y|X)$ are the conditional entropies, representing the uncertainty determined by the conditional probability.

Digital Terrestrial Television Broadcasting: Technology and System, First Edition. Edited by Jian Song, Zhixing Yang, and Jun Wang.

The channel capacity C is defined as the supremum of the mutual information over all possible choices of X:

$$C = \max_X I(X, Y) \tag{3-2}$$

The channel capacity is the maximum information that can be transmitted per unit time (or per unit symbol) by a channel. It is easy to prove that channel capacity is solely determined by the channel characteristics and independent of the information source. In other words, the channel capacity is uniquely determined for a given channel. Only when the information source satisfies certain conditions can the channel capacity be achieved [1].

In 1948, Claude Shannon published his well-known paper entitled "*A Mathematical Theory of Communication*" and gave the band-limited channel capacity under AWGN, also known as the famous Shannon formula:

$$C = W \log_2\left(1 + \frac{P_{\text{av}}}{WN_0}\right) \tag{3-3}$$

where P_{av} is the average signal power, W is the system bandwidth, and N_0 is the AWGN spectrum density. After introducing a new symbol E_b, which stands for the minimum energy required for transmitting each information bit through a band-limited channel, one can have

$$P_{\text{av}} = CE_b \tag{3-4}$$

Hence, 3-3 can be rewritten as

$$\frac{C}{W} = \log_2\left(1 + \frac{C}{W}\frac{E_b}{N_0}\right) \tag{3-5}$$

From 3-5, we get the normalized signal-to-noise ratio (SNR):

$$\frac{E_b}{N_0} = \frac{2^{C/W} - 1}{C/W} \tag{3-6}$$

which gives the minimum SNR required to reach the channel capacity C.

Figure 3-1 plots the relationship between C/W and E_b/N_0 based on 3-6. The Shannon formula indicates that the channel capacity above this curve cannot be achieved.

It is interesting to see from 3-6 that for a given channel capacity C, when W approaches infinity, i.e., C/W goes to zero, E_b/N_0 will not approach zero but will approach an asymptotic value instead: from this observation, one can have

$$\lim_{W\to\infty} \frac{E_b}{N_0} = \lim_{C/W\to 0} \frac{2^{C/W} - 1}{C/W} = \ln 2 \tag{3-7}$$

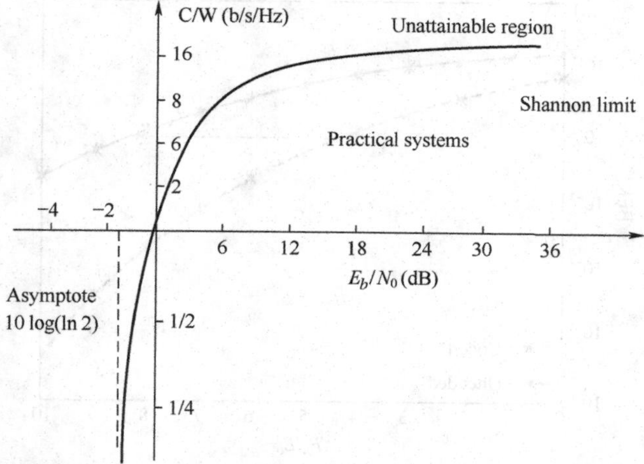

FIGURE 3-1 Relationship between C/W and E_b/N_0 based on Shannon theorem.

If the result of equation 3-7 is expressed in decibels, one can get $10 \log(\ln 2) = -1.6$ dB.

This is the so-called Shannon limit in coding theory and stands for the minimum SNR per bit (i.e., E_b/N_0) required to reach the channel capacity assuming an infinite bandwidth of additive Gaussian white noise. It gives the lower bound of the system transmission capability at the extremely low spectral efficiency.

The Shannon formula 3-3 gives the channel capacity of the bandwidth-limited channel under a given SNR, and one can make the trade-off between the two key parameters of system design, i.e., the transmission bandwidth W and the SNR. However, the Shannon formula does not specify how to obtain this compromise. In other words, this formula does not address implementation issues on how to build a capacity-approaching transmission system. In transmission system design, what needs to be considered are not only feasibility and effectiveness but also reliability and robustness. That is, whether the information can be retrieved correctly at the receiver for every transmission is also of importance. Due to the imperfect nature of the physical medium (i.e., interference and the inevitable noise) throughout the whole transmission process, correspondence between the input and output after imperfect channel transmission only has statistical meaning, and errors are therefore inevitable. The error probability depends on the channel characteristics, transmission duration, and system complexity. Therefore, a practical transmission system usually contains a channel encoding module at the transmit side and decoding modules at the receiver.

Video signals will usually be compressed to get rid of the redundancy to lower the required bit rate before transmission. In this sense, errors from transmission will have a great impact on the final image recovery, and good error correction capability for channel coding is especially important. The purpose of channel coding is to add redundant bits (called check bits) to the original information bit sequence, which satisfies certain constraints to form a codeword for the channel transmission. If errors

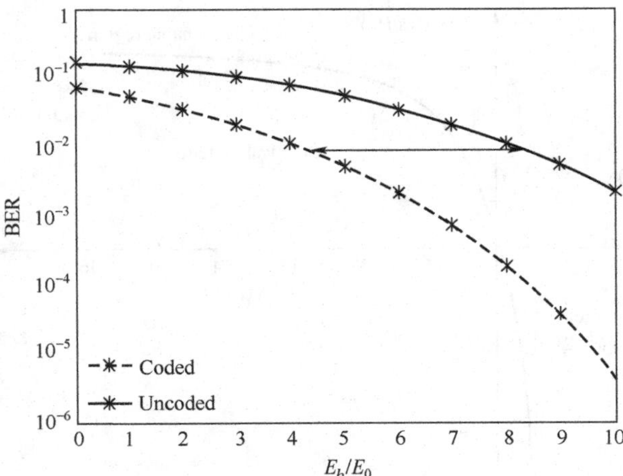

FIGURE 3-2 Coding gain diagram.

occur after transmission, the constraints between information bits and check bits will no longer hold. This constraint can be checked at the receiver so that errors can be identified and corrected. For example, if k information bits need to be transmitted, one can create a codeword of length n ($n > k$) through the coding process, which would introduce $r = n - k$ check bits. Suppose $R_c = k/n$ is the coding efficiency or code rate, the introduction of r check bits could cause the reduction of the transmission efficiency for the information. Under the same bit error rate (BER) requirement, the reduction of transmission efficiency provides a low bitwise required SNR, i.e., E_b/N_0. The E_b/N_0 difference between the coded and uncoded transmissions under the same BER is called coding gain. It should be pointed out that the E_b/N_0 value used here to evaluate the transmission performance of the system is the bitwise SNR before coding instead of after coding. Using Figure 3-2 as an example, the coding gain is around 4 dB at a BER of 10^{-2}.

The next challenge is to find a "good code" to get the highest possible coding gain with the smallest possible redundancy. Claude Shannon proposed his famous channel coding theorem: Suppose C is channel capacity. If the transmission rate $R < C$, then for any $\varepsilon > 0$ there exists a code with block length n large enough whose error probability is less than ε. If $R > C$, the error probability of any code with any block length is bounded away from zero.

This theorem indicates that reliable information transmission over an unreliable channel can be achieved under the condition that the transmission rate never exceeds the channel capability. In other words, noise impacts the transmission rate rather than the transmission accuracy of the channel.

To prove his channel coding theorem, Shannon did not focus the performance of certain good codes. Instead, he focused on the average performance of all codes from the "random coding" to avoid the tough issue of finding good codes. From this point of view, the channel coding theorem proves the existence of good codes, but it does not say how to find them. Theoretically, only when the code length n goes to infinitiy can the transmission capability of the system approach the Shannon limit. This, however, is unrealistic as the complexity of the decoding algorithm at the receiver generally increases exponentially with the coding length. Even for limited coding length, the codeword set obtained from "random coding" can still be quite large, which means much effort is needed to find good codes as the codes tend to be irregular. One can only use look-up tables when decoding, whose complexity is not acceptable. Therefore, channel coding is regularly constructed mathematically to ensure the codeword has a structure easy enough for coding and decoding. The performance of these codes usually has some differences compared with Shannon limits.

In summary, a good balance is needed between effectiveness, reliability, and computation complexity for channel code design based on actual channel conditions. Sometimes, factors such as delay and memory size should also be considered for channel code design and performance evaluation.

As discussed in Chapter 2, the severe interference in the DTTB system which operates in VHF/UHF bands comes from different sources. Typical interferences include ISI as well as the frequency-selective fading caused by the multipath effect, the "shadow effect" by terrain factors and buildings, cochannel as well as adjacent-channel interference due to the simultaneous broadcasting with analog TV signals, the dynamic multipath effect, and the Doppler effect from the relative movements between transmitter and receiver.

Unlike other broadcasting systems with much simplified channel coding due mainly to perfect channel conditions, the DTTB system needs much more powerful channel coding schemes.

3.2 ERROR CONTROL AND CLASSIFICATION OF CHANNEL CODING

Error control methods can be divided into two types. One is with feedback, including automatic repeat request (ARQ), information repeat request (IRQ), and hybrid error correction (HEC). The channel coding construction is simple but it requires a feedback channel. The other is without feedback and is called forward error correction (FEC). Here, "forward" means the decoder performs the error control (i.e., error correction) based solely on the characteristics of the code structure and without any need for feedback. FEC is more suitable for one-way transmission with no delay from feedback but suffers from code construction complications and low transmission efficiency. For future multimedia broadcasting and communication networks, the feedback method can be used for uplink as well as for those multimedia services with low data rates, insensibility to delay, and extreme sensitivity to transmission error. For the video broadcasting service which is the major service of this network, the FEC method is the only choice for error control because of the point-to-multipoint nature of the broadcasting. In this chapter, only FEC channel coding is addressed. Figure 3-3 gives a general classification of channel coding.

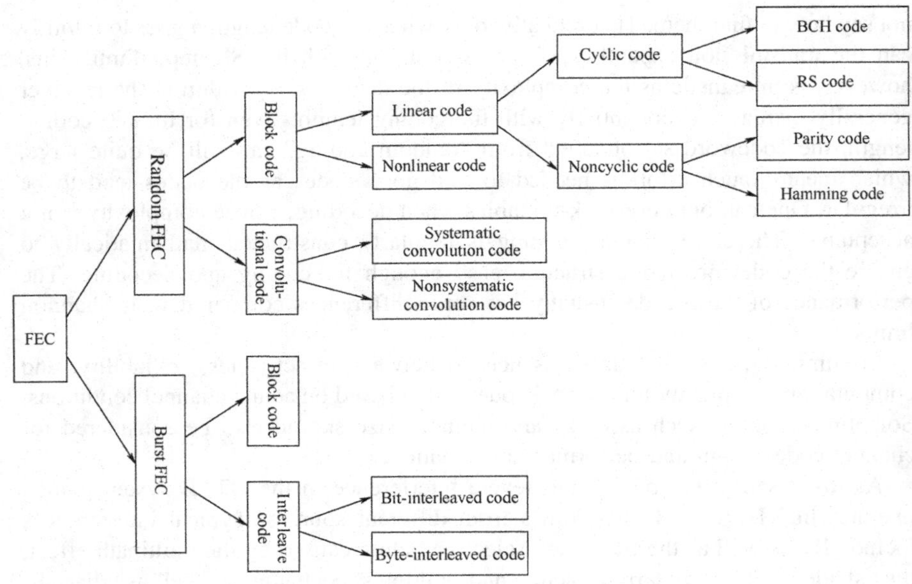

FIGURE 3-3 Channel coding classification.

There are many ways to characterize the different channel coding schemes. The most common are as follows:

1. Block and convolutional codes depend on the relationship between parity-check bits and information bits within each packet. If the parity-check bits within one packet are determined by only the information bits in the same packet, these are called block codes. If the parity-check bits in one packet not only depend on the information bits of the same packet but also are determined by the information bits in the previous packets, these are called convolutional codes.
2. Systematic and nonsystematic codes depend on whether information bits in a packet remain the same after encoding. The encoded information bits remain unchanged for the systematic code while they are changed in nonsystematic codes. For block codes, there is no performance change between the systematic and nonsystematic code, and the systematic codes are widely adopted; nonsystematic codes sometimes have better performance than their counterpart for convolutional codes.
3. Algebraic, geometric, and arithmetic codes depend on the mathematical code construction methods. Algebraic code is created using modern algebra based on the most well-established theoretical development.
4. Linear and nonlinear codes depend on the relationship between parity-check bits and information bits. If this relationship can be expressed by linear

equations, it is a linear code; otherwise, it is a nonlinear code. Linear codes make up an important family of algebraic codes.

5. Error-detecting, error-correcting, and erasure codes correcting erasure errors depend on their functionality in the system. The same code can take a different functionality when using different decoding schemes.
6. Random and burst error correction codes depend on the type of errors to be corrected. The former is mainly used to handle random errors and the latter is mainly used to deal with burst errors.
7. Binary and q-ary codes depend on the number of bits each data symbol is carrying. Usually, $q = p^m$, where p is a prime number and m is a positive integer.
8. Equal- and unequal-protection error correction codes depend on the protection capability for every information bit.

3.3 LINEAR BLOCK CODE

Linear block code based on a well-established theory has very clear mathematical structure and has been widely adopted. The concepts introduced here help lay the foundation for the study of other coding methods.

3.3.1 Basic Concept of Linear Block Code

In the linear block code, the information sequence is first divided into segments (i.e., packets) with length k. Then r check bits are derived from every k bits to form a codeword with total length of $n = k + r$, denoted by (n, k). The relationship of the information bits and parity-check bits can be characterized by linear equations, as shown in Figure 3-4.

Each code element in the linear block code may be one of the q elements in a finite Galois field $G(q)$. The code of length n has q^n possible codewords, which creates an n-dimensional linear space in $G(q)$. But the collection of q^k actual chosen code words is a k-dimensional linear subspace of the n-dimensional linear space. This linear subspace forms Abelian groups under the addition operation of the $G(q)$ field, so the linear block code is also called a group code.

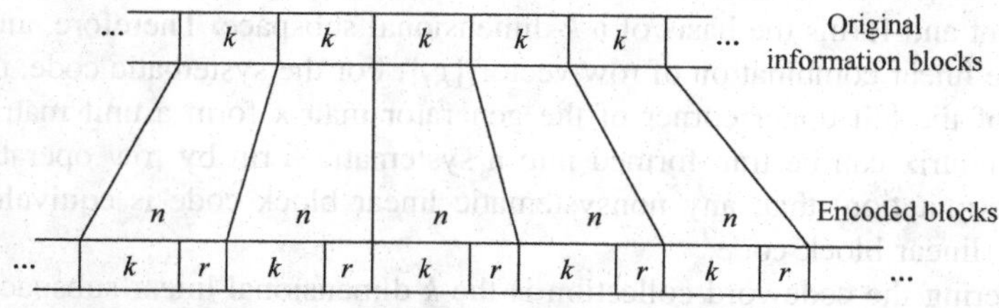

FIGURE 3-4 Schematic diagram of linear block code structure.

Several important concepts needed to discuss the linear block code are as follows:

1. *Code Distance.* Code distance is the number of different elements between two codewords C_i and C_j, denoted by $d(C_i, C_j)$. The code distance is also called the Hamming distance, which measures the distinguishable degrees between two codewords. A very useful parameter is $d_{\min}$, which is the minimum distance among all the codeword pairs, namely

$$d_{\min} = \min_{i \neq j} d(C_i, C_j) \tag{3-8}$$

Usually, the minimum code distance is used to analyze the error-detecting and error-correcting capability.

In reality, the minimum code distance only gives a theoretical bound for reliable error correction. The real error correction capability depends on the weight distribution of the codewords and decoding algorithms.

2. *Codeword Weight.* The codeword weight can be obtained by calculating the number of nonzero elements in the codeword and is denoted by $w(C_i)$. One can also define the minimum codeword weight, that is, the minimum number of nonzero elements among all the code words. It is easy to prove that the minimum code distance and the minimum code weight for the linear block code are identical.

3. *Generator Matrix and Parity-Check Matrix.* If both the input information bits and output encoded bits are expressed in vector form, one can get $X_m = [x_{m1}, x_{m2}, \ldots, x_{mk}]$ and $C_m = [c_{m1}, c_{m2}, \ldots, c_{mn}]$. Then the linear block code can be expressed in vector form, that is,

$$C_m = X_m G \tag{3-9}$$

Here

$$G = \begin{bmatrix} \leftarrow g_1 \rightarrow \\ \leftarrow g_2 \rightarrow \\ \vdots \\ \leftarrow g_k \rightarrow \end{bmatrix} = \begin{bmatrix} g_{11} & g_{12} & \cdots & g_{1n} \\ g_{21} & g_{22} & \cdots & g_{2n} \\ \vdots & \vdots & \ddots & \vdots \\ g_{k1} & g_{k2} & \cdots & g_{kn} \end{bmatrix} \tag{3-10}$$

where G is the generator matrix. The row vector $\{g_i\}$ of the generator matrix is linearly independent and forms the basis of a k-dimensional subspace. Therefore, any codeword is the linear combination of row vector $\{g_i\}$. For the systematic code, the $k \times k$ elements of the left upper corner of the generator matrix form a unit matrix. Any generator matrix can be transformed into a systematic form by row operation and column permutation; that, any nonsystematic linear block code is equivalent to a systematic linear block code.

Considering the codeword collection is the k-dimensional linear subspace of the n-dimensional linear space, it must have an r-dimensional null space in this linear space. The basis of the null space forms the parity-check matrix

$$H = \begin{bmatrix} \leftarrow h_1 \rightarrow \\ \leftarrow h_2 \rightarrow \\ \vdots \\ \leftarrow h_r \rightarrow \end{bmatrix} = \begin{bmatrix} h_{11} & h_{12} & \cdots & h_{1n} \\ h_{21} & h_{22} & \cdots & h_{2n} \\ \vdots & \vdots & \ddots & \vdots \\ h_{r1} & h_{r2} & \cdots & h_{rn} \end{bmatrix} \tag{3-11}$$

It is easy to prove the following relationship for an arbitrary codeword:

$$C_m H^{\mathrm{T}} = 0 \tag{3-12}$$

The parity-check matrix is used for the decoding. Suppose the codeword from the transmitter is C_m and the received codeword can be expressed as $R_m = C_m + E_m$. Given

$$S = R_m H^{\mathrm{T}} = (C_m + E_m)H^{\mathrm{T}} = E_m H^{\mathrm{T}} \tag{3-13}$$

This is called a concomitant formula or a corrector. If the code is designed such that each combinations of S has one-to-one correspondence to each individual error of E_m, error correction can easily be achieved.

4. *Cyclic Code.* Cyclic code is an important subset of linear block code with circular shift properties, that is, if $C = [c_{n-1}, c_{n-2}, \ldots, c_0]$ is a codeword of the cyclic code collection, its circular shift $[c_{n-2}, c_{n-3}, \ldots, c_0, c_{n-1}]$ is also a codeword of this cyclic code collection.

The special structure of a cyclic code makes it easy to be analyzed by algebra theory. The cyclic code $C = [c_{n-1}, c_{n-2}, \ldots, c_0]$ can be expressed by the polynomial

$$C(p) = c_{n-1}p^{n-1} + c_{n-2}p^{n-2} + \cdots + c_1 p + c_0 \tag{3-14}$$

One can get

$$p^i C(p) = Q(p)(p^n + 1) + C_i(p) = C_i(p) \bmod (p^n + 1) \tag{3-15}$$

The residual polynomial $C_i(p)$ represents the other cyclic codewords obtained by the circular shift. Similarly, if the information sequence can also be expressed in the form of a polynomial,

$$X(p) = x_{k-1}p^{k-1} + x_{k-2}p^{k-2} + \cdots + x_1 p + x_0 \tag{3-16}$$

one can get

$$C(p) = X(p)g(p) \tag{3-17}$$

where the $(n-k)$-order polynomial g(p) is called a generator polynomial. The generator polynomial of cyclic code is certainly the factor of the polynomial (p^n+1). Then

$$p^n + 1 = g(p)h(p) \tag{3-18}$$

where $h(p)$ is the check polynomial.

The special algebraic property of cyclic code helps construct the code according to the required error correction ability in a systematic way. Cyclic code is easy to implement by using the linear feedback shift register (LFSR) and the decoding algorithm is also quite simple. Therefore, cyclic code attracts lots of attention and has been widely used. In the following, two popular block codes (BCH and RS codes, which are subsets of cyclic code) used in DTTB systems will be discussed in detail.

3.3.2 BCH Code

BCH codes were invented in 1959 by Alexis Hocquenghem and then independently invented in 1960 by Raj Bose and D. K. Ray-Chaudhuri. The acronym BCH comprises the initials of the three inventors' names. BCH codes have a very strict algebraic structure and therefore are easy to use in both encoding and decoding. Their performance can approach the theoretical limit of finite field codes even with short- or mid-range code lengths.

Assuming that a generator polynomial $g(p)$ of a binary cyclic code has $r=(n-k)$ roots in the field $GF(2^n)$ (the performance of the code with multiple roots is usually not good and will not be discussed here), there exists a minimal polynomial of every root. The generator polynomial will be the least common multiple of minimal polynomials for all these roots.

The cyclic code generated by the polynomial whose roots are $(d-1)$ consecutive powers of an n-order element α in the field $GF(2^m)$ is a BCH code. If there is any primitive element in these roots, it is called a primitive BCH code with the features $n=2^m-1$, $d_{\min}=2t+1$, and $n-k\leq mt$. The code length for the nonprimitive BCH code is a factor of 2^m-1 and the code length of all binary BCH codes is an odd number. The cyclic Hamming code that is capable of correcting one error is a special primitive BCH code, and the famous Golay code is a nonprimitive BCH code.

The BCH decoding methods can be divided into time- and frequency-domain decoding.

The frequency-domain decoding proposed by Blahut in 1978 treats each codeword as a signal and performs the discrete Fourier transform (DFT) at the receiver. The decoding is done in the frequency domain using digital signal processing techniques, and the codeword can then be obtained after inverse discrete Fourier transform (IDFT).

Time-domain decoding is done by directly utilizing the algebraic structure of the code in the time domain, and there are many methods to do this. BCH code is a cyclic code and can be decoded using the general decoding methods for cyclic codes. More efficient decoding algorithms based on BCH code properties are needed as decoding complexity will increase rapidly with the increase of error-correcting capability of BCH code.

In 1960, Peterson put forward a decoding algorithm which is believed to have laid the foundation for BCH decoding. The syndrome calculation was used with the following steps:

1. Using the t factors of the generator polynomial $g(p)$ as the divisor to calculate the remainder of the received code polynomial and get t residues, which are called the syndromes
2. Using each syndrome to construct the decoding polynomial with its root as the error location, which is the reason the decoding polynomial is also called an error location polynomial
3. Finding the root of the error location polynomial (usually using Chien search), obtaining the error locations, which for binary code can be corrected directly

When using error location polynomials by the Peterson method, the amount of calculation and decoding time will be greatly increased with increases in the correcting ability t. In 1975, Sugiyama, Kasahara, and Msmekawa proposed Euclid's decoding algorithm. The method of using Euclid's division algorithm to calculate the error location polynomial for BCH code is similar to the continued fraction algorithm by Weldon. In 1968, Berlekamp and Massey put forward the most popular iterative decoding algorithm, with memory requirement of decoding that is smaller than Euclid's division algorithm, less computational complexity, and easy to be realized. The essence of this algorithm is to use an iterative algorithm to calculate the error location polynomial, and it just needs t iterations for binary BCH code. Details of the above algorithm will not be introduced here.

3.3.3 Reed–Solomon Code

Unlike the binary BCH code, not only the roots of the generator polynomial but also the symbols of the code elements for RS codes are from the GF(2^{m}) field. For RS code of (n, k), the input sequence is divided into groups, each including k symbols with m bits per symbol, and the generator polynomial is

$$g(x) = (x-\alpha)(x-\alpha^2)\cdots(x-\alpha^{n-k}) \tag{3-19}$$

Parameters for RS code capable of correcting t-symbol errors include code length $n=(2^m-1)$ symbols or $m\ (2^m-1)$ bits; information length k symbols or mk bits; parity check length $n-k=2t$ symbols or $m(n-k)$ bits; and minimum code distance $d_{\min}=2t+1$ symbols or $m(2t+1)$ bits.

Since the Singleton limit of block codes is $d_{\min}\leq n-k+1$, RS code belongs to those with the maximum value of minimum distance. RS code is the optimal code for the Singleton limit. This implies RS code has the best error-correcting capability, not only for random errors but also for the burst errors. The error patterns can be corrected to include:

A single burst with total length of $b_1 = (t-1)m+1$ bits
Two bursts with total length of $b_2 = (t-3)m+3$ bits

$$\vdots$$

i bursts with total length $b_i = (t-2i+1)m+2i-1$ bits.

RS code is a very popular coding scheme for various systems: optical discs, magnetic recording systems, DTV terrestrial broadcasting, etc., because of its superior error-correcting capability.

In addition to the great error-correcting capability, it is easy to choose the appropriate RS code based on the system design as both the minimum distance and weight distribution of (n, k) RS code are entirely determined by the parameters of k and n. This is another reason why RS codes are widely applied. It is worthwhile to mention that the RS code is also an algebraic geometric code which can be studied from the perspective of the algebraic geometric code family.

In 1960, Gorenstien and Zierler extended the Peterson algorithm beyond binary, i.e., q-ary, and their work can be considered as the basis of RS decoding. In fact, because RS code is a subset of the BCH code, all decoding algorithms for BCH code introduced above also apply to RS decoding in principle.

RS decoding needs one more step when the BCH decoding algorithms mentioned above are applied. After finding the root (i.e., error location) by solving the error location polynomial, one also needs to calculate the correct value by solving a set of equations. If there are many error locations and t is large, the calculation efforts will be quite huge. Therefore, the simplified Forney algorithm is usually adopted in practice to reduce computational complexity.

3.4 CONVOLUTIONAL CODES

Block codes are codewords with a total of n code symbols each consisting of k information bits and $n-k$ check bits. The check bits of each codeword are only determined by the information bits of this codeword with certain constraints. To achieve desirable error-correcting capacity while maintaining efficiency, both n and k are generally quite large, which means high memory requirements for both encoding and decoding as well as large decoding delay.

In 1955, Elias invented the convolutional code, which also converts k information bits into n bits yet both k and n are usually very small. It is especially suitable for serial information transmission with relatively small delay. Unlike the block codes, the n bits after convolutional code encoding are related not only to the k information bits at the current stage but also to the information bits in the previous $N-1$ stages, where N is the constraint length. In other words, the number of bits associated with each other

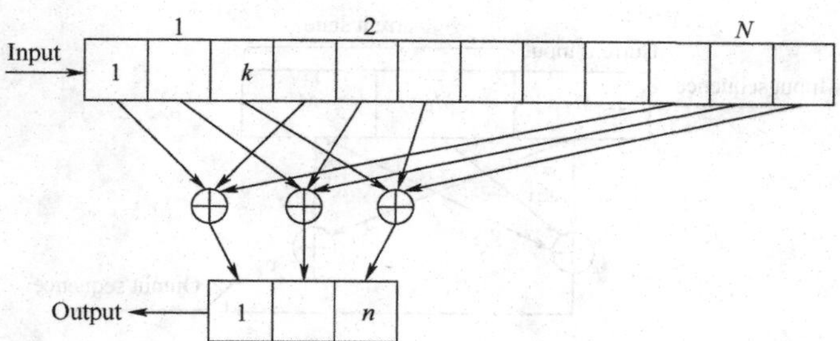

FIGURE 3-5 Binary convolutional encoders.

in the convolutional code is nN. Similarly, in the decoding process, one needs to extract the decoding information from not only the currently received bits but also those bits previously received.

The error probability exponentially decreases with the increase of N. Under the same code rate and complexity, convolutional codes in general outperform block codes by fully taking the advantage of correlations among codewords. Compared with block codes, convolutional codes do not have the vigorous mathematical structures and the analytical tools, and we usually rely on the computer to search good codes.

3.4.1 Construction and Description of Convolutional Codes

A binary convolutional encoder that includes both input and output shift registers and n modulo-2 adders is shown in Figure 3-5. The input shift registers consist of N stages, each with k registers. There are n output shift registers. Corresponding to the input of every k bits, the output is n bits. It is clear that n output bits are related not only to the current k input information bits but also to the former $(N-1)k$ input information bits. The whole encoding process can be seen as the convolution between input sequence and a binary sequence generated by LFSR, which is why it is called "convolutional code." Convolutional codes are usually noted by (n, k, N) with code rate $R_c = k/n$. Nonbinary convolutional codes can be easily derived.

Both analytical and graphical methods are used to describe convolutional codes. The analytical method includes matrix expression and generator polynomial expression, while the graphical method includes tree diagrams, state diagrams, and trellises. The convolutional code (2, 1, 3) is shown in Figure 3-6.

1. *Analytical Method.* For matrix expression, one can use the format of the generator matrix of the block code as well as the parity-check matrix to represent the convolutional code. However, due to the continuity of the information sequence, both the generator matrix and the parity matrix of the convolutional code are half (unilateral) infinite matrices. Therefore, matrix expression is not very convenient and unlikely to be used.

The generator polynomial can be expressed as follows using Figure 3-6 as an example. We define both $g_1 = [g_{10}, g_{11}, g_{12}]$ and $g_2 = [g_{20}, g_{21}, g_{22}]$, usually expressed

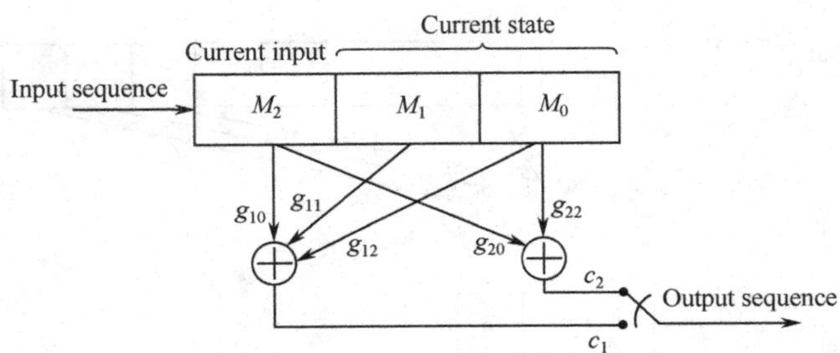

FIGURE 3-6 Encoder of convolutional code (2,1,3).

in octal, $g_1=7$, $g_2=5$. If the current information sequence in the shift register $M=[M_2, M_1, M_0]$, then

$$c_1 = [g_{10}, g_{11}, g_{12}]\begin{bmatrix} M_2 \\ M_1 \\ M_0 \end{bmatrix} \quad c_2 = [g_{20}, g_{21}, g_{22}]\begin{bmatrix} M_2 \\ M_1 \\ M_0 \end{bmatrix} \tag{3-20}$$

Define $g_1(x)=g_{10}+g_{11}\,x+g_{12}\,x^2=1+x+x^2$ and $g_2(x)=g_{20}+g_{21}\,x+g_{22}\,x^2=1+x^2$. The total generator polynomial can be expressed as

$$G(x) = \begin{bmatrix} g_{10} \\ g_{20} \end{bmatrix} + \begin{bmatrix} g_{11} \\ g_{21} \end{bmatrix} x + \begin{bmatrix} g_{12} \\ g_{22} \end{bmatrix} x^2 = \begin{bmatrix} 1 \\ 1 \end{bmatrix} + \begin{bmatrix} 1 \\ 0 \end{bmatrix} x + \begin{bmatrix} 1 \\ 1 \end{bmatrix} x^2 \tag{3-21}$$

2. *Graphical Method.* In Figure 3-6 the information bit in register M_2 represents the current input bit, while bits in registers M_0 M_1 are former $(N-1)$ input bits (usually considered as representing the encoder's current "state.") Thus, the convolutional code encoder can be viewed as a finite-state machine. Suppose state a is M_0 $M_1=00$, state b is M_0 $M_1=01$, state c is $M_0M_1=10$, and state d is M_0 $M_1=11$. One can then get three formats of the graphical method for convolutional codes.

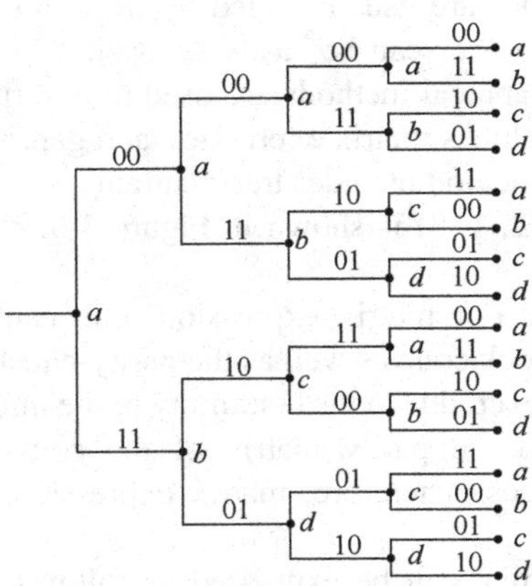

FIGURE 3-7 Tree diagram of convolutional codes.

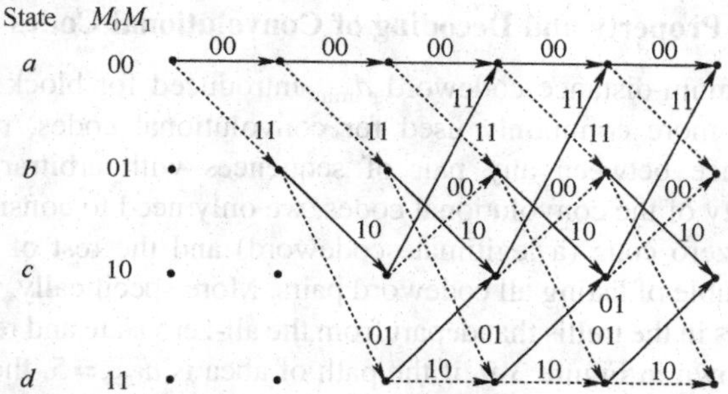

FIGURE 3-8 Trellis diagram for convolutional codes.

1. *Tree Diagrams.* As shown in Figure 3-7, if the initial state is a, when the input bit is 0, the output $c_1\, c_2 = 00$ and the state remains a; when the input bit is 1, the output $c_1\, c_2 = 11$ and state a changes to state b. One can use the upper branch to represent the case for the input bit of 0, the lower branch for the case for the input bit of 1. The labels on top of each branch represent the output of the encoder. From Figure 3-7, it is seen the tree diagram can be infinitely long when representing all possible codeword sequences. A careful look at this tree diagram shows that the structure repeats every three stages while 3 is exactly the constraint length N. For a general convolutional code (n, k, N), each tree node will have 2^k branches, and labels on top of each branch correspond to the n output symbols with number of states of $2^{k(N-1)}$.
2. *Trellis Diagram.* Merging the tree diagram gives the trellis diagram using a solid line to represent input bit 0 and a dashed line to represent input bit 1, as shown in Figure 3-8. The Trellis diagram is also semi-infinite and self-repeats after the third stage (steady state). This representation is very useful to understand the Viterbi decoding algorithm in the following.
3. *State Diagram.* Using the common representation of the finite-state machine, we can get the state diagram of a convolutional encoder as shown in Figure 3-9. State diagrams can be easily obtained using the steady-state trellis diagram.

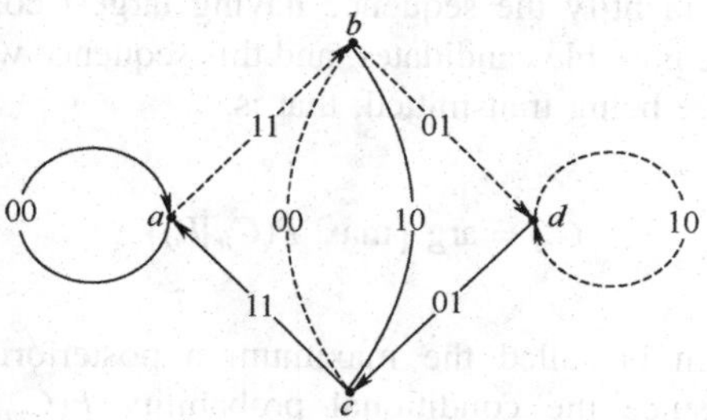

FIGURE 3-9 State diagram of convolutional codes.

3.4.2 Distance Property and Decoding of Convolutional Codes

Unlike the minimum-distance codeword $d_{\min}$ introduced for block codes, the free distance d_{free} is more commonly used for convolutional codes, representing the Hamming distance between any pair of sequences with arbitrary length. With the linear property of the convolutional codes, we only need to consider the distance between the all-zero code (a legitimate codeword) and the rest of the codewords, saving us the trouble of listing all codeword pairs. More specifically, we only need to consider the paths in the trellis that depart from the all-zero state and return to it for the first time. As shown in Figure 3-9, if the path of abca is $d_{\text{free}} = 5$, the corresponding output code word is 111011.

Proakis defines the transfer function which can be used to measure the performance of convolutional codes [3], and the general formula is

$$T(D, N, J) = \sum_d \sum_n \sum_j A_{d,n,j} D^d N^n J^j \tag{3-22}$$

This implies that the first error path of the output codeword $d_{\text{free}} = d$ will be $A_{d,n,j}$ if the code weight is n and the number of branches is j. Since it is difficult to calculate this transfer function when convolutional code parameters (such as constraint length) are large, good convolutional codes are usually obtained by computer search.

The general decoding methods for convolutional codes are algebraic decoding and probabilistic decoding. Algebraic decoding is based on the algebraic structure of the convolutional codes, and the calculation is quite simple. The probabilistic decoding considers statistical properties of the channel, and the calculation is more complicated yet has better error correction performance. The probabilistic decoding is now dominant, and the typical decoding algorithms include Viterbi decoding for relatively small values of k and sequential decoding for longer constraint length codes. This book only describes the widely used Viterbi decoding algorithm, which is based on the maximum-likelihood criterion proposed by Viterbi in 1967.

1. *Maximum-Likelihood Decoding Criterion.* Assume the code sequence to be transmitted is $C_m = (c_{m1}, c_{m2}, \ldots)$, and the received signal after the channel propagation is R (analog or digital signals, depending on the channel definition), the receiver needs to identify the sequence having largest conditional probability $P(C_m|R)$ among all the possible candidates, and this sequence will be considered the most possible sequence being transmitted, that is,

$$\tilde{C}_m = \arg\{\max_{C_m} P(C_m|R)\} \tag{3-23}$$

This decision criterion is called the maximum a posteriori (MAP) probability criterion. However, since the conditional probability $P(C_m|R)$ depends on the

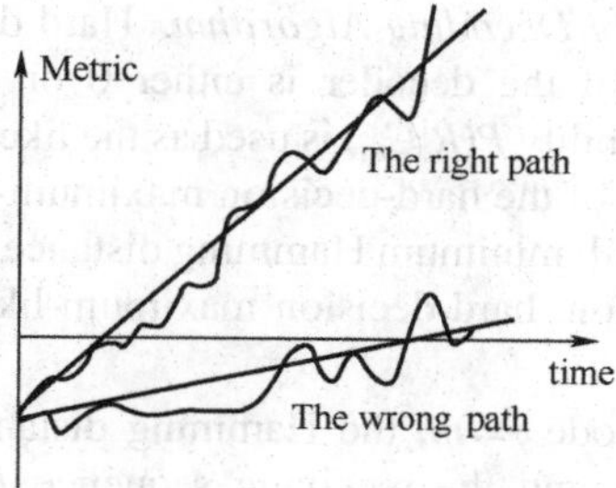

FIGURE 3-10 Logarithmic metric for right and wrong paths as function of time.

occurrence probability $P(C_m)$ of all the possible sequences C_m from the encoder and is difficult to obtain, the MAP criterion is not easy to use. According to Bayes's formula,

$$P(C_m|R) = \frac{P(R|C_m)P(C_m)}{P(R)} \tag{3-24}$$

Assume the unknown $P(C_m)$ has equal probability for all C_m. Then finding the MAP probability $P(C_m|R)$ is equivalent to finding the maximum-likelihood probability $P(R|C_m)$. One will have

$$\tilde{C}_m = \arg\{\max_{C_m} P(R|C_m)\} \tag{3-25}$$

This leads to the maximum-likelihood (ML) criterion. The maximum-likelihood probability $P(R|C_m)$ only depends on the channel characteristics and is independent of the statistics of the codewords. When the decoding is carried out along the right path, the logarithmic metric is generally a linear function of time. When the decoding is carried out along the wrong path, this logarithmic metric is also a linear function of time but with smaller slope than that of the correct path, as shown in Figure 3-10. Because of the channel noise and interference, the error path may sometimes generate larger logarithmic metric but the accumulated logarithmic metric will surely be smaller than that of the right path. This is illustrated in Figure 3-11.

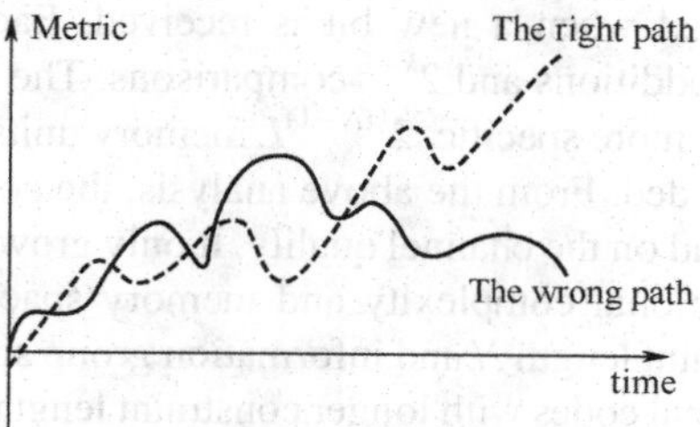

FIGURE 3-11 Fluctuations of logarithmic metric for right and wrong paths.

2. *Hard-Decision Viterbi Decoding Algorithm.* Hard decision here refers to the scenario where the input of the decoder is either 0 or 1. The logarithm of the maximum-likelihood probability $P(R|C_m)$ is used as the likelihood function. It is easy to figure out that the nature of the hard-decision maximum-likelihood decoding is to find the coding sequence with minimum Hamming distance of the received sequence. From the previous discussion, hard-decision maximum-likelihood decoding can be described as follows:

1. From a starting node $j=m$, the Hamming distance d_j between each path toward each state and the receiving sequence $R_j=(r_0, r_1, \ldots, r_{j-1})$ is calculated, known as the path metric; due to the Markov property of the finite-state machine, the nodes $j \geq m$ are only determined by the state at $j=m$ and are irrelevant to $j<m$. Therefore, for the 2^k paths that merge into each state, one only needs to retain a minimum path metric, called the surviving path. The surviving path and the path metric of each state are then stored.
2. Let $j=m+1$, add the Hamming distance between the output of the branches into each state and the corresponding receiving symbol (called branch metrics) to the metric of the surviving path with respect to these branches at node m, and get a new set of path metric d_{j+1}. Then compare and keep the surviving path and its metric. The decoding process is extended to one more branch.
3. Continue until the end of decoding and the surviving path with minimum path metric will be the decoded sequence.

The Viterbi algorithm can be summarized as sum–compare–store. Decoding is a forward process without feedback, and the implementation is quite straightforward.

When the coding sequence is long, storing the surviving path requires large memory space and decoding delay will also be quite long, and a *sliding-window Viterbi algorithm* is usually adopted in practice. This method only requires storing the latest L information bits for each surviving path at any time j. When the $(L+1)$th bit is received, it compares all the survival paths and outputs the first information bit of the surviving path with the least path metric. In this way, the decoding delay is fixed at L bits, and the algorithm can be greatly simplified at the cost of suboptimal decoding. It has been proven that the performance loss of the suboptimal algorithm can be ignored when $L \geq 5N$ (N is the constraint length).

The complexity of the Viterbi algorithm is analyzed as follows. The number of states for a convolutional code (n, k, N) is $2^{k(N-1)}$, and $2^{k(N-1)}$ operations of sum–compare–store are needed when a new bit is received. Each sum–compare–store operation consists of 2^k additions and 2^{k-1} comparisons. The $2^{k(N-1)}$ surviving paths needs to be stored. To be more specific, $2^{k(N-1)}L$ memory units in the sliding-window Viterbi algorithm are needed. From the above analysis, the complexity of the Viterbi algorithm does not depend on the channel quality. It only grows linearly with the code length, but the computational complexity and memory space required grow exponentially with the constraint length N and information group size k. It is not applicable to decode the convolutional codes with longer constraint length and larger information group size. For these types of convolutional codes, one can use the Viterbi decoding algorithm with reduced states.

3. *Soft-Decision Viterbi Decoding Algorithm.* To take full advantage of channel information to improve decoding reliability, one can use the soft-decision Viterbi decoding algorithm. In this case, the output of the demodulator which is the input of the decoder will be the "soft information," that is, either analog or multilevel (quantized) instead of 0 or 1.

Unlike the hard-decision algorithm, the path metric of the soft-decision decoding algorithm uses a "soft distance" instead of a Hamming distance, and the most commonly used is the Euclidean distance, defined as the geometric distance between the possibly sent waveform and the received waveform. When using the soft distance, the path metric is an analog variable, and computational complexity increases because special arrangements are needed to make operations of sum and comparison easy. Other than this, the soft- and hard-decision algorithms are identical in principle and structure.

In general, as hard-decision decoding cannot fully utilize the channel information in its decision process, the performance loss compared to soft-decision decoding is about 2.5 dB. As the Viterbi algorithm is a sequence-based decoding, burst decoding errors are inevitable whether the soft- or hard-decision decoding algorithm is used.

3.5 INTERLEAVING

Most channel codes are designed based on the assumption of the statistically independent error characteristic of the channel, but in reality, channel errors are a combination of burst and random errors. In this regard, directly applying channel codes with good random error-correcting capability will not guarantee satisfactory performance. Interleaving technology is therefore introduced. Interleaving is a time or frequency expansion technology which helps reduce or even destroy the time or frequency correlation of channel errors. When the interleaving is good enough, the burst errors can be converted into random errors (i.e., become totally scattered in the time or frequency domain), creating a favorable decoding condition.

Strictly speaking, interleaving is not a coding scheme: This technology by itself does not generate any redundant bits for error correction. However, if coding and interleaving are combined, the so-called interlaced code can provide better error-correcting performance. For simplicity, the time interleaver is introduced in the following. The frequency interleaver has the same design and operation principle.

3.5.1 Block Interleaving

Block interleaving is also known as matrix interleaving. As shown in Figure 3-12, the encoded codeword sequences are written in rows in an $m \times n$ matrix and then read by columns. Similarly, the de-interleaving at the receiver writes the received signals into the $m \times n$ matrix by columns and then reads the signal to the input of the decoder in rows. In this way, continuous burst errors from the channel will be separated with the duration of the m-bit by the interleaver before being sent to the decoder so that the burst errors become "random" errors at the decoder.

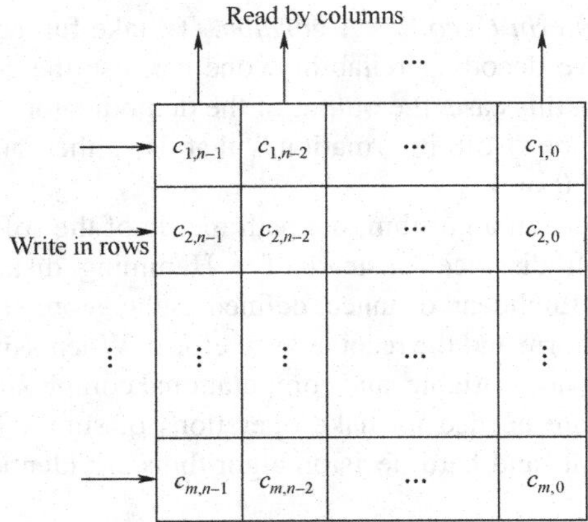

FIGURE 3-12 Schematic of block interleaving.

If an (n, k) code uses the block interleaver with an $m \times n$ matrix, each row of the interleaver stores a codeword and m is then the interleaving depth. This creates an (mn, mk) interlaced code. For this interlaced code, it is easy to prove the following properties:

1. If the original (n, k) code can correct t random errors, then the (mn, mk) interlaced code can correct single burst errors with length no more than mt.
2. If the original (n, k) code can correct t random errors, then the (mn, mk) interlaced code can correct t burst errors, and each burst errors is with length no longer than m.
3. If the original (n, k) code can correct single burst errors of length b, then the (mn, mk) interlaced code can correct single burst errors with length no more than mb.
4. As the decoding process is conducted after the de-interleaving, the total delay of interleaving and de-interleaving will be $2mn$. This long delay is not suitable for transmission systems that are sensitive to time delay.

As the matrix interleaving may happen to turn certain periodic interferences into burst errors, a variant of matrix interleaving known as random interleaving is introduced. In random interleaving, the order for the encoded sequences to be written is predetermined by a pseudorandom sequence or directly by a computer search. This interleaving technology is considerably robust to various interferences but with high implementation complexity.

3.5.2 Convolutional Interleaving

Ramsey and Forney first proposed convolutional interleaving, a process that is different from block interleaving using a matrix. Convolutional interleaving does not need to group the encoded sequence and can work continuously, which is more effective.

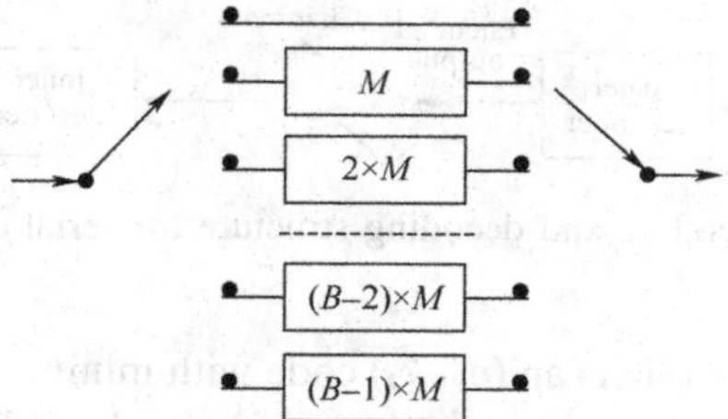

FIGURE 3-13 Schematic of convolutional interleaving.

Figure 3-13 gives the schematic diagram of the convolutional interleaver. For the input, the encoded sequence is written into the B branches in turn with control of the switch, a process that starts again when the buffer of the last branch is full. Each branch has buffer delay of M times the previous branch. As for the output, the interleaved symbols are read from B branches in turn by the switch, which is synchronous to the input control switch. Here, B is the interleaving width (branch), representing the number of interleaving branches and M is the interleaving buffer delay, as the increment of depth of the branches' delay unit. By inverting the schematic diagram of Figure 3-13, the convolution de-interleaver at the receiver can be obtained. The total delay of interleaving/de-interleaving is $M \times (B-1) \times B$.

In addition to block interleaving and convolutional interleaving, random interleaving, and ideal interleaving are also proposed. Random interleaving will provide a completely randomized interleaving pattern each time it is used. One can use the average performance value for comparison purposes. Ideal interleaving means that all burst errors can be completely converted into random errors, which is impossible to achieve in reality. This assumption is only made to facilitate the analysis.

3.6 CONCATENATION CODES

As mentioned in Section 3.1, random codes with code length n sufficiently long can approach the channel capacity. However, as n increases, both the complexity and computational effort of the decoder increase exponentially, which is unacceptable. In 1966, Forney proposed cascaded coding: The encoder, channel, and decoder can be considered as a generalized channel which has errors, and additional channel code needs to be introduced. Based on this, he proposed connecting two short codes serially to get a long code. With limited increase in complexity, error correction capabilities can be greatly improved. This concatenated code structure has been used by the U.S. National Aeronautics and Space Administration (NASA) as the signal transmission protocol for deep-space telemetry and is now widely used in DTTB and other systems.

As shown in Figure 3-14, the information sequence at the transmitter passes through two encoders (the outer and inner encoders) to get the output sequence. At the receiver, it needs two separate decodings or iterative decoding between two decoders to restore the original information. If the outer code is an (n_1, k_1) code with minimum

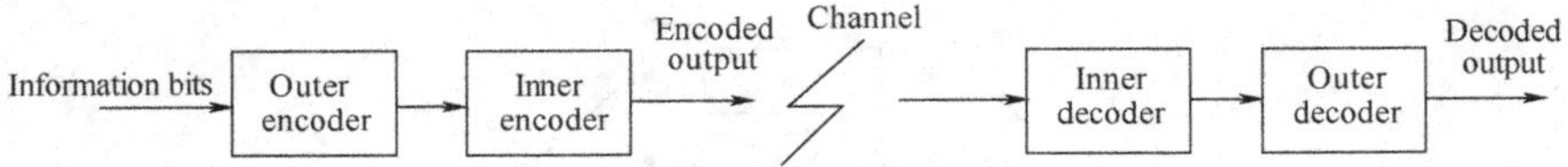

FIGURE 3-14 Encoding and decoding structure for serial concatenated code.

distance d_1 and the inner code is an (n_2, k_2) code with minimum distance d_2, then the cascaded code can be considered as an $(n_1 n_2, k_1 k_2)$ code with minimum distance $d_1 d_2$. When the channel causes a small amount of random errors, the inner decoder can take care of it. When the channel burst errors cannot be fully corrected by the inner decoder, the outer decoder helps. Thus, concatenated codes are suitable for unpredictable channels. As the errors after the inner decoder are often burst errors, it generally requires an interleaver between the inner and outer encoders and de-interleavers at the receiver. Note that adoption of the interleaving scheme is not illustrated in Figure 3-14.

There are many combinations for concatenated codes. One can use RS code as the outer code and either binary block code or convolutional code as the inner code. Convolutional codes can be used as both inner and outer codes with the inner code decoder outputting soft information to the outer code decoder. This concept of concatenation can be easily extended to multiple cascading to include more combinations, but in fact, multistage cascade is rarely used because of a "threshold effect" for concatenated codes. The threshold effect refers to the situation where the performance of the coded system is poorer than that of the uncoded system when the SNR is below a certain threshold. The threshold effect of the concatenated codes is quite obvious (the bit error rate curve is sharper). When the channel condition is good, the error rate is fairly low, and two layers of coding are sufficient. When the channel condition is bad, an additional layer of coding not only generates more errors after decoding but also causes errors to spread. In this case, multilevel coding may not be as good as single-level coding. It should be noted that cascade codes can significantly improve error correction capability, but the price paid for this capacity improvement is the low coding efficiency. From the E_b/N_0 perspective, the benefit of cascaded coding is not that great. The significant gain from cascaded coding is that when the channel condition is relatively good with high SNR, the error rate can be very low.

The encoding and decoding structures of the cascaded codes consist of the popular RS (255,239) and RS (255,223), which are illustrated in Figure 3-15 with the performance shown in Figure 3-16.

Theoretically, the minimum distance of concatenated codes is $d_1 d_2$ if the individual minimum distances of the codes are d_1 and d_2, respectively. However, a separate decoding approach cannot achieve the error correction capability determined by the minimum distance of the combined code. This is because the hard-decision results from the inner decoder cause the information loss for the outer decoder to improve the overall performance. In this regard, inner decoding should output soft instead of hard decisions so that the outer decoder can perform the soft-decision decoding. If the inner code is convolutional and the soft decision is adopted, the criterion for the inner decoder should be the minimum symbol error rate instead of

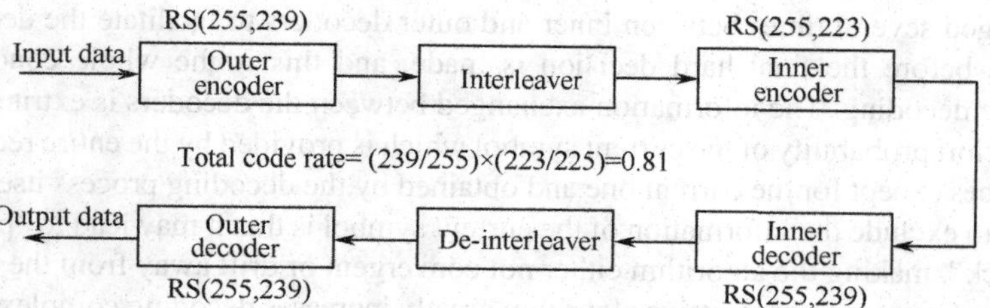

FIGURE 3-15 Encoding and decoding structure of cascaded codes consist of RS (255,239) and RS (255,223).

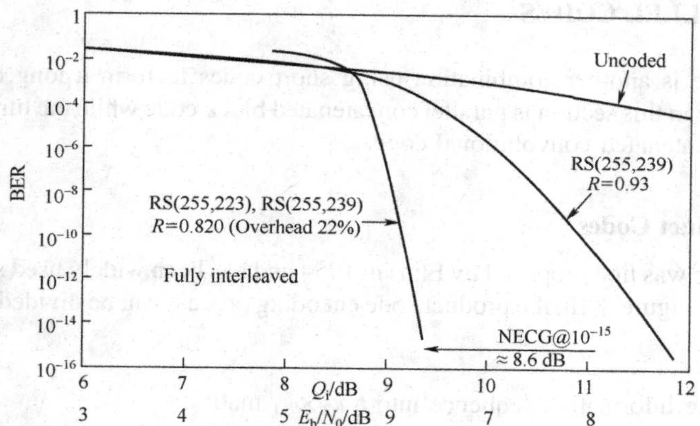

FIGURE 3-16 Performance of cascaded codes consist of RS (255,239) and RS (255,223).

the minimum bit error rate. The Viterbi algorithm is no longer optimal under this situation and the a posteriori probability of each symbol should be calculated. Readers interested in the details are referred to [3].

Recent studies show that the threshold effect of the cascading code can be greatly reduced if the iterative decoding algorithm is used. As shown in Figure 3-17, the outer decoder does not perform the one-time hard decision as the output. Instead, it sends the soft-decision information to the inner decoder. The decision information will be

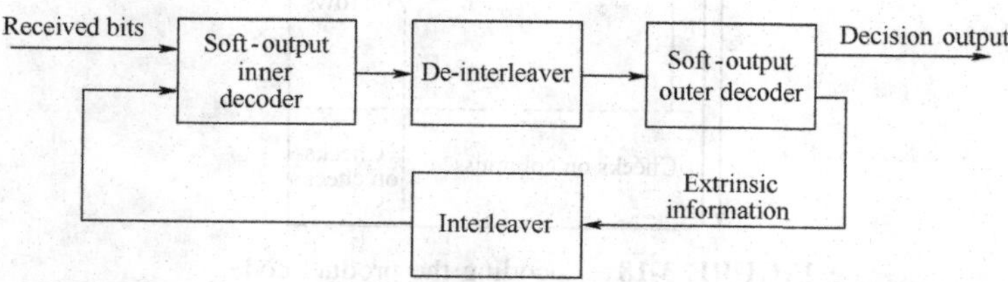

FIGURE 3-17 Iterative decoding of serial concatenated codes.

exchanged several times between inner and outer decoders to facilitate the decoding process before the final hard decision is made, and this is the whole concept of iterative decoding. The information exchanged between the decoders is extrinsic, the a posteriori probability of the current symbol which is provided by the entire receiving sequences except for the current one and obtained by the decoding process itself. The reason to exclude the information of the current symbol is that it may lead to "positive feedback", making the algorithm either not convergent or drift away from the correct solution. Of course, the iterative algorithm greatly increases decoding complexity and hardware cost. The detailed algorithm will be given in the section on turbo codes.

3.7 PARALLEL CODES

Parallel code is another combination using short codes to form a long code. The product code in this section is parallel concatenated block code while the turbo code is parallel concatenated convolutional code.

3.7.1 Product Codes

Product code was first proposed by Elias in 1954 and has been widely used since then. As shown in Figure 3-18, the product code encoding process can be divided into three steps:

1. Fill the information sequence into a $k_2 \times k_1$ matrix.
2. Encode the information symbols of each row of the matrix with a system block code V_1 of (n_1, k_1) and get a $k_2 \times n_1$ matrix.
3. Encode each column of the matrix with a system block code V_2 of (n_2, k_2) and ultimately get an $n_2 \times n_1$ matrix.

The resulting product code is an $(n_1 n_2, k_1 k_2)$ block code. Different from the serial concatenated codes described in the previous section, the two encoders of

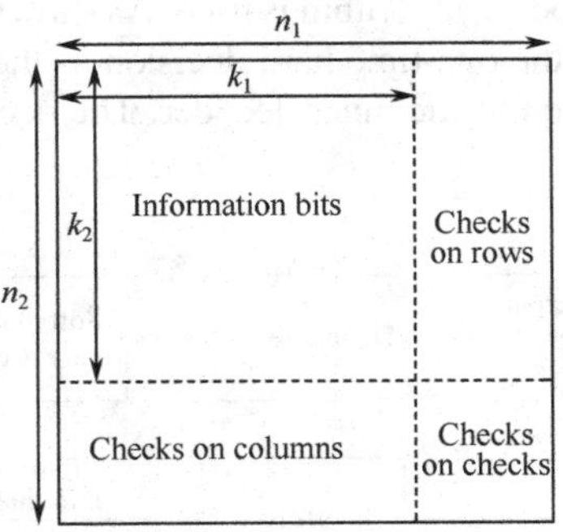

FIGURE 3-18 Encoding the product code.

the product code encode the same information sequence but arranged in different order. The two encoders work in parallel without the distinction of "inner" and "outer" and are collectively referred to as component encoders. Product code is also known as two-dimensional code, which can also be encoded by column and then by row, resulting in two identical codewords. The output of the product code could be in rows, in columns, or diagonal. The properties of the product code are as follows:

1. If the minimum distance of the system block code V_1 is d_1 and the minimum distance of the system block code V_2 is d_2, then the minimum distance of the product code is $d_1\ d_2$.
2. If system block code V_1 can correct t_1 random errors and V_2 can correct t_2 random errors, then the number of random errors the product code can correct is $b \leq \max(n_1\ t_2,\ n_2\ t_1)$.
3. If system block code V_1 can correct b_1 burst errors and V_2 can correct b_2 burst errors, then the burst errors the product code can correct is $b \leq \max(n_1\ b_2, n_2\ b_1)$.

Note that these properties only provide the lower bound of the error correction capability of the product code.

Traditional product code decoding includes two steps: first row decoding and then column decoding based on the decision of row decoding or vice versa. The complexity of this decoding method is the summation of the complexity of two component decoders, and the decoding process is relatively simple and straightforward. It also suffers from the problem that the error correction capability of product code cannot be fully utilized. Using iterative decoding mentioned in the previous section can greatly improve the error correction performance. Of course, one must present this block code in trellis format to use the iterative decoding. In 1974, Bahl, Cocke, Jelinek, and Raviv proposed the so-called BCJR iterative decoding method, solving this problem using a parity-check matrix of block codes. Assume the parity-check matrix is $H = [h_1, h_2, \ldots, h_n]$ for an (n, k) block code with h_j $(1 \leq j \leq n)$ as the $(n-k)$-dimensional column vector and $C = (c_1, c_2, \ldots, c_n)$ represents the encoded sequences, the state σ_j $(0 \leq j \leq n)$ of the block code can be defined as

$$\sigma_j = \sum_{i=1}^{j} c_i h_i = \sigma_{j-1} + c_j h_j \tag{3-26}$$

When $j = 0$, the initial state σ_0 is defined as the zero vector. When $j = n$, σ_n is also a zero vector because of the characteristics of the parity-check matrix. With this, the block code takes a trellis format and can be decoded by the following Turbo iterative algorithm.

The performance of a (128 × 128, 120 × 120) turbo product code (TPC) is shown in Figure 3-19. For comparison purpose, the performance curve of RS (255,239) is also presented in the figure.

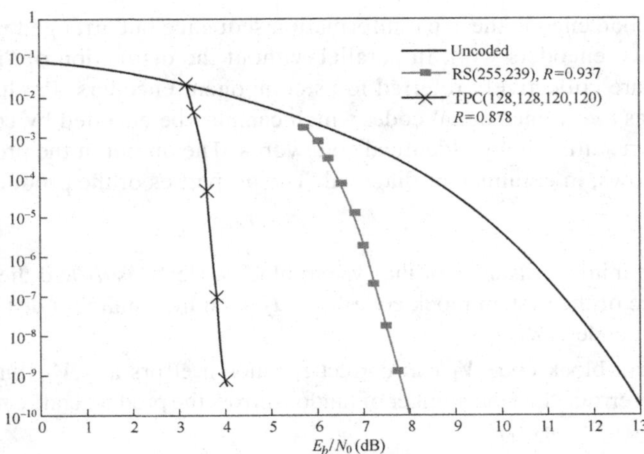

FIGURE 3-19 Comparison of TPC code (128,128,120,120) and RS codes in AWGN channel.

3.7.2 Turbo Codes and Iterative Decoding

Recall from Section 3.1, that the Shannon channel capacity can be approached if a sufficiently long random code is used. However, since traditional codes have regular instead of random algebraic structures and the code length cannot be too long given the decoding complexity and delay, there is a gap between the traditional channel coding efficiency and channel capacity. For a long time, channel capacity has only been taken as a theoretical limit, and the actual coded design and evaluation have not been targeted towards the Shannon limit.

In 1993 French professors Berrou and Glavieux and their Burmese doctoral student Thitimajshima published an article, "Near Shannon Limit Error-Correcting Coding and Decoding: Turbo Codes" [4], and introduced a new family of codes called turbo codes. By using two simple, parallel-cascaded component codes through an interleaver, a long code with pseudorandom properties is obtained. Through iterations between two soft-input and soft-output (SISO) decoders, pseudorandom decoding can be done. Using the rate 1/2 turbo code with code length of 65,532 bits as an example, it can achieve BER $\leq 10^{-5}$ at the E_b/N_0 of only 0.7 dB in an AWGN channel. Using equation 3-6, one can easily find that the theoretical E_b/N_0 value is 0 dB to reach the channel capacity in this situation. The performance of this code is far better than any other coding scheme.

The turbo code has since then received considerable attention and has had a profound impact on coding theory and research methodology. The basic turbo encoder is shown in Figure 3-20. It consists of two parallel recursive systematic convolutional (RSC) code encoders, known as component encoders. The second encoder, RSC2, is connected to the interleaver. Information sequence S_k is directly sent to the component encoder RSC1 to generate encoded sequence W_{1k}. In the meantime, S_k is rearranged by the interleaver before being sent to the component encoder RSC2 to generate another encoded sequence W_{2k}. The switch unit controls

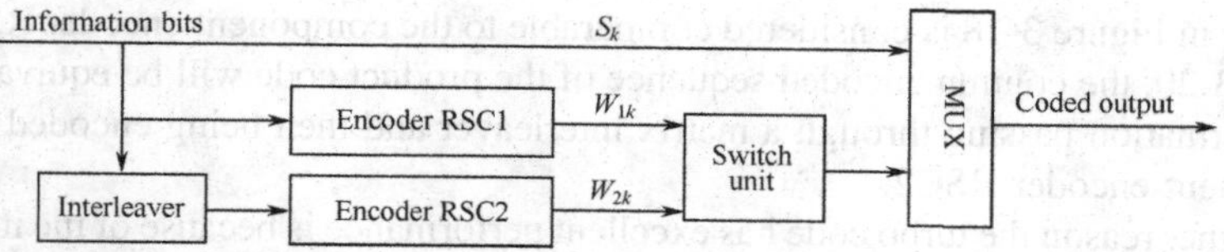

FIGURE 3-20 Schematic of turbo encoder.

which encoded sequences is selected and when to output (therefore, the code rate of the turbo code is determined), and its output is then multiplexed with the original sequence S_k to form the final output of the turbo code.

Using the RSC encoder as the turbo code encoder makes its performance better than the equivalent non-RSC encoder at any SNR for the high bit rate, and two component codes are typically identical for a simple decoder. However, using different component code encoders appropriately (so-called asymmetric structure) will have even better performance.

An example of a turbo code encoder is shown in Figure 3-21 usually with the pseudorandom matrix interleaver. The interleaver length is an important factor in determining turbo decoding performance, sometimes referred to as interleaving gain. When the interleaver is large enough, correlation between the code sequence before and after interleaving is very small. Therefore, the turbo code can have the property similar to the long random code, and it also avoids the positive feedback during iterative decoding because of the strong correlation. Meanwhile, the interleaver helps change the weight distribution of the codeword to make it as uniform as possible, which is very important to the performance. Studies show that this structure has scattered neighboring codewords. Therefore, the error probability is quite low, one of the important reasons for the turbo code gain.

Comparing Figure 3-18 with Figure 3-20, one can see that both have the same structure, with the product codes using block codes as component codes. If the row

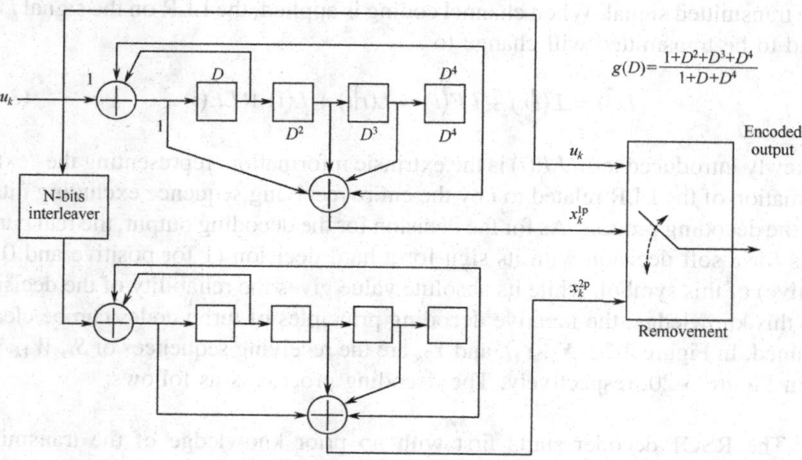

FIGURE 3-21 Example of turbo encoder.

encoder in Figure 3-18 is considered comparable to the component encoder RSC1 in Figure 3-20, the column encoded sequence of the product code will be equivalent to the information passing through a matrix interleaver and then being encoded by the component encoder RSC2.

Another reason the turbo code has excellent performance is because of the iterative decoding based on the MAP criterion, which has been discussed in Section 3.3.2. Take the binary case as an example. Assuming that the symbol i refers to the signal to be sent and is either 1 or 0 and the received signal is r, using the maximum a posteriori probability as the decision criterion, one can get

$$P(i=1|r) \underset{H_2}{\overset{H_1}{\gtrless}} P(i=0|r) \tag{3-27}$$

In other words, when the conditional probability $P(i=1|r) > P(i=0|r)$ is satisfied, the sending symbol is 1 or 0. Rearranging this expression and using the Bayes formula yield

$$\frac{P(r|i=1)P(i=1)}{P(r|i=0)P(i=0)} \underset{H_2}{\overset{H_1}{\gtrless}} 1 \tag{3-28}$$

where H_1 and H_2 represent the hypotheses of $i=1$ and $i=0$, respectively.

Taking the logarithm on both sides of 3-28, one can get log-likelihood ratio (LLR) formula:

$$\begin{aligned} L(i|r) &= \log\left(\frac{P(i=1)[P(r|i=1)]}{P(i=0)[P(r|i=0)]}\right) \\ &= \log\left(\frac{P(i=1)}{P(i=0)}\right) + \log\left(\frac{P(r|i=1)}{P(r|i=0)}\right) \\ &= L(r|i) + L(i) \end{aligned} \tag{3-29}$$

In 3-29, $L(r|i)$ only depends on the channel characteristics; $L(i)$ is the prior knowledge of the transmitted signal. When channel coding is applied, the LLR on the signal i that should to be transmitted will change to

$$L(\hat{i}) = L(i|r) + LE(\hat{i}) = L(r|i) + L(i) + LE(\hat{i}) \tag{3-30}$$

The newly introduced term $LE(\hat{i})$ is the extrinsic information, representing the "extra" information of the LLR related to i by the entire receiving sequence excluding i itself from the decoding process. As for the decision for the decoding output, the real part of $L(\hat{i})$ is for a soft decision with its sign for a hard decision (1 for positive and 0 for negative) of this symbol, while its absolute value gives the reliability of the decision. With this knowledge, the iterative decoding principles of turbo codes can be clearly explained. In Figure 3-22, X_k, Y_{1k}, and Y_{2k} are the receiving sequences of S_k, W_{1k}, and W_{2k} in Figure 3-20, respectively. The decoding process is as follows:

1. The RSC1 decoder starts first with no prior knowledge of the transmitted sequence available at this moment, namely $L(i)=0$. After calculation of $L(\hat{i})$, the extrinsic information output from the RSC1 decoder will be LE_{1k}.

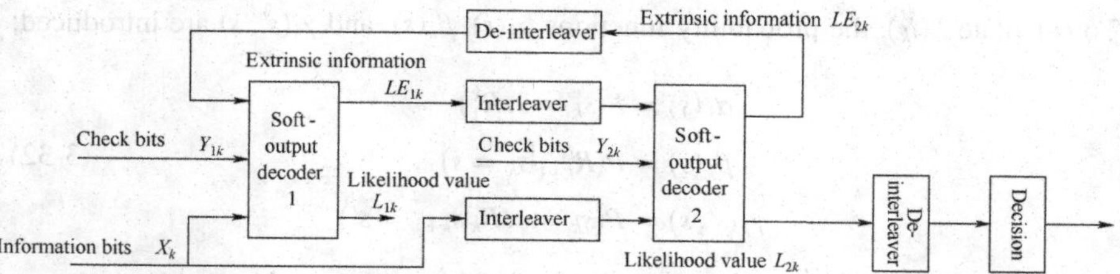

FIGURE 3-22 Iterative decoding of turbo codes.

2. Let the information sequences X_k and LE_{1k} pass through an interleaver the same as the one at the transmit side and then be sent to the RSC2 decoder.
3. The RSC2 decoder uses the extrinsic information LE_{1k} as the a priori knowledge of the transmitted sequence $L(i)$ and performs the decoding. With the output of $L(\hat{i})$ and external information LE_{2k}, one iteration is completed.
4. LE_{2k} is sent to the RSC1 decoder after de-interleaving, also taken as a priori knowledge of $L(i)$ for another iteration.

The decision could be made after several iterations or when the output is stable. The iterative process is similar to the working principles of the turbo engine, which accounts for the name turbo code.

Turbo codes decoding based on the MAP criterion need to calculate the $P(i|X_k\ Y_{1k}\ Y_{2k})$ but the complexity is too high, so $P(i|X_k\ Y_{1k}\ LE_{2k})$ and $P(i|X_k\ Y_{2k}\ LE_{1k})$ need to be separately calculated by the component decoders; through the iteration process, these two values will converge to $P(i|X_k\ Y_{1k}\ Y_{2k})$, and apparently this is the suboptimal strategy but with acceptable complexity.

Although different strategies in obtaining $P(i|X_k\ Y_{1k}\ LE_{2k})$ and $P(i|X_k\ Y_{2k}\ LE_{1k})$ come from different algorithms, they are usually based on the BCJR algorithm proposed in 1974. Assume S_k is the input and i_k the state of the encoder at the instant k and $R_1^N = (R_1, R_2, \ldots, R_N)$ is the received code word. Then

$$
\begin{aligned}
L(\hat{i}_k) &= \log \frac{P(i_k = 1|R_1^N)}{P(i_k = 0|R_1^N)} \\
&= \log \frac{\sum_s P(i_k = 1, s_k = s|R_1^N)}{\sum_s P(i_k = 0, s_k = s|R_1^N)} \\
&= \log \frac{\sum_{i_k=1(s,s')} P(s_k = s, s_{k-1} = s'|R_1^N)}{\sum_{i_k=0(s,s')} P(s_k = s, s_{k-1} = s'|R_1^N)} \\
&= \log \frac{\sum_{i_k=1(s,s')} P(s_k = s, s_{k-1} = s', R_1^N)}{\sum_{i_k=0(s,s')} P(s_k = s, s_{k-1} = s', R_1^N)}
\end{aligned}
\tag{3-31}
$$

To calculate $L(\hat{i}_k)$, the probability functions $\alpha_k(s)$, $\beta_k(s)$, and $\gamma_k(s', s)$ are introduced:

$$\begin{aligned}\alpha_k(s) &= P(s_k = s, R_1^k) \\ \beta_k(s) &= P(R_{k+1}^N | s_k = s) \\ \gamma_k(s', s) &= P(s_k = s, R_k | s_{k+1} = s')\end{aligned} \tag{3-32}$$

where $\alpha_k(s)$ and $\beta_k(s)$ are the forward and backward iteration factors, respectively, obtained as

$$\alpha_k(s) = \sum_{s'} \alpha_{k-1}(s)\gamma_k(s', s) \tag{3-33}$$

$$\beta_{k-1}(s') = \sum_s \beta_k(s)\gamma_k(s', s) \tag{3-34}$$

The initial conditions are $\alpha_0(0) = 1$ and $\alpha_0(s \neq 0) = 0$;

$$\begin{aligned}&B_N(0) = 1 \quad B_N(s \neq 0) = 0 && \text{if reset to 0 by end of process} \\ &B_N(s) = 1/2^{\nu}, \forall s && \text{if not reset to 0 and } \nu \text{ is number of states}\end{aligned} \tag{3-35}$$

Assume $\gamma_k(s', s)$ is the state transition probability expressed as

$$\gamma_k(s', s) = P(s_k = s | s_{k-1} = s')P(R_k | s_k = s, s_{k-1} = s') = P(i_k)P(R_k | i_k) \tag{3-36}$$

where $P(i_k)$ is the a priori probability of i_k and $P(R_k|i_k)$ only depends on the channel. The final result for the soft-decision output is as follows:

$$L(\hat{i}_k) = \frac{\sum_{i_k=1(s',s)} \gamma_k(s', s)\alpha_{k-1}(s')\beta_k(s)}{\sum_{i_k=0(s',s)} \gamma_k(s', s)\alpha_{k-1}(s')\beta_k(s)} \tag{3-37}$$

The widely used turbo decoding algorithms can be divided into three categories:

1. *Standard MAP Algorithm: Modified BCJR Algorithm by Normalizing both α and β.* For convolutional codes with constraint length N, each step needs about $6 \times 2^{N-1}$ multiplications and $5 \times 2^{N-1}$ additions. It suffers from high complexity and large memory requirement, and therefore is not applicable to high-speed decoding.
2. *Logarithmic-Based Algorithm such as LOG-MAP and MAX-LOG-MAP.* The LOG-MAP algorithm uses the log-likelihood ratio to represent all the likelihood values in the decoding process. Therefore, the multiplication is changed to addition. Each step requires $6 \times 2^{N-1}$ additions, $5 \times 2^{N-1}$ operations seeking the maximal value and $5 \times 2^{N-1}$ operations for the table look-up:

$$e^a + e^b = e^{\max\{a,b\} + \ln\left[1 + e^{-|a-b|}\right]} \tag{3-38}$$

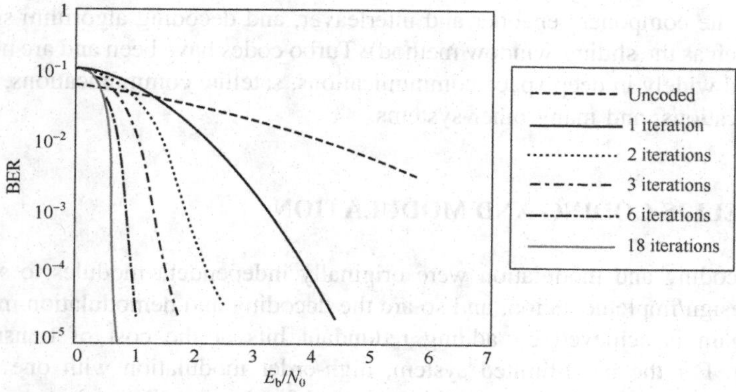

FIGURE 3-23 Decoding performance of turbo codes versus number of iterations.

Ignoring the term of the natural logarithm in 3-38, the likelihood additive operation becomes seeking for the maximum value, and that is the MAX-LOG-MAP algorithm. It greatly reduces the memory requirement and complexity with performance loss of ~0.5 dB–very suitable for hardware implementation.

3. *Soft-Output Viterbi Algorithm (SOVA): Variation and Improvement of Viterbi Algorithm.* The idea is to use the a priori information to improve the error probability of the bits on the survival paths, making it suitable for iterative decoding. The complexity of SOVA is only twice that of the standard Viterbi algorithm with performance deterioration of ~1 dB when compared with the standard MAP algorithm.

In addition to very high complexity of decoding, a long interleaver and iterative process are also needed to achieve a high coding gain. For example, the decoding result published by Berrou is only 0.7 dB from the Shannon limit using the interleaver size of 65,536 after 18 iterations [4]. This leads to huge decoding complexity, needs a large memory space, and may generate unacceptable decoding delay. To some extent, these drawbacks limit the applications of turbo codes.

Figure 3-23 shows the decoding performance versus the number of interactions for turbo codes. One can clearly see from the figure that the BER decreases significantly with the increase of number of iterations at the beginning. But when the number of iterations reaches a certain level, the gain from the increase of the iterations is not obvious and is called "floor effect." The reason is that when the number of iterations increases, the correlation between external information and internal information also increases, and the error correction capacity provided by external information becomes weaker, generating less gain by further iteration. Considering the nonnegligible complexity and decoding delay from additional iterations, a compromise among decoding performance, delay, implementation complexity, and memory size should be made in system design.

The research interest in turbo codes includes but is not limited to the weight spectrum of turbo codes, the free distance and upper bound of the BER performance, design of the component encoder and interleaver, and decoding algorithm simplification (such as the sliding-window method). Turbo codes have been and are now still being used widely in deep-space communications, satellite communications, mobile communications, and many other systems.

3.8 TRELLIS CODING AND MODULATION

Channel coding and modulation were originally independent modules to simplify system design/implementation, and so are the decoding and demodulation modules. Coding gain is achieved by adding redundant bits at the cost of transmission efficiency. For the band-limited system, high-order modulation with one symbol carrying more than one information bit is needed for the overall system transmission efficiency (i.e., to compensate for the transmission efficiency loss due to the coding). But high-order modulation is more vulnerable to noise, interference, and imperfect reception conditions, resulting in more bit errors at the demodulator. This takes away some or even most of the coding gain and makes the overall system performance unacceptable.

SOVA mentioned above can bring about 2.5 dB gain compared with the hard-decision algorithm. The demodulator does not make the decision but outputs soft information directly when SOVA is used. Decoding is done to find the one out of all the possible input sequences with minimum Euclidean distance (geometric distance) from the received sequence. In fact, SOVA actually provides better gain by combining demodulation and decoding. The problem is that the code design is based on the Hamming distance, and combining coding and modulation will not guarantee the optimal Euclidean distance, especially for higher order modulation. To achieve best performance, integrated design of channel coding and modulation is required. In 1982, Ungerboeck invented the trellis-coded modulation (TCM) technology by combining coding and modulation, with the signal space partition optimized based on the maximum Euclidean distance criterion. TCM has 3–6 dB gain in the AWGN channel with the same information rate and system bandwidth.

3.8.1 Mapping by Set Partition of TCM Codes

TCM technology is based on the set partition method in the signal space. Using the 16-QAM signal shown in Figure 3-24 as an example, the minimum Euclidean distance of this normalized 16-QAM constellation is $\Delta_0 = 2/\sqrt{10} = 0.632$. The whole constellation is divided into two subsets of class B, namely, B_0 and B_1, first, and bit y^0 is assigned to differentiate these two B-class subsets. The minimum distance among constellation points within each B-class subset will be $\Delta_1 = \sqrt{2}\Delta_0$.

Next, each subset of class B is divided into two subsets of class C and bit y^1 is assigned to differentiate these two C-class subsets. The minimum distance among constellation points within each C-class subset will be $\Delta_2 = \sqrt{2}\Delta_1$. The label of each subset is determined by the value of bits y^1y^0. Class D subsets can be obtained by repeating the previous procedure and bit y^2 is used to differentiate these two D-class

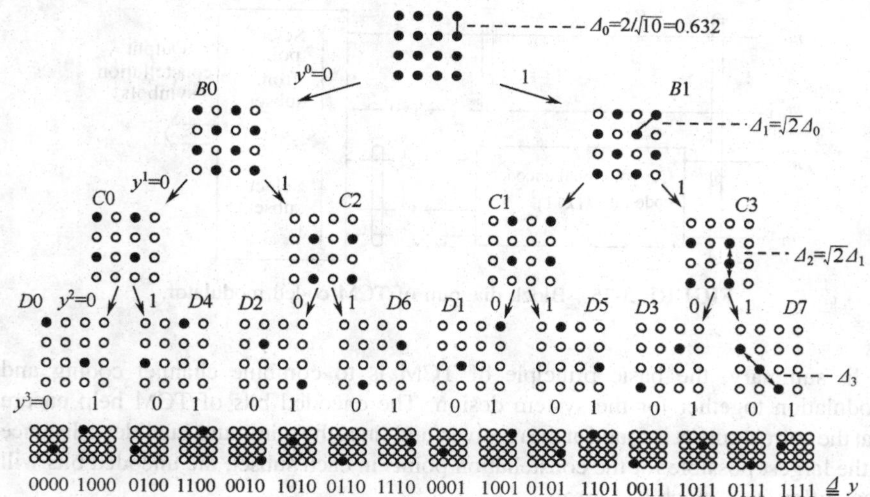

FIGURE 3-24 Set partitioning mapping for 16-QAM signal.

subsets. The minimum distance between two constellation points (differentiate by the value of y^3) within each D-class subset will be $\Delta_3 = \sqrt{2}\Delta_2$. After this, the set partitioning mapping process completes and each constellation point (i.e., modulation waveforms or symbols) is assigned to a unique code word, namely $y = y^3y^2y^1y^0$.

In short, the set partitioning mapping process is to divide one larger signal set (or subset) into two smaller subsets step by step based on the symmetry of the signal constellations. At each set partition step, one bit is assigned such that the set partition corresponds to the mapping. The minimum Euclidean distance among the constellation points in each subset increases with the number of partition settings, that is, $\Delta_0 < \Delta_1 < \Delta_2 < \cdots$. The goal is to maximize the minimum Euclidean distance among constellation points within each subset.

3.8.2 Code Construction and Basic Principles of TCM Codes

Figure 3-25 is a block diagram of the TCM coded modulator. Suppose k-bit information sequence $[m^1, m^2, \ldots, m^k]$ is to be transmitted within one interval. Then the constellation consists of 2^k points without coding. Now double the constellation points to create 2^{k+1} constellation points, and mapping by the set partition method is used to establish the relationship between the signal waveform and the codeword. Let $\tilde{k}(\leq k)$ bits from the k bits pass through a binary convolutional encoder with rate $\tilde{k}/(\tilde{k}+1)$ to get $\tilde{k}+1$ encoded bits. These $\tilde{k}+1$ bits determine one of $2^{\tilde{k}+1}$ subsets to be mapped into, and the rest $(k-\tilde{k})$ uncoded bits determine the constellation point inside this subset to be mapped. The resulting TCM coded modulator rate is $R_c = k/(k+1)$, and redundancy is introduced by doubling the constellation points without increasing the system bandwidth.

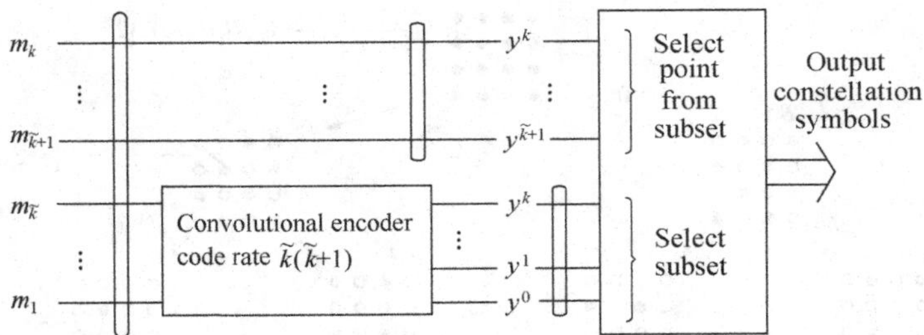

FIGURE 3-25 Block diagram of TCM coded modulator.

In summary, the basic principle of TCM is to combine channel coding and modulation together for the system design. The encoded bits of TCM help ensure that the correct subset is found at the receiver, and since the minimum Euclidean distance is the largest possible for the constellation points in each subset, the uncoded bits will also suffer from much less errors.

To facilitate coding design and decoding, TCM can be represented by a trellis format based on the states of the $\tilde{k}/(\tilde{k}+1)$ convolutional encoder. Unlike the trellis format of the ordinary convolutional codes, each state transition process has $2^{k-\tilde{k}}$ parallel branches by $k - \tilde{k}$ uncoded bits. The TCM code trellis with good performance should have the following properties:

1. All the allowed sequences should have the same probability and as regular and symmetric as possible.
2. The signals corresponding to the branches starting from the same state should be in the B-class subset, i.e., B_0 or B_1. This ensures that the distance between different branches from the same starting state are no smaller than Δ_1.
3. The signals corresponding to branches ending at the same state should be in the B-class subset. This ensures that the distance between different branches to the same state are no smaller than Δ_1.
4. The parallel state transition corresponds to the subset at the $(\tilde{k}+1)$th set partition. This ensures that the parallel transfer distance is smaller than Δ_1.

SOVA can used to decode the received TCM signals with two steps in practice:

1. This is the so-called subset decoding. Keep the constellation point nearest the received signal in each subset (that is, keep $2^{k-\tilde{k}}$ parallel branches) and remove the rest of the branches with large Euclidean distance.
2. Perform the conventional Viterbi decoding by applying those $2^{k-\tilde{k}}$ points and their corresponding Euclidean distances from step 1 to the branches in the Viterbi algorithm. Figure 3-26 shows a BER curve of a trellis coded 8PSK signal with four states under an AWGN channel. One can see that at the BER of 10^{-5} coding gain is about 3 dB. More importantly, this gain is achieved without sacrificing spectrum efficiency.

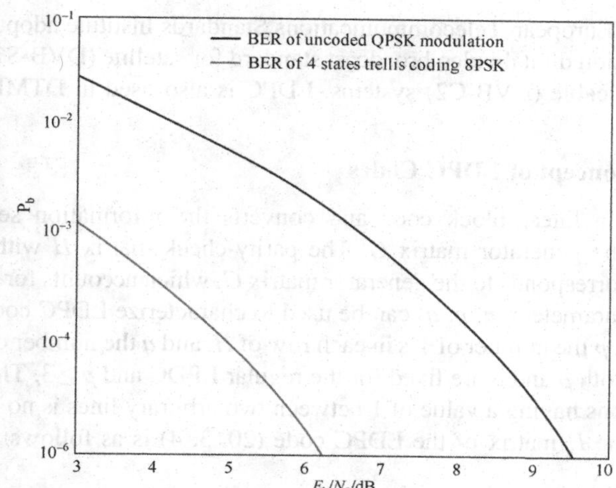

FIGURE 3-26 BER curve of trellis coded 8PSK signal with fourstates.

A considerable amount of work has been done in TCM technology and its applications. In 1989, Wei proposed the rotation-invariant code to overcome the phase ambiguity of TCM. Combinations of block codes and modulation (BCM) as well as turbo and TCM, 2D and multidimensional TCM, and TCM design for non-Gaussian channels have received extensive attention in recent years.

3.9 LOW-DENSITY PARITY-CHECK CODE

Low-density parity-check code (LDPC) is a linear code based on a sparse parity-check matrix proposed by Gallager [7]. Using a decoding algorithm that is different from those for traditional linear block code, excellent performance can be obtained.

In 1962, Gallager proved that LDPC is a good code with suboptimal minimum distance and proposed a probability-based decoding algorithm. Since then, LDPC codes have not gotten enough attention. Only after Berrou proposed turbo codes in 1993 did people start realizing that turbo code is a type of LDPC code [4]. LDPC codes began to attract people's attention again in 1996 [8], when MacKay and Neal published their work showing that LDPC also provides the Shannon limit approaching performance with iterative decoding based on the belief propagation algorithm. This discovery is considered as significant as the turbo code. Recent studies have shown that better performance than for the turbo code can be expected from optimized irregular LDPC when a belief propagation decoding algorithm is used.

The excellent performance of LDPC code makes it a good channel coding candidate for many systems, such as optical, satellite, and deep-space communications, 4G/5G mobile communication systems, high-speed digital subscriber lines, and optical and magnetic recording systems. In November 2002, the European Space Agency launched the LDPC-based adaptive coding and modulation (ACM) research project, and the European Telecommunications Standards Institute adopted LDPC as the next-generation digital video broadcast standard for satellite (DVB-S2), terrestrial (DVB-T2), and cable (DVB-C2) systems. LDPC is also used in DTMB system.

3.9.1 Basic Concept of LDPC Codes

LDPC code is a linear block code and converts the information sequence into codewords as the generator matrix G. The parity-check matrix H with almost all-zero elements corresponds to the generator matrix G, which accounts for the acronym LDPC. Three parameters (n, p, q) can be used to characterize LDPC code, whre n is the code length, p the number of 1's in each row of H, and q the number of 1's in each column of H. Both p and q are fixed for the regular LPDC and $p \geq 3$. The number of identical locations having a value of 1 between two arbitrary lines is no more than 1. For example, the H matrix of the LDPC code (20, 3, 4) is as follows:

$$H = \begin{bmatrix} 1&1&1&1&0&0&0&0&0&0&0&0&0&0&0&0&0&0&0&0 \\ 0&0&0&0&1&1&1&1&0&0&0&0&0&0&0&0&0&0&0&0 \\ 0&0&0&0&0&0&0&0&1&1&1&1&0&0&0&0&0&0&0&0 \\ 0&0&0&0&0&0&0&0&0&0&0&0&1&1&1&1&0&0&0&0 \\ 0&0&0&0&0&0&0&0&0&0&0&0&0&0&0&0&1&1&1&1 \\ 1&0&0&0&1&0&0&0&1&0&0&0&1&0&0&0&0&0&0&0 \\ 0&1&0&0&0&1&0&0&0&1&0&0&0&0&0&0&1&0&0&0 \\ 0&0&1&0&0&0&1&0&0&0&0&0&0&1&0&0&0&1&0&0 \\ 0&0&0&1&0&0&0&0&0&0&1&0&0&0&1&0&0&0&1&0 \\ 0&0&0&0&0&0&0&1&0&0&0&1&0&0&0&1&0&0&0&1 \\ 1&0&0&0&0&1&0&0&0&0&0&1&0&0&0&0&0&1&0&0 \\ 0&1&0&0&0&0&1&0&0&0&1&0&0&0&0&1&0&0&0&0 \\ 0&0&1&0&0&0&0&1&0&0&0&0&1&0&0&0&0&0&1&0 \\ 0&0&0&1&0&0&0&0&1&0&0&0&0&1&0&0&1&0&0&0 \\ 0&0&0&0&1&0&0&0&0&1&0&0&0&0&1&0&0&0&0&1 \end{bmatrix} \tag{3-39}$$

This parity-check matrix can also be presented as a Tanner graph, a bipartite graph used to state constraints or equations specifying error-correcting codes. In the Tanner graph, all nodes are divided into two groups, and the edge is used to connect two nodes of the two groups, respectively. In a Tanner graph which corresponds to a parity-check matrix H, upper nodes are information nodes corresponding to the columns in H and lower nodes are check nodes corresponding to the rows in H. There will be an edge between information node n and check node m when the element h_{mn} of H is not zero. Figure 3-27 is a Tanner graph of the LDPC code (20, 3, 4).

Usually, the number of edges connecting to a node is called the "degree," and one can see that the degree in Figure 3-27 is quite regular: The degree is always 3 associated with each information node and always 4 associated with each check node.

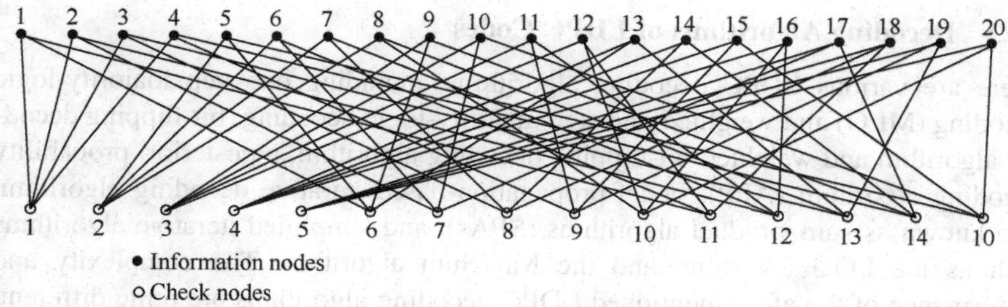

FIGURE 3-27 Tanner graph for LDPC code (20, 3, 4).

So this kind of code is a regular LDPC code. There are also irregular LDPC codes with the same code construction rule as the regular codes. The only difference is that the degree is no longer a constant. Studies show that irregular LDPC codes, even though difficult to construct, can achieve better decoding performance. There is no well-established systematic method to construct LDPC codes approaching the Shannon limit. All the existing good codes seem to be long codes with code length n generally over 1000 (typical length ~60,000).

Professor Lin Shu and his student Kou Yu [9] proposed a construction method of LDPC codes based on finite geometries. The code obtained is either circular or quasi-cyclic, and the encoding can be simply realized by shift registers. Compared with high-order matrix multiplication, this implementation is quite simple.

Richardson et al. [10] found the approximate capacity of irregular LPDC codes by optimizing the degree structure of the irregular graph. They also discuss how to determine the parity-check sparsity matrix to get an efficient encoder and how to construct code to make the coding time and code block length in line with the linear time rather than the usual square relationship.

LDPC codes are defined by the parity-check matrix H, and the generator matrix G can be obtained from H. This can be done as follows: Apply the Gaussian elimination method to H and rearrange the matrix to get

$$H' = [P|I_M] \tag{3-40}$$

Here I_M is a unity matrix with $M = N - K$ and P is an $M \times K$ matrix. The generator matrix G satisfies

$$H' \times G = 0 \tag{3-41}$$

Then, the matrix G will be

$$G = \begin{bmatrix} I_K \\ P \end{bmatrix} \tag{3-42}$$

3.9.2 Decoding Algorithms of LDPC Codes

There are various LDPC decoding algorithms, including one-step majority-logic decoding (MLG) and weighted one-step majority-logic decoding; bit-flipping decoding algorithm and weighted bit-flipping decoding algorithm; a posteriori probability decoding algorithm (APP); belief-propagation-based iterative decoding algorithm, also known as sum–product algorithms (SPAs); and simplified iterative algorithms such as the LOG algorithm and the Min-Sum algorithm. The complexity and performance of the afore-mentioned LDPC decoding algorithms are quite different. SPA provides very good performance for high-bit-rate applications, and other simplified iterative algorithms are also based on SPA. For this reason, SPA will be discussed in the following. SPA is a soft-decision algorithm. Suppose N, K, and $M = N - K$ represent the code length, information sequence, and parity bit length of LDPC, respectively. In GF(q), the generator matrix G converts the information $S = \{s_1, s_2, \ldots, s_K\}$ into the codeword $X = \{x_1, x_2, \ldots, x_N\}$. The parity-check matrix H is an $M \times N$ matrix, and in the GF(q) field, one has

$$HX^{\mathrm{T}} = 0 \tag{3-43}$$

After X passing through the generalized channel, the output will be

$$Y = \{y_1, y_2, \ldots, y_N\} = \{x_1 + n_1, x_2 + n_2, \ldots, x_N + n_N\} = X + N \tag{3-44}$$

where $N = \{n_1, n_2, \ldots, n_N\}$ is the AWGN noise vector with zero mean and variance of σ^2 for each element.

To simplify the discussion of the LDPC decoding algorithm, the GF(2) field is used as an example. In GF(2), each element of X is either 1 or 0. If a bipolar symbol waveform $\pm b$ is used to represent bit 1 or 0 for transmission over the AWGN channel, each symbol will have energy b^2, and $E_b/N_0 = b^2/2R\sigma^2$ with $R = K/N$ as the LDPC code rate. The decoding process is to find the most appropriate vector X to satisfy 3-43. The likelihood of X is the product of each component, which can be expressed as

$$\text{Likelihood}(X) = \prod_n f_n^{x_n} \tag{3-45}$$

where $f_n^{x_n}$ is the channel output likelihood. Since x_n can be 0 and 1 in the GF(2) field, f_n^1 is the probability of decoding 1 and f_n^0 the probability of decoding 0. Hence, in an AWGN channel, there is

$$f_n^1 = \frac{1}{1 + \exp\left(-2by_n/\sigma^2\right)} \quad f_n^0 = 1 - f_n^1 \tag{3-46}$$

As the length of the LDPC code is N, the likelihood(X) will be chosen from $2N$ values in GF(2), and one can denote the set of information bits that connect to check node m by

$$N(m) = \{n : H_{mn} = 1\} \tag{3-47}$$

Using Figure 3-27 as the example, one can have $N(8) = \{3, 7, 14, 18\}$. Similarly, one can define the set of check bits that connect to information bit n by

$$M(n) = \{m : H_{mn} = 1\} \tag{3-48}$$

In Figure 3-27, one can have $M(5) = \{2, 6, 15\}$.

Denote a set $N(m)/n$ as containing all the information bits that connect to check node m excluding variable node n. The algorithm describes two parts, in which quantities q_{mn} and r_{mn} associated with each of the 1's in the matrix H are updated in an iterative fashion:

q^x_{mn}: The probability that bit n of X is $x = 0$, 1 given the information obtained via check nodes other than check node m.

r^x_{mn}: The probability of check node m being statisfied if bit n of X is considered fixed at x and the other bits have a separable distribution given by the probabilities $\{q_{mn'}: n' \in N(m)\backslash n\}$.

The algorithm is described as follows:

1. *Initialization.* The variables q^0_{mn} and q^1_{mn} are initialized to the values f^0_n and f^1_{n}, respectively.
2. *Horizontal Step.* One can define

$$\delta q_{mn} = q^0_{mn} - q^1_{mn} \tag{3-49}$$

$$\delta r_{mn} = \prod_{n' \in N(m) \setminus n} \delta q_{mn'} \tag{3-50}$$

and set

$$r^0_{mn} = 1/2(1 + \delta r_{mn}) \tag{3-51}$$

$$r^1_{mn} = 1/2(1 - \delta r_{mn}) \tag{3-52}$$

3. *Vertical Step.* For each n and m and for $x = 0$, 1, one has

$$q^x_{mn} = a_{mn} f^x_n \prod_{m' \in M(n)/m} r^x_{m'n} \tag{3-53}$$

where a_{mn} is chosen such that

$$q^0_{mn} + q^1_{mn} = 1 \tag{3-54}$$

We can also update the pseudo a posteriori probabilities q_n^0 and q_n^1 as

$$q_n^x = a_n f_n^x \prod_{m \in M(n)} r_{mn}^x \tag{3-55}$$

4. The quantities are used to create a tentative bit-by-bit decoding $\tilde{x}$; if $H\tilde{x} = 0$, then the decoding algorithm stops. Otherwise, the algorithm repeats from the horizontal step. A failure is declared if the maximum number of iterations (e.g., 100) reaches without a valid decoding result.

If the belief propagation iterative decoding algorithm is applied to the regular LDPC codes, it has been proven that 5.17% of errors can be corrected at code rate $\frac{1}{2}$ and 16% of errors can be corrected at code rate $\frac{1}{4}$.

In DTV broadcasting systems and other systems, applying LDPC code also needs to address the design and implementation issue of multiple-code-rate, high-speed encoding and decoding problems. Structured LDPC codes, such as QC (quasi-cyclic)–LDPC codes, might be a good choice [12–15].

3.10 CHANNEL CODING ADOPTED BY DIFFERENT DTV BROADCASTING STANDARDS

Currently, there are four International Telecommunication Union (ITU) standards for the first generation of DTTB systems: ATSC (Advanced Television Systems Committee) [17], DVB-T (Digital Video Broadcasting-Terrestrial) [18], ISDB-T (Integrated Services Digital Broadcasting-Terrestrial) [19], and DTMB (Digital Television/Terrestrial Multimedia Broadcasting) [20].

Table 3-1 summarizes the channel coding scheme of ATSC, DVB-T, ISDB-T, and DTMB. The chapters that follow will provide detailed descriptions.

Details from Table 3-1 are as follows:

1. All DTTB standards have adopted a powerful, serial concatenation coding scheme with the pattern of outer code + outer code interleaving + inner code + inner code interleaving to deal with the very harsh transmission environment. The channel coding schemes of DVB-T and ISDB-T are essentially the same.
2. Because of its superior ability to correct burst errors, RS code has become the common choice of existing DTTB international standards. For example, the RS (207,187) code adopted by American ATSC is generated from RS (255,235) by presetting 48 information bits to zero; the RS (204,188) code in the European and Japanese plans is a shortened code of the common RS (255,239) code. Since the new shortened code is actually a subset of the original code, the minimum-distance and error correction capabilities of the shortened code are at least not worse.
3. The inner code in the European and Japanese standards is the punctuated convolutional code; it is formed by removing a certain number of bits regularly

TABLE 3-1 Comparison of Existing International Standards in Channel Coding

System	ATSC	DTMB	DVB-T	ISDB-T
Outer code	RS (207, 187, $t = 10$)	BCH(762, 752)	RS (204, 188, $t = 8$)	RS (204, 188, $t = 8$)
Interleaving of outer code	52 RS encoded blocks	None	12 Encoded RS blocks	12 Encoded RS blocks
Inner code	2/3 TCM	LDPC code rate: 0.4, 0.6, 0.8, code length 7493	Multirate punctured convolutional code rate: 1/2, 2/3, 3/4, 5/6, 7/8, constraint length 7	Multirate punctured convolutional code rate: 1/2, 2/3, 3/4, 5/6, 7/8, constraint length 7
Interleaving of inner code	12:1 Trellis code interleaving	Convolutional time interleaving and frequency interleaving	Convolutional interleaving and frequency interleaving	Convolutional interleaving, frequency interleaving, and optional time interleaving

from the convolutional encoder output. This punctuation helps adjust the transmission rate and improve the transmission efficiency easily to meet different service requirements. However, usually the error correction ability of the punctuated codes will slightly degrade compared with the original code.

4. The FEC of the DTMB system is formed by cascading the outer BCH code and the inner LDPC code. DTMB has adopted three LDPC code rates to provide different coding gains or error correction capabilities: the LDPC (7488, 3048) code with equivalent coding rate ~0.4, the LDPC (7488, 4572) code with equivalent coding rate ~0.6, and the LDPC (7488, 6096) code with equivalent coding rate ~0.8.
5. The outer code interleaving in the DTTB standards (except DTMB) is byte-based convolutional interleaving with depth determined by the RS codeword length. For inner code interleaving, ATSC's scheme is similar to convolutional interleaving while DVB-T and ISDB-T has adopted not only bit interleaving but also frequency interleaving (similar to block interleaving in the time domain) as OFDM technology is used. DTMB has adopted very long time interleaving against multipath fading. A convolutional interleaver with two modes is used across many signal frames in DTMB where the total interleaving length is 170 and 510 signal frames (the corresponding delay is ~100 and 300 ms).

6. The error correction capability of RS (207,187) for ATSC is $t = 10$ and that of RS (204,188) for DVB-T and ISDB-T systems is $t = 8$. ATSC has adopted a deeper interleaving (the delay and storage are also increasing), but the gain is only 0.3–0.5 dB. ATSC's internal code uses TCM instead of convolutional code (as used in DVB-T/ISDB-T), which brings 0.5–1 dB gain compared to that of DVB-T/ISDB-T. Added together, ATSC has the overall coding gain of 0.8–1.5 dB compared with DVB-T/ISDB-T. This has been confirmed in field tests in Brazil and Australia. The performance of the LDPC codes in DTMB has been extensively studied, and the results show excellent coding gain under different reception conditions.

Since the channel code is designed based on specific channel characteristics and the real channel condition is quite dynamic as well as changes with time, it is difficult to get a "truly" optimal code. The coding performance of an AWGN channel is usually used for comparison. Even though a wireless RF channel is normally not an AWGN channel, one can use channel estimation and compensation technologies (such as equalization) at the receiver to approximately convert the real channel condition into an AWGN channel to fully utilize its error correction capability. The synchronization (including carrier synchronization and clock synchronization) performance at the receiver also has great impact on decoding performance. Meyer et al. divide the receivers into "inner" and "outer" receivers. From a functional point of view, the inner receiver is responsible for synchronization, channel estimation, and equalization while the outer receivers performs the decoding [21]. Inner and outer receivers work independently, that is, the channel estimation and synchronization are assumed to be completed by the inner receiver before decoding is done by the outer receiver. Joint algorithms have also been widely studied with much higher implementation complexity.

3.11 SUMMARY

In this chapter, we introduced the channel coding for DTTB systems with an emphasis on the basic coding concepts and important conclusions for DTTB system design. Detailed discussion of channel coding with knowledge of finite field algebra and statistics is beyond the scope of this book, and readers interested in these topics are strongly encouraged to refer to other books dedicated to channel coding.

REFERENCES

1. S. G. Wilson, *Digital Modulation and Coding*, Upper Saddle, NJ: Prentice Hall, 1995.
2. W. E. Ryan and S. Lin, *Channel Codes: Classical and Modern*, Cambridge: Cambridge University Press, 2009.
3. J. G. Proakis, *Digital Communications*, 5th ed., Columbus, OH: McGraw-Hill, 2007.

4. C. Berrou, A. Blavieux, and P. Thitimajshima, "Near Shannon limit error-correcting coding and decoding: Turbo-codes 1," *ICC 93 Technical Program, Conference Record IEEE International Conference on Communications*, vol. 2, pp. 1064–1070, May 1993.
5. ETSI EN 302 307 V1.2.1, *Second Generation Framing Structure, Channel Coding and Modulation Systems for Broadcasting, Interactive Services, News Gathering and Other Broadband Satellite Applications (DVB-S2)*, Geneva: Digital Video Broadcasting (DVB), Apr. 2009.
6. ETSI EN 302 755 V1.3.1, *Digital Video Broadcasting (DVB); Frame Structure Channel Coding and Modulation for a Second Generation Digital Terrestrial Television Broadcasting System (DVB-T2)*, Geneva: Digital Video Broadcasting (DVB), Apr. 2012.
7. R. G. Gallager, "Low density parity check codes," *IEEE Transactions on Information Theory*, vol. 8, no. 3, pp. 208–220, Jan. 1962.
8. D. J. MacKay and R. M. Neal, "Near Shannon limit performance of low-density parity check codes," *Electronics Letters*, vol. 32, no. 8, pp. 1645–1646, Aug. 1996.
9. Y. Kou, S. Lin, and M. Fossorier, "Low density parity check codes based on finite geometries: A rediscovery and new results," *IEEE Transactions on Information Theory*, vol. 47, no. 7, pp. 2711–2736, Nov. 2001.
10. T. J. Richardson, "Design of capacity-approaching irregular low-density parity-check code," *IEEE Transations on Information Theory*, vol. 47, no. 2, pp. 619–637, Feb. 2001.
11. Communications Systems and Research Section, JPL, "Turbo code performance," Apr. 2, 2000, Available: http://www331.jpl.nasa.gov/public/TurboPerf.pdf, accessed Feb. 1, 2008.
12. Z. Li, C. Lei, L. Zeng, S. Lin, and W. Fong, "Efficient encoding of quasi-cyclic low-density parity-check codes," *IEEE Transactions on Communications*, vol. 54, no. 1, pp. 71–81, Jan. 2006.
13. Z. Yang, Q. Xie, K. Peng, and J. Fu, "A fast and efficient encoding structure for QC-LDPC codes," *Proc. ICCSC 2008: IEEE International Conference on Communications, Circuits and Systems*, pp. 16–20, 2008.
14. N. Jiang, K. Peng, J. Song, C. Pan, and Z. Yang, "High-throughput QC-LDPC decoders," *IEEE Transactions on Broadcasting*, vol. 2, no. 55, pp. 252–259, June 2009.
15. D. Niu, K. Peng, C. Pan, and Z. Yang, "Multi-rate LDPC decoder implementation for China Digital Television Terrestrial Broadcasting Standard," *Proc. ICCCAS 2007: IEEE International Conference on Communications, Circuits and Systems*, vol. 1, pp. 24–28, 2007.
16. M. S. Alencar, *Digital Television Systems*, Cambridge: Cambridge University Press, 2009.
17. Advanced Television System Committee.A/53, *ATSC Digital Television Standard*, Washington, DC: ATSC, 1995.
18. ETSI.300 744 V1.2.1, *Digital Broadcasting Systems for Television, Sound and Data Services, Framing Structure, Channel Coding and Modulation for Digital Terrestrial Television*, Sophia-Antipolis, France: ETSI, 1999.
19. ITU-R WP 11A/59, *Channel Coding, Frame Structure and Modulation Scheme for Terrestrial Integrated Service Digital Broadcasting (ISDB-T)*, Tokyo: ARIB, 1999.
20. Chinese National Standard GB 20600-2006, *Framing Structure, Channel Coding and Modulation for Digital Television Terrestrial Broadcasting System*, Aug. 2006.
21. H. Meyr, M. Moeneclaey, and S. A. Fechtel, *Digital Communication Receivers: Synchronization and Channel Estimation*, New York: Wiley, 1997.

11. Communications Systems and Research Section, JPL, "[illegible]," [illegible] Accessed [illegible] 2011.

4

MODULATION TECHNOLOGIES FOR DTTB SYSTEM

4.1 INTRODUCTION

This chapter introduces the modulation technologies used in various DTTB systems. Essentially, digital modulation is a function transformation process which changes one or more properties of a high-frequency periodic waveform known as a carrier signal with a modulating signal containing information to be transmitted. Here, channel coding is usually applied to the information to be transmitted before modulation. The properties of the carrier that can be changed include amplitude, phase and frequency. Therefore, there are three basic types of linear digital modulation: pulse amplitude modulation (PAM), also known as amplitude shift keying (ASK); phase shift keying (PSK); and frequency shift keying (FSK). All digital modulation commonly used can be considered as variations or combinations of these three basic types. Linear modulation and demodulation techniques are usually applied to the DTV broadcasting systems, that is, the modulated signal spectrum is a linear conversion of that of the baseband digital signal [1].

This chapter also introduces the basic concepts of coded modulation (CM), *a technique jointly optimizing channel coding and digital modulation to best control the errors made by the nonideal effects for transmission* [2], with emphasis on digital modulation after the discussion on channel coding in Chapter 3. Four well-known CM—schemes: multilevel coding (MLC), bit-interleaved coded modulation (BICM), and BICM with iterative demapping (BICM-ID)—will first be examined, and the CM schemes typically used in DTTB systems will then be presented. It is a difficult to

Digital Terrestrial Television Broadcasting: Technology and System, First Edition. Edited by Jian Song, Zhixing Yang, and Jun Wang.

make a clear distinction between coding and modulation when system design tends to consider a combination of both, so there will be overlap between these two chapters.

4.2 DIGITAL MODULATION

4.2.1 Signal Space and Its Representation

The linear modulation and demodulation model is shown in Figure 4-1, in which the modulation is divided into two basic functional modules: baseband processing and spectrum conversion. M-ary modulation is usually adopted in TV broadcasting systems. In this case, the baseband processing module takes $k = \log_2 M$ bits from the binary sequence $\{a_n\}$ each time to form a group, apply baseband shape filtering to each group, and then select one waveform from out of $M = 2^k$ analog carrier waveforms $\{s_i(t),\ i = 1,2, \ldots ,M\}$ based on the defined mapping relationship to complete spectrum conversion. Here, $s_i(t)$ is carefully designed to match the channel characteristics. After these operations, the transmission signal is bandwidth limited within a frequency band at the carrier frequency f_c as the center. The signal then passes through the non ideal channel suffering from the waveform distortion, noise, and interference. At the receiver, the demodulator filters the received signal to eliminate the outband noise as well as the interference and down converts the signal back to the baseband. The match filtering and decision are done in the baseband.

Following the vector space concept in linear algebra, the signal waveform set $\{s_i(t), i = 0, 1, \ldots, M-1\}$ forms a signal space. The inner product $\langle s_i(t), s_j(t)\rangle$ is defined as

$$\langle s_i(t), s_j(t)\rangle \triangleq \int_{-\infty}^{\infty} s_i(t)s_j^*(t)\, dt \tag{4-1}$$

and the norm $\|s_i(t)\|$ is defined as

$$\|s_i(t)\| \triangleq \sqrt{\langle s_i(t), s_i(t)\rangle} = \left(\int_{-\infty}^{\infty} |s_i(t)|^2\, dt\right)^{1/2} \tag{4-2}$$

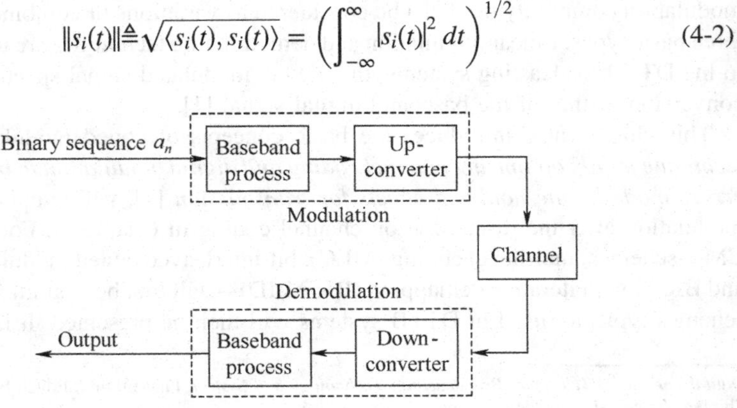

FIGURE 4-1 typical communication transceiver architecture.

Two signals are orthogonal if the inner product given by 4-1 is zero. Any two signals within the signal space satisfy the triangle inequality

$$\left\|s_i(t)+s_j(t)\right\| \le \|s_i(t)\| + \left\|s_j(t)\right\| \tag{4-3}$$

and the Cauchy–Schwarz inequality

$$\left|\langle s_i(t), s_j(t)\rangle\right| \le \|s_i(t)\| \cdot \left\|s_j(t)\right\| \tag{4-4}$$

For a physically implementable modulation, $s_i(t)$ should be deterministic with finite energy, that is,

$$E_{si} = \int_{-\infty}^{\infty} |s_i(t)|^2\, dt < \infty \tag{4-5}$$

Using the Gram–Schmidt algorithm, a set of standard orthogonal basis functions $\{\varphi_j(t), j = 1, \ldots, N\}$ satisfying the following condition can be obtained from the waveform set $\{s_i(t), i = 0, 1, \ldots, M-1\}$ where N stands for the dimension of the set $\{s_i(t)\}$ and one has $N \le M$. The orthogonality is defined as

$$\int_{-\infty}^{\infty} \varphi_i(t)\varphi_j(t)\, dt = \begin{cases} 0 & i \ne j \\ 1 & i = j \end{cases} \tag{4-6}$$

and each element in $\{s_i(t)\}$ can be obtained by using a linear combination of them, namely,

$$s_i(t) = \sum_{j=1}^{N} s_{ij}\varphi_j(t) \tag{4-7}$$

In this way, a one-to-one relationship between M signal waveforms $\{s_i(t)\}$ and M vectors of N dimensions is established, that is, $s_i(t)$ can be equivalently expressed by a point $s_i = [s_{i1}, s_{i2}, \ldots, s_{iN}]$ in an N-dimensional signal space. In fact, a linearly modulated signal is usually characterized by two orthonormal basis function sets under rectangular pulse conditions whereby

$$\varphi_1(t) = \sqrt{\frac{2}{T_s}} \cos 2\pi f_c t \tag{4-8}$$

and

$$\varphi_2(t) = -\sqrt{\frac{2}{T_s}} \sin 2\pi f_c t \tag{4-9}$$

where T_s is the duration of the signal waveform $s_i(t)$ and f_c is the carrier frequency.

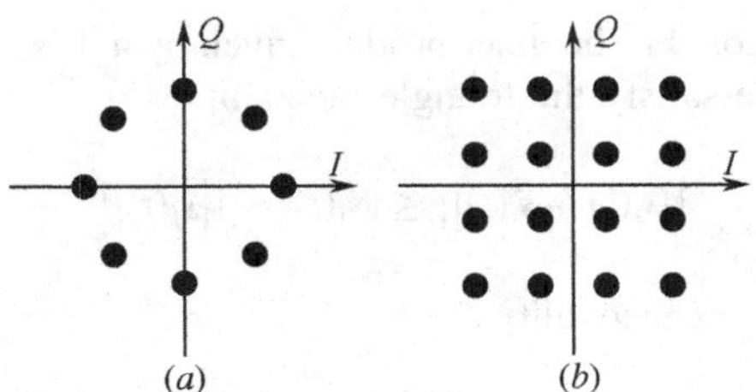

FIGURE 4-2 Signal constellations of (*a*) 8PSK and (*b*) 16QAM.

Thereafter, waveforms from linear modulation can be expressed in a two-dimensional plane known as the constellation. The signal constellations of 8PSK and 16QAM modulations are shown in Figure 4-2 whereby the horizontal component (x component) is the in-phase (I) component and the vertical component (y component) is the quadrature (Q) component.

As the modulated digital signal is generally a bandpass signal, it can be further expressed as

$$s_i(t) = \Re\{s_{Li}(t) \exp(j2\pi f_c t)\} \tag{4-10}$$

where the low-pass signal $s_{Li}(t)$ is the complex envelope or the equivalent low-pass signal of $s_i(t)$ and $j = \sqrt{-1}$. Here, $\Re(\cdot)$ denotes the operation to obtain the real part from a complex-valued signal and $s_{Li}(t)$ will be given as

$$s_{Li}(t) = x_i(t) + jy_i(t) = a_i(t) \exp(j\theta_i(t)) \tag{4-11}$$

where the baseband components $x_i(t)$ and $y_i(t)$ can be considered as amplitude modulation signals separately applied to in-phase carrier cos $2\pi f_c t$ and quadrature carrier sin $2\pi f_c t$, and one has

$$a_i(t) = \sqrt{x_i^2(t) + y_i^2(t)} \quad \theta_i(t) = \arctan \frac{y_i(t)}{x_i(t)} \tag{4-12}$$

Since the constellation points also reflect the complex envelope distribution of the waveform set, the bandpass signal $s_i(t)$ can also be expressed as

$$\begin{aligned} s_i(t) &= x_i(t) \cos 2\pi f_c t - y_i(t) \sin 2\pi f_c t \\ &= a_i(t) \cos [2\pi f_c t + \theta_i(t)] \end{aligned} \tag{4-13}$$

Thus the Fourier transform of $s_i(t)$ is

$$S_i(f) = \int_{-\infty}^{\infty} s_i(t) \exp(-j2\pi f t)\, dt = \frac{1}{2}\left[S_{Li}(f - f_c) + S_{Li}(f + f_c)\right] \tag{4-14}$$

where $S_{Li}(f)$ is the Fourier transform of the complex envelope $s_{Li}(t)$. Similarly, the energy of $s_i(t)$ is

$$E_{si} = \int_{-\infty}^{\infty} |s_i(t)|^2 dt = \frac{1}{2}\int_{-\infty}^{\infty} |S_{Li}(t)|^2 dt \tag{4-15}$$

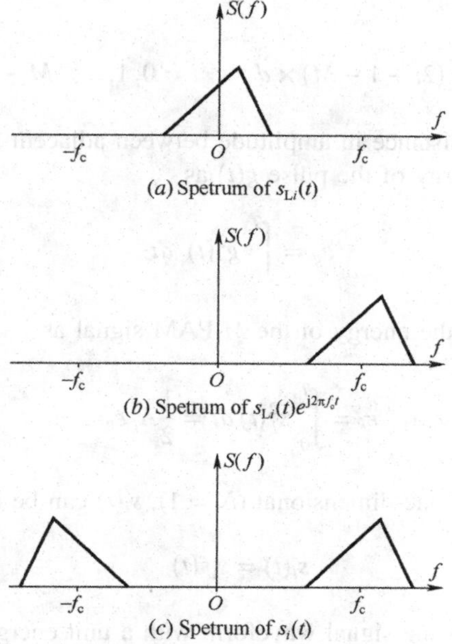

FIGURE 4-3 Illustration of signal spectrum throughout modulation.

The spectrum of the low-pass complex signal $s_{Li}(t)$, the modulated complex signal $s_{Li}(t)\exp(j2\pi f_c t)$, and the corresponding real signal $s_i(t)$ is illustrated in Figures 4-3*a*, *b*, and c respectively.

4.2.2 Typical Digital Modulations

As discussed above, a passband signal has an equivalent baseband representation, while a digital baseband waveform can be written as a linear combination of a set of basis functions. These waveforms may differ in amplitude, phase, frequency, or combinations of them. The modulations that differ in amplitude and phase are widely used in digital video broadcasting systems, including the terrestrial, satellite and cable systems, and will be discussed in this chapter, starting with digital PAM.

4.2.2.1 Pulse Amplitude Modulation In digital PAM, the signal waveform $s_i(t)$ is represented as

$$\begin{aligned} s_i(t) &= \Re[A_i g(t)\exp(j2\pi f_c t)] \\ &= A_i g(t)\cos 2\pi f_c t \qquad i = 0,\ldots,M-1 \qquad 0 \le t \le T_s \end{aligned} \tag{4-16}$$

where $\{A_i, 0 \le i < M\}$ represents the amplitude of $M = 2^m$ possible symbols and $g(t)$ is a real-valued shaping signal that influences the spectrum of the transmitted signal. For the traditional uniformly spaced M-PAM, the signal amplitude of A_i takes the discrete levels as

$$A_i = (2i + 1 - M) \times d \qquad i = 0, 1, \ldots, M - 1 \tag{4-17}$$

where $2d$ denotes the distance in amplitude between adjacent signals.

By denoting the energy of the pulse $g(t)$ as

$$\varepsilon_g = \int_0^{T_s} g^2(t)\, dt \tag{4-18}$$

it is clear that one has the energy of the M-PAM signal as

$$\varepsilon_i = \int_0^{T_s} s_i^2(t)\, dt = \frac{1}{2} A_i^2 \varepsilon_g \tag{4-19}$$

Since PAM signals are one-dimensional ($N = 1$), $s_i(t)$ can be expressed as

$$s_i(t) = s_i f(t) \tag{4-20}$$

where $f(t)$ is defined as the signal waveform with a unit energy given by

$$f(t) = \sqrt{\frac{2}{\varepsilon_g}} g(t) \cos\, 2\pi f_c t \tag{4-21}$$

and, for $i = 0, 1, \ldots, M - 1$, s_i is represented by

$$s_i = A_i \sqrt{\frac{\varepsilon_g}{2}} \tag{4-22}$$

The corresponding signal diagrams of 2/4/8-ary PAM are shown in Figure 4-4, wherein the *Gray labeling* is also depicted. The bit-to-symbol mapping function, i.e., labeling, is

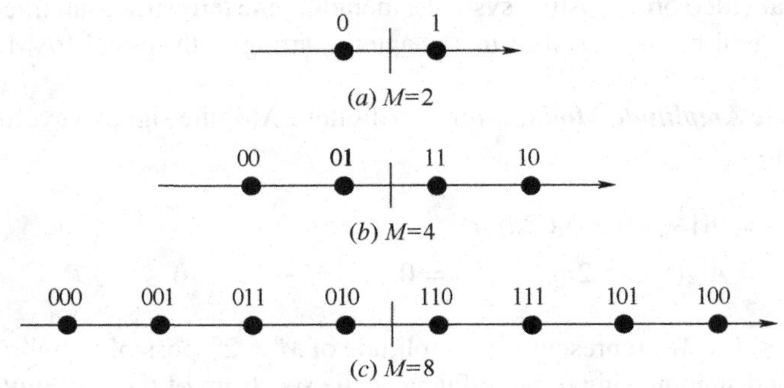

FIGURE 4-4 Signal space diagram for digital M-PAM with Gray labeling.

crucial to the error performance of the transmission. There are plenty of labeling functions associated with m information bits to M possible signal amplitudes. The preferred labeling is the one where any two adjacent constellation points have their corresponding bit labels differ only in one bit, as most likely the decision symbol errors occur among the adjacent constellation points. With this labeling, the bit errors can be minimized after demodulation and can be taken care of by the decoder. Nevertheless, this is based on the assumption that no a priori information about the transmitted bits is available at the demodulator, i.e., there is no feedback from the decoder to the demodulator, such as in traditional BICM receivers. If iteration or feedback is employed, the Gray labeling may not be the optimal and that will be demonstrated later.

The error performance associated with modulation is quite relevant to the Euclidean distance between any pair of signal points. The Euclidean distance here is defined as the distance between two points defined as the square root of the sum of the squares of the differences between the corresponding coordinates of the points, which may be calculated as

$$\begin{aligned} d_{mn}^{(e)} &= \sqrt{(s_m - s_n)^2} \\ &= \sqrt{\frac{\varepsilon_g}{2}}|A_m - A_n| \\ &= d\sqrt{2\varepsilon_g}|m - n| \end{aligned} \tag{4-23}$$

Hence, the minimum Euclidean distance, i.e., the distance between a pair of adjacent signal points, is

$$d_{\min}^{(e)} = d\sqrt{2\varepsilon_g} \tag{4-24}$$

4.2.2.2 *Phase Shift Keying* The PSK signal waveform $s_i(t)$ is represented by

$$\begin{aligned} s_i(t) &= \Re[g(t)\exp(j\theta_i)\exp(j2\pi f_c t)] \\ &= g(t)\cos\theta_i \cos 2\pi f_c t - g(t)\sin\theta_i \sin 2\pi f_c t \end{aligned} \tag{4-25}$$

for $i = 0, 1 \ldots, M-1, 0 \le t \le T_s$, and $\theta_i = 2\pi(i-1)/M$ or $\theta_i = \pi(2i-1)/M$. Here, $\{\theta_i\}$ is the set of M possible phases of the carrier to identify every m-bit information where $M = 2^m$. For instance, the quaternary PSK (QPSK) is widely used in digital transmission systems. The commonly used QPSK signal constellation is shown in Figure 4-5 and that for 8PSK with Gray labeling is shown in Figure 4-6.

Again, by denoting the energy of the pulse signal $g(t)$ as ε_g, we note that the PSK signal waveforms have equal energy shown as

$$\varepsilon = \int_0^{T_s} s_i^2(t)\,dt = \frac{\varepsilon_g}{2} \tag{4-26}$$

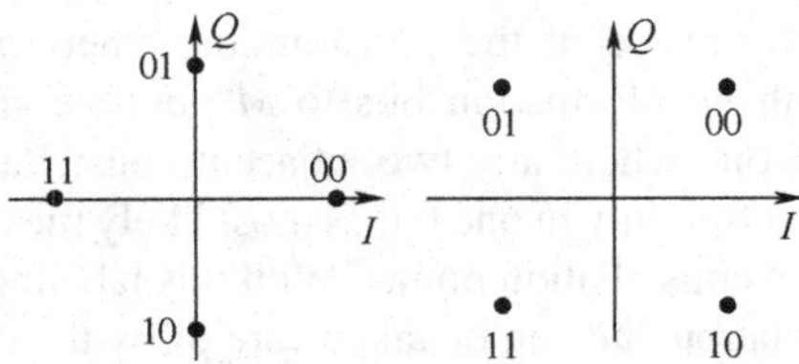

FIGURE 4-5 Signal space diagram of commonly used QPSK with Gray labeling.

Furthermore, PSK waveform $s_i(t)$ can be taken as the combination of two orthonormal functions,

$$s_i(t) = s_{i1}\varphi_1(t)s_{i2}\varphi_2(t) \tag{4-27}$$

where the orthonormal functions $\varphi_1(t)$ and $\varphi_2(t)$ are, respectively, presented as

$$\varphi_1(t) = \sqrt{\frac{2}{\varepsilon_g}}g(t)\cos\ 2\pi f_c t \tag{4-28}$$

and

$$\varphi_2(t) = -\sqrt{\frac{2}{\varepsilon_g}}g(t)\sin\ 2\pi f_c t \tag{4-29}$$

The two-dimensional vectors $s_i = [s_{i1}, s_{i2}]$ are given by

$$\mathbf{s}_i = \left[\sqrt{\frac{\varepsilon_g}{2}}\cos\ \theta_i, \sqrt{\frac{\varepsilon_g}{2}}\sin\ \theta_i\right] \qquad i = 0, 1, \ldots, M-1 \tag{4-30}$$

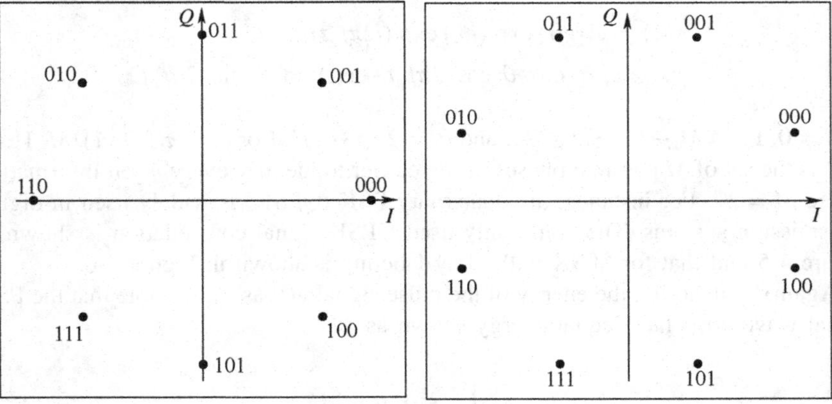

FIGURE 4-6 Signal space diagram of 8PSK with Gray labeling.

The Euclidean distance between the pair of signal points s_m and s_n is

$$\begin{aligned} d_{mn}^{(e)} &= \| (s_m - s_n \| \\ &= \{\varepsilon_g[1 - \cos(2\pi(m-n)/M)]\}^{1/2} \end{aligned} \tag{4-31}$$

The minimum Euclidean distance between adjacent signal phases for $|m-n| = 1$ may be written as

$$d_{\min}^{(e)} = \sqrt{\varepsilon_g \left(1 - \cos\frac{2\pi}{M}\right)} \tag{4-32}$$

4.2.2.3 Quadrature Amplitude Modulation For bandwidth-limited systems, high-throughput transmission requires high-order constellation modulation, i.e., one symbol carries more than one information bit. As discussed earlier, it is quite straightforward to consider a combination of amplitude and phase modulations together to fully utilize the resource of the carrier. With this consideration, quadrature amplitude modulation (QAM) is introduced, which has been widely used in digital broadcasting systems to deliver higher bit rates within a limited bandwidth.

The common M-ary QAM signal constellation is shown in Figure 4-7 with the waveform expressed as

$$\begin{aligned} s_i(t) &= \Re[(A_{ic} + jA_{is})g(t)\exp(j2\pi f_c t)] \\ &= A_{ic}g(t)\cos 2\pi f_c t - A_{is}g(t)\sin 2\pi f_c t \\ i &= 0, \ldots, M-1 \qquad 0 \le t \le T_s \end{aligned} \tag{4-33}$$

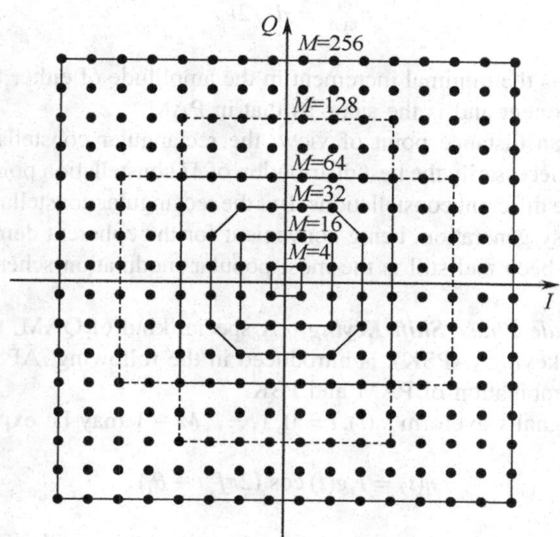

FIGURE 4-7 Signal space diagram of M-ary QAM constellation.

where A_{ic} and A_{is} carrying the information to be transmitted denote the carrier amplitudes of the in-phase and quadrature components respectively. In addition, the QAM signal can also be expressed as

$$s_i(t) = V_i g(t) \cos\left(2\pi f_c t + \theta_i\right) \tag{4-34}$$

where $V_i = \sqrt{A_{ic}^2 + A_{is}^2}$, $\theta_i = \arctan\left(A_{is}/A_{ic}\right)$.

The QAM signal waveform can also be taken as a combination of amplitude– and phase-modulated signals.

By denoting the energy of $g(t)$ as ε_g and defining the orthonormal functions $\varphi_1(t)$ and $\varphi_2(t)$ as 4-28 and 4-29, the QAM signal waveform $s_i(t)$ can be represented by a two-dimensional vector

$$\mathbf{s}_i = [s_{i1}, s_{i2}] = [A_{ic}\sqrt{\varepsilon_g/2}, A_{is}\sqrt{\varepsilon_g/2}] \tag{4-35}$$

The Euclidean distance between any two constellation points for the QAM signal can be calculated by

$$d_{mn}^{(e)} = |\mathbf{s}_m - \mathbf{s}_n| = \sqrt{\frac{\varepsilon_g}{2}\left[(A_{mc} - A_{nc})^2 + (A_{ms} - A_{ns})^2\right]} \tag{4-36}$$

When the signal amplitudes take the discrete values of $\{(2i + 1 - M)d,\ i = 0, 1, \ldots, M - 1\}$, i.e., the signal space diagram is rectangular, as shown in Figure 4-7. The minimum Euclidean distance is

$$d_{\min}^{(e)} = d\sqrt{2\varepsilon_g} \tag{4-37}$$

where d represents the minimal increment in the amplitude of either the in-phase or quadrature component and is the same as that in PAM.

From Euclidean distance point of view, the rectangular constellation shown in Figure 4-7 is not necessarily the best distribution of M constellation points, and system design can choose different constellations. But the rectangular constellation has unique advantages of easy generation, being convenient for the coherent demodulation, etc. Therefore, it has been and still is the most popular modulation scheme.

4.2.2.4 Amplitude Phase Shift Keying A special kind of QAM, namely, amplitude phase shift keying (APSK), is introduced in the following. APSK can also be regarded as a combination of PAM and PSK.

The APSK signal waveform $s_i(t), i = 0, 1, \ldots, M - 1$ may be expressed as

$$s_i(t) = r_i g(t) \cos\left(2\pi f_c t + \theta_i\right) \tag{4-38}$$

where r_i and θ_i represent the amplitude and phase of $s_i(t)$, respectively. The APSK signal waveform $s_i(t)$ can also be written as a two-dimensional vector $\mathbf{s}_i$ as shown in 4-35.

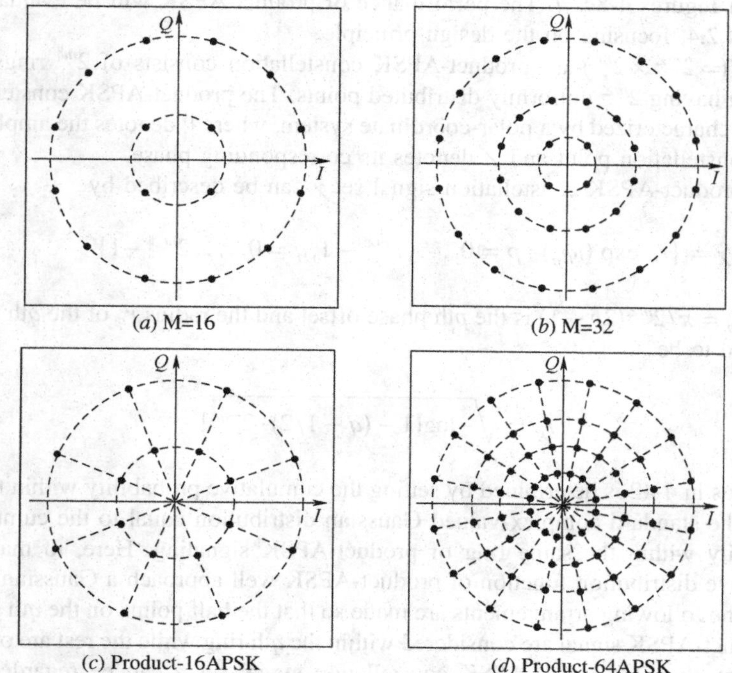

FIGURE 4-8 Signal space diagrams of M-ary APSK constellations of (a) $M = 16$ and (b) $M = 32$ used by DVB-S2 and M-ary product-APSK constellations of (c) $M = 16$ and (d) $M = 64$.

However, different from the traditional rectangular QAM, the APSK $\{\mathbf{s}_i\}$ locates on several rings, each with uniformly distributed signal points. Figures 4-8*a*, *b* 16- and 32-ary APSK constellations used in DVB-S2.

The Euclidean distance between any pair of constellation points for the APSK signal can be calculated by

$$d_{mn}^{(e)} = |\mathbf{s}_m - \mathbf{s}_n| = \sqrt{\frac{\mathcal{E}_g}{2}\left[r_m^2 + r_n^2 - 2r_m r_n \cos(\theta_m - \theta_n)\right]} \tag{4-39}$$

The minimum Euclidean distance $d_{\min}^{(e)}$ is readily evaluated as

$$d_{\min}^{(e)} = \min_{m,n,m \neq n} d_{mn}^{(e)} \tag{4-40}$$

However, APSK $d_{\min}^{(e)}$ does not have a closed-form expression as it depends on many factors such as the number of rings, the number of points on each ring, and the phase offset of each ring.

A special APSK called product-APSK will be discussed in detail as it can be taken as the amplitude–phase product of a classical PSK and a pseudo nonuniform PAM, as shown in Figures 4-8*c*, *d*. The performance of product-APSK will be evaluated in Section 4.2.4, focusing on the design principle.

An $M(=2^{m\angle} \times 2^{m\|})$-ary product-APSK constellation consists of $2^{m\|}$ rings with each ring having $2^{m\angle}$ uniformly distributed points. The product-APSK constellation could be characterized by a polar-coordinate system, where $||$ denotes the amplitudes of the constellation point and $\angle$ denotes its corresponding phase.

The product-APSK constellation signal set χ can be described by

$$\chi = \{r_q \exp(j\varphi_p) : p = 0, \ldots, 2^{m\angle} - 1; q = 0, \ldots, 2^m \| -1\} \tag{4-41}$$

where $\varphi_p = \pi/2^{m\angle}(2p+1)$ is the pth phase offset and the radius r_q of the qth ring is suggested to be

$$r_q = \sqrt{-\log[1 - (q + 1/2) \cdot 2^{-m\|}]} \tag{4-42}$$

The radius in 4-42 is determined by setting the cumulative probability within the qth ring of the standard complex-valued Gaussian distribution equal to the cumulative probability within the same ring of product-APSK signaling. Here, to make the cumulative distribution function of product-APSK well approach a Gaussian distribution, the following arrangements are made so that the half points on the qth ring of the product-APSK signal are considered within the qth ring while the rest are outside.

In fact, the $(2^{m\angle} \times 2^{m\|})$-APSK constellation signal set χ can be regarded as a product of the $2^{m\angle}$-PSK set $\mathcal{P} = \{\exp(j\varphi_p)\}$ and the pseudo $2^{m\|}$-PAM set $\mathcal{A} = \{r_q\}$, i.e., $\chi = \mathcal{P} \times \mathcal{A}$. Additionally, one can define the set of phases as $\mathcal{P}^{\angle} = \{\varphi_p\}$.

The minimum Euclidean distance for product-APSK may be evaluated as

$$d_{\min}^{(e)} = \min_{1 \le q < 2^{m\|}} \{2r_0 \sin(\pi/2^{m\angle}), r_q - r_{q-1}\} \tag{4-43}$$

In addition, for an $M = 2^m$-ary product-APSK, usually one has [3]

$$m_{\|} = \begin{cases} m/2 - 1, m_{\angle} = m/2 + 1 & \text{for even } m \\ (m-1)/2, m_{\angle} = (m+1)/2 & \text{for odd } m \end{cases} \tag{4-44}$$

4.2.3 The Power Spectrum of Modulated Signal

Excluding the PAM signal, modulated signals such as PSK and APSK can be taken as QAM signals. The modulator and the power spectrum of a general QAM signal will be presented in the following.

4.2.3.1 Modulator Figure 4-9 is a block diagram of the QAM modulator. A binary bit sequence is mapped onto a symbol sequence, i.e., every m bits are mapped onto a symbol belonging to the M-QAM constellation set. The symbol sequence is usually

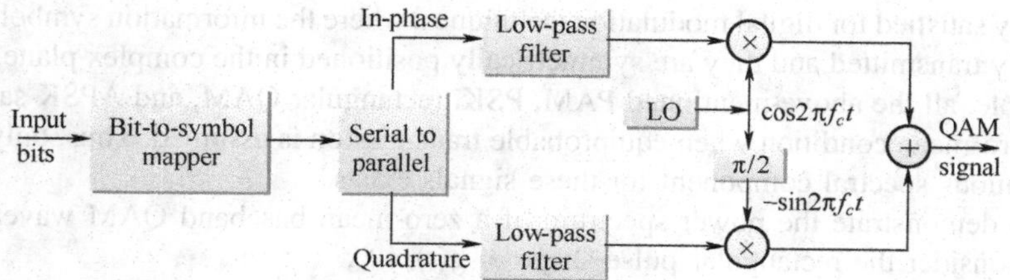

FIGURE 4-9 Schematic diagram of QAM modulator.

separated into the I- and Q-component streams. After passing through the low-pass filter separately, the two streams are modulated by cos $2\pi f_c t$ and -sin $2\pi f_c t$ (both from the same local oscillator, LO), respectively. The QAM signal $s_i(t)$ of 4-33 then can be obtained by combining the outcome of these two modulated streams. Clearly, a squared M-QAM quadrature modulation can be taken as the sum of $\sqrt{M}$-PAM of two orthogonal carriers.

4.2.3.2 Power Spectrum The QAM waveform $s(t)$ can be expressed as

$$\begin{aligned} s(t) &= \Re[x \cdot g(t) \exp(j2\pi f_c t)] \\ &= \Re[v(t) \exp(j2\pi f_c t)] \end{aligned} \tag{4-45}$$

where x is the random process for QAM and $v(t)$ is the equivalent low-pass signal. The autocorrelation function of $s(t)$ is

$$\phi_{ss}(\tau) = \Re[\phi_{vv}(\tau) \exp(j2\pi f_c \tau)] \tag{4-46}$$

where $\phi_{vv}(\tau)$ represents the autocorrelation function of $v(t)$. Thus, the desired power spectrum of $s(t)$ can be readily expressed as

$$\Phi_{ss}(f) = \frac{1}{2}\left[\Phi_{vv}(f - f_c) + \Phi_{vv}(-f - f_c)\right] \tag{4-47}$$

where $\Phi_{vv}(f)$ denotes the power spectrum of $v(t)$.Here, $\Phi_{vv}(f)$ can be written as

$$\Phi_{vv}(f) = \frac{\sigma_x^2}{T_s}|G(f)|^2 + \frac{\mu_x^2}{T_s}\sum_{n=-\infty}^{\infty}\left|G\left(\frac{n}{T_s}\right)\right|^2 \delta\left(f - \frac{n}{T_s}\right) \tag{4-48}$$

where $G(f)$ is the power spectrum of $g(t)$ and $\delta(\cdot)$ is the Dirac function; σ_x^2 denotes the variance of the random process x and μ_x is the mean value. Equation (4-48) consists of two terms with different spectral components. The first term is the continuous spectrum whose shape depends only on the spectrum of $g(t)$. The second term consists of discrete components with the frequency spaced $1/T_s$. It is worth emphasizing that if $\mu_x = 0$, then the second discrete term would be zero. This condition is usually satisfied for digital modulation techniques, where the information symbols are equally transmitted and they are symmetrically positioned in the complex plane. For example, all the above-mentioned PAM, PSK, rectangular QAM, and APSK satisfy the zero-mean condition when equiprobable transmission is assumed. Thus, only the continuous spectral component for these signals exists.

To demonstrate the power spectrum of a zero-mean baseband QAM waveform $v(t)$, consider the rectangular pulse shape of $g(t)$, i.e.,

$$g(t) = \begin{cases} A & 0 \le t \le T_s \\ 0 & \text{otherwise} \end{cases} \tag{4-49}$$

The Fourier transform of $g(t)$ is

$$G(f) = AT_s \frac{\sin \pi f T_s}{\pi f T_s} \exp(-j\pi f T_s) \tag{4-50}$$

Hence

$$|G(f)|^2 = (AT_s)^2 \left(\frac{\sin \pi f T_s}{\pi f T_s} \right)^2 \tag{4-51}$$

and

$$\Phi_{vv}(f) = A^2 \sigma_x^2 T_s \left(\frac{\sin \pi f T_s}{\pi f T_s} \right)^2 \tag{4-52}$$

The spectrum of $\Phi_{vv}(f)$ is illustrated in Figure 4-10. Note that $\Phi_{vv}(f)$ is inversely proportional to the square of the frequency f.

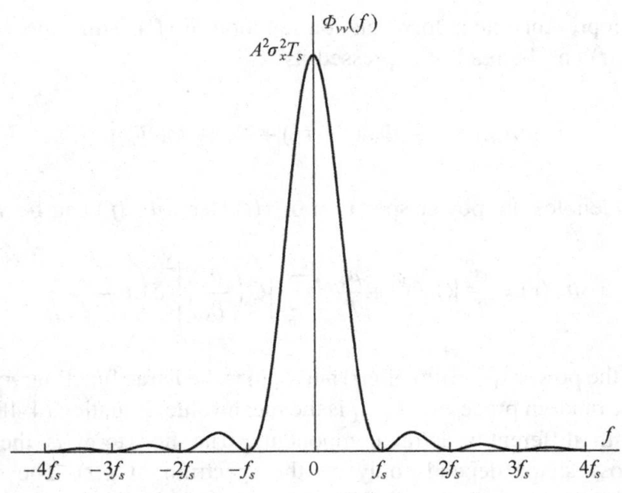

FIGURE 4-10 Power spectrum of baseband QAM waveform having rectangular pulse shape with $f_s = 1/T_s$.

As discussed above, the shape of the power spectrum of the QAM waveform $s(t)$ and its equivalent low-pass waveform $v(t)$ are determined by the pulse shape function $g(t)$. Furthermore, the above signal design neglects the bandwidth limitation of a practical channel. Indeed, the signal design for a bandwidth-limited channel is quite important, which is relevant to the design of $g(t)$. The goal of a typical signal design is to make sure that the sampled data symbols at time $t = nTs, n \in \mathbb{N}$, are inter symbol interference (ISI) free. This condition is known as the Nyquist pulse-shaping criterion for zero ISI, which is stated as follows.

Nyquist Theorem: The necessary and sufficient condition for $x(t)$ to satisfy the sampling ISI free, i.e.,

$$x(nT_s) = \begin{cases} 1 & n = 0 \\ 0 & n \neq 0 \end{cases} \tag{4-53}$$

means that its Fourier transform $X(f)$ satisfies

$$\sum_{m=-\infty}^{\infty} X\left(f + \frac{m}{T_s}\right) = T_s \tag{4-54}$$

The proof can be found in numerous textbooks such as [4].

Suppose the channel bandwidth is W, and $X(f) = 0$ for $|f| > W$, i.e.,

$$X(f) = \begin{cases} T_s & \text{abs}(f) \leq W \\ 0 & \text{otherwise} \end{cases} \tag{4-55}$$

The highest symbol rate $1/T_s$ when zero ISI is guaranteed will be $1/T_s = 2W$, and this is the well-known Nyquist rate. The corresponding pulse waveform in the time domain will be

$$x(t) = \frac{\sin\,(\pi t/T_s)}{\pi t/T_s} = \text{sinc}\left(\frac{\pi t}{T_s}\right) \tag{4-56}$$

There are two problems when choosing the sinc function as the pulse shape waveform in practice. The first is that $\text{sinc}(\pi t/T_s)$ is non causal and not practically implementable. To solve this, a delay version, i.e., $\text{sinc}[\pi(t - t_0)/T_s]$, is used to make sure that $\text{sinc}[\pi(t - t_0)/T_s] \approx 0$ for $t < 0$, whereby the sampling time also should be delayed by t_0. The second problem is that the tails of $\text{sinc}(\pi t/T_s)$ decay inversely with t, which is too slow, and a small timing error in sampling would result in a very high ISI. In fact, as the series on the order of $\{1/t\}$ are not absolutely summable, the resulting ISI may not converge.

The most popular pulse shape in practice is the raised-cosine spectrum, which is readily shown as

$$X_{\text{rc}}(f) = \begin{cases} T_s & |f| < \dfrac{1-\beta}{2T_s} \\ \dfrac{T_s}{2}\left\{1 + \cos\left[\dfrac{\pi T_s}{\beta}\right]\left(|f| - \dfrac{1-\beta}{2T_s}\right)\right\} & \dfrac{1-\beta}{2T_s} \leq |f| \leq \dfrac{1+\beta}{2T_s} \\ 0 & \text{otherwise} \end{cases} \tag{4-57}$$

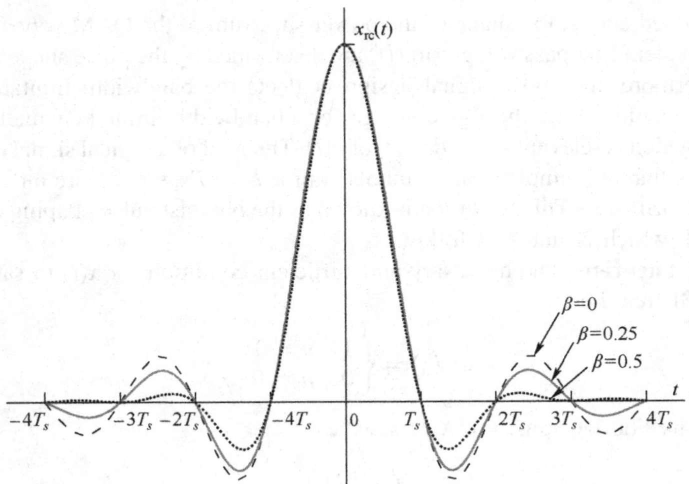

FIGURE 4-11 Time-domain pulse having raised-cosine spectrum.

where $\beta \in [0, 1]$ is called the roll-off factor. In this way, the symbol rate $1/T_s$ could be $2W/(1+\beta)$. It is clear that for $\beta = 0$ the raised-cosine spectrum in 4-57 is degenerated to the rectangular spectrum in 4-51. The corresponding time-domain pulse $x_{rc}(t)$ can be written as

$$x_{rc}(t) = \text{sinc}\left(\frac{\pi t}{T_s}\right)\frac{\cos\,(\pi\beta t/T_s)}{1-4\beta^2t^2/T_s^2} \tag{4-58}$$

Now the tail of $x_{rc}(t)$ decays inversely proportional to t^3 for $\beta > 0$, and consequently the ISI caused by timing error on sampling is controlled. Figure 4-11 illustrates the time-domain pulse having a raised-cosine spectrum for $\beta = 0, \frac{1}{4}, \frac{1}{2}$, respectively.

It is more common to use shape pulse $g(t)$ whose Fourier magnitude spectrum is the square root of the raised cosine waveform in 4-57 for the ideal-channel assumption of $C(f) = 1$ for $|f| \le W$, and there will be

$$x_{rc}(f) = G_T(f)G_R(f) \tag{4-59}$$

where $G_T(f)$ and $G_R(f)$ are the frequency responses of the transmitter and receiver filters and $g(t)$ is the inverse Fourier transform of $G_T(f)$. To satisfy 4-57, it is natural to let

$$|G_T(f)| = \sqrt{|X_{rc}(f)|} \tag{4-60}$$

In addition, it is notable that the rectangular and sinc functions in this case are a Fourier transform pair.

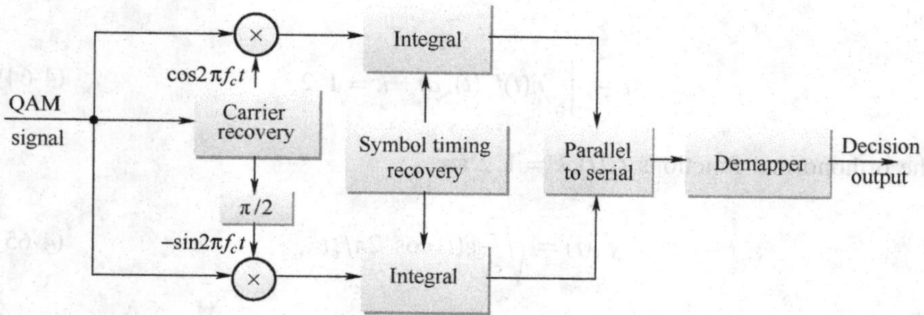

FIGURE 4-12 Schematic diagram of coherent QAM demodulator.

4.2.4 Demodulation and Performance Evaluation

4.2.4.1 Demodulator Figure 4-12 depicts a block diagram for the coherent QAM demodulator using correlation. "Coherent" demodulation means that a demodulator is required to locally generate the carrier exactly at the same frequency and phase as that at the transmitter to complete the demodulation process and restore the original information. In Figure 4-12, the carrier recovery module obtains the in-phase and the quadrature carriers from the received signal and completes the down conversion by multipliers. The integral correlator (or matched filter) is used to remove the noise and unwanted signal components, and the original information stream can then be restored after parallel-to-serial conversion and demapping. The detailed demodulation procedure is provided below.

If a QAM signal $s_i(t)$ is transmitted over an AWGN channel, the received signal $r(t)$ will be

$$r(t) = s_i(t) + n(t) \tag{4-61}$$

where $n(t)$ denotes the AWGN process with zero mean and the power spectral density $\Phi_{nn}(f) = N_0/2$. Based on the observation of $r(t)$ over time interval $[0, T_s]$ and denoting the corresponding m-bit vector as $\mathbf{b}_i$, the demodulator is to make estimation on $\mathbf{b}_i$, say $\hat{\mathbf{b}}_i$ after a hard decision, or from the soft decision representing the reliability of the estimation of $\mathbf{b}_i$ usually expressed by the log-likelihood ratio (LLR) $\mathbf{L}_i$.

The demodulator first converts $r(t)$ into a two-dimensional vector $\mathbf{r} = [r_1, r_2]$ by using correlation as shown in Figure 4-12. That is,

$$r_k = s_{ik} + n_k \quad k = 1, 2 \tag{4-62}$$

where

$$s_{ik} = \int_0^{T_s} s_i(t) f_k(t)\, dt \quad k = 1, 2 \tag{4-63}$$

and

$$n_k = \int_0^{T_s} n(t) f_k(t)\ dt \quad k = 1, 2 \tag{4-64}$$

The orthonormal functions $f_k(t), k = 1, 2$ are

$$f_1(t) = \sqrt{\frac{2}{\varepsilon_g}} g(t) \cos\ 2\pi f_c t \tag{4-65}$$

$$f_2(t) = -\sqrt{\frac{2}{\varepsilon_g}} g(t) \sin\ 2\pi f_c t \tag{4-66}$$

where $\varepsilon_g = \int_0^{T_s} |g(t)|^2\ dt$ is the energy of $g(t)$. If $g(t)$ is a rectangular function, i.e., $g(t)$ equals A for $0 \le t \le T_s$ and 0 for other values of t, then $\varepsilon_g = A^2 T_s$ and

$$f_1(t) = \sqrt{\frac{2}{T_s}} \cos\ 2\pi f_c t \tag{4-67}$$

$$f_2(t) = -\sqrt{\frac{2}{T_s}} \sin\ 2\pi f_c t \tag{4-68}$$

The signal components s_{i1} and s_{i2} are deterministic. The noise components n_1 and n_2, are still Gaussian as any linear transform of a Gaussian variable is still a Gaussian variable. It is clear that both n_1 and n_2 have zero mean and the variance of $N_0/2$. They are statistically independent of each other since

$$\begin{aligned} \mathbb{E}(n_1 n_2) &= \int_0^{E_s} \int_0^{E_s} \mathbb{E}[n_1(t) n_2(\tau)] f_1(t) f_2(\tau)\, dt\, d\tau \\ &= \frac{N_0}{2} \int_0^{E_s} f_1(t) f_2(t)\, dt \\ &= 0 \end{aligned} \tag{4-69}$$

Hence the conditional probability density functions of the random variables $\mathbf{r} = [r_1, r_2]$ can be written as

$$\begin{aligned} p(\mathbf{r}|\mathbf{s}_i) &= p(r_1|s_{i1}) p(r_2|s_{i2}) \\ &= \frac{1}{\pi N_0} \exp\left[-\frac{(r_1 - s_{1i})^2 + (r_2 - s_{2i})^2}{N_0} \right] \end{aligned} \tag{4-70}$$

Because the QAM signal waveform $s_i(t)$ can be expressed as

$$\begin{aligned} s_i(t) &= A_{ic} g(t) \cos\ 2\pi f_c t - A_{is} g(t) \sin\ 2\pi f_c t \\ i &= 0, \ldots, M-1 \quad 0 \le t \le T_s \end{aligned} \tag{4-71}$$

then

$$\mathbf{s}_i = [s_{i1}, s_{i2}] = \left[A_{ic}\sqrt{\tfrac{1}{2}\varepsilon_g}, A_{is}\sqrt{\tfrac{1}{2}\varepsilon_g}\right] \tag{4-72}$$

In the above derivation, the property of $\int_0^{T_s}|g(t)|^2\cos(4\pi f_c t + \theta)\,dt = 0$ is used. This is true when the frequency of the carrier is much higher than the symbol rate $1/T_s$ regardless of the phase θ. By comparing 4-72 with 4-35, it is clear that they have exactly the same expression, which indicates that the received signal $\mathbf{r} = [r_1, r_2]$ can be viewed as the transmitted signal disturbed by Gaussian noise with zero mean and variance of $N_0/2$ in each dimension. Consequently, by denoting a random vector $\mathbf{s} \in \{\mathbf{s}_i\}$ as the transmitted signal, the received random vector $\mathbf{r}$ after demodulation can be written as

$$\mathbf{r} = \mathbf{s} + \mathbf{n} \tag{4-73}$$

where $\mathbf{n} = [n_1, n_2]$ denotes the noise shown in 4-62. The signal-to-noise ratio (SNR) in each dimension can be calculated as

$$\text{SNR}_1 = \frac{\mathbb{E}[s_1^2]}{\mathbb{E}[n_1^2]} = \frac{2\varepsilon_1}{N_0} \tag{4-74}$$

$$\text{SNR}_2 = \frac{\mathbb{E}[s_2^2]}{\mathbb{E}[n_2^2]} = \frac{2\varepsilon_2}{N_0} \tag{4-75}$$

where the transmitted signal energy in each dimension is

$$\varepsilon_1 = \frac{\varepsilon_g}{2}\sum_i \Pr(\mathbf{s}_i)A_{ic}^2 \tag{4-76}$$

$$\varepsilon_2 = \frac{\varepsilon_g}{2}\sum_i \Pr(\mathbf{s}_i)A_{is}^2 \tag{4-77}$$

In most cases, the signal $\{\mathbf{s}_i\}$ is transmitted with equal probability, i.e., $\Pr(\mathbf{s}_i) = 1/M, \forall 0 \le i < M$, and with equal transmit power in each dimension, i.e., $\varepsilon_1 = \varepsilon_2$. Hence, by denoting the average energy of the transmitted signal as $\varepsilon = \varepsilon_1 + \varepsilon_2 = 2\varepsilon_1$, one has

$$\text{SNR} = \frac{\mathbb{E}[\|\mathbf{s}\|^2]}{\mathbb{E}[\|\mathbf{n}\|^2]} = \frac{\varepsilon}{N_0} \tag{4-78}$$

The discrete-time channel model 4-73 is useful for evaluating the digital modulation performance. After $\mathbf{r}$ is received, the task left is to decide which is the symbol transmitted and the corresponding bits. This classic decision by directly outputting the estimated bits is called a *hard decision*. Modern communication systems are likely to calculate the soft information for each bit expressed in the format of log-likelihood ratio (LLR), which is called the *soft decision*, and leave the bit hard decision to the channel decoder.

4.2.4.2 Error Probability Performance We start with the error performance evaluation for the hard decision by evaluating the symbol error rate (SER). The SER for PAM, PSK, QAM, and product-APSK will be derived.

1. *SER for PAM.* Recall that the M-ary PAM signal waveform can be represented as M points of one dimension with values

$$s_i = A_i\sqrt{\tfrac{1}{2}\varepsilon_g} \quad i = 0, 1, \ldots, M-1 \tag{4-79}$$

where ε_g denotes the energy of $g(t)$. The amplitude values are

$$A_i = (2i + 1 - M)d, \quad i = 0, 1, \ldots, M-1 \tag{4-80}$$

The Euclidean distance between adjacent points, i.e., the minimum Euclidean distance, is

$$d_{\min}^{(e)} = d\sqrt{2\varepsilon_g} \tag{4-81}$$

Assuming the equiprobable signal, the average energy of the transmit signal is

$$\varepsilon = \tfrac{1}{6}(M^2 - 1)d^2\varepsilon_g \tag{4-82}$$

The received signal over AWGN channels can be expressed as

$$r = s_i + n \tag{4-83}$$

where n is a zero-mean $N_0/2$-variance Gaussian noise. Hence, the SER of PAM signals over AWGN channels can be readily derived as

$$\begin{aligned} P_s &= \frac{M-1}{M}\Pr\left(|r - s_i| > d\sqrt{\frac{\varepsilon_g}{2}}\right) \\ &= \frac{2(M-1)}{M}Q\left(\sqrt{\frac{d^2\varepsilon_g}{N_0}}\right) \end{aligned} \tag{4-84}$$

where

$$Q(x) = \frac{1}{\sqrt{2\pi}}\int_x^{\infty}\exp\left(\frac{-x^2}{2}\right)dx \tag{4-85}$$

Equivalently, by substituting for $d^2\varepsilon_g$ in 4-84, SER can be written as

$$P_s = \frac{2(M-1)}{M} Q\left(\sqrt{\frac{3}{(M^2-1)}\frac{2\varepsilon}{N_0}}\right) \tag{4-86}$$

Clearly, it is a function of average symbol SNR $2\varepsilon/N_0$.

2. *SER for PSK.* Recall that the M-ary PSK signal waveform can be represented as M two-dimensional points with values of

$$\mathbf{s}_i = \left[\sqrt{\varepsilon}\cos\,\theta_i, \sqrt{\varepsilon}\sin\,\theta_i\right] \quad i = 0, 1, \ldots M-1 \tag{4-87}$$

where $\theta_i = 2\pi_i/M$ or $\theta_i = (2i+1)\pi/M$ and ε denotes the PSK symbol energy which equals to $\varepsilon_g/2$ and ε_g denotes the energy of $g(t)$.

Owing to the circular symmetric property of PSK signals, the SER of PSK can be calculated as the error probability assuming transmitting a signal with zero phase, i.e., $\theta = 0$, and an error occurs when the phase of the received signal is beyond the interval $[-\pi/M, \pi/M)$. By denoting the phase of the received signal as a random variable Θ, the PDF (power density function) of Θ can be determined by

$$\begin{aligned} p_\Theta(\theta) &= \frac{1}{2\pi}\exp\left(-\gamma\sin^2\theta\right)\int_0^\infty r\exp\left[-(r-\sqrt{2\gamma}\cos\,\theta)^2/2\right]dr \\ &= \frac{1}{2\pi}\exp\left(-\gamma\right) + \frac{\sqrt{\gamma}}{2\sqrt{\pi}}\cos\,\theta\exp\left(-\gamma\sin^2\theta\right)\left[1 + \operatorname{erf}\left(\sqrt{\gamma}\cos\,\theta\right)\right] \end{aligned} \tag{4-88}$$

where $\gamma = \varepsilon/N_0$ denotes the symbol SNR and

$$\operatorname{erf}(x) \triangleq \frac{2}{\sqrt{\pi}}\int_0^x \exp\left(-t^2\right)dt \tag{4-89}$$

Hence, the SER for M-PSK can be written as

$$P_s = 1 - \int_{-\pi/M}^{\pi/M} p_\Theta(\theta)\,d\theta \tag{4-90}$$

Clearly, 4-90 does not have a closed-form expression except for $M = 2, 4$, which are the same as 2PAM and 4QAM, respectively. For $M > 4$, the SER can be obtained by numerical integration on 4-90 based on the PDF shown in Figure 4-13.

3. *SER for Square QAM.* A square QAM signal can be regarded as two PAM signals on phase-quadrature carriers. Thereby, they are easily demodulated and are employed in most practical communication and broadcasting systems. Hence, the SER of square QAM can be derived as

$$P_s = 1 - \left(1 - P_s^{\sqrt{M}\text{-PAM}}\right)^2 \tag{4-91}$$

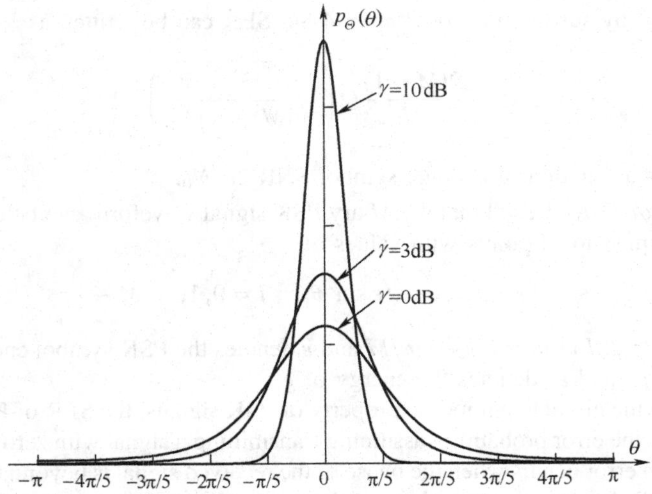

FIGURE 4-13 PDF $p\Theta(\theta)$ for SNRs of 0, 3, and 10 dB.

where $P_s^{\sqrt{M}-\text{PAM}}$ denotes the SER of and $\sqrt{M}$-ary PAM whose symbol energy is half that of the corresponding QAM, and therefore one has

$$P_s^{\sqrt{M}-\text{PAM}} = 2\left(1 - \frac{1}{\sqrt{M}}\right) Q\left(\sqrt{\frac{3}{M-1}\frac{\varepsilon}{N_0}}\right) \tag{4-92}$$

where ε denotes the symbol energy of the QAM signal.

4. *SER for Product-APSK.* Recall from Section 4.2.2.4 that an $M(= 2^{m\angle} \times 2^{m\|})$-ary product-APSK constellation consists of $2^{m\|}$ rings, wherein each ring possesses $2^{m\angle}$ uniformly distributed points, the product-APSK constellation signal set χ can be described by

$$\chi = \{r_q \exp{(j\varphi_p)} : p = 0, \ldots, 2^{m\angle} - 1; q = 0, \ldots, 2^{m\|} - 1\} \tag{4-93}$$

where $\varphi_p = (\pi/2^{m\angle})(2p + 1)$ denotes the pth phase shift. Assuming the equiprobable signal transmission, the symbol energy of product-APSK may be evaluated as

$$\varepsilon = \frac{1}{2^{m\|}} \sum_{q=0}^{2^{m\|}-1} r_q^2 \tag{4-94}$$

Similar to the PSK signal, product-APSK signal is also circular symmetric, and hence one can also safely assume a signal with a zero phase is transmitted when evaluating the SER. We let $x_\| \in \{r_q\}$ be the amplitude of the transmitted APSK and $y_\|$ and $y_\angle$, respectively, be the amplitude and phase of the received signal. We can write the joint

PDF of $y_{\|}$ and $y_{\angle}$ as

$$pY_{\|}, Y_{\angle}(y_{\|}, y_{\angle}) = \frac{y_{\|}}{\pi N_0} \exp\left(-\frac{y_{\|}^2 + x_{\|}^2 - 2x_{\|}y_{\|} \cos\, y_{\angle}}{N_0}\right) \tag{4-95}$$

An error occurs if the phase $y_{\angle}$ is outside of interval $[-\pi/2^{m\angle}, \pi/2^{m\angle})$ or the amplitude $y_{\|}$ is beyond

$$(r_q^-, r_q^+) = \begin{cases} ((r_{q-1} + r_q)/2, (r_q + r_{q+1})/2) & \text{if } 1 \le q < 2^{m\|} - 1 \\ (0, (r_0 + r_1)/2) & \text{if } q = 0 \\ ((r_{q-1} + r_q)/2, \infty) & \text{if } q = 2^{m\|} - 1 \end{cases} \tag{4-96}$$

when $x_{\|} = r_q$. Now, the SER for the product-APSK may be determined as

$$P_s = 1 - \frac{1}{2^{m\|}} \sum_{q=0}^{2^{m\|}-1} \int_{r_q^-}^{r_q^+} \int_{-\pi/2^{m\angle}}^{\pi/2^{m\angle}} pY_{\|}, Y_{\angle}(y_{\|}, y_{\angle})\, dy_{\angle}\, dy_{\|} \tag{4-97}$$

Similar to the SER for PSK, it is clear that 4-97 cannot ben expressed in a closed form, expect for a few special cases, e.g., for $m_{\|} = 1$ and $m_{\angle} = 1, 2$, or for $m_{\angle} = 1$, which degrade to BPSK/QPSK or non uniformly spaced PAM.

5. *SER Comparison of Square QAM and Product-APSK.* After discussing the SER performance of each modulation scheme individually, the SER performance between square QAM and product-APSK can be compared in the following. Square QAM is widely used in most practical systems, much more often than PAM and PSK because the performance of QAM is much better at the same spectrum efficiency with its demodulation complexity as low as that of PAM and PSK. Nevertheless, it is interesting that product-APSK may have a better performance than square QAM at the practical coding rates while keeping a similar low complexity. Here, the performance of square QAM and product-APSK is compared in SER first and then by mutual information in the following section. The complexity analysis will cover each specific coded modulation scheme, as the demodulation process from the received signal to the demapper shown in Figure 4-14 is identical for both QAM and APSK, and the only difference is the demapper, which depends on the specific coded modulation schemes.

Figure 4-14 provides the SER performance comparison among 16/64/256-ary square QAM and product-APSK, where the parameters of product-APSK are determined using 4-42 to 4-44. The theoretical analysis is based on 4-95 and 4-97, and Monte Carlo simulation is carried out to verify the analysis. It is interesting that at low to medium SNR or equivalently for SER higher than 0.3, product-APSK is better than square QAM, while at high SNR it is much worse than its counterpart. This is because at high SNR, the SER performance is dominated by the minimum Euclidean distance, and the minimum Euclidean distance of product-APSK is obviously smaller than that

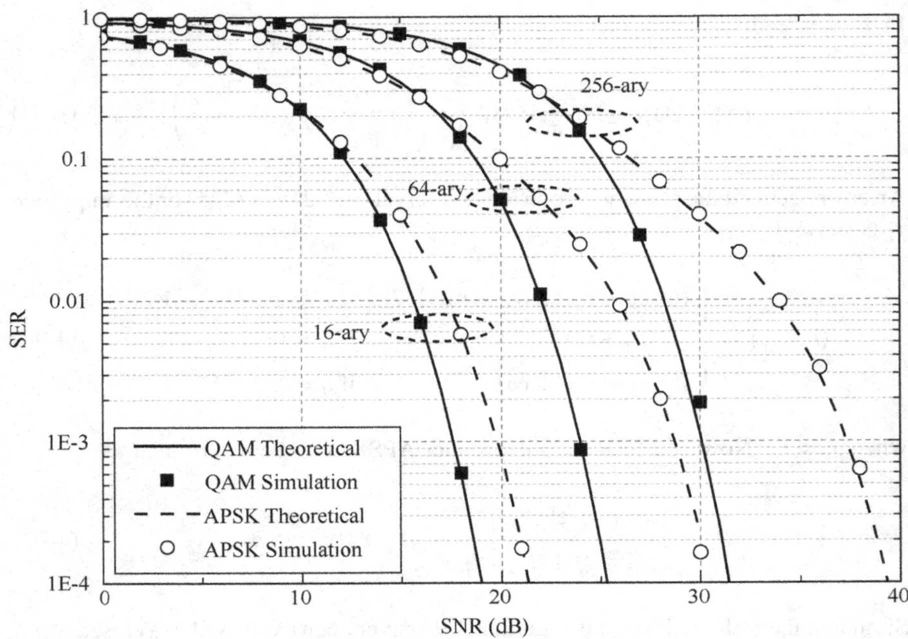

FIGURE 4-14 SER comparison of square QAM and product-APSK.

of square QAM. But when channel coding is considered, practical systems may work at the SNR region whereby product-APSK is better than QAM.

4.2.4.3 Performance in Terms of Mutual Information Mutual information of the transmitted symbol X and the received symbol Y is defined in 3-1, and compared with the SER or Euclidean distance spectrum, mutual information is more effective for analyzing practical coded systems. This can be explained from the perspective of an information-theory-based framework for telecommunication system design and optimization. The fundamental objective of the telecommunication system is to reliably deliver as much information as possible at a certain cost. There exists a channel capacity for an arbitrary channel, namely, the maximum information that can be reliably transmitted, which equals to the maximum mutual information between the input and output signals of the channel. However, the signal distribution for the capacity achievement is usually unrealizable in practice. Therefore, one must find a feasible distribution with the mutual information close to the channel capacity and use coding techniques to approach such mutual information.

The mutual information between the channel input signal X from an M-ary constellation set χ at equal probability and the output signal Y can be evaluated as

$$I(X;Y) = \log_2 M - \mathbb{E}_{x,y} \log_2 \left[\frac{\sum_{\hat{x} \in \chi} p_{Y|X}(y|\hat{x})}{p_{Y|X}(y|x)} \right] \qquad (4\text{-}98)$$

This mutual information is called average mutual information for coded modulation (CM-AMI) as it quantizes the maximum information that can be reliably transmitted for any scheme associated with a specific constellation. CM-AMI is also referred to as constellation-constrained capacity [5] or discrete-input, continuous-output memoryless channel (DCMC) capacity [6].

For transmissions over complex-valued AWGN channels, CM-AMI can be further written as

$$I(X;Y) = \log_2 M - \frac{1}{M}\sum_{x\in\chi} \mathbb{E}_w \log_2\left[\sum_{\hat{x}\in\chi} \exp\left(-\frac{|x-\hat{x}|^2 + 2\Re((x-\hat{x})w)}{N_0}\right)\right] \tag{4-99}$$

where w denotes one realization of the AWGN random process with zero mean and variance N_0. Unfortunately, 4-99 still cannot be expressed in a close form even for the simplest case such as BPSK. Nevertheless, for an extremely high SNR, one has $|x-\hat{x}| \gg 2|w|$ for $x \neq \hat{x}$, and consequently one can approximately rewrite $I(X;Y)$ as

$$\begin{aligned} I(X;Y) &\approx \log_2 M - \frac{1}{M}\sum_{x\in\chi} \log_2\left[\sum_{\hat{x}\in\chi} \exp\left(-\frac{|x-\hat{x}|^2}{N_0}\right)\right] \\ &\approx \log_2 M - \frac{1}{M}\sum_{x\in\chi} \log_2\left[1 + n_x \exp\left(\frac{-d_x^2}{N_0}\right)\right] \qquad (4\text{-}100) \\ &\approx \log_2 M - \frac{N_{\min}}{M\ln 2} \exp\left(\frac{-d_{\min}^2}{N_0}\right) \end{aligned}$$

where n_x denotes the number of neighboring points with d_x away from x, while $d_{\min}$ and $N_{\min}$ represent the minimum Euclidean distance and the number of pairs, respectively, by assuming $(x, \hat{x})$ and $(\hat{x}, x)$ are two different pairs. Based on 4-100, it is clear that CM-AMI under extremely high SNR is dominated by the minimum Euclidean distance. For low to medium SNR, the numeric results of the SNR gap between CM-AMI of either product-APSK or square QAM to the channel capacity with Gaussian inputs are provided in Figure 4-15. The parameters of product-APSK are determined according to Section 4.2.2.4. A similar conclusion to that of SER comparison in Figure 4-14 can be drawn: At low to medium SNRs (corresponding to low to medium AMI), APSK is better than its QAM counterpart, while at very high SNR, APSK is worse because of the smaller minimum Euclidean distance.

4.2.5 Variations of Digital Modulations

Besides the standard modulation schemes, one can have different variations for different requirements. We introduce two types of variations associated with PSK and QAM signals in the following.

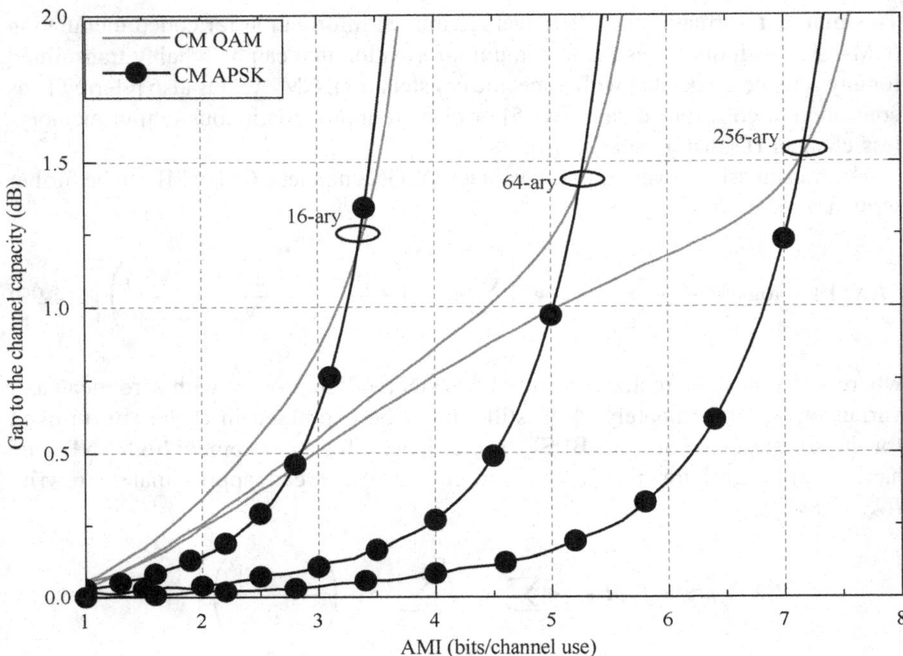

FIGURE 4-15 Gap between the CM-AMI and capacity for AWGN channels, whereby channel capacity $C = \log_2(1 + \mathrm{SNR})$ that is achieved by Gaussian inputs.

4.2.5.1 Variations of PSK Signals Two variations of PSK are briefly introduced here which can provide a sharper spectrum (faster decay around the cut-off frequency) than conventional PSK modulations.

1. *OQPSK.* In conventional QPSK modulation schemes, both I and Q components are synchronous. When two signals change their values simultaneously, a 180° phase shift between the adjacent QPSK symbols occurs and the signal envelope crosses the zero. If the amplifier at the transmitter is not ideally linear, it will cause greater spectrum expansion and the side lobes rising, which could cause interference to adjacent channels.
Offset QPSK (OQPSK) technology helps solve this problem. Its implementation is very simple: Just let either the Q or I signal components delay by half the symbol period (i.e., $T_s/2$). With this arrangement, the phase angle of the modulated signal changes once every $T_s/2$, but since only one of the two signal components changes each time, the phase angle change can only be ±90°. This much smaller phase change greatly reduces the spectrum expansion problem to the adjacent channels. It should be noted that even though the phase change rate is twice as much as that of the QPSK signal, the spectrum of the OQPSK signal is the same as the QPSK signal as it is still essentially a superposition of two orthogonal BPSK signals with a period of T_s, and the BER performance of coherent demodulation in the AWGN channel remains the same.

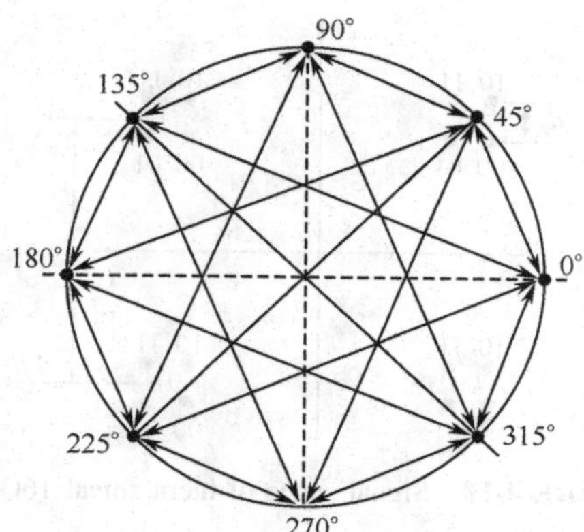

FIGURE 4-16 The π/4-QPSK constellation

2. *$\pi/4$-QPSK.* From the maximal phase change point of view, $\pi/4$-QPSK is a compromise between QPSK and OQPSK with the maximum phase change of $\pm 135°$, and its constellation diagram is shown in Figure 4-16. The connection in the diagram represents the possible phase change of the modulated signal. It can be seen that the constellation distribution of the $\pi/4$-QPSK signal is the same as the 8PSK signal, but the phase change of the $\pi/4$-QPSK signal between adjacent symbols is not completely random and actually switches back and forth between the two QPSK constellations given in Figure 4-5. The $\pi/4$-QPSK signal is generated by introducing the mapping relationship with memory to the conventional QPSK modulator. The BER performance of coherent demodulation in an AWGN channel is also the same as that of the conventional QPSK signal. The most attractive feature of $\pi/4$-QPSK is that the demodulation can be noncoherent, greatly simplifying receiver design. It also outperforms the OQPSK signal in multipath fading channels.

4.2.5.2 Variations of QAM Signals There also exists offset modulation for the QAM (OQAM) signal, and its principles and generation methods are the same as OQPSK. Hence we introduce another variation of the QAM signal below.

Since the QAM signal can be considered a linear combination of multilevel QPSK signals, this feature can be applied to the digital transmission, and hierarchical transmission services can therefore be provided. Figures 4-17 and 4-18 provide the typical nonuniform 16QAM and 64QAM hierarchical constellations, respectively. Using Figure 4-17 as an example the modulated signal is divided into two layers (corresponding to two priorities), that is, the QPSK signal of layer 1 (high priority, HP) and QPSK signal of layer 2 (low priority, LP). The transmitter completes one layer of QPSK mapping and then the other layer of QPSK mapping on top of the previous QPSK signal. The information used for two mappings usually comes from different sources with different channel coding to provide different levels of error protection capability. The receiver can receive either all or only the high-priority streams based on the user's requirements and the transmission performance.

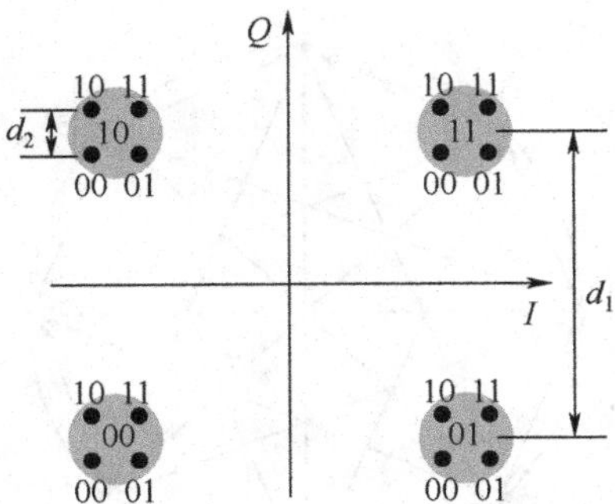

FIGURE 4-17 Signal space of hierarchical 16QAM.

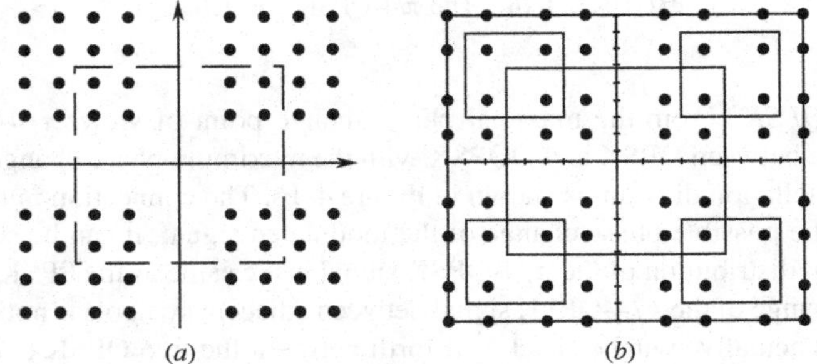

FIGURE 4-18 Signal space of hierarchical 64QAM.

Furthermore, MLC is one of the most powerful and applicable coded modulation schemes and was originally proposed by Imai and Hirakawa [7]. For readers interested in the details, please refer to [8,9] for the basic concepts of MLC with its encoder/decoder, design rules, and performance.

4.3 BIT-INTERLEAVED CODED MODULATION

4.3.1 BICM System Model

Bit-interleaved coded modulation (BICM), which simply consists of a channel code, a bit wise interleaver, and a bit-to-symbol constellation mapper is proposed by Zehavi, [10]. Figure 4-19 illustrates the BICM encoder. The source bits **u** are encoded, bit wise interleaved, and then mapped onto constellation symbols **x** for

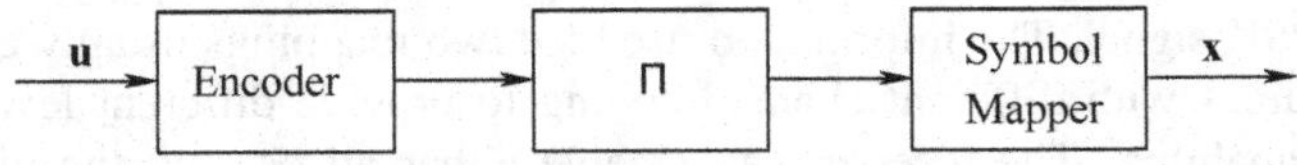

FIGURE 4-19 BICM encoder.

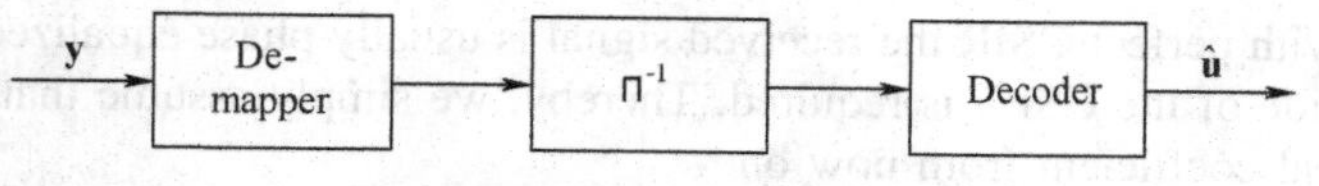

FIGURE 4-20 BICM decoder.

transmission. At the receiver side, the received signal symbols **y** are demapped and de-interleaved before passing to the channel decoder, which estimates the bits as $\hat{\mathbf{u}}$, as shown in Figure 4-20.

Compared with MLC or TCM (introduced in, Section 3.8), BICM avoids the multiple coding in MLC and the complexity of the demapper is lower than that of TCM.

Moreover, the error performance of BICM is usually better than MLC or TCM under fading channels. First, to achieve a targeted transmission rate by using MLC, one needs several near-capacity code components with different rates. For different channels, the required rates are usually different. Therefore, it is hard to obtain a near-capacity MLC simultaneously for various channels. Second, TCM tries to maximize the free Euclidean distance, which is good for AWGN channels [11]. However, for fading channels, a larger Hamming distance rather than the Euclidean distance is much preferred.

4.3.2 BICM Design and Performance Evaluation

The BICM scheme is simple not only for implementation but also for design methodology. Suppose Gray labeling $\mu : \mathbf{b} \mapsto x \in \chi$ is used. Here **b** is an m-bit-long coded bit vector, x is the mapped constellation symbol and χ denotes the constellation set with size of 2^m. At the receiver after passing through a memoryless channel: $x \to y$ and y is the received signal. A demapper is utilized to calculate the soft information of each bit, expressed in the log-likelihood ratio (LLR) form. For the log-MAP (*maximum a posteriori*) noniterative demapping, the extrinsic LLR of the ith $(0 \leq i < m)$ bit L_i is calculated as

$$L_i = \log \frac{\Pr(b_i = 0|y)}{\Pr(b_i = 1|y)} = \log \frac{\sum_{x \in \chi_i^{(0)}} p(y|x)}{\sum_{x \in \chi_i^{(1)}} p(y|x)} \tag{4-101}$$

wherein b_i denotes the ith bit of **b** and $\chi_i^{(b)}$ denotes the constellation subset with the ith bit being $b \in \{0, 1\}$. A memoryless fading channel can be modeled as $y = hx + n$, where h is the fading coefficient and n is the AWGN with zero mean and variance N_0. The conditional PDF $p(y|x)$ when perfect channel state information (CSI) h is available at the receiver side (CSIR) can be written as

$$p(y|x) = \frac{1}{\pi N_0} \exp\left(-\frac{\|y - hx\|^2}{N_0}\right) \tag{4-102}$$

Note that with perfect CSIR the received signal is usually phase equalized, and only the amplitude of the CSI h is required. Thereby, we simply assume that h is a non negative real coefficient from now on.

According to the approximation of the Jacobian logarithm that $\log(e^x + e^y) \approx \max(x, y)$, the log-MAP demapping can be approximated by

$$L_i \approx \log \frac{\max_{x \in \chi_i^{(0)}} p(y|x)}{\max_{x \in \chi_i^{(1)}} p(y|x)} \tag{4-103}$$

which is called the max-log-MAP demapper.

The exponential operations can be avoided by using the max-log-MAP demapper since one has $\log(p(y|x)) \propto -\|y - hx\|^2 / N_0$ by ignoring the common coefficient of $1/(\pi N_0)$, with which the complexity can be significantly reduced, compared to the log-MAP demapper.

Each bit level $b_i, 0 \leq i < m$, to the received signal y forms an equivalent binary channel. Therefore, the achievable throughput of BICM can be written as

$$\sum_{i=0}^{m-1} I(B_i; Y) \tag{4-104}$$

This is called "BICM capacity" in [5] and "BICM-AMI" in [3]. It is notable that BICM-AMI is exactly the same as PID-AMI. One has the following theorem associated with BICM-AMI.

Theorem 4-1 BICM-AMI Associated with LLR of Log-MAP Demapping Output

If L_i is the non iterative log-MAP demapping output, i.e.,

$$L_i = \log \frac{\sum_{x \in \chi_i^{(0)}} p(Y|x, H)}{\sum_{x \in \chi_i^{(1)}} p(Y|x, H)} \tag{4-105}$$

where H is the CSIR, one has

$$I(B_i; Y|H) = I(B_i; L_i) = 1 + \mathbb{E}_{l_i}[f(l_i)] \tag{4-106}$$

in bits per channel use, with

$$f(l_i) = \frac{e^{l_i}}{1 + e^{l_i}} \log_2 \frac{e^{l_i}}{1 + e^{l_i}} + \frac{1}{1 + e^{l_i}} \log_2 \frac{1}{1 + e^{l_i}}$$

and l_i a realization of L_i.

Proof: For non iterative log-MAP demapping, one has

$$L_i = \log \frac{\Pr(B_i = 0|Y, H)}{\Pr(B_i = 1|Y, H)} \tag{4-107}$$

Furthermore, since $\Pr(B_i = 0|Y,H) + \Pr(B_i = 1|Y,H) = 1$, one has

$$\begin{aligned}\Pr(B_i = b|Y,H) &= \frac{\exp[L_i(1-b)]}{1+\exp(L_i)} \\ &= \Pr(B_i = b|L_i), \forall b \in \{0,1\}\end{aligned} \tag{4-108}$$

Hence, $H(B_i|Y,H) = H(B_i|L_i)$. As $I(B_i;Y|H) = H(B_i) - H(B_i|Y,H)$ and $I(B_i;L_i) = H(B_i) - H(B_i|L_i)$, one has $I(B_i;Y|H) = I(B_i;L_i)$, where $H(\cdot)$ and $H(\cdot|\cdot)$ denote the entropy and conditional entropy functions, respectively.

Apparently, since the input binary signal $B_i \in \{0,1\}$ is uniformly distributed, one has $H(B_i) = 1$ bit, and $H(B_i|L_i) = -\mathbb{E}_{b_i,l_i}[\log_2(\Pr(b_i|l_i))] = -\mathbb{E}_{l_i}[f(l_i)]$.

Theorem 4-1 not only shows that the non iterative MAP demapper is optimal in terms of BICM-AMI but also provides an alternative way to calculate BICM-AMI. In addition, since the demapping output L_i is a function of Y and H and H is independent of B_i, $I(B_i;L_i) \le I(B_i;Y,H) = I(B_i;Y|H)$ according to the data processing theorem. Theorem 4-1 shows that the equality holds for the log-MAP demapper. For other demappers such as the max-log-MAP demapper, the equality usually does not hold. However, since the demapping output is passed to the decoder without any feedback for conventional noniterative demapping, the AMI between the transmitted bits and the demapping outputs becomes the new bound of the information rate for possible error-free transmission, regardless of the different demappers. Thereby, $\sum_{i=0}^{m-1} I(B_i;L_i)$ is named the *BICM-AMI associated with a specific demapper*, e.g., BICM-AMI with max-log-MAP demapping (max-log-MAP BICM-AMI). Please note that based on Theorem 4-1 the BICM-AMI with log-MAP demapping (log-MAP BICM-AMI) is exactly the true BICM-AMI. Monte-Carlo-based 4-106 can be used to calculate BICM-AMI. However, since 4-108 may not hold for other demappers, 4-106 is unable to be used for calculating the max-log-MAP BICM-AMI. At this time, the histogram method can be employed, i.e., statistically to obtain the PDF $p(l_i)$ and the conditional PDF $p(l_i|B_i)$ and then determine the AMI according to $I(B_i;L_i) = h(L_i) - h(L_i|B_i)$, where $h(\cdot)$ and $h(\cdot|\cdot)$ denote the differential and conditional differential entropies, respectively.

Due to the noniterative demapping, the channel between the input signal X of 2^m-ary and the output signal Y is often modeled as m different channels with binary input B_i and the output of L_i for $i \in \{0,1,\ldots,m-1\}$. However, after the bit-level de-interleaving, the channel decoder may not need to treat its input LLR L differently associated with different bit levels and the decoder may mix the m parallel channels into only one channel.

Theorem 4-2 Mixing Channel Reduces AMI

If the decoder mixes the m BICM parallel channels into one, i.e., it observes the transmitted bit B and the demapper's output L, but does not differentiate the bit level L, then

$$m \times I(B;L) \le \sum_{i=0}^{m-1} I(B_i;L_i) \tag{4-109}$$

where $m \times I(B;L)$ is called the *mixed BICM-AMI*.

Proof: Intuitively, the LLR L corresponds to a specific bit level. However, if the decoder does not use this information, the AMI usually decreases.

Without loss of generality, assume $m = 2$ (i.e., two parallel channels with channel 0, $B_0 \to L_0$, and channel 1, $B_1 \to L_1$, with equal probability). A Random variable Z is used to denote which channel is used, i.e., $Z = 0/1$ denotes channel 0/1 is used, and Pr $(Z = 0) =$ Pr $(Z = 1) = 1/2$. Then, $\frac{1}{2}[I(B_0; L_0) + I(B_1; L_1)] = I(B; L|Z)$, which is equivalent to $I(B; L) \leq I(B; L|Z)$. As $I(B; Z) = 0$, the transmitted bit B is independent of the channel, and $I(B; L|Z) = I(B; L, Z) = I(B; L) + I(B; Z|L) \geq I(B; L)$.

BICM design can be done by not differentiating the LLR of different bit levels: Choosing a near-capacity channel code with a random bit interleaver is enough for designing a BICM scheme. By considering the difference of each level, the performance can be further improved. This is because (1) mixed BICM-AMI is normally lower than the true BICM-AMI and (2) a near-capacity code usually assumes that its input LLRs are independently and identically distributed (i.i.d.), which cannot be satisfied in BICM schemes when using high-order constellations. The technique considering the difference of the LLRs is called bit mapping or DEMUX [12]. A Bit-mapping design associated with LDPC and Gray-APSK is investigated in [13]. The performance of BICM will be detailed together with BICM-ID in the next section.

4.3.3 BICM-ID System Model

The iterative version of BICM, namely, BICM-ID, was proposed by Li and Ritcey [14] and ten Brink et al. [15]. Conventionally, the transmitter of BICM-ID is exactly the same as that of BICM. The difference is that in the BICM-ID scheme the decoder's output *extrinsic* information is fed back to the demapper after re-interleaving, and iterative demapping and decoding are thus performed, as shown in Figure 4-21.

BICM-ID can be regarded as a serially concatenated turbo code, whereby the demapper is a unity-rate inner code, and convolutional code is traditionally used as the outer code because of its simple trellis representation to ensure soft-in, soft-out decoders, just like these for parallel turbo codes. Nevertheless, the convolutional code is not capacity approaching. Even if perfect a priori information is input to the demapper, one cannot expect perfect extrinsic information to produced by the demapper. This leads to an inevitable high *error floor*. There are mainly two ways to overcome this error floor: One is to use a capacity-achieving outer channel code such as LDPC or turbo codes and the other is to use the technique called *doping*.

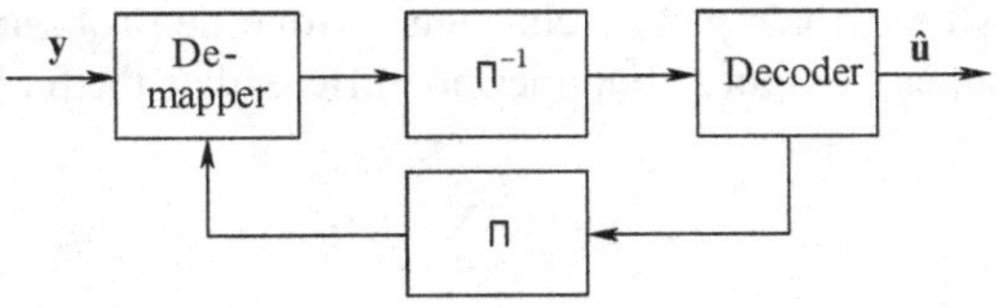

FIGURE 4-21 Conventional BICM-ID decoder.

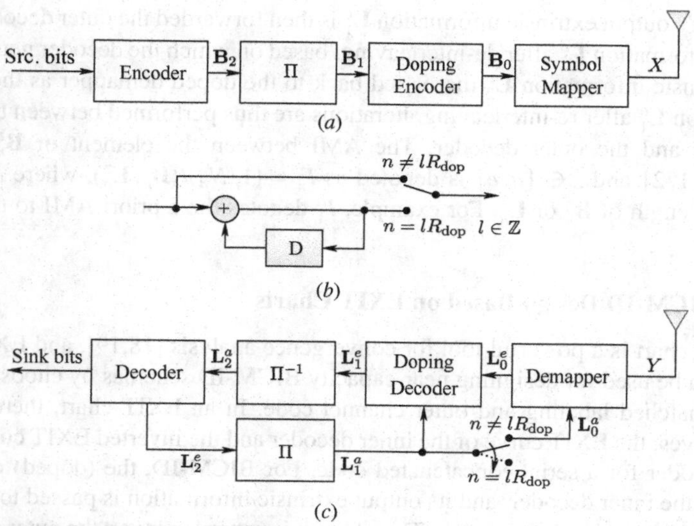

FIGURE 4-22 BICM-ID with doping: (*a*) transmitter with doping, (*b*) doping encoder, and (*c*) serial detection at receiver.

4.3.4 BICM-ID with Doping: Design and Performance Evaluation

Consider a BICM-ID scheme with doping [16], also referred to unity-rate precoding [6], as shown in Figure 4-22. The doping technique was originally proposed by ten Brink for serially concatenated codes [17] and later introduced into BICM-ID by Pfletschinger and Sanzi [16] and Tee et al. [6]. The doping code is a low-complexity unity-rate two-state recursive systematic convolutional (RSC) code, where every (R_{dop})-th information bit is replaced by the RSC encoded bits, and R_{dop} is called the *doping rate* [16]. This unity-rate RSC code introduces dependencies between adjacent bits. If perfect a priori information is input to the doping decoder, perfect extrinsic information can be obtained at its output, allowing the doped demapper's EXIT curve to reach the point of (1, 1) in the EXIT chart, and perfect convergence can be achieved. Hence, the high error floor can be successfully removed for an infinite block length, or lowered down to the satisfactory level for a finite block length.

At the BICM-ID transmitter, the source bits are encoded to $\mathbf{B}_2$, which is reformed into $\mathbf{B}_1$ after bit interleaving. Then, $\mathbf{B}_1$ is doping encoded into $\mathbf{B}_0$, which is mapped onto constellation symbols X for transmission. The average symbol energy of X is denoted as $E_s = \mathbb{E}\left[\|X\|^2\right]$.

The receiver usually uses *serial detection* rather than TCM detection to lower the complexity while maintaining a similar performance. The procedure of serial detection is illustrated in Figure 4-22*c*. With channel observation Y and a priori information $\mathbf{L}_1^a$, the doped demapper, namely, the demapper plus the doping decoder, generates its output extrinsic information $\mathbf{L}_1^e$, wherein the a priori and extrinsic information associated with the demapper is denoted by $\mathbf{L}_0^a$ and $\mathbf{L}_0^e$, respectively. The doped demapper's output extrinsic information $\mathbf{L}_1^e$ is then forwarded the outer decoder as the a priori information $\mathbf{L}_2^a$ after de-interleaving, based on which the decoder generates its own extrinsic information $\mathbf{L}_2^e$ that is fed back to the doped demapper as the a priori information $\mathbf{L}_1^a$ after re-interleaving. Iterations are thus performed between the doped demapper and the outer decoder. The AMI between the element of $\mathbf{B}_n^z$ and $\mathbf{L}_n^z$ $\forall n \in \{0, 1, 2\}$ and $z \in \{a, e\}$ is denoted as $I_n^z = (1/N_n)I(\mathbf{B}_n^z; \mathbf{L}_n^z)$, where N_n represents the length of $\mathbf{B}_n^z$ or $\mathbf{L}_n^z$. For example, I_1^a denotes the a priori AMI to the doped demapper.

4.3.5 BICM-ID Design Based on EXIT Charts

An EXIT chart is a powerful tool for convergence analysis [18,19], and EXIT curve fitting can be used for designing near-capacity BICM-ID schemes by choosing a pair of well-matched labeling and outer channel code. In an EXIT chart, there are two EXIT curves: the EXIT curve of the inner decoder and the inverted EXIT curve of the outer decoder for a serial-concatenated code. For BICM-ID, the (doped) demapper serves as the inner decoder, and its output extrinsic information is passed to the outer decoder as the a priori information. Therefore, it is natural to invert the outer decoder's EXIT curve and view it together with that of the inner decoder.

First, a unity-rate inner code is preferred because the inner code should maximize its degree of freedom for the CM-AMI. The bit-to-symbol mapper can be regarded as a unity-rate inner code, and the doping code also has a unity rate. The area theorem associated with the EXIT chart for a serially concatenated code is depicted in Figure 4-23, whereby the inner rate is unity. For successful detection, one should have an open EXIT

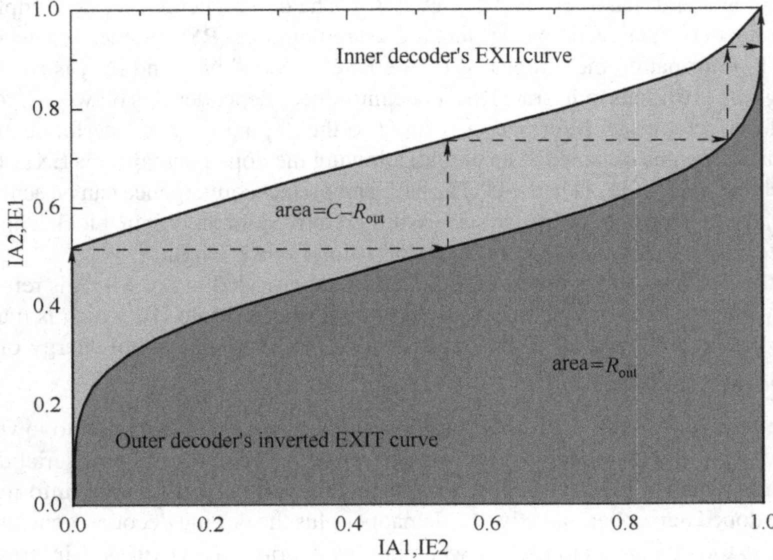

FIGURE 4-23 EXIT chart for serially concatenated code where inner code has unity rate.

tunnel, i.e., $R_{out} \leq C$. This exactly matches the famous Shannon channel coding theorem that the code rate must be no more than the capacity to achieve reliable transmission.

Second, find a labeling which allows the (doped) demapper's EXIT curve to fit the outer decoder's inverted EXIT curve. The basic idea of the adaptive binary switching algorithm (ABSA) is to adaptively change the cost function (CF) for binary switching algorithm (BSA) search according to the EXIT chart analysis. If the (doped) demapper's EXIT curve is above the outer decoder's inverted EXIT curve, i.e. there is an open EXIT tunnel, then the tentative labeling will be the desired one. Otherwise, one needs to figure out at which particular a priori information I_A points: this condition not satisfied and adjust the CF accordingly and then re start the BSA search. Unlike the conventional BSA, ABSA does not curtail its action upon finding the minimum CF and rather treats this as an intermediate stage in finding a specific labeling having an EXIT curve which closely matches the outer channel decoder's inverted EXIT curve. This is beneficial because the distance of the BER curve from the DCMC capacity is proportional to the area within the open EXIT tunnel; hence near capacity design should target a small EXIT tunnel area.

ABSA is an effective way to obtain a well-matched pair of bit-to-symbol labeling and outer channel code. Further improvements by EXIT curve match can be done by periodically renewing the initial labeling to avoid a local trap during the search or selecting a particular initial labeling (i.e., Gray labeling) in finding the match of the inverted EXIT curve for a capacity-approaching code (i.e., turbo or LDPC code). For the details, please refer to [20,21].

4.3.6 BICM-ID with LDPC Coding

As indicated above, BICM-ID using more powerful channel codes such as turbo or LDPC codes is also of practical interest. In fact, the serially concatenated outer convolutional code and the inner doping code can be regarded as a serial turbo code, while LDPC can also be regarded as a serial concatenation of a parity-check outer code with a repetition inner code [19].

BICM-ID with LDPC coding [bit-interleaved LDPC coded modulation (BILCM-ID) with iterative detection] is a good choice in practice. Basically, the joint modulation and LDPC code design is a good approach [22], but the joint design with the main focus on iterative detection is complicated. The following simple design also yields excellent performance: Simply choose a good LDPC code optimized for binary input channels and find a good labeling function, usually Gray labeling for this LDPC code, wherein the bit-mapping technique could also be used. This simple design could generate satisfactory performance for both BILCM-ID and conventional non iterative BICM. It is really an important feature as traditional non iterative BICM remains to keep receiving complexity low.

The BILCM-ID encoder is simple as it is exactly the same as BILCM. Information source bits are LDPC encoded, bit wise interleaved, and then mapped onto constellation symbols for transmission. The BILCM-ID decoder should be carefully designed as different decoder architectures result in quite different performance

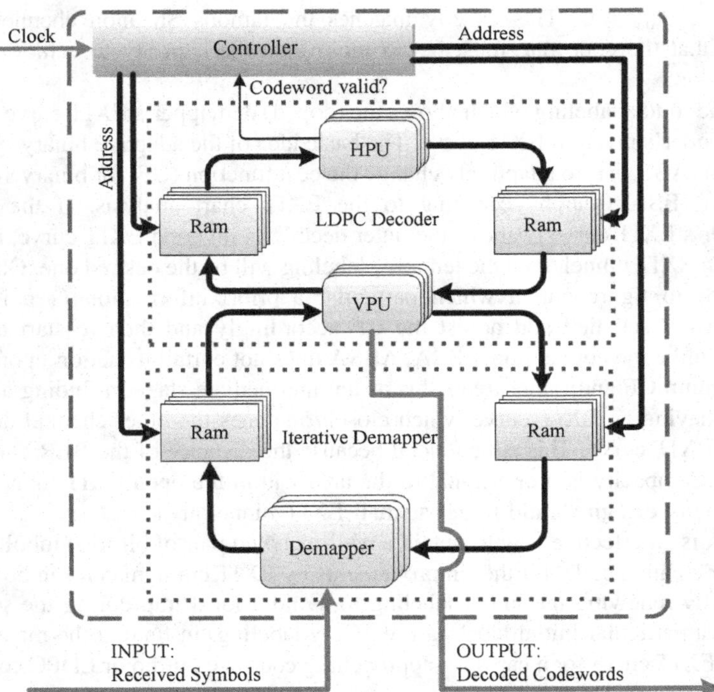

FIGURE 4-24 Shuffle detection architecture for BILCM-ID.

and complexity. A good architecture that achieves good performance at acceptable complexity and flexible detection provides an excellent trade-off. Figure 4-24 is the shuffle detection architecture for BILCM-ID, which is flexible for hardware implementation. There are multiple horizontal process units (HPUs) and vertical process units (VPUs) for the LDPC decoding which process the horizontal (minimization) and vertical (summation) computation, respectively. Furthermore, the VPUs also process iteratively with multiple symbol demappers. For a single iteration between the VPU and demapper, several iterations are usually performed between the VPU and HPU. This is flexible for achieving a good trade-off between complexity and performance. For instance, more iteration between the VPU and demapper results in better performance but with increased complexity.

4.4 MULTICARRIER MODULATION

For broadband digital radio communication systems, a major limitation of high-speed information transmission is the frequency-selective fading caused by the multipath effect of the channel. With this fading, some frequency components of the signal suffer from serious attention, which is called frequency selective. The conventional method for dealing with frequency-selective fading on a single-carrier modulated signal is to use an equalizer at the receiver or a combination of spread–spectrum modulation and RAKE receive, yet faced with the significant implementation complexity and performance limitations. Multicarrier modulation (MCM) technology, therefore, has become very popular recently due to the rapid development of digital integrated circuit technologies.

In principle, MCM transmits high-speed, single–stream data (broadband) by dividing it into many parallel and low-bit-rate streams (i.e., substreams or subchannels) first and then using these substreams to modulate corresponding carriers (subcarriers). The bandwidth of signals carrying substreams should be less than the channel coherence bandwidth, which can effectively reduce the impact of frequency-selective fading and achieve frequency diversity. The most commonly used MCM method is the orthogonal frequency division multiplexing (OFDM) technology, which will be described in detail.

In a traditional frequency division multiplexing (FDM) system carrying different information sequences in different frequency bands, no bandwidth overlapping among different bands is strictly required. To ensure good signal separation by the filter at the receiver, a protective band (or guard band) is usually provided between adjacent frequency bands. As a result, the spectrum efficiency of a conventional FDM system is far less than that of the single-carrier system. Unlike traditional FDM, OFDM technology uses a multicarrier, parallel transmission scheme with the permission of band overlapping yet certain rules must be satisfied.

4.4.1 Principle of Orthogonal Frequency Division Multiplexing [23]

Figure 4-25 presents a typical OFDM system. At the transmitter, a high-speed baseband stream is converted into N parallel substreams of low speed, and these substreams are then modulated by N subcarriers. Each subcarrier carries different or the same constellations, such as PAM, QPSK, and MQAM.

Assuming the symbol period of each parallel signal is T_s, the subcarrier spacing $\Delta f = 1/T_s$. Without loss of generality, a subcarrier can be expressed as

$$\Psi_k(t) = e^{j2\pi f_k t} \quad k = 0, 1, 2, \ldots, N-1 \tag{4-110}$$

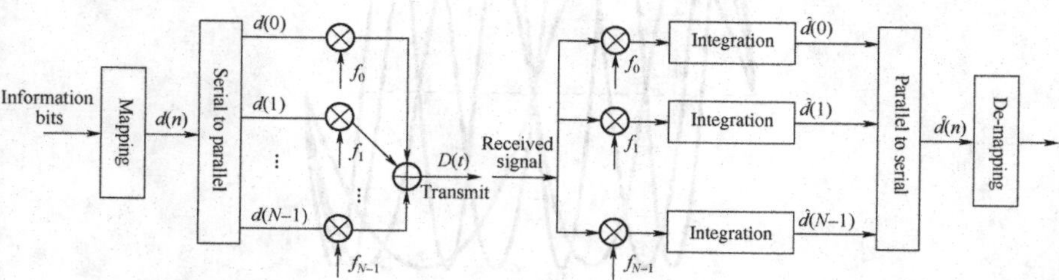

FIGURE 4-25 Block diagram of typical OFDM system.

where $f_k = f_0 + k\ \Delta f = f_0 + k/T_s$. The output signal of the transmitter is

$$D(t) = \mathrm{Re}\left[\sum_{k=0}^{N-1} d(k)\psi_k(t)\right] = \mathrm{Re}\left[\sum_{k=0}^{N-1} d(k)e^{j2\pi(f_0+k/T_s)t}\right] \tag{4-111}$$

The orthogonality among all subcarriers within the symbol period T_s is easy to prove in the following:

$$\int_{\tau}^{\tau+T_s} \psi_k(t)\psi_l^*(t)\,dt = \begin{cases} 0 & (k \neq l) \\ T_s & (k = l) \end{cases} \tag{4-112}$$

where τ can be an arbitrary value. Assume both ideal channel and perfect synchronization. The following can be derived based on the orthogonality condition of equation 4-112:

$$\hat{d}(k) = \frac{1}{T_s}\int_{0}^{T_s} \left(\sum_{k=0}^{N-1} d(k)\psi_k(t)\right)\psi_l^*(t)\,dt = d(k) \tag{4-113}$$

It is clear from 4-113 that the receiver can correctly recover the signal from each subcarrier individually because of the orthogonality. That is, the data symbol will not be affected by other carriers.

One can easily understand the orthogonality condition for OFDM systems in the time domain using the example given in Figure 4-26. Assume an OFDM symbol of four subcarriers with identical amplitude and phase, each subcarrier contains an integer number of cycles within one OFDM symbol period, and there is only one cycle difference between adjacent subcarriers to ensure the orthogonality.

The orthogonality concept of OFDM also can be explained in the frequency domain. According to 4-111, the spectrum of the OFDM symbol can be obtained by the convolution between the spectrum of a rectangular pulse with a period of T_s and a group of δ functions centered at subcarrier frequencies. The fast Fourier transform

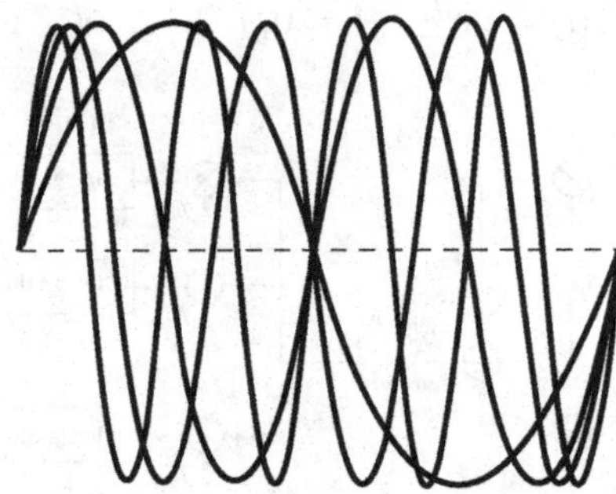

FIGURE 4-26 OFDM symbol containing four subcarriers.

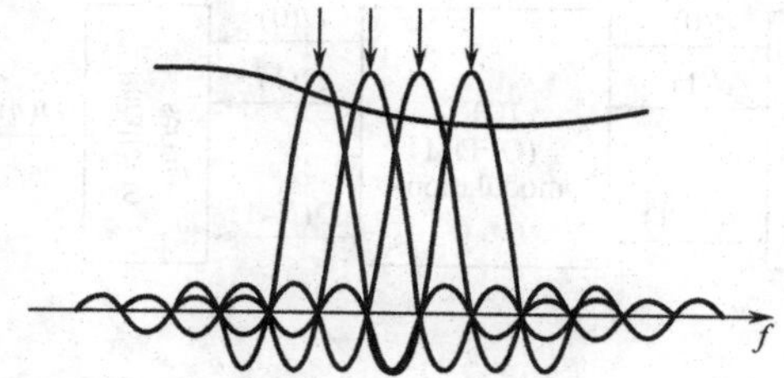

FIGURE 4-27 Spectrum of subcarriers in OFDM signal.

FFT of a rectangular pulses a, sampling function, i.e., $\mathrm{sinc}(\pi f T_s)$, and all zeros are at the frequency of an integer number of $1/T_s$, as shown in Figure 4-27. At the central frequency point of each subchannel (usually it is the frequency corresponding to the peak of the subchannel frequency response), the contribution from the spectrum of any other subchannel is exactly zero. In the demodulation, by obtaining the signal value at the peak of each subchannel, one can extract the symbol for each subchannel without interference from other subchannels. The spectrum of the OFDM signal actually satisfies the Nyquist criterion without distortions among symbols, that is, there is no interference among subchannels in the frequency domain.

4.4.2 Implementation of OFDM with Discrete Fourier Transform

The system structure illustrated in Figure 4-25 cannot be directly applied in practice since each carrier needs a modem. When the number of subcarriers N is large, the implementation complexity for an OFDM system will become unacceptable. This problem can be solved by applying the discrete Fourier transform(DFT) to OFDM systems.

For 4-111, if f_0 is regarded as the only carrier of the modulated signal, then the complex envelope of $D(t)$ (equivalently to the low-pass band form) can be expressed as

$$D_L(t) = \sum_{k=0}^{N-1} d(k)e^{j2\pi(k/T_s)t} \tag{4-114}$$

The value of the optimal Nyquist sampling point will be

$$D(n) = D_L(t)|_{t=nT_s/N} = \sum_{k=0}^{N-1} d(k)e^{j2\pi kn/N} \qquad n = 0, 2, \ldots, N-1 \tag{4-115}$$

This is exactly the inverse discrete Fourier transform (IDFT) on sent stream $\{d\,(k),$ $k=0, 1, \ldots, N\}$ to be transmitted. At the receiver, the DFT operation is applied to D (n) to recover the original data symbol $d(k)$,

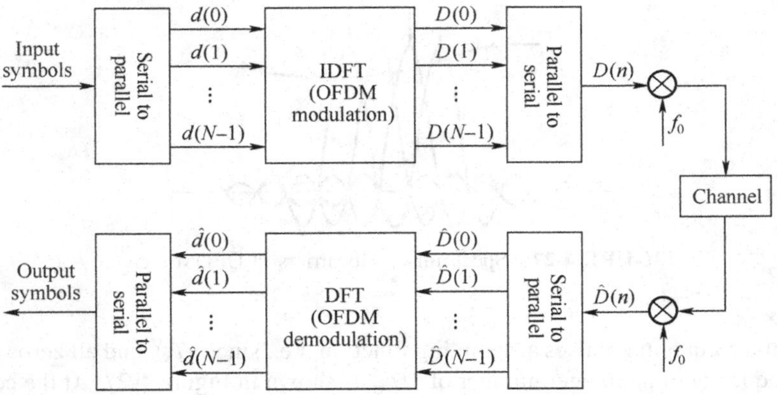

FIGURE 4-28 OFDM implementation using DFT.

$$d(k) = \frac{1}{N}\sum_{n=0}^{N-1} D(n)e^{-j2\pi nk/N} \quad k = 1, 2, \ldots, N \tag{4-116}$$

Based on this, the OFDM system in Figure 4-25 can be converted into its equivalent form shown in Figure 4-28, which is implementation friendly. The rationale is to convert the conventional FDM process usually implemented at the carrier frequency to the baseband digital preprocessing. The frequency-domain data symbols $d(k)$ are converted into time-domain data symbols $D(n)$ by an N-point IDFT operation and then sent to the channel after MCM. The coherent demodulation is applied the received signal at receiver, and then the N-point DFT operation is performed on the baseband signal to get the transmitted data symbol $d(k)$.

In practice, DFT is generally implemented using the FFT algorithm. With this conversion, the traditional single-carrier model can still be used in the RF part of the OFDM system. This greatly avoids the inter-modulation interference among all subcarriers, multicarrier synchronization and other issues, which greatly simplifies the system structure while maintaining the advantages of MCM systems. In addition, it is easy to perform the data recovery using digital signal processing algorithms at the receiver, and this has become the trend in digital receiver design and development.

4.4.3 Guard Interval and Cyclic Prefix of OFDM

A major advantage for adopting OFDM technology is that it can effectively combat multipath delay spread. By assigning the input data stream in parallel to the N parallel subchannels, each OFDM symbol period is N times that of the original data symbol duration. Therefore, the ratio of the multipath delay spread to symbol duration is also reduced to 1/N. In OFDM systems, the guard interval (GI) is inserted between adjacent OFDM symbols to eliminate ISI as much as possible. When padding is considered, the length of the OFDM symbol becomes $T_s + T_g$ with T_s representing the

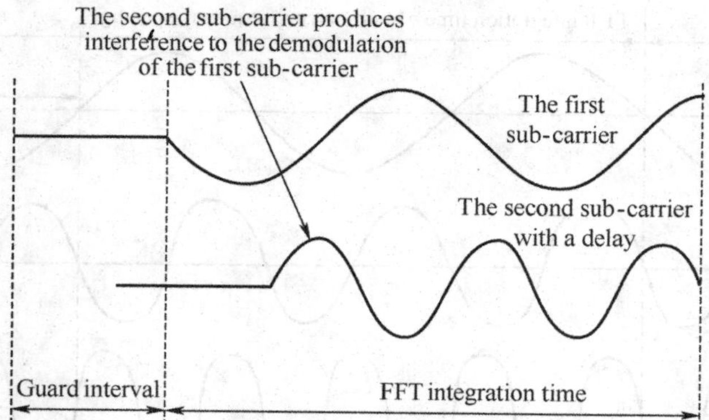

FIGURE 4-29 Effect of multipath channel on OFDM symbol with zero guard interval.

data block (or FFT/IFFT block) duration. The length of the guard interval T_g should generally be greater than the maximum channel delay spread such that the multipath components of the current OFDM symbol will have no interference on the following OFDM symbols. Instead of giving the actual time duration for the GI, the ratio between T_g and T_s is more commonly used as a very important system parameter.

Theoretically, one can choose a blank guard interval (i.e., zero padding), and in this case, however, the orthogonality among subcarriers will be destroyed, and the inter channel interference (ICI) shown in Figure 4-29 will appear, which gives an example of the delayed signal of the first and second subcarriers. It can be seen that within FFT operation time the differences among all subcarriers in terms of number of cycles are no longer integers, and the orthogonality illustrated in Figure 4-26 is no longer satisfied. In this case, the other subcarrier will introduce the interference on the current subcarrier modulation without resorting to the complicated algorithms at the receiver. To simultaneously overcome both ISI from the multipath effect and ICI from the zero padding, an effective method is introduced to perform the period expansion for the OFDM symbol with its original width T_s. The expansion is done by filling GI with the part of the OFDM signal as shown in Figure 4-30. The GI with duration T_g will be padded by the last portion (with duration T_g) of the OFDM signal, and this is called a cyclic prefix (CP).

In practice, OFDM symbols should have its GI padded by CP before being sent to the channel. At the receiver, after the CP part of the received OFDM symbol is removed, Fourier transform will be performed on the remaining part of the OFDM symbol with duration T_s, and followed by demodulation. Adding CP to the OFDM symbol ensures that the differences among all subcarriers in terms of number of cycles are still integers within $T_s + \mathrm{T}_g$. As long as the delay spread is less than T_g, ICI will not appear. Figure 4-31 presents an example of the impact of the multipath on the OFDM symbol assuming three subcarriers and two paths. When the multipath delay is less than T_g, the difference in number of cycles among all subcarriers within the FFT integration time (IFFT block length) remains an integer due to the introduction of CP.

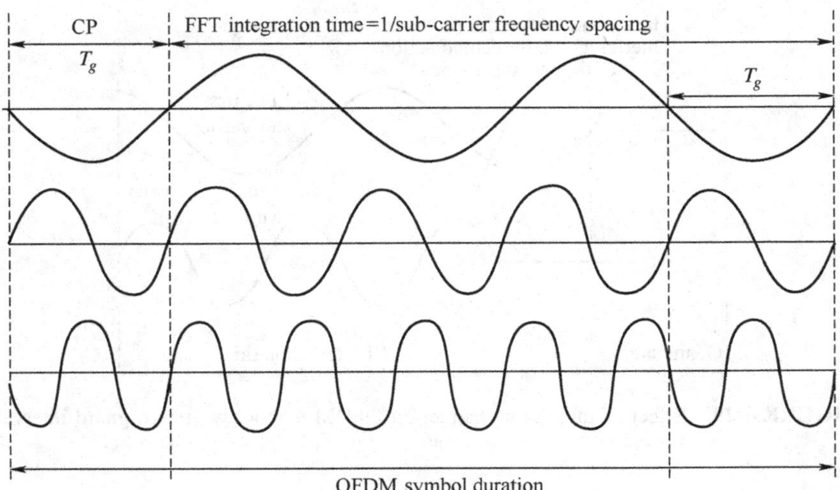

FIGURE 4-30 OFDM symbol with cyclic prefix.

Without CP, the resulting signal after passing through a channel will be the linear convolution between the IFFT block and the channel impulse response. With CP, however, the resulting signal after passing through a channel will be the circular convolution between the IFFT block and the channel impulse response if the length of channel impulse response is no more than the length of CP and the observation is within the IFFT block duration:

$$D(t) = \sum_{i=0}^{\infty} \sum_{k=0}^{N-1} \text{Re}\ \{d_i(k) \exp\ [j2\pi f_k(t - iT'_s)]\} g(t - iT'_s) \tag{4-117}$$

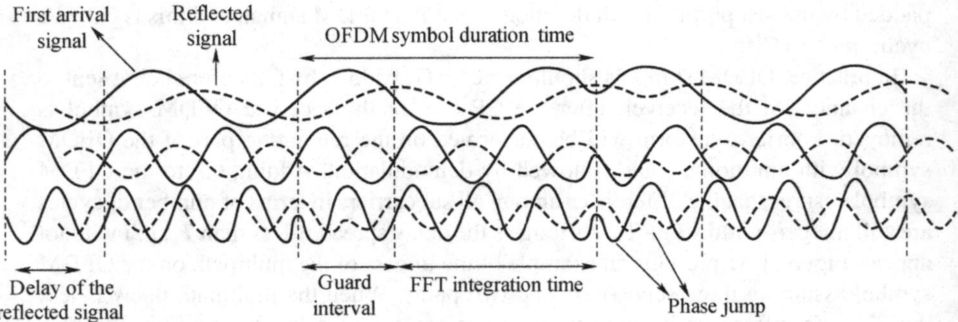

FIGURE 4-31 Impact of multipath on OFDM signal.

where the variable i is the index of the OFDM symbols and $g(t)$ is the rectangular pulse waveform, that is,

$$g(t) = \begin{cases} 1 & -T_g \leq t \leq T_s \\ 0 & t < -T_g,\ t > T_s \end{cases} \tag{4-118}$$

Under the static frequency-selective fading channel, the channel impulse response is

$$h(t) = \sum_{l=1}^{M_1+M_2} h_l \delta(t - \tau_l) \tag{4-119}$$

where h_l is the complex envelope and τ_l is the delay of the lth path. Assuming every echo delay is shorter than the OFDM symbol length and among all echoes there are M_1 echoes with delays smaller than T_s and M_2 echoes with delays larger than T_s,

$$\begin{cases} 0 \leq \tau_l \leq T_g & l = 1, \ldots, M_1 \\ T_g < \tau_l < T' = T_s + T_g & l = M_1 + 1, \ldots, M_1 + M_2 \end{cases} \tag{4-120}$$

The input signal at the receiver after multipath channel should be

$$\hat{D}(t) = \int_0^{\infty} D(t - \tau) h(\tau)\, d\tau + n(t) \tag{4-121}$$

where $n(t)$ is the complex AWGN. For the ith OFDM symbol, the decision after DFT is

$$\begin{aligned}
\hat{d}_i(k) &= \frac{1}{T_s} \int_{iT'}^{iT'+T_s} \hat{D}(t) e^{-j2\pi f_k (t - iT')}\, dt \\
&= \left(\sum_{l=1}^{M_1} h_l e^{-j2\pi f_k \tau_l} + \sum_{l=M_1+1}^{M_1+M_2} \frac{T - \tau_l + T_g}{T} h_l e^{-j2\pi f_k \tau_l} \right) d_i(k) \\
&\quad - \sum_{l=M_1+1}^{M_1+M_2} \sum_{\substack{m=0 \\ m \neq k}}^{N-1} \frac{\tau_l - T_g}{T} h_l e^{-j2\pi f_m \tau_l - j\frac{\pi(m-k)(\tau_l - T_g)}{T}} \operatorname{sinc}\left(\frac{\pi(m-k)(\tau_l - T_g)}{T} \right) d_i(m) \\
&\quad + \sum_{l=M_1+1}^{M_1+M_2} \sum_{m=0}^{N-1} \frac{\tau_l - T_g}{T} h_l e^{-j2\pi f_k (\tau_l - T) - j\frac{\pi(m-k)(\tau_l - T_g)}{T}} \operatorname{sinc}\left(\frac{\pi(m-k)(\tau_l - T_g)}{T} \right) d_{i-1}(m) + n_i(k)
\end{aligned} \tag{4-122}$$

where the first term is the desired signal, the second is the ICI, the third is the inter period interference of the OFDM (including the ISI of the same subcarrier, that is, $m = k$, and crosstalk of different subcarriers, that is, $m \neq k$); and the last is random noise. It can be seen from 4-122 that for those M^1 echoes with their delays shorter than T_g, the impact is on the complex coefficient of the decision result and will not change the result itself. When all the echoes have their delays shorter than T_g and assuming $M = M_1 + M_2$, 4-122 can be simplified to

$$\hat{d}_i(k) = \left(\sum_{l=1}^{M} h_l e^{-j2\pi f_k \tau_l} \right) d_i(k) + n_i(k) \tag{4-123}$$

The subcarrier spacing of the OFDM signal is usually chosen to be less than the channel coherence bandwidth so that the fading on each subchannel can be considered flat. That is,

$$H_k = \sum_{l=1}^{M} h_l e^{-j2\pi f_k \tau_l} \quad k = 0, 1, \ldots, N-1 \tag{4-124}$$

Then we have

$$\hat{d}_i(k) = H_k d_i(k) + n_i(k) \tag{4-125}$$

Obviously, H_k in 4-124 is the transfer function of each subchannel. If H_k can be estimated accurately at the receiver, the original information from the transmitter can be correctly recovered.

4.4.4 Frequency Domain Property

In practice, the number of OFDM subcarriers N is usually large. Based on the law of large numbers, the statistical characteristics of the time-domain signal from the linear superposition of N independent subcarriers will approach a Gaussian distribution when N is large. The power spectrum of the OFDM signal consisting of N subcarriers with a subcarrier power spectrum envelope as the sampling function is approximately rectangular even without passing through a filter, which is similar to the spectrum of the band-limited AWGN (refer to Figure 4-27). If M-ary modulation is applied to each subcarrier, the overall spectrum efficiency for an ideal OFDM signal is about $\log_2 M$ bits(s·Hz), reaching the "theoretical maximum spectrum efficiency." In comparison, the maximum spectrum efficiency of a real single-carrier system is only 80–90% of the theoretical maximum due to the implementation restrictions of the filter.

The introduction of GI, however, will inevitably reduce the spectrum efficiency of the actual system. For a multipath channel with defined delay, the spectrum efficiency that can be achieved by an OFDM system will be

$$\eta_{W\ \text{actual}} = \eta_{W\ \text{ideal}} \frac{T}{T + T_g} \tag{4-126}$$

From 4-126, it is seen that by increasing T_s or increasing the subcarrier number N under the fixed bandwidth, the spectrum efficiency of the system can be improved.

However, since the computational complexity and hardware resource requirement of DFT will increase rapidly with the increase of N, it is not always true that the higher the number of subcarriers, the better. In addition, the smaller the subcarrier spacing (with the increased N under a fixed system bandwidth), the more sensitive the OFDM system will be to spectrum expansion and the carrier phase noise caused by time-selective fading and the Doppler effect. In this case, orthogonality among all subcarriers may no longer hold. Therefore, the trade-off between these two requirements should be considered in practice. Also, N should be the product of the smallest possible prime numbers, preferably an integer power of 2, to use the highly efficient FFT butterfly algorithm.

Many factors other than multipath fading can cause the loss of orthogonality and lead to the ICI, including the synchronization error from either a local carrier or sampling clock at the receiver and the local oscillator phase noise. In general, OFDM modulation is more sensitive to nonlinear factors than single-carrier modulation.

4.4.5 General Comparison between OFDM and Single-Carrier Modulation System

For a single-carrier modulation system, the traditional method to overcome frequency-selective fading is equalization at the receiver. In this section, the system performance and implementation complexity for both single-carrier and OFDM systems, both uncoded, will be addressed.

The performance issue is discussed first. Two channel models will be used, one without severe frequency-selective fading, labeled channel 1, the other with deep fading, labeled channel 2. The frequency response with frequency normalized to the system bandwidth of these two channels is shown in Figure 4-32. For a single-carrier modulation system, a 31-tap decision feedback equalizer is used at the receiver for the QPSK modulated signal while for OFDM systems the number of subcarriers is 256 and each subcarrier is also modulated by QPSK. A bit-loading algorithm is used to

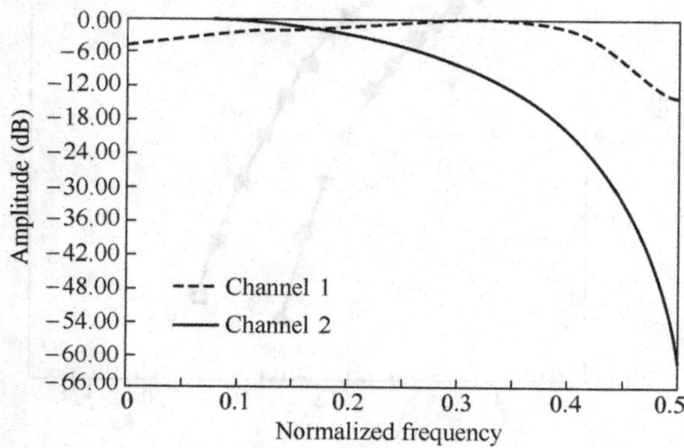

FIGURE 4-32 Frequency response of channels 1 and 2.

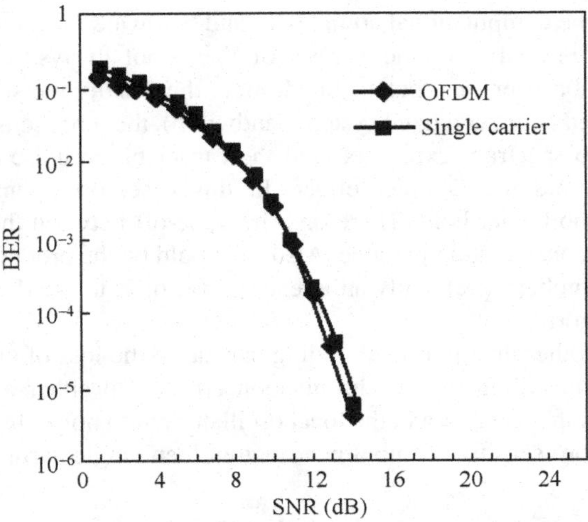

FIGURE 4-33 Performance of both OFDM and single-carrier systems for channel 1.

optimize the number of bits carried by each subcarrier. The BER simulation results for channels 1 and 2 are shown in Figures 4-33 and 4-34, respectively. It can be seen that the performance of the two transmission systems are similar for channel 1 and OFDM shows better performance than that of the single-carrier system for channel 2.

Now, the implementation complexity issue is addressed. In OFDM systems, the implementation complexity mainly depends on the IFFT/FFT operations. Most of the implementation complexity of the single-carrier systems comes from the equalizer. In

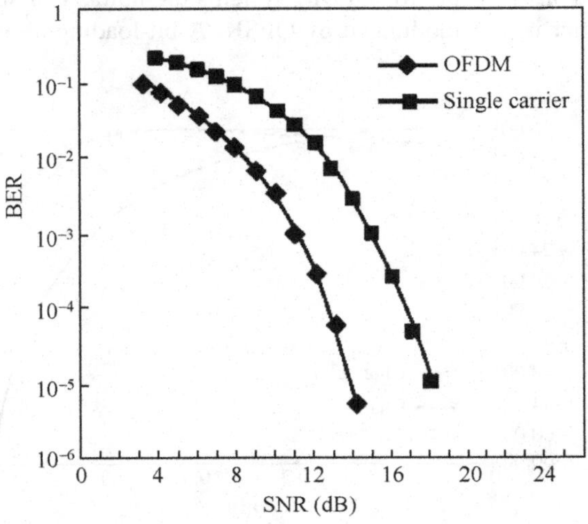

FIGURE 4-34 Performance of both OFDM and single-carrier systems for channel 2.

general, when the product of system bandwidth (roughly inverse to system symbol duration) and channel delay increases, the number of equalizer taps for the single-carrier system will increase significantly, and the complexity could be higher than that of the OFDM system.

4.5 DESIGN CONSIDERATIONS OF DTTB MODULATION

4.5.1 Modulation Scheme Determination

Choosing which digital modulation scheme to use for DTTB systems usually involves consideration of the following aspects.

4.5.1.1 Transmission Rate The transmission rate is used to measure the transmission capacity of the system, and the net bit rate (payload data rate) the system is able to offer is of most interest. The relationship of the bit rate R_b (in units of bits/s), defined as the number of binary bits transmitted per unit time, and the baud rate, R_s (in units of baud/s), representing the number of modulated symbols transmitted per unit time for M-ary modulation, is given as

$$R_b = R_s \log_2 M \tag{4-127}$$

In fact, the bit rate and baud rate give the system transmission capacity at different stages during the signal transmission for convenience. For example, during the coding/decoding stage of either the source or channel at the transceiver, the information is usually expressed in binary form, measured as the bit rate. Between the stages after the mapping at the modulator and before the demapping at the demodulator, the information is generally in the form of M-ary symbols, measured as the baud-rate. In addition, the payload data rate is used to assess the transmission capacity of the system, which means the "pure" information rate after the deduction of all the overhead from the channel coding, the training information (as the known sequence) for the synchronization in the transmitted signals, the padding, etc. It is usually measured in bits per second.

4.5.1.2 Spectrum Efficiency The spectrum efficiency η, also known as the spectrum utilization efficiency, is used to measure the effectiveness of the system utilizing the given spectrum and is defined as the information transmission rate of the unit bandwidth (in units of bits/s/Hz). If the bandwidth of the transmission channel is W, then

$$\eta = \frac{R_b}{W} \tag{4-128}$$

For the modulation schemes with $\eta > 1$, it is commonly referred to as bandwidth-efficient modulation. Otherwise, it would be considered power-efficient modulation. From Shannon's channel capacity formula, the theoretical upper bound of the spectrum efficiency for AWGN channels can be given as

$$\eta_{\max} = \frac{C}{W} = \log_2\left(1 + \frac{S}{N}\right) \tag{4-129}$$

where C is the channel capacity and S/N is the SNR.

Basically, the bandwidth W is relevant to the baud rate R_s. For an ideal rectangular filter, one has $W = R_s$ for complex-valued symbols such as QAM. Nevertheless, to satisfy the Nyquist criterion, ones needs to use the SRRC filter shown in Section 4.2.3.2, and therefore, $W = (1+\beta)R_s$ with β as the roll-off factor of SRRC. It is notable that in Section 4.2.3.2 one has $W = R_s/2$ for real-valued PAM. However, the QAM signal must be transmitted via a double sideband. Therefore, QAM and PAM have the same bandwidth efficiency when the bandwidth is a bandpass signal.

4.5.1.3 Error Performance The bit error rate (BER) and symbol error rate (SER) are mainly used to measure the transmission reliability of the system. BER is the ratio of the number of bits received in error at the receiver to the total number of transmitted bits. For the signal using constellation points of more than 2, the receiver's decision is on the symbol basis, so the SER is more commonly used, which is the ratio of symbols received in error to total transmitted symbols.

The SNR to achieve an error-free transmission is lower bounded by the Shannon limit. Approaching the Shannon limit is one of the most important targets of coded modulation design. That is the goal of several famous coded modulation schemes proposed in the last several decades.

4.5.2 Modulation Schemes in Typical DTTB Standards

For the four International Telecommunication Union (ITU) standards (labeled Systems A–D) of the generation of DTTB systems [24–27], the modulation schemes are summarized into Table 4-1. Refer to the following chapters for detailed descriptions.

4.6 SUMMARY

Technically speaking, both single-carrier modulation and multicarrier modulation have their own advantages and disadvantages in delivering the very important system performance. The final selection of the modulation scheme should come from a comprehensive understanding as well as thorough consideration of the technical features of each scheme, system Performance requirements, and application scenarios. Lots of effort must be made with various trade-offs to pick the most appropriate DTTB system.

Finally, it is worth pointing out that receiver will performance improve continuously, and with the introduction of new technologies/algorithms, some conclusions we made above might need to be modified. For example, some have recently proposed that the method of frequency-domain equalization or time-domain pre filtering can be used to allow single-carrier modulation to handle the long-delay echoes well [28].

TABLE 4-1 Modulation Scheme of Existing International Standards

System	ATSC	DVB-T	ISDB-T	DTMB
Modulation	Single-carrier 8-VSB	COFDM QPSK, 16QAM, and 64QAM	BST-OFDM DQPSK, QPSK, 16QAM, and 64QAM	TDS-OFDM QPSK, 16QAM, 32QAM, and 64QAM
System Bandwidth	6 MHz	8 MHz	6 MHz	8 MHz
Similar payload rate at 6 MHz	19.4 Mbits/s	17.9 Mbits/s at code rate = 2/3, 64QAM, GI = 1/32	17.7 Mbits/s at code rate = 2/3, 64QAM, GI = 1/32	18.3 Mbits/s at code rate = 0.6, 64QAM, PN = 420
Spectrum efficiency	3.2 bits/s·Hz	3.0 bits/s·Hz	3.0 bits/s·Hz	3.0 bits/s·Hz
RF test system E_b/N_0 threshold at threshold of visibility	10.1 dB	13.1 dB	13.2 dB	10.4 dB

Recent research work on a peak-to-average power ratio (PAPR) reduction scheme may alleviate the impact of the PAPR issue for OFDM signals [29].

REFERENCES

1. S. G. Wilson, *Digital Modulation and Coding*, Upper Saddle River, NJ: Prentice Hall, 1995.
2. J. Massey,"Coding and modulation in digital communications," in *Proc. of International Zurich Seminar on Digital Communications*, 1974.
3. Z. Liu, Q. Xie, K. Peng, and Z. Yang, "APSK constellation with Gray mapping," *IEEE Communications Letters*, vol. 15, no. 12, pp. 1271–1273, Dec. 2011.
4. J. G. Proakis, *Digital Communications* 5th ed., Columbus, OH: McGraw-Hill, 2007.
5. G. Caire, G. Taricco, and E. Biglieri, "Bit-interleaved coded modulation," *IEEE Transactions on Information Theory*, vol. 44, no. 3, pp. 927–946, May 1998.
6. R. Y. Tee, R. G. Maunder, and L. Hanzo, "EXIT-chart aided near capacity irregular bit-interleaved coded modulation design," *IEEE Transactions on Wireless Communications*, vol. 8, no. 1, pp. 32–37, Jan. 2009.
7. H. Imai and S. Hirakawa, "A new multilevel coding method using error correcting codes," *IEEE Transactions on Information Theory*, vol. IT-23, pp. 371–377, May 1977.
8. U. Wachsmann, R. F. H. Fischer and J. B. Huber, "Multilevel codes: Theoretical concepts and practical design rules", *IEEE Transactions on Information Theory*, vol. 45, no. 5, pp. 1361–1391, July 1999.
9. J. Hou, P. H. Siegel, L. B. Milstein, and H.D. Pfister, "Capacity-approaching bandwidth-efficient coded modulation schemes based on low-density parity-check codes," *IEEE Transactions on Information Theory*, vol. 49, no. 9, pp. 2141–2155, Sept. 2003.
10. E. Zehavi, "8-PSK trellis codes for a Rayleigh channel," *IEEE Transactions on Communications*, vol. 40, no. 5, pp. 873–884, May 1992.

11. G. Ungerbock, "Channel coding with multilevel/phase signals," *IEEE Transactions on Information Theory*, vol. 28, no. 1, pp. 55–67, Jan. 1982.
12. Y. Li and W. E. Ryan, "Bit-reliability mapping in LDPC-coded modulation systems," *IEEE Communications Letters*, vol. 9, no. 1, pp. 1–3, Jan. 2005.
13. T. Cheng, K. Peng, J. Song, and K. Yan, "EXIT-aided bit mapping design for LDPC coded modulation with APSK constellations," *IEEE Communications Letters*, vol. 16, no. 6, pp. 777–780, June 2012.
14. X. Li and J. A. Ritcey, "Bit-interleaved coded modulation with iterative decoding using soft feedback," *Electronics Letters*, vol. 34, no. 10, pp. 942–943, May 1998.
15. S. ten Brink, J. Speidel, and R.-H. Yan,"Iterative demapping and decoding for multilevel modulation," in *IEEE Globecom*, 1998, pp. 579–584.
16. S. Pfletschinger and F. Sanzi, "Error floor removal for bit-interleaved coded modulation with iterative detection," *IEEE Transactions on Wireless Communications*, vol. 5, no. 11, pp. 3174–3181, Nov. 2006.
17. S. ten Brink,"Code doping for triggering iterative decoding convergence," in *Proc. IEEE International Symposium on Information Theory*, 2001, p. 235.
18. S. ten Brink, "Convergence behavior of iteratively decoded parallel concatenated codes," *IEEE Transactions on Communications*, vol. 49, no. 10, pp. 1727–1737, Oct. 2001.
19. A. Ashikhmin, G. Kramer, and S. ten Brink, "Extrinsic information transfer functions: model and erasure channel properties," *IEEE Transactions on Information Theory*, vol. 50, no. 11, pp. 2657–2673, Nov. 2004.
20. Z. Yang, Q. Xie, K. Peng, and J. Song, "Labeling optimization for BICM-ID systems," *IEEE Communications Letters*, vol. 14, no. 11, pp. 1047–1049, Nov. 2010.
21. Q. Xie, J. Song, Z. Yang, and L. Hanzo, "EXIT-chart-matching aided near-capacity coded-modulation design and a BICM-ID design example for both Gaussian and Rayleigh channels," *IEEE Transactions on Vehicular Technology*, vol. 63, no. 3, pp. 1216–1227, Nov. 2012.
22. B. M. Hochwald and S. ten Brink, "Achieving near-capacity on a multiple-antenna channel," *IEEE Communications Letters*, vol. 51, no. 3, pp. 389–399, Mar. 2003.
23. M. Singh and R. Heath, *OFDM: Principles of Multicarrier Modulation*, Upper Saddle River, NJ: Prentice Hall, 2008.
24. Advanced Television System Committee.A/53, *ATSC Digital Television Standard*, Washington, DC: ATSC, 1995.
25. ETSI.300 744, *Digital Broadcasting Systems for Television, Sound and Data Services, Framing Structure, Channel Coding and Modulation for Digital Terrestrial Television*, Sophia-Antiplis, France: ETSI, 1999.
26. ITU-R WP 11A/59, *Channel Coding, Frame Structure and Modulation Scheme for Terrestrial Integrated Service Digital Broadcasting (ISDB-T)*, Tokyo: ARIB, 1999.
27. J. Song, Z. Yang, K. Gong, C. Pan, J. Wang, and Y. Wu, "Technical review on Chinese digital terrestrial television broadcasting standard and measurements on some working modes," *IEEE Transactions on Broadcasting*, vol. 53, no. 1, pp. 1–7, Mar. 2007.
28. N. Benvenuto, R. Dinis, D. Falconer, and S. Tomasin, "Single carrier modulation with nonlinear frequency domain equlization: An idea whose time has come again", *Proceedings of the IEEE*, vol. 98, no. 1, pp. 69–96, 2010.
29. ETSI EN 302 755 V1.3.1, *Digital Video Broadcasting (DVB); Frame Structure Channel Coding and Modulation for a Second Generation Digital Terrestrial Television Broadcasting System (DVB-T2)*, Geneva: Digital Video Broadcasting (DVB), Apr. 2012.

5

FIRST-GENERATION DTTB STANDARDS

5.1 GENERAL INTRODUCTION

Currently, four first-generation DTTB standards are officially approved by the ITU: ATSC (Advanced Television Systems Committee) [1], DVB-T (Digital Video Broadcasting-Terrestrial) [2], ISDB-T (Terrestrial Integrated Service Digital Broadcasting) [3], and DTMB (Digital Terrestrial/Television Multimedia Broadcasting) [4]. The channel coding, modulation techniques, and operating mode supported by these standards will be introduced and discussed in detail in this chapter.

5.1.1 ATSC Standard

The ATSC DTV standard was originally designed for terrestrial broadcast and cable distribution systems for fixed outdoor reception of HDTV programs. The system transmits high-quality video, audio, and auxiliary data in a 6 MHz channel and is able to deliver approximately 19 Mbps payload data rate in a 6 MHz terrestrial broadcast channel using eight-level vestigial sideband modulation (8-VSB). The system consists of a source coding and compression subsystem, a service multiplexing and transmission subsystem, and an RF transmission subsystem. Images with a variety of quality can be provided by means of 18 kinds of video formats (SD or HD, progressive scanning or interlacing, as well as different frame rates).

Digital Terrestrial Television Broadcasting: Technology and System, First Edition. Edited by Jian Song, Zhixing Yang, and Jun Wang.

5.1.2 DVB-T Standard

The DVB-T system is designed for indoor and outdoor fixed reception as well as for supporting the portable reception of services of digital video and audio as well as multimedia programs.

The payload data rate in an 8-MHz channel ranges from 4.98 to 31.67 Mbps, depending on the selection of channel coding parameters, modulation types, and guard interval. DVB-T is an OFDM system using cascade error-correcting codes. The system can support different network structures, including the large-scale single-frequency network (SFN) and single transmitter network, while maintaining high spectral efficiency. With respect to the source coding, the DVB standard requires that DTV systems use a unified MPEG-2 compression algorithms and MPEG-2 transport stream (TS) and multiplexing methods.

5.1.3 ISDB-T Standard

The ISDB-T system is designed for indoor and outdoor fixed, portable, and mobile broadcast reception of integrated services, including audio, video, graphics, and text from LDTV to HDTV. The ISDB system uses segment-OFDM technology (having 13 segments for 6 MHz bandwidth) and was originally developed for 6 MHz bandwith with a net bit rate ranging from 280.85 to 1787.28 kbps for each segment. The overall data throughput is between 3.65 and 23.23 Mbps, which is sufficient to carry various types of service streams, including HDTV. Those segments in the frequency spectrum can be flexibly combined and matched so that it is possible to carry various kinds of services in the same operating band, in particuler mobile DTV services.

5.1.4 DTMB Standard

DTMB is designed to support the various requirements for broadcasting services such as HDTV, SDTV, and multimedia data broadcasting. It can provide large-area coverage including the SFN and support both fixed (indoor and outdoor) and mobile reception. DTMB adopts both single- and multicarrier modulation with the unique frame structure called time-domain synchronous OFDM (TDS-OFDM) and using low-density parity code (LDPC) as the inner code. By combining all the technical features together, DTMB has been proven to be able to provide fast system synchronization, better receiving sensitivity, excellent system performance against the multipath effect, high spectrum efficiency, and flexibility for the multimedia service support.

5.2 INTRODUCTION TO ATSC STANDARD

The ATSC DTV standard describes a system which is designed within a single 6-MHz channel for transmitting high-quality video, audio, and auxiliary data. The system can reliably support a bit rate of about 19 Mbps in a 6-MHz terrestrial broadcast channel.

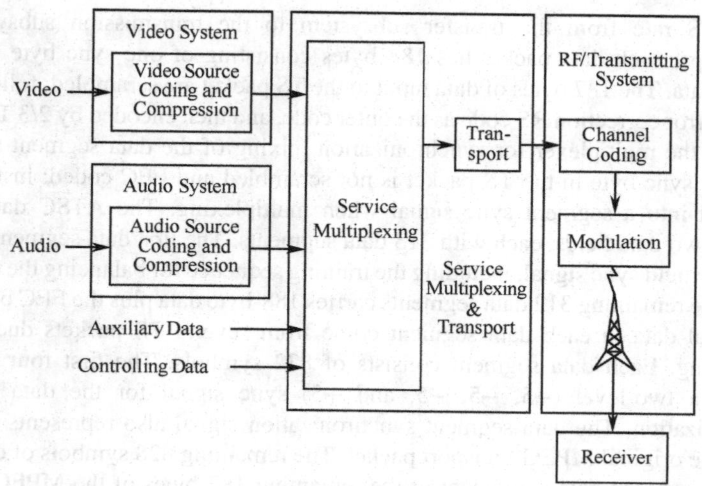

FIGURE 5-1 ATSC DTTB system.

The block diagram of the ATSC system is shown in Figure 5-1, which consists of the source coding and compression, service multiplexing and transmission, and RF transmission subsystems. The source coding and compression are used for data compression of video, audio, and auxiliary data. The MPEG-2 video stream is used for video coding in HDTV systems, and the Dolby AC-3 digital audio compression standard is used for audio coding. The service multiplexing and transmission subsystem is designed to pack the video, audio, and auxiliary data from the respective data stream packets and multiplex them as a single data stream. The subsystem uses the MPEG-2 transport stream. Interoperability with the various digital media and computer interface is fully considered in the transmission scheme. The RF transmission subsystem performs the channel coding and modulation, providing two modes, terrestrial broadcast mode (8-VSB) and high-data-rate mode (16-VSB), in which the latter is used for wired broadcasting.

This section describes the operating principle of the 8-VSB mode within 6 MHz bandwidth where the schematic of channel coding and the transmission system is shown in Figure 5-2.

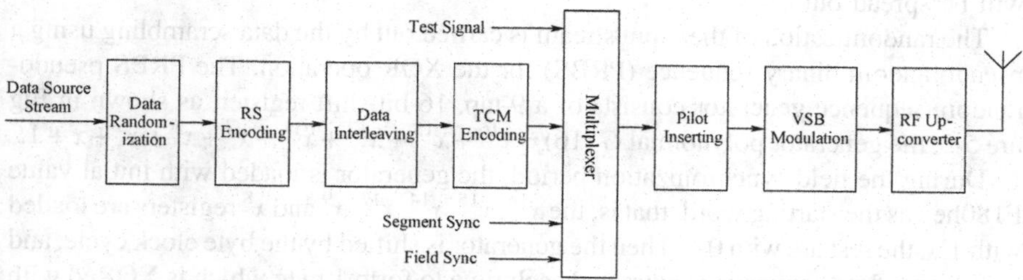

FIGURE 5-2 ATSC channel coding and transmission system.

The TS rate from the transfer subsystem to the transmission subsystem is 19.39 Mbps; each data packet has 188 bytes consisting of one sync byte and 187 bytes of data. The 187 bytes of data input to the TS packet are scrambled, followed by forward error correction RS code as the outer code, and then encoded by 2/3 TCM and output to the multiplexer for synchronization mixing of the data segment and data field. The sync byte in the TS packet is not scrambled and FEC coded; instead it is converted into a segment sync signal when multiplexing. The ATSC data frame includes two data fields, each with 313 data segments. The first data segment in each field is the field sync signal, including the training sequence for balancing the receiver; each of the remaining 312 data segments carries 188-byte data plus the FEC overhead. The actual data in each data segment come from several TS packets due to data interleaving. Each data segment consists of 832 symbols. The first four symbols transmit a two-level (+5, −5, −5, and +5) sync signal for the data segment synchronization. The data segment synchronization signal also represents the sync byte of the original MPEG transport packet. The remaining 828 symbols of each data segment carry the data equivalent of the remaining 187 bytes of the MPEG packet. The 20 parity bytes are added to 187 bytes of data via the RS encoding and then encoded by 2/3 TCM code, so that every 2 bits will become 3 bits and then be mapped to the 8-VSB signal: every 2 bits is mapped into an 8-VSB symbol. Therefore, the sync byte occupies four symbols (end synchronization), and the 207-byte data encoded by the RS code occupy 828 symbols. In this way, each data segment just carries one TS packet ($207 \times 8/2 = 828$).

5.2.1 Scrambler

Data scrambling is also known as data randomization or energy dispersal. A long string of consecutive 0's or 1's may appear in the source code stream, which will cause some difficulty in restoring the bit timing information at the receiver. In addition, if the stream is interrupted during DTV broadcasting or the stream format does not match with the MPEG-2 TS structure, the modulator will transmit unmodulated carrier signal, causing a large energy concentrated on the carrier spectrum, and this may jeopardize the safety of broadcast transmitters. To eliminate these two cases, the randomizing or scrambling treatment can be made to a baseband signal, so that it has a pseudo-random-like characteristics. In this way, the length of consecutive 0 or 1 codes will be shortened, and the spectrum of the modulated signal will be spread out.

The randomization of the input stream is carried out by the data scrambling using a pseudorandom binary sequence (PRBS) for the XOR operation. The PRBS pseudo-random sequence generator consists of a 9-tap, 16-bit shift register, as shown in Fig ure 5-3, the generator polynomial $G(16) = x^{16} + x^{13} + x^{12} + x^{11} + x^{7} + x^{6} + x^{3} + x + 1$.

During the field synchronization period, the generator is loaded with initial value F180hex as the starting word, that is, the $x^{16}, x^{15}, x^{14}, x^{13}, x^{9}$ and x^{8} registers are loaded with 1's, the rest are with 0's. Then the generator is shifted by the byte clock cycle, and an 8-bit random sequence is extracted each time to form 1 byte which is XORed with

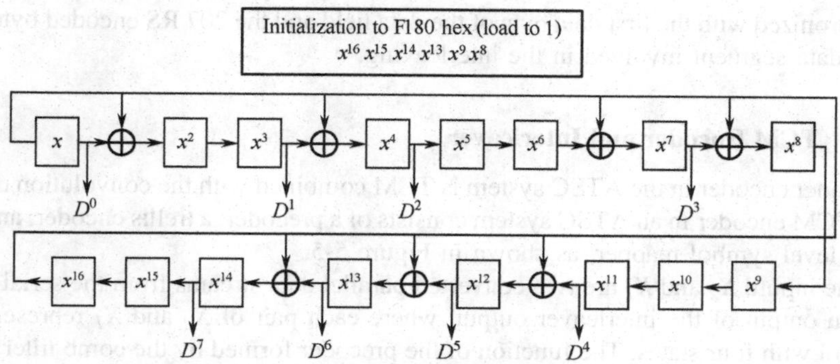

FIGURE 5-3 Pseudorandom sequence generator.

the input TS information bytes. Again, only 187 bytes of data are scrambled, instead of the segment sync, field sync, and 20 RS parity byte.

5.2.2 RS Encoding and Data Interleaving

ATSC uses RS (207,187, $t = 10$) code, which is a shortened RS (255,235) code. Twenty bytes of errors can be detected and 10 bytes of errors can be corrected. To get the RS (207,187) code, 48 bytes of 0's are added in front of 187 information bytes to make 235 information bytes, and then 20 check bytes are generated by the RS encoder and attached to the information bytes, and the desired RS code can be obtained after deleting all those 0's.

The interleaver between the RS and trellis encoders is a convolutional interleaver with 52 data segments, and the interleaving depth is one-sixth of the data field (4 ms), as shown in Figure 5-4. Only data bytes are interleaved. The interleaving is

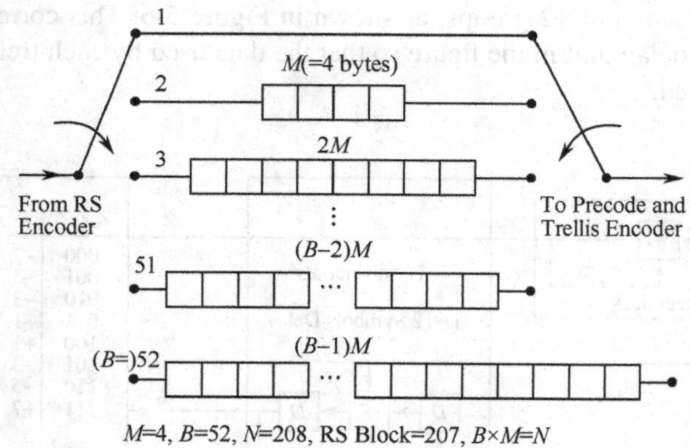

FIGURE 5-4 ATSC convolutional interleaver.

synchronized with the first data byte of the data field and the 207 RS encoded bytes of each data segment involved in the interleaving.

5.2.3 TCM Encoder and Interleaver

The inner encoder in the ATSC system is TCM combined with the convolution code. The TCM encoder in an ATSC system consists of a precoder, a trellis encoder, and an eight-level symbol mapper, as shown in Figure 5-5.

The inputd X_2 and X_1 are two converted parallel data streams from the serial data stream output of the interleaver output, where each pair of X_2 and X_1 represents a symbol with four states. The function of the precoder formed by the comb filter is to reduce the cochannel interference from NTSC signals. Input Y_2 is directly marked as Z_2 via the trellis encoder. The Z_1 Z_0 bit pair is output after Y_1 goes through the convolutional encoder with a coding rate of 1/2, forming a four-level symbol set (00,01,10,11); the plus or minus of the level is determined by the value of Z_2, and the mapping relationship is shown in the symbol-mapping table.

It is seen that a fourlevel state of original X_2 and X_1 is changed to the eight-level state of Z_2 Z_1 Z_0 by the TCM encoder. When the balanced amplitude modulation mode is used for the carrier, the modulated carrier has eight different oscillatory waves (± 1, ± 3, ± 5, ± 7). Therefore, the TCM encoding doesnot affect the information rate and the required channel bandwidth of the modulated carrier, except for doubling the number of levels of the carrier's amplitude and halving the spacing. The reduction of the spacing increases the demodulation error possibility but those errors can be corrected by the TCM decoder.

A trellis encoder helps correct random errors, but its capability to handle impulsive interference and burst errors is poor. To improve this performance, 12 identical trellis encoders in parallel are used to form a trellis code interleaver. It uses the intrasegment symbols for interleaving, each segment consisting of 828 symbols; the interleaving is accomplished by encoding symbols (0, 12, 24, 36, . . .) as the first group, symbols (1, 13, 25, 37, . . .) as the second group, symbols (2, 14, 26, 38, . . .) as the third group, and so on, for a total of 12 groups, as shown in Figure 5-6. This corresponds to 12 symbols in the delay unit in the figure so that the data used by each trellis encoder is exactly staggered.

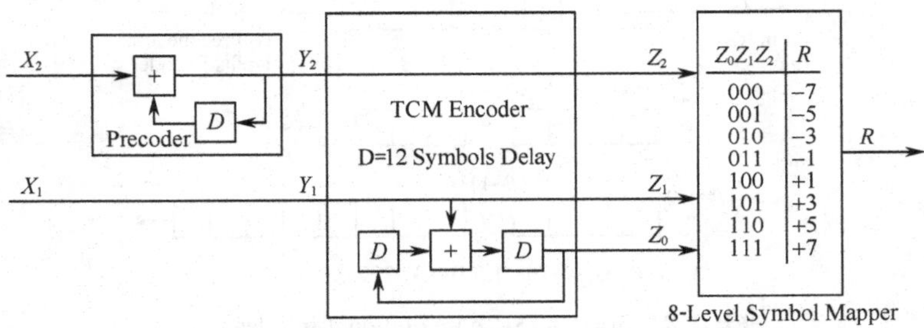

FIGURE 5-5 Operating principle of TCM trellis encoder.

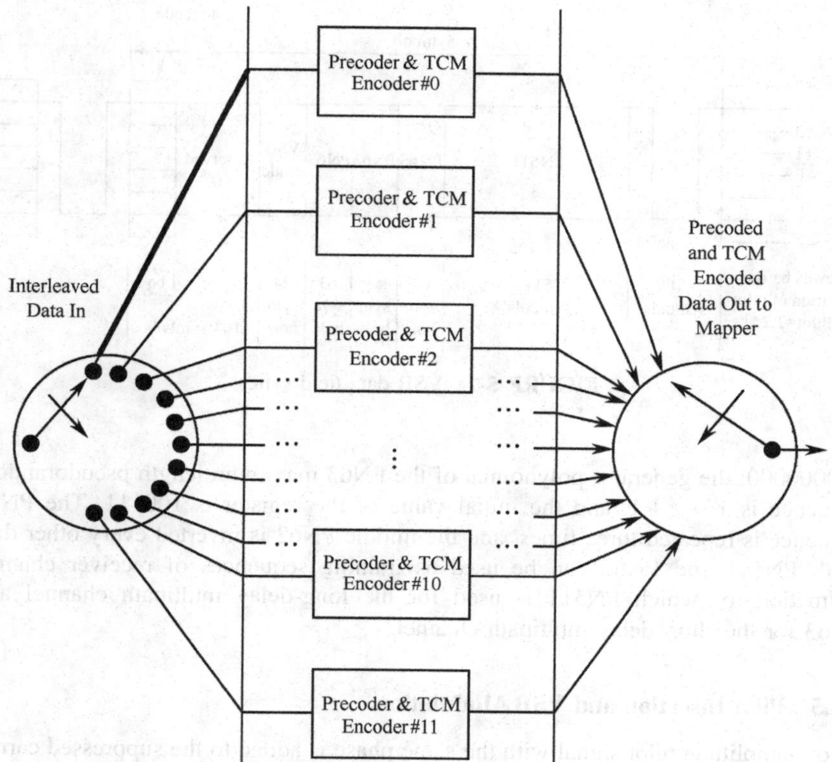

FIGURE 5-6 TCM interleaver.

5.2.4 Multiplexing

The TCM coded data constitute segments (832 symbols each) into symbols of eight levels, and the segment further forms the field (313 segments) and frames. Data symbols, after the data segment synchronization and data field sync inserted by the multiplexer, are output. The segment synchronization as shown in Figure 5-7 consists of four two-level symbols at the beginning of each segment. After complete data segments are obtained, the field structure will be applied. The first segment of every field is the field synchronization segment. Like other data segments, field sync signal includes 832 symbols, including (1) 4 segment sync symbols of 1001, like the general segment synchronization signal; (2) 511 symbols for PN511, where PN511 is the maximum-length pseudorandom sequence by a ninth-order shift register; (3) 63 symbols for PN63, where PN63 is the maximum-length pseudorandom sequence by a sixth-order shift register; (4) 24 symbols for the VSB mode, that is, the 8-VSB or 16-VSB mode; and (5) 104 reserved symbols, including 12 precoded symbols.

The generator polynomial of the PN511 maximum-length pseudorandom sequence is $x^9+x^7+x^6+x^4+x^3+x+1$, and the initial value of the register is

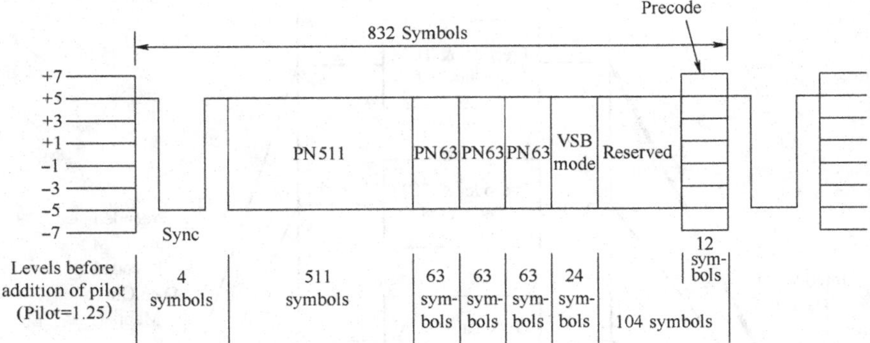

FIGURE 5-7 VSB data field sync.

010000000; the generator polynomial of the PN63 maximum length pseudorandom sequence is $x^6 + x + 1$, and the initial value of the register is 100111. The PN63 sequence is repeated three times, and the middle PN63 is inverted every other data field. PN511 and PN63 can be used as training sequences of receiver channel estimation, of which PN511 is used for the long-delay multipath channel and PN63 for the short-delay multipath channel.

5.2.5 Pilot Insertion and VSB Modulation

A low-amplitude pilot signal with the same phase is added to the suppressed carrier frequency. This signal can be generated by adding a negligible digital DC component (1.25) to the level (± 1, ± 3, ± 5, and ± 7) of each symbol (including the data signal and synchronization signal). Since the data are random, it is considered that the eight-level symbols are having equiprobability, and the average power of the eight-level data signal is 21 dB. After the pilot signal is superimposed, the total average signal power will be 22.5 dB.

After adding synchronization and a pilot signal, the symbol rate is 10.76 MS/s (where MS means one million symbols), and then it enters the VSB modulator. The modulator completes the spectrum shaping based on a square-root raised-cosine filter with linear phase. The raised-cosine filter is used by cascading the pulse-shaping filter at the transmitter and the match filter at the receiver. Except for the transition bands at both ends of the spectrum, the frequency response of the square-root raised-cosine filter throughout the bandwidth is relatively flat. The transition band at each side occupies 310 kHz, and the 3 dB bandwidth is $6 - 0.31 \times 2 = 5.38$ MHz; thus the baseband symbol rate is exactly $5.38 \times 2 = 10.76$ MS/s.

The operating principles of the modules in the high-data-rate mode are basically the same as that of the 8-VSB mode. The main difference is that the 16-VSB modulation is used and each symbol is expressed by 16 levels. This mapping is achieved by using 4 bits per symbol, unlike the 8-VSB, using 3-bit mapping. The data segment is shown in Figure 5-8.

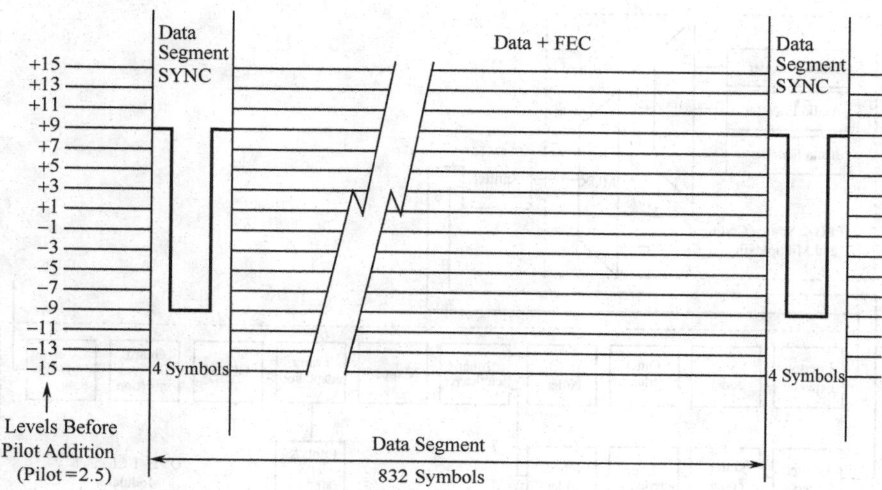

FIGURE 5-8 16VSB data segment structure.

Meanwhile, the 2/3 TCM encoding is removed in the high-data-rate mode; thus the transmission rate is increased to 38.57 Mbps, and the SNR threshold for receiving is increased to 28.3 dB. Most modules in the high-data-rate mode are identical or similar to the 8-VSB system, so further description will not be provided here.

5.3 INTRODUCTION TO DVB-T STANDARD

DVB-T uses the COFDM (coded orthogonal frequency division multiplexing) modulation technology and defines 2 K and 8 K operating modes with 1705 (2 K) and 6817 (8 K) subcarriers, respectively. The 2 K mode is suitable for single-transmitter operation and small-scale SFNs with limited transmission distance while the 8 K mode is suitable for single-transmitter operation and both small- and large-scale SFNs. System design is able to adapt to all channel conditions. It can deal with not only a Gaussian channel but also both the Ricean and Rayleigh channels and can handle high-level (0-dB), long-delay static and dynamic multipath echoes well. This system can reliably overcome the interference caused by the delay signal, including echoes reflected by the terrain and buildings or signals transmitted by remote transmitters in the SFN environment.

DVB-T has many parameters to accommodate a wide range of carrier-to-noise ratio and channel characteristics. It allows fixed, portable, or mobile reception, but the payload data rate has to be compromised. Those parameters allow broadcasters to select the most appropriate mode. Different combinations of QAM modulation and different inner code rates can be used to for trade-off between bit rate and robustness. The system also supports two-level hierarchical channel coding and modulation, including uniform and multiresolution constellation. In this case, the system block

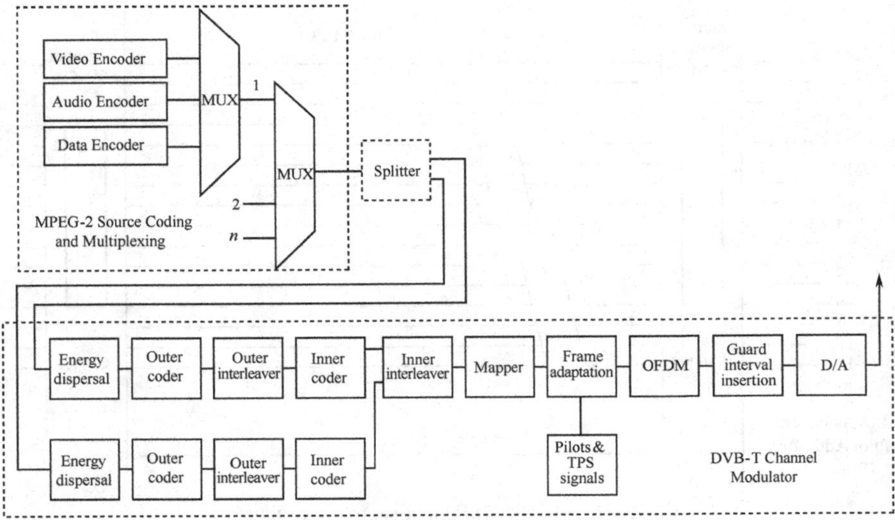

FIGURE 5-9 DVB-T system block diagram.

diagram should include the modules shown in the dashed block in Figure 5-9. The splitter separates the incoming transport stream into two independent MPEG transport streams, corresponding to high-priority and low-priority systems, respectively. These two bit streams are mapped onto the signal constellation by the mapper and modulator. To ensure that the signal transmitted by such a hierarchical system can be handled by a simple receiver, and the hierarchical nature is restricted to hierarchical channel coding and modulation without the use of a hierarchical source code. The program service could thus be simulcasted as a low-bit-rate, better-protected version and another version of high data rate and less protection. In other words, entirely different programs could be transmitted on separate transport streams with different protection. In either case, the receiver requires those blocks for the inverse operation: inner de-interleaver, inner decoder, outer de-interleaver, outer decoder, and demultiplexer.

The DVB-T transmitting system is shown in Figure 5-9, which includes channel coding, modulation, and pilot and TPS parts. The channel coding includes data scrambling, outer coding, outer interleaving, inner coding, and inner interleaving. The modulation includes constellation mapping and OFDM modulation, of which the former supports QPSK, 16QAM, and 64QAM, while the mapping could be either uniform or nonuniform. The OFDM modulation is generally achieved using inverse fast Fourier transform (IFFT). Depending on the number of IFFT points when implementing the OFDM, there are two different working modes in the DVB-T system, i.e., 2 K mode and 8 K mode. Some reference signals are inserted in the OFDM symbols, including continuous pilot, scattered pilot, and transmission parameter signal (TPS). The pilot helps the receiver achieve carrier synchronization and channel state estimation, and the TPS mainly provides the system mode and parameters to the receiver. The detailed operating principles are discussed below.

5.3.1 Channel Coding

5.3.1.1 Scrambler The input TS stream of a DVB-T system is 188-byte-long TS packets. In the DVB-T, a large TS packet is formed by every eight TS packets, and then the randomization is carried out to the input stream. The principle of the scrambler is exactly the same as that of ATSC. The polynomial $1+x^{14}+x^{15}$ is generated by using a pseudorandom binary sequence (PRBS) consisting of 15 shift registers. Initialization is done for every TS large packet. To detect the start position of the TS large package, the sync byte of the first TS packet in the TS large package is inverted as 0xB8, and the random-sequence generator functions from the inverted sync byte and deals with $8\times 188-1=1503$ bytes before reinitialization. The sync bytes of the seven TS packets are involved in the calculations, but the output is still 0x47. The enable signal is used to control the descrambling of these sync bytes. The randomized treatment of the transport packet is shown in Figure 5-10.

The randomization process will continue when there is no input or when the input is not compliant with the MPEG-2 TS format. The null packet can be added simply by inserting the sync byte followed by all zeros, and it will be identified and removed at the receiver. The null packet insertion helps adjust the rate of the input stream to a constant matching to the transmitter channel coding and modulation clock, which ensures that the continuous data stream is finally generated by the transmitter.

5.3.1.2 Outer Encoding and Outer Interleaving The outer code is RS (204,188, t=8), which is RS (255,239, t=8) shortened code. The primitive polynomial of coding circuit is

$$G(x) = x^8 + x^4 + x^3 + x^1 + 1 \tag{5-1}$$

The structure of the RS coded error protection packet for the fixed-length 188-byte MPEG transport stream data frame is shown in Figure 5-11.

DVB employs the cascaded coding scheme with the RS outer code and the convolutional inner code. The convolutional code has a strong error correction capability but suffers from error propagation. If the resultant burst errors from the convolutional decoder are beyond the error correction capability of the RS code, severe BER performance degradation would occur. Data interleaving is therefore introduced between outer and inner encoders, helping to convert the burst errors into random errors for better error correction performance by the RS code.

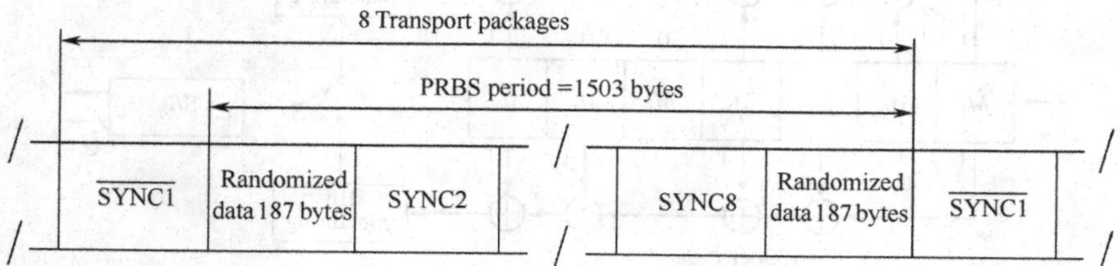

FIGURE 5-10 Scrambling processing of transport packet.

FIGURE 5-11 RS (204,188) error protection packet.

The DVB system has a convolutional interleaver (outer interleaver) with $I = 12$, $M = 17$ after the RS encoder, and its structure is identical to that of the ATSC byte interleaver. The interleaved data bytes are constituted by the error correction packet and separated by an inverted or noninverted MPEG-2 sync byte (to keep the 204-byte cycle). The interleaver, consisting of $I = 12$ branches, is cyclically connected to the input byte stream by the input switch. Each branch j is a first-in, first-out (FIFO) shift register with depth $j \times M$ cells, where $M = 17 = N/I$ and $N = 204$. The cells of the FIFO register contains a byte with input and output switching synchronously. For synchronization purpose, the SYNC and $\overline{\text{SYNC}}$ bytes will always be routed to the branch 0 of the interleaver (corresponding to no delay).

The de-interleaver is similar to the interleaver in principle with the branch index reversed (i.e. $j = 0$ corresponds to the maximum delay). The de-interleaver synchronization can be carried out by routing the first recognized sync byte (SYNC or $\overline{\text{SYNC}}$) in the 0 branch.

5.3.1.3 Inner Encoding The inner code comes from a range of punctured convolutional codes, based on a mother code with 1/2 code rate and 64 states. This will allow the selection of the most appropriate level of error correction for a given service or data rate in either the nonhierarchical or hierarchical transmission mode. The generator polynomials of the mother code are $G_1 = 171$oct for X output and $G_2 = 133$ oct for Y output. The structure of inner coding is shown in Figure 5-12. The puncturing patterns are shown in Table 5-1.

5.3.1.4 Inner Interleaving The inner interleaving consists of bitwise interleaving followed by symbol interleaving. Both bitwise and symbol interleaving processes are block based.

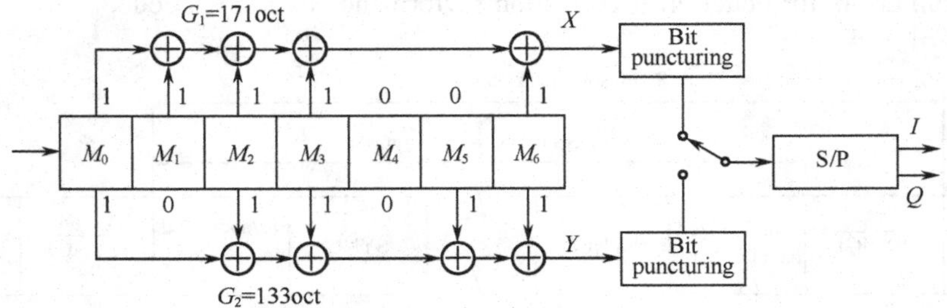

FIGURE 5-12 (2, 1, and 7) generation of punctured convolutional code.

TABLE 5-1 DVB-T Punctured Convolutional Code

Code Rates r	Puncturing Pattern	Transmitted Sequence (After Parallel-to-Serial Conversion)
1/2	X: 1 Y: 1	$X_1\ Y_1$
2/3	X: 1 0 Y: 1 1	$X_1\ Y_1\ Y_2$
3/4	X: 1 0 1 Y: 1 1 0	$X_1\ Y_1\ Y_2\ X_3$
5/6	X: 1 0 1 0 1 Y: 1 1 0 1 0	$X_1\ Y_1\ Y_2\ X_3\ Y_4\ X_5$
7/8	X:1 0 0 0 1 0 1 Y: 1 1 1 1 0 1 0	$X_1\ Y_1\ Y_2\ Y_3\ Y_4\ X_5\ Y_6\ X_7$

Bitwise interleaving is carried out first. The bit stream for which the inner codes are punctured is fed to the bit interleaver, and then it is divided into v substreams depending on the constellation and transmitted into the v-way bit interleaver in the shift register. Here, $v=2$ for QPSK, $v=4$ for 16QAM, $v=6$ for 64QAM. In the nonhierarchical mode, the single input stream is demultiplexed into v substreams. In the hierarchical mode, the high-priority stream is demultiplexed into two substreams, and low-priority stream is demultiplexed into $v-2$ substreams. This applies to both uniform and nonuniform QAM modes.

Bitinterleaving is performed only on the useful data. The block size is the same for each interleaver, but the interleaving sequence is different in each case. With bit interleaving block size of 126 bits, the block interleaving process is repeated exactly 12 times per OFDM symbol of useful data in the 2 K mode and 48 times per symbol in the 8 K mode.

For each bit interleaver, the input bit vector is defined by

$$B(e) = (b_{e,0}, b_{e,1}, b_{e,2}, \ldots, b_{e,125}) \qquad e = 1, 2, \ldots, v-1$$

The interleaved output vector $A(e) = (a_{e,0}, a_{e,1}, a_{e,2}, \ldots, a_{e,125})$ is defined by

$$a_{e,W} = b_{e,H_e(W)} \qquad W = 0, 1, 2, \ldots, 125$$

where $H_e(W)$ is a permutation function which is different for each interleaver, as defined in Table 5-2.

The outputs of the v bit interleavers are grouped to form the digital data symbols such that each symbol of v bits will consist of exactly one bit from each v interleaver. Hence, the output from the bitwise interleaver is a v bit word y' that has output I_0 as its most significant bit, i.e.,

$$y'_W = (a_{0,W}, a_{1,W}, \ldots, a_{v-1,W})$$

TABLE 5-2 Permutation Function of Bit Interleaver

Each Bit Interleaver	Permutation Function
I_0	$H_0(W) = W$
I_1	$H_1(W) = (W + 63) \bmod 126$
I_2	$H_2(W) = (W + 105) \bmod 126$
I_3	$H_3(W) = (W + 42) \bmod 126$
I_4	$H_4(W) = (W + 21) \bmod 126$
I_5	$H_5(W) = (W + 84) \bmod 126$

Next, symbol interleaving (frequency interleaving) is performed to the codeword with the purpose of mapping these v-bit-long words onto the 1512 (2 K mode) or 6048 (8 K mode) active carriers per OFDM symbol. The symbol interleaver acts on blocks of 1512 (2 K mode) or 6048 (8 K mode) data symbols.

Therefore, in the 2 K mode, 12 groups of 126 data words from the bit interleaver are sequentially read into a vector $Y' = (y'_0, y'_1, y'_2, \ldots, y'_{1511})$. Similarly, in the 8 K mode, a vector $Y' = (y'_0, y'_1, y'_2, \ldots, y'_{6047})$ is formed by 48 groups of 126 data words.

The interleaved vector $Y = (y_0, y_1, y_2, \ldots, y_{N_{\max}-1})$ is defined by

$$y_{H(q)} = y'_q \quad \text{for even } q = 0, \ldots, N_{\max} - 1$$
$$y_q = y'_{H(q)} \quad \text{for odd } q = 0, \ldots, N_{\max} - 1$$

where $N_{\max} = 1512$ in the 2 K mode and $N_{\max} = 6048$ in the 8 K mode.

The symbol index is used to define the position of the current OFDM symbol in the OFDM frame, where $H(q)$ is a permutation function generated by a random sequence. See the ETS 300 744 standard for detailed description.

Like y', y is also constituted by v bits:

$$y_{q'} = (y_{0,q'}, y_{1,q'}, \ldots, y_{v-1,q'})$$

where q' is the symbol index at the output of the symbol interleaver. The value of y is used to map the data into the signal constellation. The data are fed to the corresponding symbol interleaver for interleaving by means of the controller according to the parity of the current symbol marked.

5.3.2 Modulation

5.3.2.1 Constellation Mapping The DVB-T system uses OFDM technology for transmission. All data carriers in an OFDM frame use the constellations of QPSK, 16QAM, 64QAM, nonuniform 16QAM/64QAM, all with Gray mapping.

For nonuniform modulation, the real part $\mathrm{Re}\{z\}$ and the imaginary part $\mathrm{Im}\{z\}$ of the complex number z after mapping are related to the nonuniform factor α. If $\alpha = 2$, the uniform 16QAM mapped values +3, +1, −1, and −3 will change to the nonuniform values +4, +2, −2, and −4, respectively; uniform 64QAM mapped values +7, +5, +3, +1, −1, −3, −5, and −7 change to the nonuniform values +8, +6, +4, +2, −2, −4, −6, and −8, respectively. If $\alpha = 4$, the nonuniform 16QAM mapped values are +6, +4, −4, and −6, respectively; the nonuniform 64QAM mapped values are +10, +8, +6, +4, −4, −6, −8, and −10.

For nonhierarchical transmission, the data stream at the input of the inner interleaver consists of v bit words that are mapped onto a complex number z. For example, the constellation point at the top left, i.e., 1000, represents $y_{0,q'} = 1, y_{1,q'} = y_{2,q'} = y_{3,q'} = 0$. For the hierarchical 16QAM, the high-priority bits are $y_{0,q'}$ and $y_{1,q'}$ bits of the inner interleaver output words, and the low-priority bits are the $y_{2,q'}$ and $y_{3,q'}$ bits of the inner interleaver output words. If the 16QAM constellation is decoded as if it were QPSK, the high-priority bits $y_{0,q'}$ and $y_{1,q'}$ will be deduced. To decode the low-priority bits, the full constellation must be examined and the appropriate bits $(y_{2,q'}, y_{3,q'})$ extracted from $y_{0,q'}, y_{1,q'}, y_{2,q'}$, and $y_{3,q'}$.

For the hierarchical 64 QAM, the high-priority bits are $y_{0,q'}$ and $y_{1,q'}$ bits of the inner interleaver output, and the low priority bits are the $y_{2,q'}, y_{3,q'}, y_{4,q'}$, and $y_{5,q'}$ bits of the inner interleaver output. If the constellation is decoded as if it were QPSK, the high-priority bits $y_{0,q'}$ and $y_{1,q'}$ will be deduced. To decode the low-priority bits, the full constellation must be examined and the appropriate bits $(y_{2,q'}, y_{3,q'}, y_{4,q'}$, and $y_{5,q'})$ extracted from $y_{0,q'}, y_{1,q'}, y_{2,q'}, y_{3,q'}, y_{4,q'}$, and $y_{5,q'}$.

5.3.2.2 ***OFDM Frame Structure*** The transmitted signal is organized in frames, each of duration T_F, and consists of 68 OFDM symbols (numbered from 0 to 67). Each symbol is constituted by a set $K = 6817$ subcarriers in the 8 K mode and $K = 1705$ subcarriers in 2 K mode. The duration of a 1 symbol is T_s, where T_s consists of two parts, a useful part with duration T_u and guard interval Δ. The guard interval consists of a cyclic continuation of the useful part, T_u, and is inserted before it. Four values of guard intervals may be used according to Table 5-3, where the different values are given both in multiples of the elementary period $T = 7/64$ μs and in microseconds.

Since the OFDM signal has many separately modulated carriers, each symbol can be divided into small cells, each corresponding to a subcarrier within the OFDM symbol. In addition to the data subcarrier, each OFDM symbol includes a scattered pilot, a consecutive pilot, and TPS subcarriers. These pilots are used for frame synchronization, frequency synchronization, time synchronization, channel estimation, and transmission mode recognition and can also be used to track phase noise.

TABLE 5-3 Symbol Duration and Guard Interval

Mode	8 K Mode			
Guard interval Δ/T_u	1/4	1/8	1/16	1/32
Duration of symbol part T_u	8192 × T (896 μs)			
Duration of guard interval Δ	2048 × T (224 μs)	1024 × T (112 μs)	512 × T (56 μs)	256 × T (28 μs)
Symbol duration T_s	10240 × T (1120 μs)	9216 × T (1008 μs)	8704 × T (952 μs)	8448 × T (924 μs)

TABLE 5-4 OFDM Parameters for 8 K and 2 K Mode

OFDM parameter	8 K Mode	2 K Mode
Number of carrier K	6817	1705
Value of carrier number K_{min}	0	0
Value of carrier number K_{max}	6816	1704
Duration T_u	896 μs	224 μs
Carrier spacing $1/T_u$	1116 Hz	4464 Hz
Bandwidth	7.61 MHz	7.1 MHz

The subcarriers are indexed by $k \in [K_{min}, K_{max}]$, $K_{min} = 0$, $K_{max} = 1704$ in the 2 K mode, and $K_{max} = 6816$ in the 8 K mode. If the carrier K_{min}, K_{max} are determined by $(K-1)/T_u$, the spacing between adjacent subcarriers is $1/T_u$. The OFDM parameters for the 8 K and 2 K modes are given in the Table 5-4.

The parameters given in Table 5-4 are for channel spacing at 8 MHz. For the 7 MHz channel, the system clock can be changed from $1/T = 64/7$ MHz to an accurate 8.0 MHz, so that all system parameters are scaled down. After the channel bandwidth is changed from 8 to 7 MHz, the frame structure and the coding, mapping, and interleaving method remain the same, but the signal bandwidth will be 7/8 of the original one, and the data transmission capability of the system is reduced.

5.3.2.3 Pilot Subcarrier Not all subcarriers in an OFDM symbol of the DVB-T system are used for data transmission, some of which are the pilots. There are 177 continuous pilots in the 8 K mode and 45 in the 2 K mode which are inserted in fixed positions of all OFDM symbols. "Continuous" here means that the pilots appear at the same subcarrier position of each OFDM symbol. In each OFDM symbol, a scattered pilot appears every 12 subcarriers and repeats its pattern every four OFDM symbols. TPS is defined over 68 consecutive OFDM symbols (or one OFDM frame). Four consecutive frames constitute an OFDM superframe. The reference sequence corresponding to the TPS carrier in the first symbol of each OFDM frame is used to initialize the TPS modulation on each TPS carrier. Each OFDM symbol conveys a TPS bit, and each TPS block (corresponding to one OFDM frame) contains 68 bits. The 68 bits are constantly transmitted at the cycle of the OFDM frame. The same symbol in each OFDM frame transmits one TPS bit. The 68 bits are defined as follows: (1) initialization bit (bit s_0); (2) 16 synchronization bits (bits s_1–s_{16}); (3) 37 information bits (bits s_{17}–s_{53}); and (4) 14 redundancy bits for error protection (bits s_{54}–s_{67}). Of 37 information bits (bits s_{17}–s_{53}), 23 bits are used at present. The remaining 14 bits are reserved for future use and must be set to zero.

Every TPS carrier is modulated using DBPSK. DBPSK is initialized at the beginning of each TPS block. The first initialization bit (bit s_0) is the initial bit for the differential modulation. Unlike other pilots, the TPS subcarrier is transmitted at the normalized power level, i.e., $E(C_{i,k}C^*_{i,k}) = 1$.

To improve the error resilience of TPS, 53 bits, including TPS synchronization and information (bits s_1–s_{53}), are encoded by 14 parity bits of the BCH (67, 52, $t = 2$) punctured code, derived from primitive systematic BHC (127, 113, $t = 2$) code. The

TABLE 5-5 Definitions of Bits in TPS Block

Bit Number	Definition	Format
s_0	Initialization	Initial bit for differential BPSK modulation
s_1–s_{16}	Synchronization word	Synchronization word for TPS
s_{17}–s_{22}	Length indicator	Indicate number of used bits of TPS, 010111 is used (23 TPS bit is transmitted)
s_{23}, s_{24}	Frame number	Four frames constitute one superframe; frames inside superframe numbered 1–4.
s_{25}, s_{26}	Constellation	QPSK, 16QAM, or 64QAM
s_{27}, s_{28}, s_{29}	Hierarchy information	Hierarchy information specifies whether transmission is hierarchical and, if so, what α value is: $\alpha=1$, $\alpha=2$, or $\alpha=4$
s_{30}, s_{31}, s_{32}	HP stream code rate	HP level code rate, 1/2, 2/3, 3/4, 5/6, or 7/8
s_{33}, s_{34}, s_{35}	LP stream code rate	LP level code rate, 1/2, 2/3, 3/4, 5/6, or 7/8
s_{36}, s_{37}	Guard interval	Value of guard interval, 1/32, 1/16, 1/8, or 1/4
s_{38}, s_{39}	Transmission mode	2 K mode or 8 K mode
s_{40}–s_{53}	Reserved	All 0
s_{54}–s_{67}	TPS error protection	BCH error protection bits

code generator polynomial $H(x)=x^{14}+x^9+x^8+x^6+x^5+x^4+x^2+x+1$. The punctured BCH codes can be obtained by adding 60 bits (all set to 0) in front of the information bit for the input of the BCH (127, 113, $t=2$) encoder. After the BCH coding, these zeros will be removed for codeword length of 67 bits.

The bits in the TPS block are defined in Table 5-5.

Each OFDM superframe, structure contains an integer number of RS packets (204 bytes) to avoid padding, and this has nothing to do with the constellation, guard interval length, coding rate, or channel bandwidth. See Table 5-6 for details. The first data byte transmitted on each OFDM superframe is one of the synchronization bytes for the MPEG packet.

TABLE 5-6 Number of RS Packet per OFDM Superframe in Combinations of Various Modes

Code Rate	QPSK		16-QAM		64-QAM	
	2 K Mode	8K Mode	2 K Mode	8 K Mode	2 K Mode	8K Mode
1/2	252	1 008	504	2 016	756	3 024
2/3	336	1 344	672	2 688	1 008	4 032
3/4	378	1 512	756	3 024	1 134	4 536
5/6	420	1 680	840	3 360	1 260	5 040
7/3	441'	1 764	882	3 528	1 323	5 292

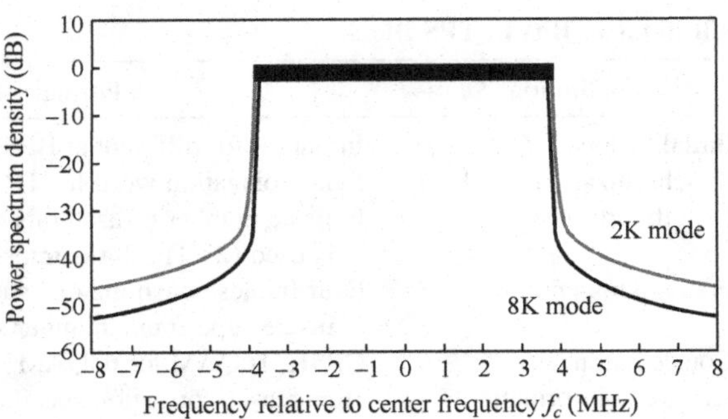

FIGURE 5-13 Theoretical spectrum for DVB-T transmission signal at $\Delta = T_u/4$ guard interval.

5.3.2.4 OFDM Modulation The OFDM symbol consists of equally spaced orthogonal carriers. The amplitude and phase of each subcarrier vary with the symbol, like the mapping process described above.

Usually, OFDM modulation is done by IFFT transformation. At the frequency $f_k = f_c + k'/T_u$ $[k' = k - (K_{\max} + K_{\min})/2,\ K_{\max} \leq k \leq K_{\min}]$, the power spectral density $P_k(f)$ is defined as:

$$P_k(f) = \left[\frac{\sin\ \pi(f - f_k)T_s}{\pi(f - f_k)T_s}\right]^2 \tag{5-2}$$

The entire power spectral density for the modulated data unit carrier is the sum of the total power spectral density for these carriers. Theoretically the power spectrum of the DVB transmitted signal is shown in Figure 5-13. Since the OFDM symbol duration is larger than the carrier spacing, the main lobe of each carrier's power spectrum is less than 2 times the carrier spacing. Thus, the spectral density is nonconstant within the 7.608258 MHz bandwidth of the 8 K mode or 7.611607 MHz bandwidth of the 2 K mode.

5.4 INTRODUCTION TO ISDB-T STANDARD

In the DVB-T system, hierarchical transmission can be achieved to support both high- and low-priority streams (see Figure 5-14) by changing the modulation (16/64QAM-QPSK), and it is not really layered transmission on separate channels. To further expand the carrying capacity for integrated services of DTV terrestrial transmission systems, the whole system bandwidth is divided into 13 subbands with the same bandwidth (called segment), i.e. segmented OFDM. The bandwidth of each segment is determined by 1/13 of the channel bandwidth used; the values are 428.57 and 571.428 kHz relative to 6 and 8 MHz operating bandwidth, so the modulation mode is called band-segmented transmission (BST) OFDM. The ISDB-T provides

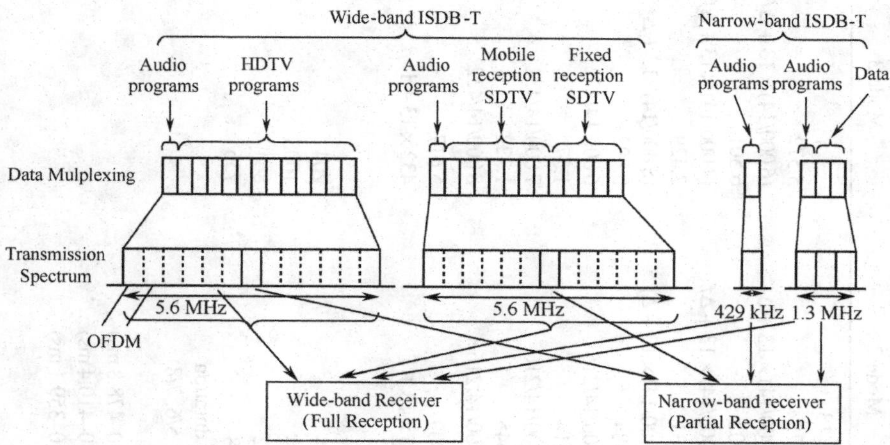

FIGURE 5-14 ISDB-T hierarchical reception and partial reception.

hierarchical transmission characteristics by specifying each OFDM segment as an independent data channel and using different carrier modulation scheme and inner coding rate on different segments. Each data segment has its own error protection scheme (inner coding rate and time interleaving depth) and modulation type (QPSK, DQPSK, 16QAM, or 64QAM), and each segment can then meet different service requirements. The analog bandwidth of ISDB-T is divided into two types: broadband ISDB-T (ISDB-Tw, composed of 13 OFDM segments, occupying 5.6 MHz) and narrowband ISDB-T (ISDB-Tn), using one or three OFDM segments, occupying bandwidth 430 kHz (3000/7 kHz) or 3×430 kHz = 1.3 MHz. The broadband ISDB-T is used for broadband services such as HDTV, SDTV, and other high-speed multimedia services, and the parameters are shown in Table 5-7. The narrowband ISDB-T is suitable for digital sound broadcasting, all kinds of low-speed data transmission such as low-frame-rate graphic information, procedures, and configuration information, and narrowband services such as data download; the parameters are shown in Table 5-8.

The spectral segments are independent each other. The segments can be flexibly combined into different OFDM segments based on service flow, and a variety of error protection levels for each segment can be achieved by different system parameter combinations, ensuring independent transmission. One ISDB-T channel can provide the service layers of A, B, and C to adapt to different transmission environments and service requirements, such as fixed reception, portable reception, mobile reception, and SFN. Under normal circumstances, layer A corresponds to digital audio or data broadcasting, layer B corresponds to mobile reception for SDTV and multimedia services, and layer C corresponds to the fixed reception for HDTV, SDTV, and multimedia services. Only layer operates on segment 0, which is at the center of the spectrum, and individual reception using a narrowband receiver is allowed. This is the partial-reception mode specific to the ISDB-T standard. The operating principle of

TABLE 5-7 ISDB-Tw Parameters

ISDB-Tw mode		Mode 1	Mode 2	Mode 3
Number of OFDM segment	6, 7, 8 MHz		$N = 13$	
Bandwidth (kHz)	6 MHz	5.575	5.573	5.572
		$(6\,000/14) \times 13 + \Delta f$	$(6\,000/14) \times 13 + \Delta f$	$(6\,000/14) \times 13 + \Delta f$
	7 MHz	6.504	6.502	6.501
		$(7\,000/14) \times 13 + \Delta f$	$(7\,000/14) \times 13 + \Delta f$	$(7\,000/14) \times 13 + \Delta f$
	8 MHz	7.433	7.431	7.429
		$(8\,000/14) \times 13 + \Delta f$	$(8\,000/14) \times 13 + \Delta f$	$(8\,000/14) \times 13 + \Delta f$
Carrier spacing (kHz)	6 MHz	3.968	1.984	0.992
		6 000/14/108	6 000/14/216	6 000/14/432
	7 MHz	4.629	2.314	1.57
		7 000/14/108	7 000/14/216	7 000/14/432
	8 MHz	5.291	2.645	1.322
		8 000/14/108	8 000/14/216	8 000/14/432
Total carriers	6, 7, 8 MHz	1,405	2,809	5,617
		$108 \times 13 + 1$	$216 \times 13 + 1$	$432 \times 13 + 1$
Mode of modulation	6, 7, 8 MHz	QPSK, 16QAM, 64QAM, DQPSK		
Number of symbols in each frame	6, 7, 8 MHz	204	204	204
Symbol duration	6 MHz	252	504	1008
	7 MHz	216	432	864
	8 MHz	189	378	756
Guard interval	6, 7, 8 MHz	1/4, 1/8, 1/16, 1/32 of the symbol duration		
Inner code	6, 7, 8 MHz	Convolutional code (1/2, 2/3, 3/4, 5/6, 7/8)		
Outer code	6, 7, 8 MHz	RS (204, 188)		
Interleaving	6 MHz	Frequency and time interleaving (0–478.8 ms)		
	7 MHz	Frequency and time interleaving (0–410.4 ms)		
	8 MHz	Frequency and time interleaving (0–359.1 ms)		
Information bit rate	6 MHz	3.651–23.234		
	7 MHz	4.259–27.107		
	8 MHz	4.868–30.979		

TABLE 5-8 ISDB-Tn Parameters

ISDB-Tn mode	Mode 1	Mode 2	Mode 3
Number of OFDM segments	$N_s = 1$ (1 segment), $N_s = 3$ (3 segments)		
Bandwidth	430 kHz (1 segment) or 1.3 MHz (3 segments)		
Carrier spacing/kHz	3.968	1.984	0.992
Total carriers	109 (1 segment) 325 (3 segments) $108 \times N_s + 1$	217 (1 segment) 649 (3 segments) $216 \times N_s + 1$	433 (1 segment) 1297 (3 segments) $432 \times N_s + 1$
Mode of modulation	QPSK, 16QAM, 64QAM, DQPSK		
Number of symbols in each frame	204		
Symbols duration	252	504	1 008
Guard interval	1/4, 1/8, 1/16, 1/32 of the symbol duration		
Inner code	Convolutional code (1/2, 2/3, 3/4, 5/6, 7/8)		
Outer code	RS (204, 188)		
Information bit rate (Mbits/s)	0.28085–1.873 0.842–5.361		

partial reception is shown in Figure 5-15. In the case of partial reception, the sampling frequency can be reduced to 1/13 of all channel reception, and the implementation complexity and power consumption will be decreased accordingly, facilitating the use of hand-held devices. The Japanese mobile TV standard is achieved using segment 0, called a one-segment.

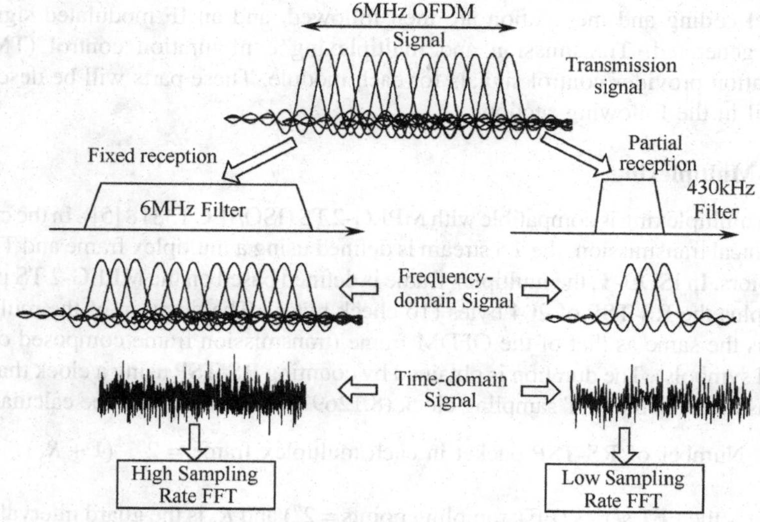

FIGURE 5-15 Principle of partial reception.

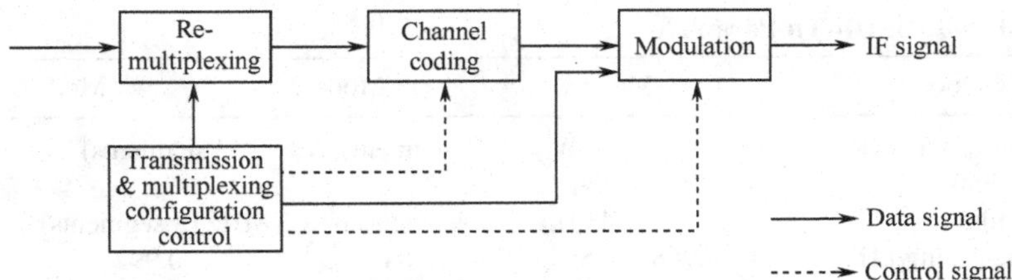

FIGURE 5-16 Basic structure of ISDB-T transmission system.

In should be noted that the modulation mode and system parameters for each segment are not specifically designated in the ISDB-T standard, even if individual reception is allowed for segment 0. This provides maximal flexibility for a variety of service combinations. However, considering that higher receiver sensitivity is required for narrowband digital broadcasting services in practice, it is generally preferred to use DQPSK modulation and a best error protection code rate to support layer A with maximum receiving robustness.

The narrowband ISDB-T mode only uses single or triple OFDM segments (segment 0 or segments 0–2) for transmission. This is considered the scaled-down version of the ISDB-T system, of which segment 0 is in the center of the three-segment mode of 1.29 MHz. The first-level transport layer can be formed by DQPSK modulation with best robustness for the individual reception; the second-level transport layer can be formed by other modulations combined with error protection parameters for segments 1 and 2 on each side.

The multiple services to be supported by ISDB-T should be in MPEG-2 TS format and the basic structure of the ISDB-T transmission system is shown in Figure 5-16. The input TSs are multiplexed and adjusted to match the transmission parameters. Channel coding and modulation are then followed, and an IF modulated signal is finally generated. Transmission and multiplexing configuration control (TMCC) information provides control signals for each module. These parts will be described in detail in the following sections.

5.4.1 Multiplexing

System multiplexing is compatible with MPEG-2 TS (ISO/IEC 13818 [5]). In the case of hierarchical transmission, the TS stream is defined using a multiplex frame and TMCC descriptors. In ISDB-T, the multiplex frame is defined based on the MPEG-2 TS packet (TSP) plus the RS-TSP of 204 bytes (16 check bytes). The duration of the multiplex frame is the same as that of the OFDM frame (transmission frame composed of 204 OFDM symbols). The duration is obtained by counting RS-TSP using a clock that four times as much as the FFT sampling clock (8.12693 MHz), which can be calculated as

$$\text{Number of RS-TSP packet in each multiplex frame} = 2^{n-1}(1 + R_y)$$

where n is the FFT series (FFT sampling points $= 2^n$) and R_y is the guard interval ratio, such as 1/4, 1/8, and so on.

TABLE 5-9 Number of Transmission RS-TSPs Included in One Multiplex Frame

	Number of Transmission TSPs			
Mode	Guard Interval Ratio 1/4	Guard Interval Ratio 1/8	Guard Interval Ratio 1/16	Guard Interval Ratio 1/32
Mode 1	1280	1152	1088	1056
Mode 2	2560	2304	2176	2112
Made 3	5120	4608	4352	4224

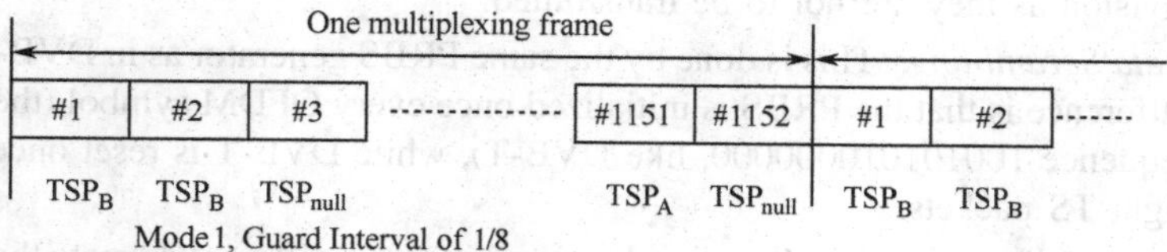

FIGURE 5-17 Diagram of transport stream TS.

The number of transmission RS-TSPs included in each multiplex frame is shown Table 5-9.

The values in Table 5-9 are larger than the actual number of transmission TSPs included in an OFDM frame. The redundant part is padded with RS-TSPs full of zeros for the rate matching. Figure 5-17 is an example of TS using the modulation mode 1, with the guard interval of 1/8 and containing layers A, B, and C.

5.4.2 Channel Coding

A channel coding module receives the RS-TSP in multiplex frame format, and the functional block diagram is shown in Figure 5-18.

The ISDB-T channel coding block is basically the same as that of the DVB-T, which uses the RS (204,188) and convolutional coding, including data scrambling, interleaving, and other modules. The biggest difference is that the ISDB-T channel

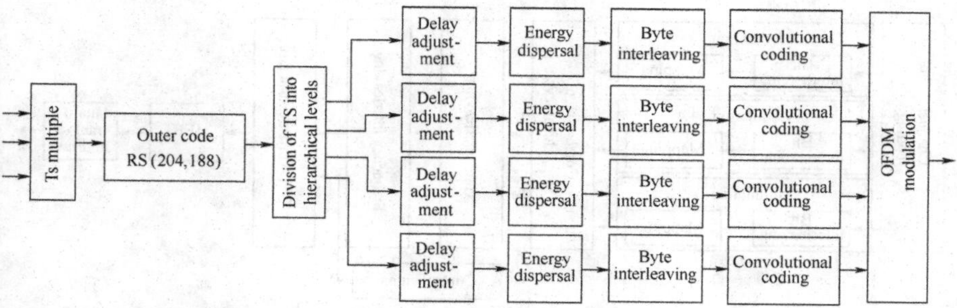

FIGURE 5-18 Block diagram of ISDB-T channel coding.

coding is carried out individually in the three types of layers, which include the following:

1. *Outer Code Encoding.* Like DVB-T, FEC of RS (204, 188) is applied to correct up to 8-byte random errors. However, in DVB-T, the randomization (data scrambling) is done before the outer code encoding.
2. *TS Stream Segmentation.* The TS data stream from the outer code is segmented in units of 204 bytes. If the hierarchy is not defined, the segmentation is not required. It should be noted that the inserted null TSPs are removed in the division as they are not to be transmitted.
3. *Data Scrambling.* This is done by the same PRBS generator as in DVB-T. The difference is that the PRBS is initialized once every OFDM symbol (the initial sequence 100101010000000, like DVB-T), while DVB-T is reset once every eight TS packets.
4. *Delay Adjustment.* The byte interleaving will cause different delay to the layers using different channel coding and modulation mode, so a delay adjustment module is inserted between the randomization and byte interleaving to compensate for the data stream delay from different layers, including de-interleaving at the receiver. The amount of delay adjustment depends on the number of TSPs in each data stream.
5. *Bytewise Interleaving.* The bytewise interleaver is a Forney convolutional interleaver of $I = 12$, $M = 17$ bytes, like the convolutional interleaver in DVB systems.

5.4.3 Constellation Mapping and Modulation

The block diagram of an ISDB-T modulation circuit is shown in the Figure 5-19.

5.4.3.1 Bit Interleaving and Constellation Mapping The DQPSK, QPSK, 16QAM, or 64QAM constellation mappings are uniformly used in ISDB-T for the subcarriers. Due to the segmented structure and hierarchy partition, different modulation modes can be applied to 13 different segments, namely, QPSK (DQPSK),

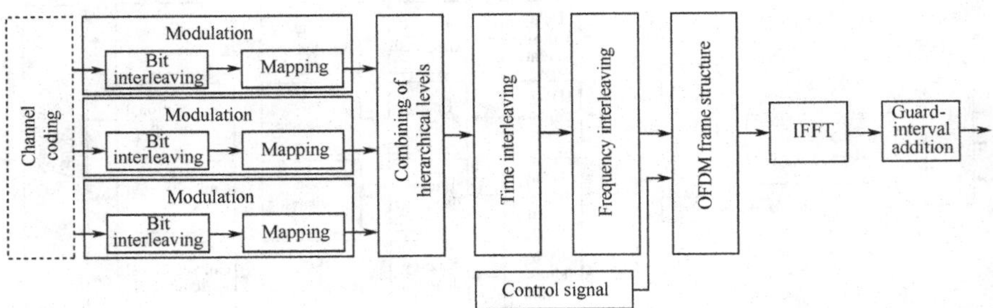

FIGURE 5-19 Block diagram of ISDB-T modulation.

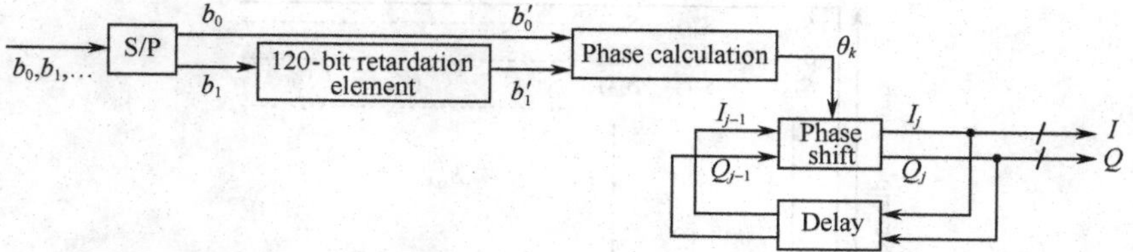

FIGURE 5-20 Block diagram of π/4 - shift DQPSK modulation.

16QAM, and 64QAM can coexist, and thus there are independent bit interleaving and mapping modules.

DQPSK bit interleaving and mapping are shown in Figure 5-20. The serial bit stream (b_0, b_1, . . .) of incoming one-way channel coding changes into two-way parallel streams. To improve the anti-interference ability, a delay of 120 bits is added to the second branch (the one corresponding to b_1), equivalent to a simple bitinterleaving for two bit streams. Then the complex constellation points (vector points) expressed by I and Q components are derived by π/4 rotation of DQPSK modulation to two bit streams. The phase shift for mapping (θ_j) and modulated constellation points are shown in Figure 5-21.

The differential operation is carried out as

$$\begin{bmatrix} I_j \\ Q_j \end{bmatrix} = \begin{bmatrix} \cos\,\theta_j & -\sin\,\theta_j \\ \sin\,\theta_j & \cos\,\theta_j \end{bmatrix} \begin{bmatrix} I_{j-1} \\ Q_{j-1} \end{bmatrix} \tag{5-3}$$

The QPSK, 16QAM, and 64QAM bit interleaving and mapping methods are similar to DQPSK. The incoming encoded data stream will be changed into two, four, and six parallel data streams and constellation points are then performed. Each data stream has different bit delay. The QPSK, 16QAM and 64QAM mapped constellation points are similar to DVB-T uniform constellation points using Gray mapping.

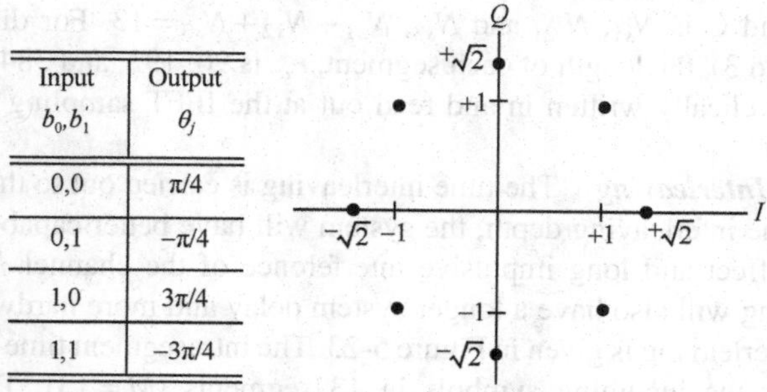

Input b_0, b_1	Output θ_j
0,0	π/4
0,1	−π/4
1,0	3π/4
1,1	−3π/4

FIGURE 5-21 Phase extraction and π/4 DQPSK constellation.

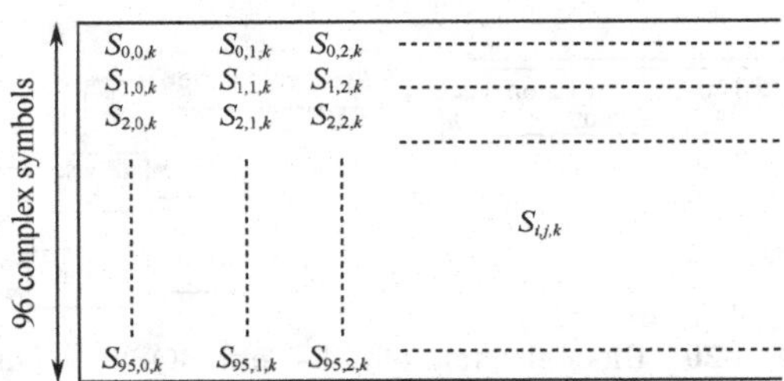

FIGURE 5-22 Structure of data segment in mode 1.

5.4.3.2 Synthesis of Hierarchical Data Flow and Rate Conversion The 13 OFDM segments defined in ISDB-T can have up to three types of transport layers (each layer contains its own modulation mode, code rate, etc.). Now, 13 segments of hierarchical data streams will be combined into one data stream, and different rates from different levels need to be converted into the same transmission rate.

The synthesis of the hierarchical data stream is carried out based on data segments. Here the meaning of the data segment is as follows: The synthesized output data (complex constellation point) are read at the IFFT sampling clock, so the synthesized output data will match the subcarriers in the OFDM segments when performing the OFDM modulation. After this matching, one OFDM segment in a given mode is defined as a data segment, and all the data segments will constitute an OFDM data symbol.

Thus, the layers of input complex constellation points $S_{i,j,k}$ are sequentially arranged in the form of a data segment structure. The data segment structure of mode 1 is shown in Figure 5-22 (the structures of mode 2 and mode 3 are exactly the same as that of mode 1, but the number of data contained in each column is different). It can be seen from the figure that there are 96 data subcarriers per OFDM segment in the case of mode 1 (see Table 5-12 below), corresponding to 96 complex constellation points in each column of the figure. In Figure 5-22, i in the $S_{i,j,k}$ denotes the subcarrier number, k denotes numbers 0–12 of the OFDM segment, and j represents the number of 204 OFDM symbols in an OFDM frame.

The complex data at the same level for 13 OFDM segments are written into the appropriate data segment memory. Assuming the number of OFDM segments of layers A, B, and C is N_{s1}, N_{s2}, and N_{s3}, $N_{s1}+N_{s2}+N_{s3}=13$. For different modes (modes 1, 2, and 3), the length of each segment, n_c, is 96, 192, and 384, respectively. The data are cyclically written in and read out at the IFFT sampling clock.

5.4.3.3 Time Interleaving The time interleaving is carried out to the above layer. With larger time interleaving depth, the system will have better capability under the time-varying effect and long impulsive interference of the channel. The increased time interleaving will also have a longer system delay and more hardware overhead.

The time interleaving is given in Figure 5-23. The intrasegment time interleaving is carried out for the incoming symbols in 13 segments ($M=13$). The interleaver performs input/output operations bit by bit to data segments based on the IFFT clock,

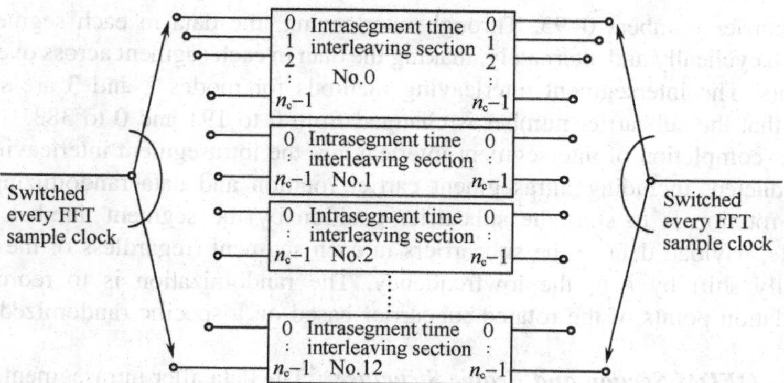

FIGURE 5-23 Time interleaving structure.

and the input/output of a data symbol is completed in each cycle. To ensure the independence of OFDM segments in the ISDB-T system, the data segment interleaving corresponding to each OFDM segment should be independent from each other, and the interleaving structure of each segment should be consistent. The interleaving parameters depend only on the parameters of the transport layer involved in the segment (transmission mode, modulation, receiving mode, etc.).

5.4.3.4 Frequency Interleaving The frequency-domain interleaving is carried out between different OFDM segments within the same OFDM symbol versus the frequency-selective fading due to the multipath channel. The structural block diagram of frequency interleaving is shown in Figure 5-24.

The time-interleaved symbol data stream is then divided into partial reception, differential modulation, and coherent modulation parts in the segment divider according to segment number and the different modulation mode. As only one OFDM segment is used for partial reception, the intersegment interleaving is not required, while it is required for the symbol data stream of the other two. For the symbol data stream in the same modulation mode, the intersegment interleaving should be carried out to differential modulation (DQPSK) and coherent modulation (QPSK, 16QAM, 64QAM) parts, respectively. Taking mode 1 as an example, before intersegment interleaving, the constellation point data in each segment are arranged

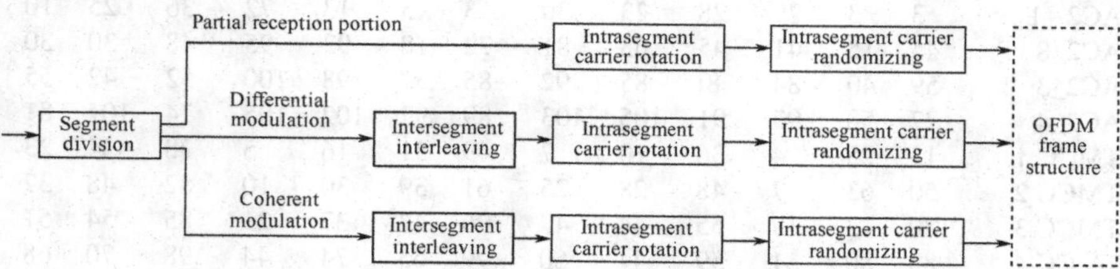

FIGURE 5-24 Frequency interleaving structure.

by subcarrier numbers 0–95. Through interleaving, the data in each segment are arranged cyclically and alternately, making the data in each segment across over the n segments. The intersegment interleaving methods for modes 2 and 3 are similar, except that the subcarrier number is changed from 0 to 191 and 0 to 383.

After completion of intersegment interleaving, the intrasegment interleaving will be conducted, including intrasegment carrier rotation and data randomizing. The carrier rotation is to shift the subcarrier position by the segment number k. For example, payload data of the subcarriers in each segment (regardless of the mode) cyclically shift by k to the lowfrequency. The randomization is to reorder the constellation points of the rotated subcarrier based on a specific randomized table.

5.4.3.5 OFDM Section and Frame Structure The data after intrasegment carrier rotation and randomization will form the OFDM frame including 204 OFDM symbols and pilot carriers. The added pilot signals include scattered pilot (SP), continuous pilot (CP), transmission and multiplexing configuration control (TMCC), and auxiliary channels AC1 and AC2: CP, AC1, AC2, and TMCC can be used for frequency and time synchronization; SP is mainly used for channel estimation; TMCC is also used for the transmission multiplexing configuration control signal to tell the receiving terminal how to interpret the received data, similar to TPS signal in DVB-T.

The channel estimation is not required for differential modulation, so there is no need for SP. In this mode, the number of carriers in each segment is 108, 216, or 432, and the number of data subcarriers is 96,192, and 384, respectively. In mode 1, the pilot assignment of OFDM segments is shown in Table 5-10.

The SP for four adjacent OFDM symbols for the coherent demodulation completes a cycle whenever an SP is inserted every 12 subcarriers in the OFDM. In mode 1, for example, each OFDM segment has 108 subcarriers, so it requires 108/12 = 9 SP subcarriers, and the other three pilot symbols include two AC1 and a TMCC. The positions of the AC1 and TMCC subcarriers for each segment are shown in Table 5-11.

TABLE 5-10 CP, AC, and TMCC Carrier Distribution in Model 1

Segment no.	11	9	7	5	3	1	0	2	4	6	8	10	12
CP	0	0	0	0	0	0	0	0	0	0	0	0	0
AC1_ 1	10	53	61	11	20	74	35	76	4	40	8	7	98
AC1_2	28	83	100	101	40	100	79	97	89	89	64	89	101
AC2_ 1	3	3	29	28	23	30	3	5	13	72	36	25	10
AC2_8	45	15	41	45	63	81	72	18	93	95	48	30	30
AC2_3	59	40	84	81	85	92	85	57	98	100	52	42	55
AC2_4	77	58	93	91	105	103	89	92	102	105	74	104	81
TMCC 1	13	25	4	36	10	7	49	31	16	5	78	34	23
TMCC 2	50	63	7	48	28	25	61	39	30	10	82	48	37
TMCC 3	70	73	17	55	44	47	96	47	37	21	85	54	51
TMCC 4	83	80	51	59	47	60	99	65	74	44	98	70	68
TMCC 5	87	93	71	86	54	87	104	72	83	61	102	101	105

TABLE 5-11 AC and TMCC Assignment in Coherent Modulation

Segment no.	11	9	7	5	3	1	0	2	4	6	8	10	12
AC1_ 1	10	53	61	11	20	74	35	76	4	40	8	7	98
AC1_ 2	28	83	100	101	40	100	79	97	89	89	64	89	101
TMCC 1	70	25	17	86	44	47	49	31	83	61	85	101	23

5.4.3.6 OFDM Modulation The total number of carriers in 6 MHz RF bandwidth is selectable from 1405, 2809, or 5617 in an ISDB-T system. All data subcarriers in DVB-T are uniformly using the same constellation. Due to the segmented structure and hierarchy partition in ISDB-T, different modulation modes can be applied to carriers within each of 13 OFDM segments, i.e., QPSK (DQPSK), 16QAM, and 64QAM can coexist. Meanwhile, the pilots are added to each OFDM segment to form the OFDM segments with pilots. See Table 5-12 for the specific parameters. In ISDB-T, the output data are segmented and multiplexed to 13 OFDM data segments, which are zero padded and followed by IFFT to achieve OFDM modulation. Then the guard interval is added between IFFT output data to prevent the interference between OFDM symbols from multipath transmission; the guard interval can be 1/4, 1/8, 1/16, and 1/32 of the IFFT length, exactly the same as that in the DVB-T system.

The formed transmission spectrum is shown in Figure 5-25. The 13 effective OFDM segments in the figure are divided into three modes. The partial reception layer uses segment number 0, and the differential modulation and coherent modulation layers are configured according to the OFDM segment number. In addition, half a segment is reserved on both sides of 13 segments as a guard band, so that there are always 14 segments occupying 6 MHz of the entire channel bandwidth, and thus each OFDM segment has a bandwidth of 429 kHz, and 13 effective segments occupy the bandwidth of about 5.6 MHz. A continuous pilot (rightmost spectrum line in the figure) is placed on the high end of the spectrum for receiver synchronization.

5.4.3.7 Modulation of the Pilot Signal DBPSK modulation (differential binary phase-shift-keying modulation) is applied to pilot subcarriers:

1. *Scattered Pilot SP.* SP is generated by modulating the W_i produced by a PRBS, where the subscript i corresponds to the carrier number i. The PRBS generator polynomial is $g(x)=x^{11}+x^9+1$. The initial value for each register in the corresponding PRBS generator circuit has a different definition for different modes and segments. The modulated complex pilot signal is derived by the resulting W_i mapping. In ISDB-T and DVB-T, different modulated data symbols are normalized so that the average transmitting power of different modulation symbols is 1. The pilot symbols here are sent using a higher rate of 4/3: $W_i = 1$, corresponding (I, Q) is $(-4/3,0)$; $W_i = 0$, corresponding (I, Q) is $(+4/3,0)$.

TABLE 5-12 ISDB-T Segment Parameters (MHz)

Mode	1		2		3	
Bandwidth			3000/7 = 428.57 ··· kHz			
Spacing between carrier frequencies	250/63 = 3.968 ··· kHz		125/63 = 1.9841 ··· kHz		125/126 = 0.99206 ··· kHz	
			Number of carriers			
Total count	108	108	216	216	432	432
Data	96	96	192	192	384	384
SP*1	9	0	18	0	36	0
CP*1	0	1	0	1	0	1
TMCC*2	1	5	2	10	4	20
AC1*3	2	2	4	4	8	8
AC2*3	0	4	0	9	0	19
Carrier modulation scheme	QPSK 16QAM 64QAM	DQPSK	QPSK 16QAM 64QAM	DQPSK	QPSK 16QAM 64QAM	DQPSK
Symbols per frame			204			
Effective symbol length	252 μs		504 μs		1008 μs	
Guard interval	63 μs (1/4), 31.5 μs (1/8), 15.75 μs (1/16), 7.875 μs (1/32)		126 μs (1/4), 63 μs (1/8), 31.5 μs (1/16), 15.75 μs (1/32)		252 μs (1/4), 126 μs (1/8), 63 μs (1/16), 31.5 μs (1/32)	
Frame length	64.26 ms (1/4), 57.834 ms (1/8), 54.621 ms (1/16), 53.0145 ms (1/32)		128.52 ms (1/4), 115.668 ms (1/8), 109.242 ms (1/16), 106.029 ms (1/32)		257.04 ms (1/4), 231.336 ms (1/8), 218.484 ms (1/16), 212.058 ms (1/32)	
IFFT sampling frequency			512/63 = 8.12698 ··· MHz			

Segment No.11	Segment No.9	Segment No.7	Segment No.5	Segment No.3	Segment No.1	Segment No.0	Segment No.2	Segment No.4	Segment No.6	Segment No.8	Segment No.10	Segment No.12
Coherent modulation segment	Coherent modulation segment	Coherent modulation segment	Differential modulation segment	Differential modulation segment	Differential modulation segment	Partial reception segment	Differential modulation segment	Differential modulation segment	Differential modulation segment	Coherent modulation segment	Coherent modulation segment	Coherent modulation segment

FIGURE 5-25 Examples of segment number and segment assignment in transmission spectrum.

2. *Continuous Pilot CP.* Like the SP, the CP is generated by modulating W_i. The difference is that the phase of a continuous pilot is constant in all OFDM symbols.
3. *Transmission and Multiplexing Configuration TMCC.* A differential modulation is applied to the TMCC. The first bit B_0 of TMCC data is taken as a reference bit. The $B'_1 - B'_{203}$ are derived by means of differential encoding for TMCC information data B_1–B_{203}:

$$B'_0 = W_i \qquad \text{(initialization bit for DBPSK modulation)}$$
$$B'_k = B'_{k-1} \oplus B_k \qquad (k = 1, 2, \ldots, 203)$$

Differential coded bits $B'_i = 0, 1$ are mapped to complex values of (4/3, 0), (−4/3,0).
4. *Auxiliary Channel AC.* The modulation mode of AC data is the same as that of TMCC. In the case of no auxiliary information, each B_k is filled with 1.

5.4.4 TMCC Information

Similar to the TPS in the DVB-T, the TMCC is used to tell the receiving terminal how to interpret the data received, which includes the following information: indicate whether the transmission system is narrowband transmission or broadband transmission; indicate that transmission parameters will be changed in several frames; indicate how many OFDM subbands are used in the transmission system; indicate whether the intermediate OFDM subband is used for partial reception; and indicate the service-level configuration in full-band receiving, including layers, information coding of each layer, and carrier modulation parameters.

TMCC includes 204 bits (B_0–B_{203}), defined as follows: 1 initialization bit (bit B_0); 16 sync bits (bits B_1–B_{16}); 3 segment description bits (bits B_{17}–B_{19}); 102 TMCC information bits (bits B_{20}–B_{121}); and 82 redundant check bits (bits B_{122}–B_{203}).

The first initialization bit (bit B_0) is the initial bit of the differential modulation, and the amplitude and phase are determined by W_i. Sixteen sync bits are filled with w_0 and w_1 in sequence (bitwise negation to w_0) with the frame, where $w_0 = [0011010111101110]$ and $w_1 = [1100101000010001]$. Three segment description bits should be set to 111 in the differential modulation segment and 000 in the coherent modulation segment. To protect the TMCC information, the CDSC (184,102) code is applied to 102 TMCC information bits (bits B_{20}–B_{121}) for encoding; the code is the shortened code of source code CDSC (273,191). Of 102 TMCC information bits 87 bits are currently used, and the remaining 15 bits are reserved for future extensions.

By selecting different modulation, inner code rate, and guard interval, the system uses data rates provided by 13 OFDM segments, as shown in Table 5-13. For a data rate provided by an OFDM segment, the corresponding data rate is simply divided by 13.

TABLE 5-13 Data Rate of ISDB-T System

Carrier Modulation	Convolutional Code	Number of TSPs Transmitted (Modes 1–3)	Data Rate (kbps) Guard Ratio: 1/4	Guard Ratio: 1/8	Guard Ratio: 1/16	Guard Ratio: 1/32
DQPSK, QPSK	1/2	12/24/48	280.85	312.06	330.42	340.43
	2/3	16/32/64	374.47	416.08	440.56	453.91
	3/4	18/36/72	421.28	468.09	495.63	510.65
	5/6	20/40/80	468.09	520.10	550.70	567.39
	7/8	21/42/84	491.50	546.11	578.23	595.76
16QAM	1/2	24/48/96	561.71	624.13	660.84	680.87
	2/3	32/64/128	748.95	832.17	881.12	907.82
	3/4	36/72/144	842.57	936.19	991.26	1021.30
	5/6	40/80/160	936.19	1040.21	1101.40	1134.78
	7/8	42/84/168	983.00	1092.22	1156.47	1191.52
64QAM	1/2	36/72/144	842.57	936.19	991.26	1021.30
	2/3	48/96/192	1123.43	1248.26	1321.68	1361.74
	3/4	54/108/216	1263.86	1404.29	1486.90	1531.95
	5/6	60/120/240	1404.29	1560.32	1652.11	1702.17
	7/8	63/126/252	1474.50	1638.34	1734.71	1787.28

5.5 INTRODUCTION TO DTMB STANDARD

In August 2006, the Chinese DTTB standard DTMB (Digital Terrestrial/Television Multimedia Broadcasting) was approved as the national mandatory standard with the label GB20600-2006. The full name of this standard is "Framing Structure, Channel Coding and Modulation for Digital Television Terrestrial Broadcasting System" [4,10]. There was an one-year probationary period before being implemented officially in August 2007. The DTMB standard has the features of fast acquisition and robust synchronization, high spectrum efficiency, good mobile performance, large broadcast coverage, easy multiservice broadcasting and so on.

For the first time, time-domain synchronous orthogonal frequency division multiplexing (TDS-OFDM) digital transmission was introduced. With the time–frequency combination processing capability of TDS-OFDM [6,7,8,9], the system can achieve fast synchronization and channel estimation. A superframe structure of date and time synchronization is also proposed, and each 500 μs data could have a unique address, enabling the support of multimedia broadcasting and power saving. The key for DTMB lies in the well-designed frame header and body structure, good error correction capability, flexible singnal processing in time and frequency domains, etc.

1. *TDS-OFDM technology:* The greatest difficulty for the DTTB system lies in the frequency-selective fading caused by the multipath. The OFDM technique

has a unique advantage versus frequency-selective fading. However, to maintain the orthogonality between subcarriers, the synchronization requirement is very strict. The TDS-OFDM technique can achieve fast signal capture and robust synchronous tracking simply and conveniently through both time- and frequency-domain processing and facilitate power saving and portable reception.

2. *PN padding technology for guard interval:* DTMB takes the PN sequence as a guard interval to avoid the interblock interference and also service as the training sequence for the channel synchronization as well as estimation. Therefore, the spectrum efficiency of the system is improved by 10%.
3. The frame header of TDS-OFDM is a self-protected unique PN sequence with strong ability to adapt to the fast moving and combat multipath interference. The multilevel header joint processing is also allowed in the time domain to adapt to a complex interference environment caused by large coverage, improving the robustness of synchronization.
4. *Cascaded error correction code structure using BCH as outer code and LDPC as inner code:* This helps significantly improve the C/N threshold of the DTMB standard. Because of this low C/N, DTMB will introduce less interference to existing analog TV services.
5. *Frame structure synchronized to absolute time:* The structure synchronous to the calender day facilitates the automatic waking function and other functions. As PN sequence could be different for each OFDM symbol, each OFDM block can be uniquely found, and this helps reduce the power consumption.
6. *Single- and multi-carrier integration:* The selection of subcarrier parameters under the TDS-OFDM frame structure is done under the same system platform with identical frame structure, scrambling, error correction coding, system clock, time-domain interleaving, modulation, and signal bandwidth.

The block diagram of a transmitter for the DTMB system is shown in Figure 5-26. The input data stream goes through a scrambler (to perform the data randomization), FEC coding, constellation mapping, and time-domain interleaving to form the basic data blocks with length of 3744 data symbols. The basic data block and the system information of 36 symbols are combined (multiplexed) when passing through the module of the frame body data processing to form the frame body. The time-domain

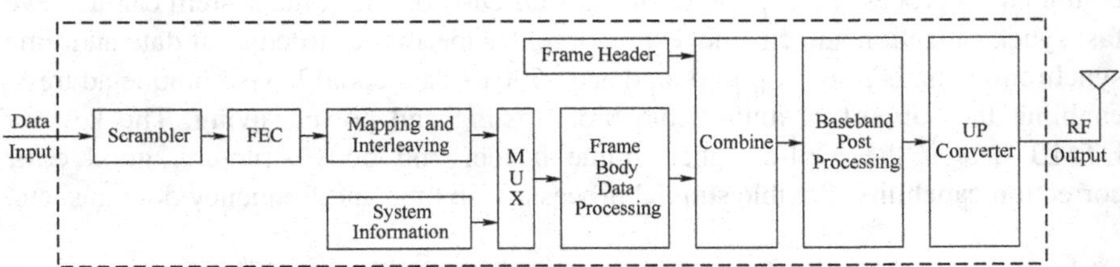

FIGURE 5-26 Block diagram of DTMB transmitter.

signal of the frame body and the corresponding frame header (i.e., PN sequence or PN sequence with cyclic expansion) are combined (multiplexed) to form the signal frame which the basic building block of the frame structure of the DTMB signal, and is then sent to the baseband postprocessing module for pulse shaping to minimize the interference to adjacent channels. The postprocessed signal will finally be converted to the RF signal by the frequency upconversion and be ready for transmission from the TV tower.

5.5.1 Major System Parameters

The major parameters of the DTMB system are listed in Table 5-14.

TABLE 5-14 Major Parameters of DTMB System for 8-MHz Channel Spacing

Item	Definition	Explanation
Data		
Data format	MPEG-2 interface	188 bytes/packet
Payload	4.8–32.4 Mbps	8 MHz bandwidth
Outer code	BCH (762,752)	
Inner code	LDPC (7488,3008/4512/6016)	Three code rates of around 0.4, 0.6, and 0.8
Time interleaving	Convolutional interleaving with interleaving width of 52	Mode 1: depth = 240 symbols Mode 2: depth = 720 symbols
Frequency interleaving	FFT block interleaving	Block size = 3780
TPS		
TPS pilot number	36 symbols	36 bits
TPS modulation mode	4QAM with same I and Q components	
Carrier mode	Multicarrier/single carrier	IDFT parameter: $C = 3780$ for multicarrier mode, $C = 1$ for single-carrier mode
Frame body duration	500 μs	For 8-MHz-bandwidth systems
Guard interval (PN sequence insertion)	420, 595, 945 symbols	1/9, 1/6, 1/4 of frame body
Constellation mapping	4QAM-NR 4QAM, 16QAM, 32QAM, or 64QAM	
Multicarrier parameters		
Carrier interval	2.0 kHz	
RF characteristics		
Effective bandwidth	7.56 MHz	
Channel shaping filter	Square-root raised-cosine filter	Roll-off factor $\alpha = 0.05$

5.5.2 Input Data Format

DTMB takes the MPEG-2 interface as the data input format and also leaves flexibility to accommodate other compression standards.

5.5.3 Scrambler

The input data stream needs to be scrambled to ensure the randomness of the transmitted data to facilitate the signal processing at the receiver. At the transmitter, the binary information sequence is first converted into a sequence with pseudorandom property by randomization, limiting the number of consecutive 0's or 1's. The reverse process to restore the original information sequence from this "scrambled" sequence must be done at the receiver, which is called "descrambling." The scrambling code in DTMB is a maximum-length pseudorandom binary sequence with generator polynomial

$$G(x) = 1 + x^{14} + x^{15} \tag{5-4}$$

The initial phase of the sequence is 100101010000000. Figure 5-27 shows the block diagram of the scrambler.

It should be noted that every bit of the input sequence is scrambled in DTMB while the MPEG-2 TS packet header is not scrambled in DVB-T.

5.5.4 FEC Coding

The forward error correction coding of the DTMB system is formed by cascading the outer code (BCH code) and inner code (LDPC code). Although LDPC code has the best overall coding efficiency and performance, it may suffer from the errorfloor problem with not enough length at low BER. That is, the BER of LDPC is not monotonically decreased with the increase of SNR but approaches to a very small constant. The BCH code is therefore used in DTMB for two purposes: (1) to lower the errorfloor down to 10^{-12}, meeting the requirement for HDTV reception, and (2) for the rate adaptation so that there is always an integer number of MPEG-2 packets within each signal frame regardless of the code rates and constellation mapping options.

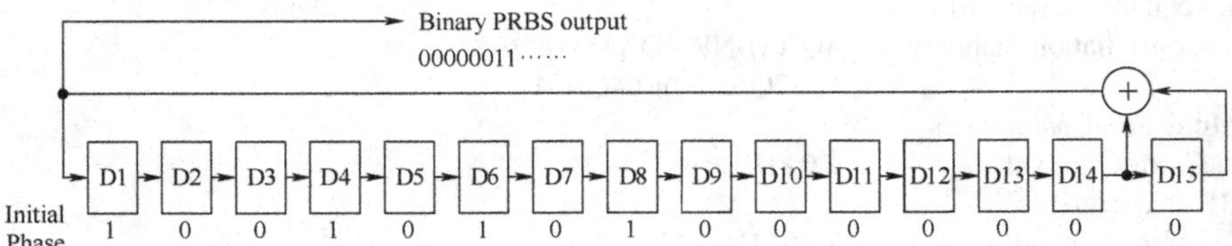

FIGURE 5-27 Scrambler in DTMB system.

752 information bits	10 encoded bits

FIGURE 5-28 Output sequence structure of BCH encoder.

The BCH outer code of DTMB is BCH (762,752), which is shortened from BCH (1023, 1013) systematic code with generator polynomial as

$$G_{\mathrm{BCH}}(x) = 1 + x^3 + x^{10} \tag{5-5}$$

To make up 1013 bits as the input to the BCH (1023, 1013) encoder, 261 bits of 0s are prefixed to the 752 bits of the scrambled data. BCH codes with 762 bits long can finally be obtained by deleting the first 261 bits of 0's from the output of the BCH (1023, 1013) encoder. One-bit burst error can be corrected by BCH code of (1023, 1013), and Figure 5-28 shows the sequence structure for the output from the BCH encoder. The generated 10 parity-check bits are attached to the input information.

DTMB has adopted three LDPC codes to provide different coding gain or error correction capability: LDPC (7493, 3048) code with equivalent coding rate $R = 0.4$, LDPC (7493, 4572) code with equivalent coding rate $R = 0.6$, and LDPC (7493, 6096) code with equivalent coding rate $R = 0.8$.

The generator matrix and parity-check matrix of LDPC codes used in DTMB are given in the Appendix of GB 20600-2006 [4]. The structure of the generator matrix G_{qc} for LDPC codes is

$$G_{qc} = \begin{bmatrix} G_{0,0} & G_{0,1} & \cdots & G_{0,c-1} & I & O & \cdots & O \\ G_{1,0} & G_{1,1} & \cdots & G_{1,c-1} & O & I & \cdots & O \\ \vdots & \vdots & G_{i,j} & \vdots & \vdots & \vdots & \ddots & \vdots \\ G_{k-1,0} & G_{k-1,1} & \cdots & G_{k-1,c-1} & O & O & \cdots & I \end{bmatrix} \tag{5-6}$$

where I is a $b \times b$ unity matrix, O is the $b \times b$ zero matrix, $G_{i,j}$ is a $b \times \mathrm{b}$ cyclic matrix, with $0 \le \mathrm{i} \le \mathrm{k}-1$, $0 \le \mathrm{j} \le \mathrm{c}-1$.

The parameter definitions are described as follows:

1. *FEC (7488, 3048) Code with Coding Rate ~0.4.* Four packets of 752-bit-long information sequences are sent to the BCH (762, 752) outer encoder and an encoded bit sequence 3048-bits long can be obtained. This outer encoder output is then sent to the LDPC (7493, 3048) encoder, and the 7488 encoded bits from the LDPC encoder can be obtained by removing the first 5 parity-check bits of the output from the LDPC (7493, 3048) encoder. The parameters of generator matrix G_{qc} of LDPC (7493, 3048) codes are $k = 24$, $c = 35$, and $b = 127$.
2. *FEC (7488, 4512) Code with Coding Rate ~0.6.* Six packets of 752 bit-long information sequences are sent to the BCH (762, 752) outer encoder and the encoded bit sequence 4572 bit-long can be obtained. This outer encoder output

is then sent to the LDPC (7493, 4572) encoder, and the 7488 encoded bits from the LDPC encoder can be obtained by removing the first 5 parity-check bits of the output from the LDPC (7493, 4572) encoder. The parameters of generator matrix G_{qc} of LDPC (7493, 4572) codes are $k=36$, $c=23$, and $b=127$.

3. *FEC (7488, 6016) Code with Coding Rate ~0.8.* Eight packets of 752 bit-long information sequences are sent to the BCH (762, 752) outer encoder and the encoded bit sequence 6096 bits long can be obtained. This outer encoder output is then sent to the LDPC (7493, 6096) encoder, and the 7488 encoded bits from the LDPC encoder can be obtained by removing the first 5 parity-check bits of the output from the LDPC (7493, 6096) encoder. The parameters of generator matrix G_{qc} of LDPC (7493, 6096) codes are $k=48$, $c=11$, and $b=127$.

The matrix $G_{i,j}$ given in Appendix A of GB 20600-2006 only provides the first row. Each row of $G_{i,j}$ is a 1-bit cyclically right-shifted version of the previous row, and the first row is a 1-bit cyclically right-shifted version of the last row. Each column of the matrix is constructed by cyclically down-shifting 1 bit from its left column, and the first column is the 1-bit cyclically down-shifted version of the last column of the matrix.

The codeword structure after the LDPC encoder is the information bits followed by the parity check bits. The relationship of the input information sequence, BCH parity check bits, and LDPC parity check bits after the FEC module with code rate ~0.4 is shown in Figure 5-29.

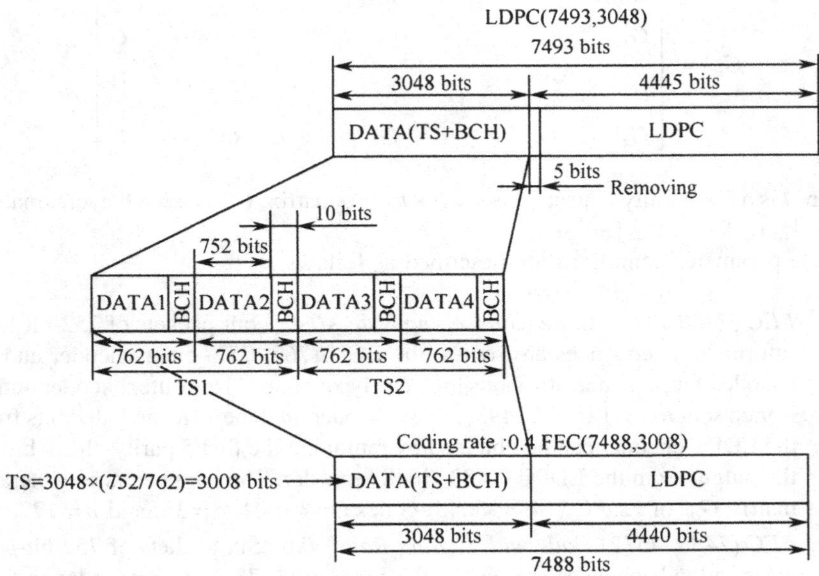

FIGURE 5-29 Output sequence structure from FEC encoder.

5.5.5 Constellation Mapping

DTMB supports constellation mapping of 64QAM, 32QAM, 16QAM, 4QAM, and 4QAM-NR (a combination of 4QAM with Nordstrom–Robinson code of rate 1/2). The appropriate power normalized factors are applied to different mapping schemes to keep the average output power roughly the same.

The bit sequence from the FEC module needs to be converted into a uniform nQAM (n is the number of constellation points) data symbol stream (the first bit of the FEC codeword serves as the LSB of the symbol word) before further processing. With this arrangement, each data symbol can carry more information bits, achieving higher spectrum efficiency.

The DTMB constellation mapping includes 64QAM, 32QAM, 4QAM, 16QAM, and 4QAM-NR in the following:

1. *64QAM Mapping.* For 64QAM, each data symbol carries 6 bits. The encoded information bits from the FEC coding are divided into groups of 6 bits ($b_5\ b_4\ b_3\ b_2\ b_1\ b_0$) which consist of in-phase component $I = b_2\ b_1\ b_0$ and quadrature component $Q = b_5\ b_4\ b_3$ for the constellation mapping symbol. The values of I and Q corresponding to coordinates of constellation points could be −7, −5, −3, −1, 1, 3, 5, and 7, respectively, as shown in Figure 5-30.
2. *32QAM Mapping.* For 32QAM, each data symbol carries 5 bits. The encoded information bits from the FEC coding are divided into groups of 5 bits ($b_4\ b_3\ b_2\ b_1\ b_0$). The values of I and Q corresponding to coordinates of constellation points could be −7.5, −4.5, −1.5, 1.5, 4.5, and 7.5, respectively, as shown in Figure 5-31.
3. *16QAM Mapping.* For 16QAM, each data symbol carries 4 bits. The encoded information bits from the FEC coding are divided into groups of 4 bits ($b_3\ b_2\ b_1\ b_0$) which consist of in-phase component $I = b_1\ b_0$ and quadrature component $Q = b_3\ b_2$ for the constellation mapping symbol. The values of I and Q corresponding to coordinates of constellation points could be −6, −2, 2, and 6, respectively, as shown in Figure 5-32.
4. *4QAM Mapping.* For 4QAM, each data symbol carries 2 bits. The encoded information bits from the FEC coding are divided into groups of 2 bits ($b_1\ b_0$) which consist of in-phase component $I = b_0$ and quadrature component $Q = b_1$ for the constellation mapping symbol. The values of I and Q corresponding to coordinates of constellation points could be either −4.5 or 4.5, respectively, as shown in Figure 5-33.
5. *4QAM-NR Mapping.* The 4QAM-NR is realized by applying NR encoding to the FEC output before 4QAM mapping. The detailed procedure is that an FEC encoded information sequence will first go through bitwise convolutional interleaving, similar to the time-domain interleaving method to be introduced later, and every 8-bit encoded bits will be converted into 16 bits through NR coding. Every 2 NR encoded bits are then sent to the 4QAM mapper. Using the relationship in the following with additions and multiplications all modulo-2

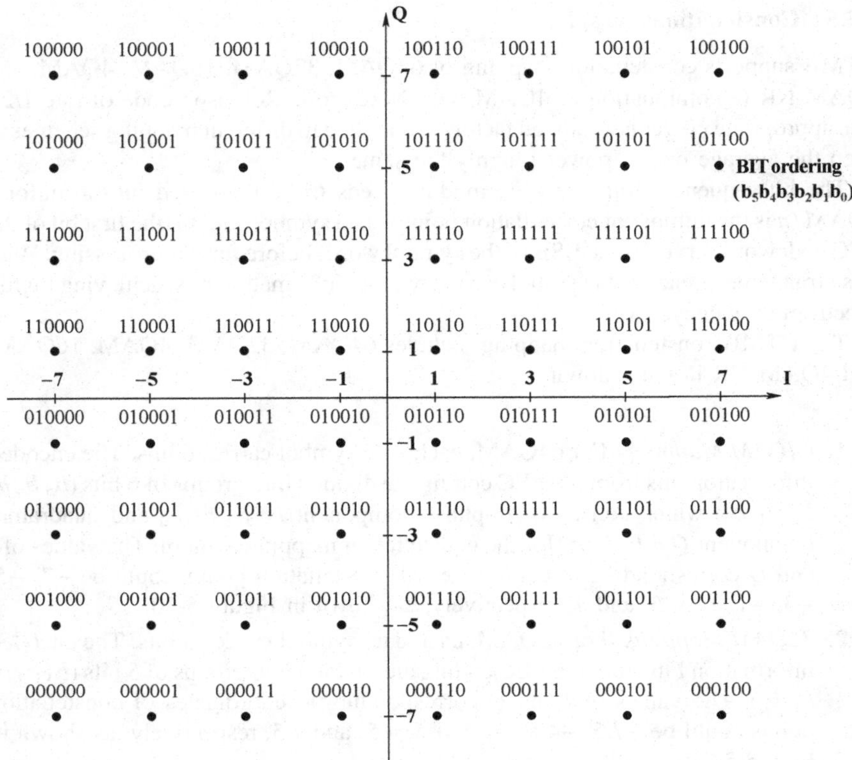

FIGURE 5-30 64QAM mapping.

operations, every 8 FEC encoded bits of $x_0\,x_1\,x_2\,x_3\,x_4\,x_5\,x_6\,x_7$ are converted into 16 bits of $x_0\,x_1\,x_2\,x_3\,x_4\,x_5\,x_6\,x_7\,y_0\,y_1\,y_2\,y_3\,y_4\,y_5\,y_6\,y_7$ (where $y_0\,y_1\,y_2\,y_3\,y_4\,y_5\,y_6\,y_7$ are the binary derivative bits) by the NR encoder [4]:

$$\begin{aligned}
y_0 &= x_7 + x_6 + x_0 + x_1 + x_3 + (x_0 + x_4)(x_1 + x_2 + x_3 + x_5) + (x_1 + x_2)(x_3 + x_5)\\
y_1 &= x_7 + x_0 + x_1 + x_2 + x_4 + (x_1 + x_5)(x_2 + x_3 + x_4 + x_6) + (x_2 + x_3)(x_4 + x_6)\\
y_2 &= x_7 + x_1 + x_2 + x_3 + x_5 + (x_2 + x_6)(x_3 + x_4 + x_5 + x_0) + (x_3 + x_4)(x_5 + x_0)\\
y_3 &= x_7 + x_2 + x_3 + x_4 + x_6 + (x_3 + x_0)(x_4 + x_5 + x_6 + x_1) + (x_4 + x_5)(x_6 + x_1)\\
y_4 &= x_7 + x_3 + x_4 + x_5 + x_0 + (x_4 + x_1)(x_5 + x_6 + x_0 + x_2) + (x_5 + x_6)(x_0 + x_2)\\
y_5 &= x_7 + x_4 + x_5 + x_6 + x_1 + (x_5 + x_2)(x_6 + x_0 + x_1 + x_3) + (x_6 + x_0)(x_1 + x_3)\\
y_6 &= x_7 + x_5 + x_6 + x_0 + x_2 + (x_6 + x_3)(x_0 + x_1 + x_2 + x_4) + (x_0 + x_1)(x_2 + x_4)\\
y_7 &= x_0 + x_1 + x_2 + x_3 + x_4 + x_5 + x_6 + x_7 + y_0 + y_1 + y_2 + y_3 + y_4 + y_5 + y_6
\end{aligned} \tag{5-7}$$

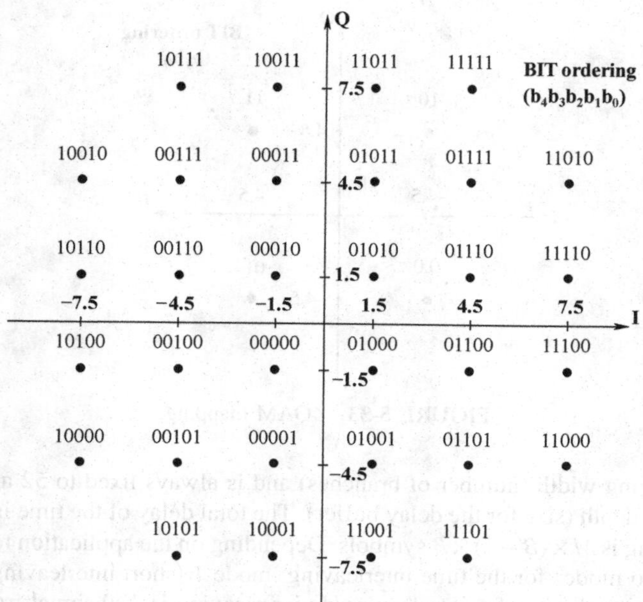

FIGURE 5-31 32QAM mapping.

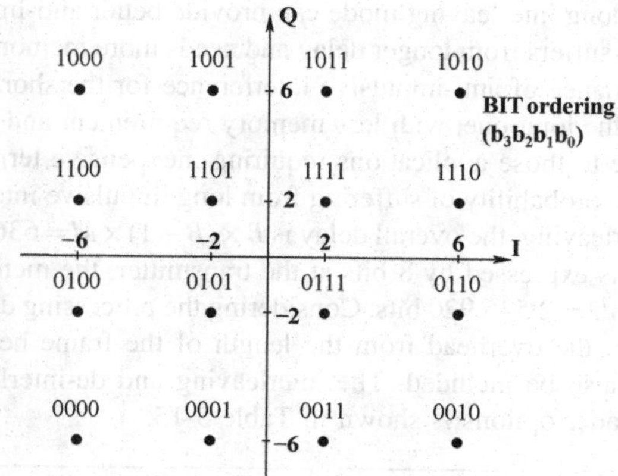

FIGURE 5-32 16QAM mapping.

5.5.6 Interleaving

DTMB uses time-domain interleaving to improve the system performance versus impulsive interference and time-selective fading. Time-domain interleaving is carried out among data symbols with the length of multiple consecutive signal frames. Symbol-based convolutional interleaving is shown in Figure 5-34, where B denotes

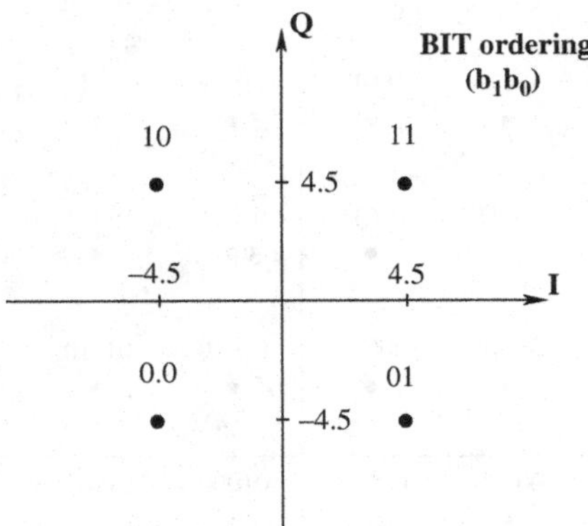

FIGURE 5-33 4QAM mapping.

the interleaving width (number of branches) and is always fixed to 52 and M is the interleaving depth (size for the delay buffer). The total delay of the time interleaving/de-interlacing is $M \times (B-1) \times B$ symbols. Depending on the application requirement, there are two modes for the time interleaving: mode 1 (short interleaving): $M=240$ symbols, the total delay for interleaving/de-interleaving is 170 signal frames; mode 2 (long interleaving): $M=720$ symbols, the total delay for interleaving/de-interleaving is 510 signal frames.

Obviously, the long interleaving mode can provide better anti-impulsive interference capability but suffers from longer delay and needs more memory resource. Even though the performance of anti-impulsive interference for the short interleaver will not be as good as the long one, with less memory requirement and short processing delay, it is suitable to those applications requiring inexpensive terminals and delay sensitivty while the probability of suffering from long impulsive interference is quite low. For short interleaving, the overall delay is $B \times (B-1) \times M = 636{,}480$ symbols. If each data symbol is expressed by 8 bits at the transmitter, the memory size will be $B \times (B-1) \times M \times 8/2 = 2{,}545{,}920$ bits. Considering the processing delay of interleaving/de-interleaving, the overhead from the length of the frame header and system information must also be included. The interleaving and de-interleaving delay for different frame header options is shown in Table 5-15.

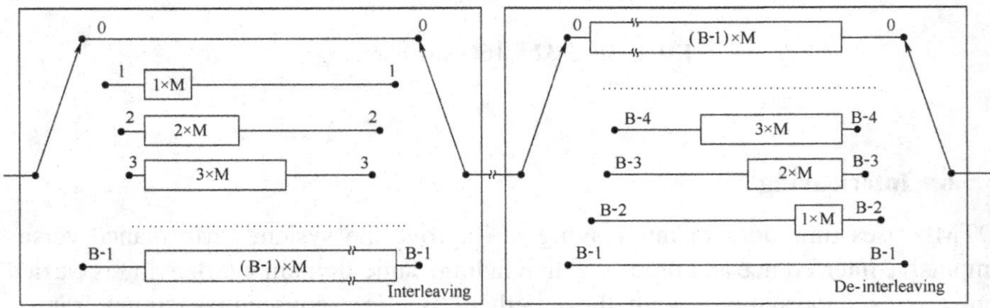

FIGURE 5-34 Convolutional interleaving among data symbols.

TABLE 5-15 Interleaving/De-Interleaving Processing Delay

Interleaving Depth M	Frame Header Mode	Interleaving Processing Delay (ms)
240	420	94.44
	595	98.38
	945	106.25
720	420	288.33
	595	295.14
	945	318.75

5.5.7 System Information

The system information of DTMB containing 36 data symbols provides parameters for each signal frame, including constellation mapping mode, LDPC code rate, interleaving mode, and single- or multicarrier mode. DTMB uses 6 system information bits (s_5 s_4 s_3 s_2 s_1 s_0) to indicate the 64 possible different system modes, and the spectrum spreading is applied for error protection during the transmission. Among the 6 system information bits, s_5 is the MSB (reserved), s_4 denotes interleaving mode with 0 for mode 1 (short interleaving mode) and 1 for mode 2 (long interleaving mode), and s_3 s_2 s_1 s_0 are used to indicate the code rate and constellation mapping mode, which are defined in Table 5-16.

The generic frame body structure is shown in Figure 5-35. The time-domain interleaving is not required for the system information symbols.

TABLE 5-16 System Information Definition for Bits $s_3s_2s_1s_0$

$s_3s_2s_1s_0$	Definition
0000	Head frame indicator of superframe with odd frame number
0001	4QAM, FEC code rate 1
0010	4QAM, FEC code rate 2
0011	4QAM, FEC code rate 3
0100	Reserved
0101	Reserved
0110	Reserved
0111	4QAM-NR, FEC code rate 3
1000	Reserved
1001	16QAM, FEC code rate 1
1010	16QAM, FEC code rate 2
1011	16QAM, FEC code rate 3
1100	32QAM, FEC code rate 3
1101	64QAM, FEC code rate 1
1110	64QAM, FEC code rate 2
1111	64QAM, FEC code rate 3

4 symbols to indicate whether single- or multi-carrier modulation is used	32 symbols to indicate the constellation mapping, FEC code rate, interleaving mode, and etc.	3744 data symbols

System Information (36 symbols) +Data (3744 data symbols)

FIGURE 5-35 Structure of frame body.

5.5.8 Signal Frame Structure

The DTMB signal has a hierarchical frame structure as shown in Figure 5-36 with the basic unit of the signal frame. In the following, each layer will be introduced taking a top-down approach:

1. *Calendar Day Frame.* It lasts exactly 24 h and consists of 1440-min frames. The calendar day frame is synchronous to the calendar day (starting from 00:00:00 AM local time).
2. *Minuteframe.* It lasts exactly 1 min with 480 superframes.
3. *Super Frame.* Every superframe lasts 125 ms. With this arrangement, it is easy to calibrate the time using the timing system (e.g., GPS). The first signal frame in the superframe is the superframe header. Depending on guard interval length, the superframe contains different signal frames. But there are always an integer number of MPEG-2 TS packets per second for DTMB system.
4. *Signal Frame.* Signal frame is the basic unit of the DTMB frame structure. A signal frame consists of two parts: the frame header (FH) and the frame body

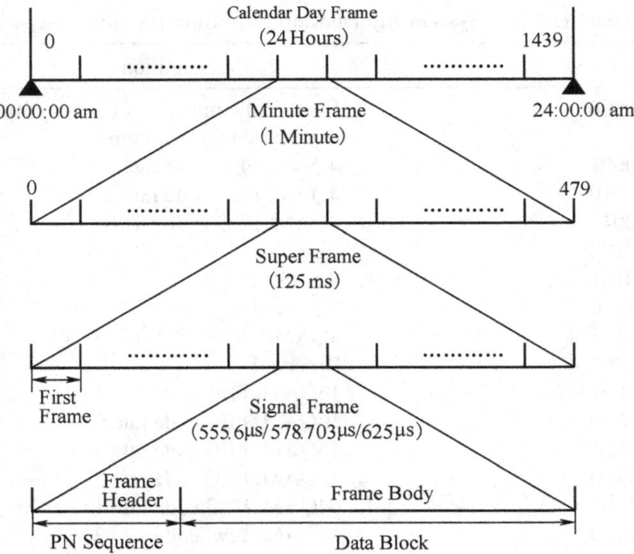

FIGURE 5-36 Hierarchical frame structure of DTMB signal.

Frame Header option 1 (420 symbols) (55.6μs)	Frame Body (System Information and data) (3780 symbols) (500μs)

(a) Signal Frame structure 1

Frame Header option 2 (595 symbols) (78.7μs)	Frame Body (System Information and data) (3780 symbols) (500μs)

(b) Signal Frame structure 2

Frame Header option 3 (945 symbols) (125μs)	Frame Body (System Information and data) (3780 symbols) (500μs)

(c) Signal Frame structure 3

FIGURE 5-37 Signal frame structure with different FH options.

(FB). Both FH and FB have the same baseband symbol rate of 7.56 MS/s. FH is formed by the PN sequence using 4QAM modulation with the same I and Q components. The duration of FB is fixed to 500 μs at the system bandwidth of 8 MHz. DTMB defines three options for the FH length and the corresponding signal frame structures to meet the different application requirements, as shown in Figure 5-37. For the signal frame structure with lengths of 4200, 4375, and 4725 symbols, respectively, each superframe can consist of 225, 216, and 200 signal frames, respectively.

5.5.9 Frame Header (FH)

DTMB provides 3 FH modes for different applications:

1. The sequence for PN420 is based on mapping of bit 0 to symbol +1 and bit 1 to symbol −1 to the cyclically extended eighth-order m sequence from LFSR. PN420 consists of a preamble of 82 symbols, an m sequence of 255-symbol-long PN255), and a postamble of 83 symbols as shown in Figure 5-38. The initial state of the LFSR determines the phase of the PN sequence. Each FH of

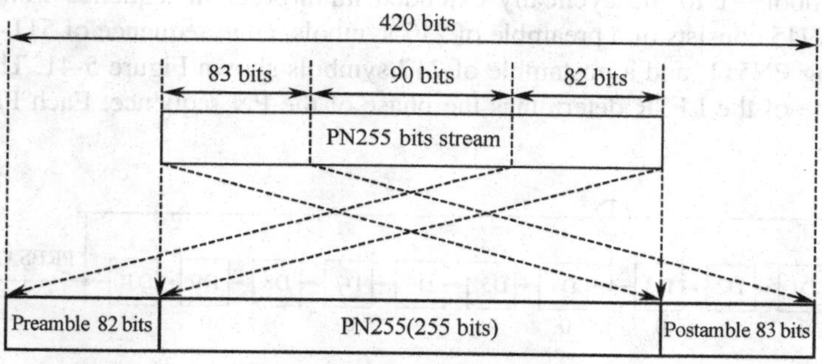

FIGURE 5-38 Schematic diagram of PN420 structure.

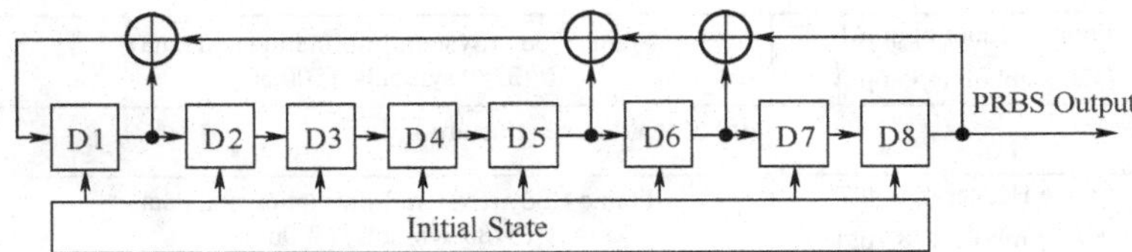

FIGURE 5-39 Eighth-order *m*-sequence generator.

the signal frame in a superframe uses either the identical PN or different PN sequences to identify each signal frame.

The generator polynomial of the LFSR that generates the PN255 sequence is defined as

$$G_{255}(x) = 1 + x + x^5 + x^6 + x^8 \tag{5-8}$$

The 255 PN 420 sequences can be generated using the LFSR in Figure 5-39. Based on the initial state of the LFSR, the 225 PN420 sequences with different phases can be obtained by computer searching to minimize the correlation between any two adjacent PN sequences in DTMB. The LFSR is reset to the initial phase at the beginning of each superframe. The average power for PN420 is twice as much that of the frame body signal.

2. The sequence for PN595 is generated by truncating the first 595 chips of the 1023-chip-long *m* sequence generator. The generator polynomial of this *m* sequence is

$$G_{1023}(x) = 1 + x^3 + x^{10} \tag{5-9}$$

The initial phase is 0000000001, which is reset at the beginning of each signal frame. Figure 5-40 gives the structure of this of the 10th-order LFSR.

The average signal power of the FH mode 2 is the same as that of the frame body signal.

3. The sequence for PN945 is based on mapping bit 0 to symbol +1 and bit 1 to symbol −1 to the cyclically extended ninth-order m-sequence from LFSR. PN945 consists of a preamble of 217 symbols, an *m* sequence of 511-symbol-long PN511, and a postamble of 217 symbols shown Figure 5-41. The initial state of the LFSR determines the phase of the PN sequence. Each FH of the

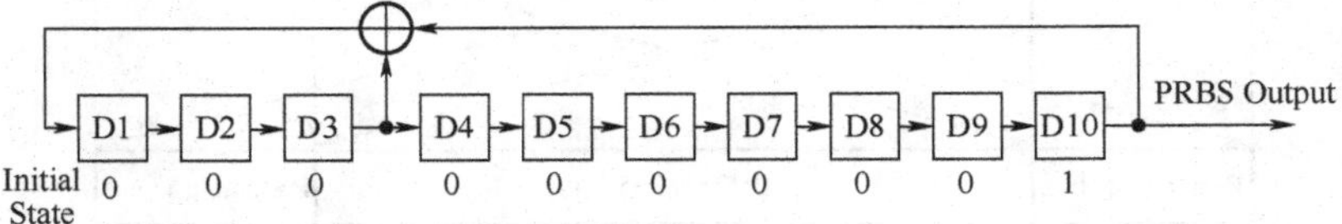

FIGURE 5-40 Tenth-order *m*-sequence generator.

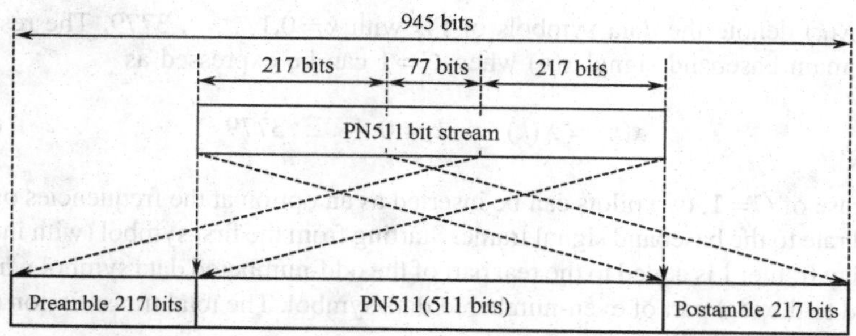

FIGURE 5-41 Schematic diagram of PN945 structure.

signal frame in a superframe uses either the identical PN or different PN sequences to identify each signal frame.The generator polynomial of the LFSR that generates the PN511 sequence is defined as

$$G_{511}(x) = 1 + x^2 + x^7 + x^8 + x^9 \tag{5-10}$$

The 511 PN 945 sequences can be generated using the LFSR in Figure 5-42. Based on the initial state of the LFSR, the 200 PN945 sequences with different phases can be obtained by computer searching to minimize the correlation between any two adjacent PN sequences in DTMB. The LFSR is reset to the initial phase at the beginning of each superframe. The average power for PN945 is twice as much as that of the frame body signal.

5.5.10 Frame Body Data Processing

The DTMB system provides two modulation options, $C = 1$ for single-carrier modulation and $C = 3780$ for multicarrier modulation, and the difference is whether the IDFT processing is performed and the IDFT is done for the multicarrier modulation mode. Except for the IDFT unit, all other functional blocks at the transmitter are exactly the same. At the receiver, the demodulator can automatically determine which modulation is used.

The data symbols after the time interleaver and the system information are multiplexed to form the frame body by the frame body data processing module and then FB will be modulated with the number of C subcarriers.

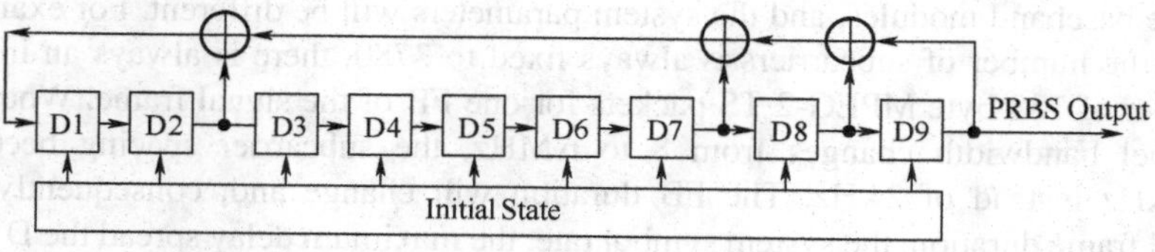

FIGURE 5-42 Ninth-order m sequence generator.

Let $X(k)$ denote the data symbols of FB with $k = 0, 1, \ldots, 3779$. The resulting time-domain baseband signal $x(n)$ when $C = 1$ can be expressed as

$$x(n) = X(k) \qquad k = 0, 1, \ldots, 3779 \tag{5-11}$$

In the case of $C = 1$, two pilots can be inserted as an option at the frequencies of ± 0.5 symbol rate to the baseband signal frame. Starting from the first symbol (with index 0) of the day frame, 1 is added to the real part of the odd-numbered data symbol while -1 is added to the real part of even-numbered data symbol. The total transmission power of the pilots is -16 dBc.

In the case of $C = 3780$, the subcarrier spacing is 2 kHz for the 8 MHz channel spacing. After frequency-domain interleaving on the data symbols $X(k)$, the time-domain signal $x(n)$ is given by the formula.

$$x(n) = \frac{1}{\sqrt{C}} \sum_{n=0}^{C-1} X(k) e^{j2\pi nk/C} \qquad (n = 0, 1, \ldots, 3779) \tag{5-12}$$

5.5.11 Baseband Signal Post Processing

The baseband pulse shaping is done by the following square-root raised-cosine (SRRC) filter with roll-off factor $\alpha = 0.05$:

$$H(f) = \begin{cases} 1 & (|f| \le f_N(1-\alpha)) \\ \left[\frac{1}{2} + \frac{1}{2}\cos\frac{\pi}{\alpha f_N}\left(\frac{|f| - f_N(1-\alpha)}{2}\right)\right]^{1/2} & (f_N(1-\alpha) < |f| \le f_N(1+\alpha)) \\ 0 & (|f| > f_N(1+\alpha)) \end{cases} \tag{5-13}$$

where $f_N = 1/(2T_s) = R_s/2$ is the Nyquist frequency with T_s the symbol period (1/7.56 μs in 8 MHz bandwidth) of input signal and $R_s = 1/T_s$ the symbol rate.

The frequency response of this SRRC is shown in Figure 5-43. For the implementation of the SRRC filter, a high-order FIR filter is needed.

5.5.12 RF Output Interface

In principle, DTMB can support different channel bandwidths without modification on the baseband modules, and the system parameters will be different. For example, since the number of subcarriers is always fixed to 3780, there is always an integer number of 188-byte MPEG-2 TS packets for one FB of the signal frame. When the channel bandwidth changes from 8 to 6 MHz, the subcarrier spacing becomes 1.75 kHz instead of 2 kHz. The FB duration will change and, consequently, the signal frame duration, the system symbol rate, the maximum delay spread the DTMB can handle, etc., eventually, the system data throughput will also be different.

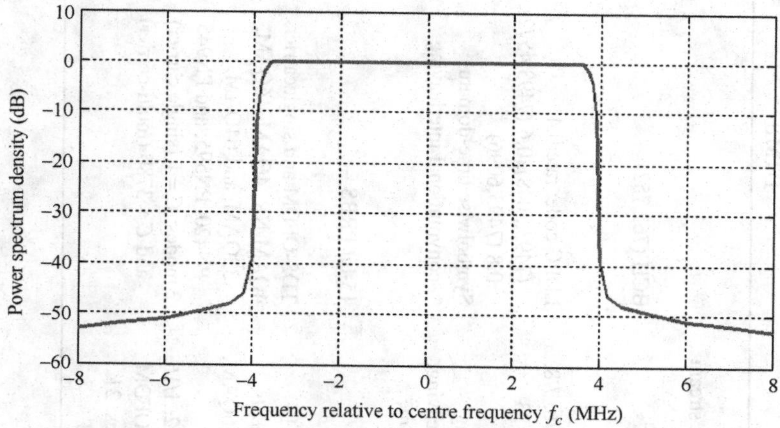

FIGURE 5-43 Frequency response of SRRC filter for DTMB system.

TABLE 5-17 Payload Data Throughput (Mbps)

		Signal Frame with 4200 Symbols	
FEC rate	0.4	0.6	0.8
Mapping			
4QAM-NR	—	—	**5.414**
4QAM	**5.414**	**8.122**	**10.829**
16QAM	**10.829**	**16.243**	**21.658**
32QAM	—	—	**27.072**
64QAM	**16.243**	**24.365**	**32.486**
		Signal Frame with 4375 Symbols	
FEC rate	0.4	0.6	0.8
Mapping			
4QAM-NR	—	—	**5.198**
4QAM	**5.198**	**7.797**	**10.396**
16QAM	**10.396**	**15.593**	**20.791**
32QAM	—	—	**25.989**
64QAM	**15.593**	**23.390**	**31.187**
		Signal Frame with 4725 Symbols	
FEC rate	0.4	0.6	0.8
Mapping			
4QAM-NR	—	—	**4.813**
4QAM	**4.813**	**7.219**	**9.626**
16QAM	**9.626**	**14.438**	**19.251**
32QAM	—	—	**24.064**
64QAM	**14.438**	**21.658**	**28.877**

Note: The mode with dash means this mode is not included in DTMB standard.

TABLE 5-18 Main Characteristics of Four Standards

Systems	ATSC	DVB-T	ISDB-T	DTMB
Interface				
Transport stream	ISO/IEC 13818-1 (MPEG-TS) transport stream			
Transmission system				
Outer coding	R-S (207,187, $t = 10$)	R-S (204, 188, $t = 8$)		BCH (762,752)
Outer interleaver	52 R-S block interleaver	12 R-S block interleaver		
Inner coding	Rate 2/3 trellis code	Punctured convolutional code: rate 1/2, 2/3, 3/4, 5/6, 7/8, constraint length 7, polynomials (octal) = 171, 133		LDPC code: rate 0.4 (7493,3048), 0.6 (7493,4572), 0.8 (7493,6096)
Inner interleaver	12 to 1 trellis code interleaver	Bitwise interleaving and frequency interleaving	Bitwise interleaving, frequency interleaving, and selectable time interleaving	Symbolwise time-domain convolutional interleaving
Data randomization	16-bit PRBS	15-bit PRBS	15-bit PRBS	15-bit PRBS
Modulation	8-VSB and 16-VSB	COFDM QPSK, 16QAM, and 64QAM Guard interval 1/32, 1/16, 1/8, and 1/4 of OFDM symbol; 2 modes: 2 k and 8 k FFT	BST-OFDM with 13 frequency segments DQPSK, QPSK, 16QAM, and 64QAM Guard interval 1/32, 1/16, 1/8, and 1/4 of OFDM symbol; 3 modes: 2 k, 4 k, and 8 k FFT	TDS-OFDM and singlecarrier 4QAM-NR, 4QAM, 16QAM, 32QAM, and 64QAM, PN420, PN595, and PN945 2 modes: $C = 1$ (single carrier) and $C = 3780$ (multi-carrier)

5.5.13 System Payload Data Throughput

The DMTB system, if used with channel bandwidth of 8 MHz, can support the payload data rate from 4.813 to 32.486 Mbps depending on different signal frame lengths, inner code rates, and constellation mapping modes. The payload data throughput is calculated as

$$\text{Payload data throughput} = \frac{3744}{\text{FH length} + 3780} \times R_i \times R_m \times 7.56\,\text{Mbps} \quad (5\text{-}14)$$

where FH length = 420/595/945 is the length of FH, R_i is the overall code rate, with values of 3008/7488 (~0.4), 4512/7488 (~0.6), 6016/7488 (~0.8), R_m is the modulation efficiency or the number of bits per data symbol, i.e., 2 (for 4QAM), 4 (for 16QAM), 5 (for 32QAM), and 6 (for 64QAM). According to (5-14), the payload data throughput of the DTMB system of different modes can be summarized as in Table 5-17.

5.6 SUMMARY

Digital terrestrial television broadcasting (DTTB) is an important part of broadcasting TV systems. DTTB not only overcomes the shortcomings of analog terrestrial TV, such as vulnerability to interference, poor image quality, and ghosting, but also greatly improves the spectrum efficiency. DTTB can also provide stable quality of signal reception for fixed, mobile, and portable terminals. In this chapter, the frame structure, channel coding, modulation technique, and operating mode of the four first-generation international digital terrestrial TV broadcasting standards were introduced in detail and the main characteristics of the four standards are summarized in Table 5-18.

REFERENCES

1. Advanced Television System Committee A/53, *ATSC Digital Television Standard*, Washington, DC: ATSC, 1995.
2. ETSI.300 744, *Digital Broadcasting Systems for Television, Sound and Data Services, Framing Structure, Channel Coding and Modulation for Digital Terrestrial Television*, European: Sophia-Antipolis, France, 1999.
3. ITU-R WP 11A/59, *Channel Coding, Frame Structure and Modulation Scheme for Terrestrial Integrated Service Digital Broadcasting (ISDB-T)*, Tokyo: ARIB, 1999.
4. Chinese National Standard GB 20600-2006, *Framing Structure, Channel Coding and Modulation for Digital Television Terrestrial Broadcasting System*, Aug. 2006.
5. Advanced Television Systems Committee A/54, *A Guide to Use of the ATSC Digital Television Standard, with Corrigendum No. 1*, Washington, DC: ATSC, 2006.
6. L. Yang, K. Gong, and Z. Yang, "Filling method of guard interval in orthogonal frequency division multiplexing system," China patent 01124144.6, Feb. 6, 2002.

7. L. Yang and Z. Yang,"Digital Terrestrial multimedia broadcasting system," China patent 00123597.4, Mar. 21, 2001.
8. L. Yang and Z. Yang, "Digital terrestrial multimedia/television broadcasting (DMB-T) transmission system," *Modern TV Technology*, no. 4, pp. 33–36, Apr. 2001.
9. L. Yang, Z. Yang, and Y. Wu, "A new terrestrial digital multimedia/television broadcasting transmission systems," *TV Technology*, no. 1, pp. 12–16, Jan. 2002.
10. J. Song et al., "Technical review on Chinese digital terrestrial television broadcasting standard and measurements on some working modes," *IEEE Transactions on Broadcasting*, vol. 53, no. 1, pp. 1–7, Mar. 2007.
11. W. Zhang, Y. Guan, W. Liang, D. He, F. Ju, and J. Sun, "An introduction of the Chinese DTTB standard and analysis of the PN595 working modes," *IEEE Transactions on Broadcasting*, vol. 53, no. 1, pp. 8–13, Mar. 2007.

6

SECOND-GENERATION DTTB STANDARDS

6.1 INTRODUCTION TO SECOND-GENERATION DIGITAL VIDEO BROADCASTING

The DVB-T2 system is the second-generation DVB-T system developed by the Digital Video Broadcasting (DVB) organization [1]. It offers efficient and reliable audio, video, and data transmission services for fixed, portable, and mobile devices using the latest modulation and coding techniques. The DVB-T2 is not originally designed to replace DVB-T in the short term, and both will coexist in the market for services of different needs in the foreseeable future. The DVB organization has defined a series of business requirements as the development framework for DVB-T2 (T2 for short also used throughout this chapter). These requirements can be summarized as follows:

1. T2 transmission must be able to use existing domestic receive antennas and be compatible with existing transmitting facilities (this requirement limits the use of the MIMO technique, which would involve both new receiving and transmitting antenna).
2. T2 should primarily target services to fixed and portable receivers.
3. T2 should provide a minimum of 30% capacity increase over DVB-T working in the same planning constraints and conditions.

Digital Terrestrial Television Broadcasting: Technology and System, First Edition. Edited by Jian Song, Zhixing Yang, and Jun Wang.

4. T2 should provide improved single-frequency network (SFN) performance compared with DVB-T.
5. T2 should have a mechanism to provide service-specific robustness; namely, T2 should be able to provide different levels of robustness for different services. For example, within a single 8-MHz channel, T2 should be able to provide some services for the roof antenna and other services for portable receivers.
6. T2 should provide for a good bandwidth and frequency flexibility.
7. There should be a mechanism defined, if possible, to reduce the peak-to-average-power ratio of the transmitted signal in order to reduce transmission loss.

Like the DVB-T standard, DVB-T2 also uses OFDM technology to pass signals via multiple subcarriers. Meanwhile, DVB-T2 is a flexible standard with a number of different modes. DVB-T2 uses the same error-correcting codes as DVB-S2, i.e., cascaded LDPC and BCH codes, with excellent performance in noise and interference resistance. Further, combined with the characteristics of terrestrial broadcasting channels, the new bit interleaving and constellation mapping techniques are introduced into DVB-T2 based on the guard interval COFDM technique used for DVB-T.

Table 6-1 compares DVB-T2 and DVB-T in terms of the forward error correction code, modulation mode, guard interval mode, FFT size, scattered and continuous pilots, etc.

In addition to the items listed in the table, DVB-T2 also provides some new features, including the following:

1. T2 frame structure containing a header with special (short) identifying symbol, which can be used for fast channel scanning and signal capture, and carries some basic system parameters
2. Constellation rotation, a form of signal space diversity used to improve receiving performance for high-rate coding mode
3. Expertise for reducing the peak-to-average-power ratio of transmitted signal
4. Future expansion frame (FEF), one type of frame to expand the signal, the nonstandardized part ignored by the first-generation receiver, which could ensure that the next generation would be compatible with future upgrades

TABLE 6-1 Comparison of DVB-T2 and DVB-T

	DVB-T	DVB-T2
FEC code	Convolutional code + RS code 1/2, 2/3, 3/4, 5/6, 7/8	LDPC + BCH 1/2, 3/5, 2/3, 3/4, 4/5, 5/6
Modulation mode	QPSK, 16QAM, 64QAM	QPSK, 16QAM, 64QAM, 256QAM
Guard interval	1/4, 1/8, 1/16, 1/32	1/4, 19/256, 1/8, 19/128, 1/16, 1/32, 1/128
FFT size	2 k, 8 k	1 K, 2 K, 4 K, 8 K, 16 K, 32 K
Scattered pilots	8% of the total carriers	1%, 2%, 4%, 8% of the total carriers
Continuous pilots	2.6% of total carriers	0.35% of total carriers

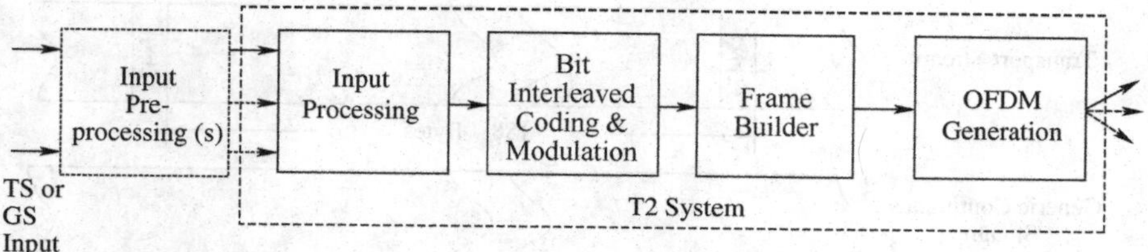

FIGURE 6-1 Top-level model of DVB-T2 system.

6.1.1 System Structure

The top-level model of the DVB-T2 system is shown in Figure 6-1 [2]. The DVB-T2 system process can be divided into input processing, bit-interleaved coding and modulation, framing, and OFDM symbol generation modules. The specific content of the modules will be given below.

The whole system input can be one or more MPEG transport stream(s) and (or) generic stream(s). The input preprocessor is not part of the DVB-T2 system, which includes a service splitter, or demultiplexer for transport stream for separating the services to be transmitted into the DVB-T2 system inputs, which are one or multiple logical data stream(s). Each logical data stream is carried by a physical layer pipe (PLP). The DVB-T2 system completes the transmission of multiple PLPs, and the preprocessor outputs correspond to the PLPs.

If there are multiple physical layer pipes (PLPs), data transmission will be flexibly time divided in the physical layer, and the corresponding parameter range is provided, thus allowing a trade-off between time diversity and power consumption reduction.

The multi-PLPs and time division technique of the DVB-T2 allow the different levels of coding, modulation, and time-domain interleaving depth to be applied to different PLPs, providing different robustness for each service. The receiver can also concentrate its decoding resources on one PLP containing the required data. Especially the limited buffer for time de-interleaving can support larger interleaving depth compared to the single PLP mode, since the de-interleaver is only processing the data corresponding to the required PLPs. With the single PLP, the time interleaving depth is around 70 ms, whereas with multiple PLPs this can be extended to the full-frame duration (150–250 ms) or for lower data rate services where it can be extended over multiple frames.

In the SISO transmission mode (single transmit antenna), the DVB-T2 physical layer output is the RF signal modulated on one RF channel, like DVB-T. In the MISO transmission mode (dual antennas), DVB-T2 uses the modified Alamouti coding [3], where the physical layer output can separate out the second output signal, which is transmitted by the second antenna.

6.1.2 Input Processing

In order to improve the system's multiservice applications, the input data format for each DVB-T2 PLP supports transport stream, generic continuous stream, generic fixed-length packetized stream, and generic stream encapsulated, as shown in

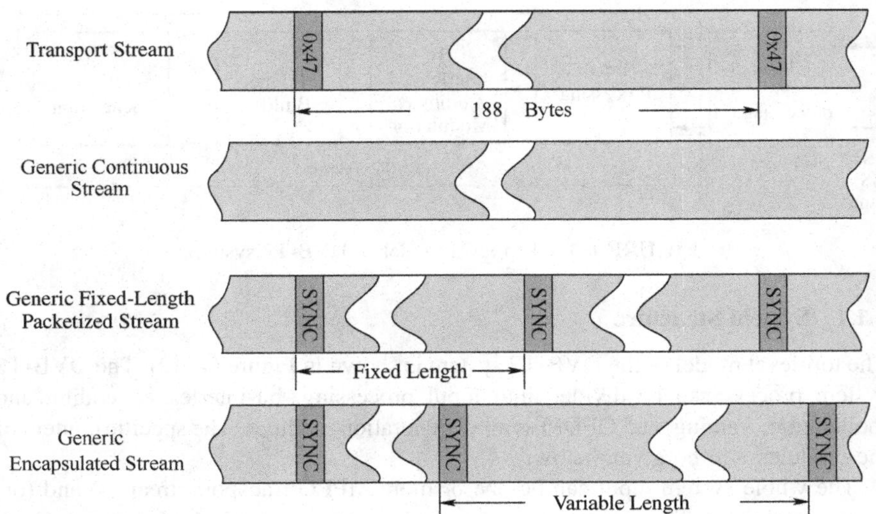

FIGURE 6-2 PLP input stream format.

Figure 6-2 and explained in the following, while the original DVB-T system only supports the transport stream format.

1. *Transport Stream (TS).* Shall be characterized by user packets (UP) of fixed length O-UPL = 188 bits (one MPEG packet), the first byte being a 0x47. It shall be signaled in the BBHEADER TS/GS field. In the case of transport stream, the data packet rate is a constant value, some of which correspond to the service data packet and the other part to the filled empty packet.
2. *Generic Continuous Flow (GCS, Variable-Length Packet Stream Where Modulator Is Not Aware of Packet Boundaries).* Shall be characterized by a continuous bit stream and shall be signaled in the BBHEADER by TS/GS field and UPL = 0_D. A variable-length packet stream where the modulator is not aware of the packet boundaries or a constant-length packet stream exceeding 64 kbits shall be treated as a GCS and shall be signaled in the BBHEADER by TS/GS field as a GCS and UPL = 0_D.
3. *Generic Fixed-Length Packetized Stream (GFPS).* The purpose of this format is compatible with DVB-S2, which may be replaced by GSE in the future. It shall be a stream of constant-length UPs, with length O-UPL bits (maximum O-UPL value 64 K), and shall be signaled in the baseband header TS/GS field. O-UPL is the original user packet length. UPL is the transmitted user packet length, as signaled in the BBHEADER TS/GS field.
4. *Generic Stream Encapsulated (GSE).* Shall be characterized by variable-length packets or constant-length packets, as signaled within GSE packet headers and shall be signaled in the BBHEADER by TS/GS field.

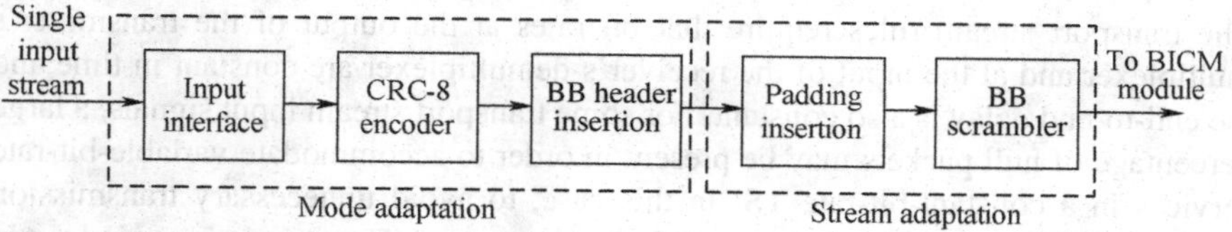

FIGURE 6-3 Block diagram of input processing module in input mode A (i.e. single PLP).

The input to the DVB-T2 system shall consist of one or more logical data streams. One logical data stream is carried by one PLP. Each PLP employs a static configured modulation encoding mode, and the transmission rate remains constant after being set. The input pattern A processing mode is used for the single PLP, while the input mode B processing mode is used for multiple PLP. The specific structures for input processing are shown in Figures 6-3 and 6-4. The input processing comprises a mode adaptation module and a stream adaptation module. The mode adaptation modules, which operate separately on the contents of each PLP, slice the input data stream into data fields and insert a baseband header at the start of each data field, which, after stream adaptation, will form baseband (BB) frames that will be further forwarded to the channel coding module.

The baseband frame header comprises the data stream type and processing mode signaling, which support two modes, the normal mode (NM) and the high-efficiency mode (HEM), where the NM is compatible with DVB-T mode adaptation, whereas in HEM, further stream-specific optimizations may be performed to reduce signaling overhead. The mode adaptation module also includes three optional submodules, providing the functions of input streams synchronization, null packet deletion, and CRC-8 encoder. The data processing in the DVB-T2 modulator may produce variable transmission delay on the user information. The input stream synchronization module (optional) provides suitable means to guarantee constant-bit-rate (CBR) and constant end-to-end transmission delay for any input data format, and this process will also allow synchronization of multiple input streams traveling in independent PLPs, since the reference clock and the counter of the input stream synchronizers are the same.

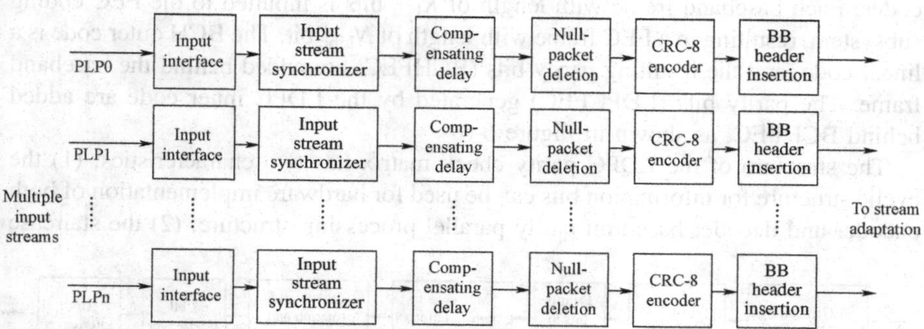

FIGURE 6-4 Block diagram of input processing module in input mode B (i.e., multiple PLPs).

The transport stream rules require that bit rates at the output of the transmitter's multiplexer and at the input of the receiver's demultiplexer are constant in time and the end-to-end delay is also constant. For some transport stream input signals, a large percentage of null packets may be present in order to accommodate variable-bit-rate services in a constant-bit-rate TS. In this case, to avoid unnecessary transmission overhead, TS null packets will be identified and removed. The process is carried out in a way that the removed null packets can be reinserted in the receiver in the exact place where they were originally, thus guaranteeing a constant bit rate and avoiding the need for time-stamp updating. The user packets without sync byte are encoded by the CRC-8 encoder to yield 8 parity bits which are added to the end of the user packet for transmission together. CRC parity bits are used for error detection at the user packet level for only the normal mode and transport stream packet input format at the receiving end.

6.1.3 Bit-Interleaved Coding and Modulation

The baseband frame formed by input processing enters into the bit-interleaved coding and modulation module. The DVB-T2 bit-interleaved coding and modulation scheme is shown in Figure 6-5.

DVB-T2 has LDPC codes of two lengths (standard block code length $N_{\text{ldpc}} =$ 64,800, short block code length $N_{\text{ldpc}} = 16{,}200$). There are six different code rates: 1/2, 3/5, 2/3, 3/4, 4/5, and 5/6 code rates. In addition to the bit interleaver and demultiplexer designed for 16QAM/64QAM/256QAM, the parity column winding interleaver is a new structure which can improve system performance in a terrestrial environment, such as 0-dB echo. Performance of short-block-code words is worse than standard-block-code words, which can be used for low-bit-rate transmission applications at a short delay.

6.1.3.1 FEC Encoding and Constellation Mapping The FEC encoding part includes outer coding (BCH), inner coding (LDPC), and bit interleaving. The input stream is formed by the baseband frames, and the output stream is formed by the FEC frames. The data packets of the input stream need to be adapted to the length of LDPC codes. Each baseband frame with length of K_{bch} bits is inputted to the FEC coding subsystem, resulting in a FEC frame with length of N_{ldpc} bit. The BCH outer code is a linear code, and the resulting parity bits (BCHFEC) are added behind the baseband frame. The parity bits (LDPCFEC) generated by the LDPC inner code are added behind BCHFEC, as shown in Figure 6-6.

The structure of the LDPC parity check matrix has two characteristics: (1) the cyclic structure for information bits can be used for hardware implementation of both encoder and decoder based on partly parallel processing structure; (2) the staircase

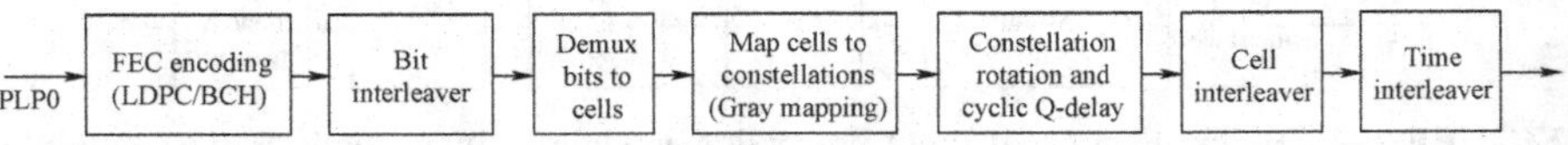

FIGURE 6-5 Block diagram of bit-interleaved coding and modulation scheme.

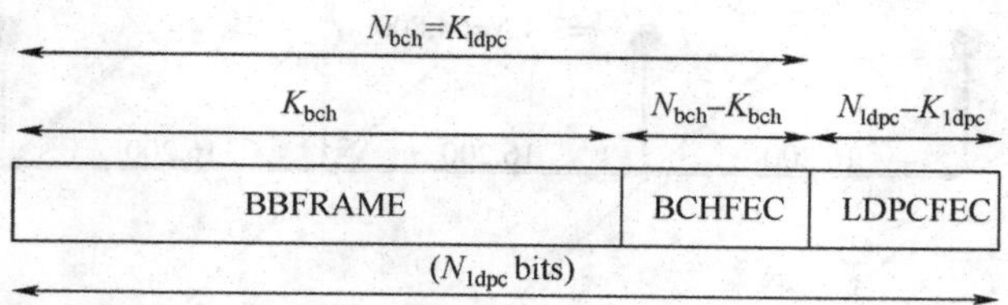

FIGURE 6-6 Data format before bit comparison.

structure for parity bits can be used to generate parity bits by an accumulator. The LDPC codes in DVB-T2 are irregular LDPC codes, with different protection levels for each code bit, depending on the column weight of check matrix. The protection level among bits in a multilevel constellation symbol is not uniform. The performance of LDPC codes with multilevel constellation depends highly on the correspondence between code bits and constellation bits, and both the bit interleaver and the demultiplexer are therefore required.

The bit interleaver is a block interleaver applied to each LDPC code word. In DVB-T2, each 2^m-QAM constellation corresponds to the bit interleaver having $N_c = 2m$ columns (except for 256QAM with a short code which uses a bit interleaver with $m = 8$ columns). The LDPC encoded bits are written columnwise and read out rowwise. The demux part demultiplexes $2m$ bits in the same row to create two constellation symbols (except for the 256QAM), as shown in Figure 6-7.

The interleaver module has introduced a new feature: the parity and column twist (PCT) interleaver. The check matrix of LDPC codes has the diagonal structure depicted in Figure 6-8 after parity interleaving. If a block interleaver is directly applied, there could be many constellation symbols having multiple code bits

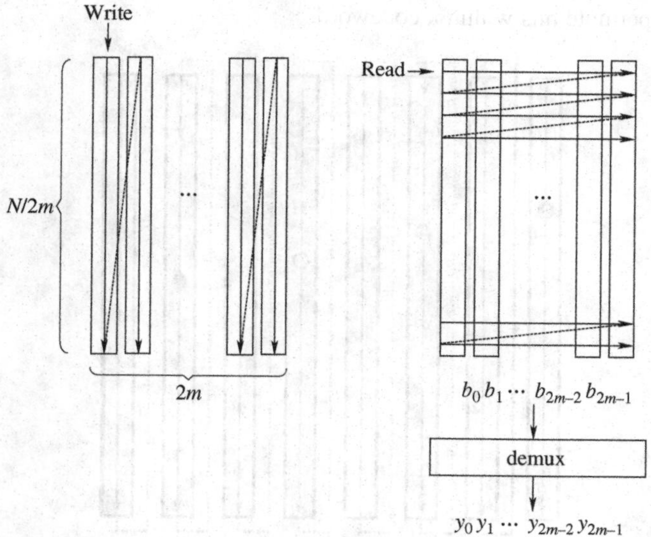

FIGURE 6-7 Bit interleaver.

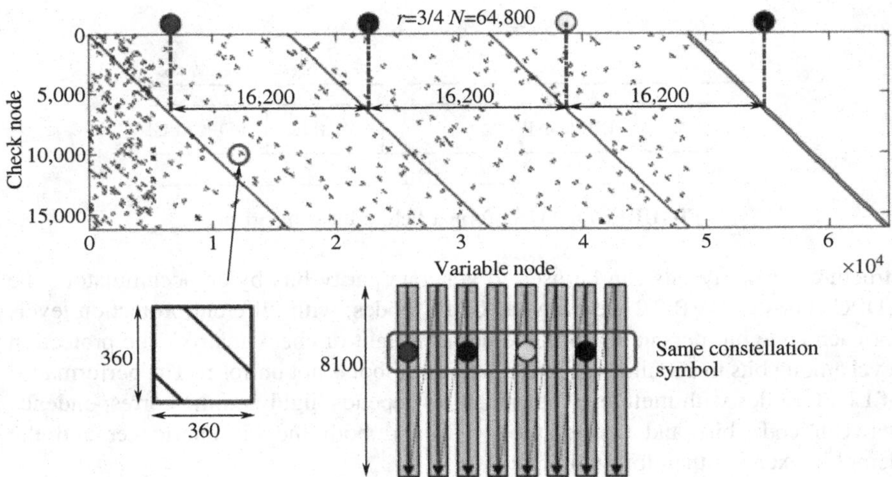

FIGURE 6-8 Check matrix of rate 3/4 and bit interleaver.

connected to the same check node (as illustrated in the figure), causing performance degradation under the erasure channel.

To avoid performance degradation, the parity interleaver interleaves the parity bits in such a way that the parity part of the check matrix has the same structure as the information part, as shown in Figure 6-8. The starting position of the written-in data in each column is different so that the code bits connected to the same check node will not be included in the same constellation symbol (see the difference between Figure 6-8 and Figure 6-9). The two interleavers can be taken as a single interleaver because both interleavers permute bits within a codeword.

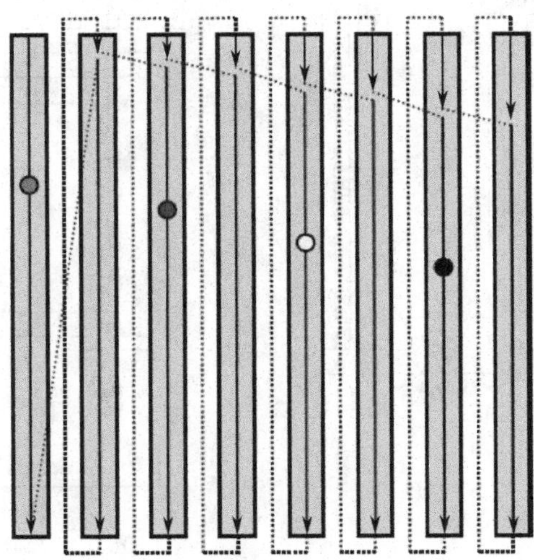

FIGURE 6-9 Column twist bit interleaver.

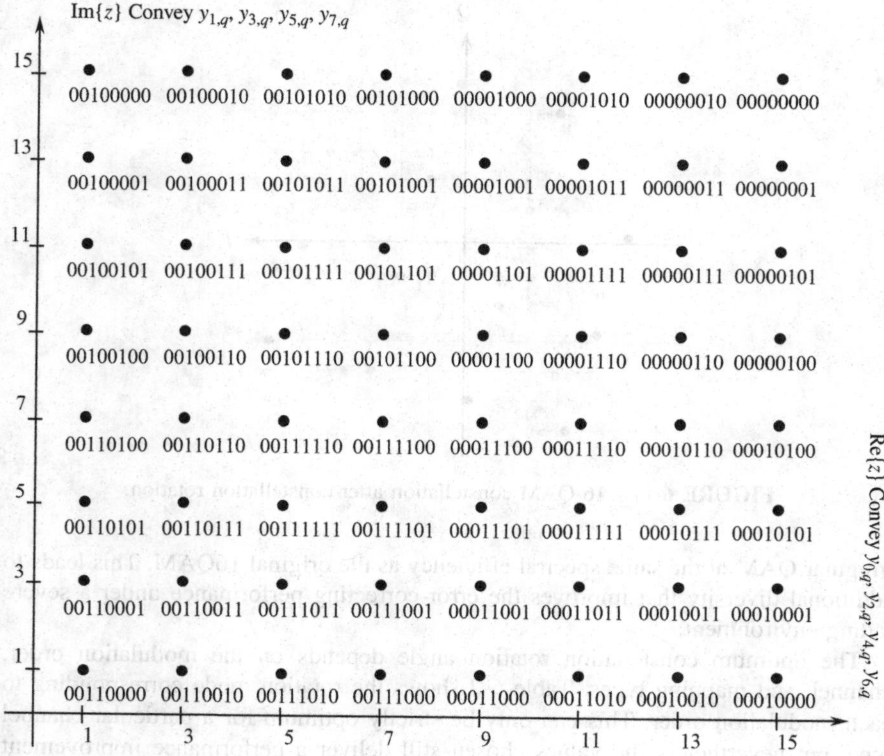

FIGURE 6-10 256QAM constellation mapping.

The demultiplexer is used to optimize the performance for each QAM constellation mapping in an AWGN channel. For QPSK modulation, no bit interleaver is used because simulations show no performance improvement over the AWGN channel and no significant performance improvement over the erasure channel. Other than the constellation mapping of QPSK, 16QAM, 64QAM as used for DVB-T, Gray mapping of 256QAM is added in DVB-T2. The constellation points within the first quadrant are shown in Figure 6-10 for simplicity.

6.1.3.2 Constellation Rotation Constellation rotation can be carried out after the mapping, which comprises rotation to the QAM constellation followed by component axis interleaving. After constellation rotation, the projections of the constellation points will have more values along each axis. For 16QAM, instead of having $2^{m/2} = 4$ projections on each axis for the original constellation points, the constellation could now have $2^m = 16$ projections, as shown in Figure 6-11. Before the cell interleaving and time-domain interleaving, axis (or I/Q) interleaving is achieved by the cyclic delay. The resulting "virtual" constellation after rotation and cyclic delay in the case of a 16QAM are shown in Figure 6-12. It is equivalent to sending a high-order

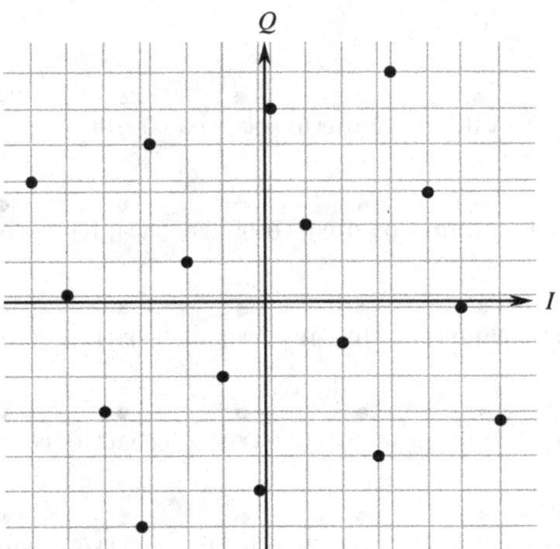

FIGURE 6-11 16-QAM constellation after constellation rotation.

irregular QAM at the same spectral efficiency as the original 16QAM. This leads to additional diversity that improves the error-correcting performance under a severe fading environment.

The optimum constellation rotation angle depends on the modulation order, channel, and mapping types. Table 6-2 shows the rotation mode corresponding to each modulation order. This can only be strictly optimum for a particular channel type, but nevertheless the values chosen still deliver a performance improvement (compared with nonrotated constellations) for all encountered channel models

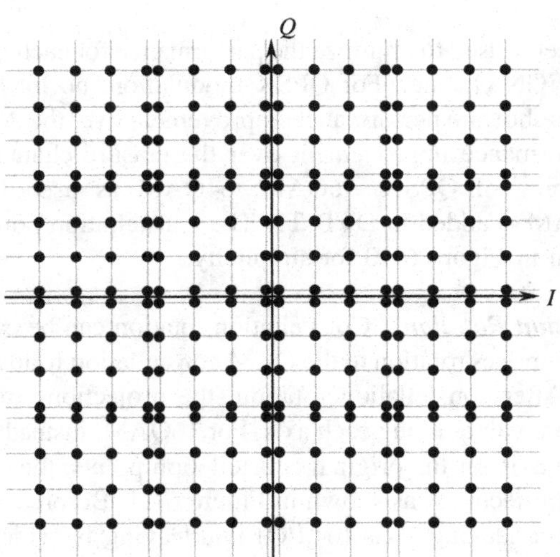

FIGURE 6-12 16-QAM constellation after constellation rotation and cyclic delay.

TABLE 6-2 Rotation Angle Corresponding to Each Constellation Type

Constellation	QPSK	16QAM	64QAM	256QAM
Rotated degree	29.0	16.8	8.6	Atan(1/16)

ranging from the classical fading channel (Rayleigh) to the severe fading channel (Rayleigh with erasures).

6.1.3.3 Interleaving Further interleaving is carried out to constellation points after constellation rotation, including a cell interleaver and a time interleaver.

The cell interleaver is a pseudorandom permutation of the cells in the FEC codeword, different for each FEC codeword of one time interleaver block, to ensure the uncorrelated distribution of channel distortions and interference for each cell corresponding to the FEC codeword at the receiver. The memory requirement of the cell interleaver depends on the time interleaver, and thus efficient implementation of the time interleaver leads to memory efficiency improvement in the cell interleaver.

The time interleaver operates at the PLP level. The time interleaving parameters may be different for different PLPs in a T2 system. The FEC blocks from the cell interleaver for each PLP will be grouped into interleaving frames (which are mapped onto one or more T2 frames). Each interleaving frame will contain a dynamically variable whole number of FEC blocks.

Each interleaving frame is either mapped directly onto one T2 frame or spread out over several T2 frames. Each interleaving frame is also divided into one or more TI blocks. The TI blocks within an interleaving frame can contain a slightly different number of FEC blocks. Therefore, there are three modes for the time interleaving for each PLP: (1) Each interleaving frame contains a TI block mapped directly to a T2 frame, as shown in Figure 6-13. (2) Each interleaving frame contains a TI block and is mapped to multiple T2 frames. As shown in Figure 6-14, one interleaving frame is mapped to two T2 frames. This gives greater time diversity for low-data-rate services. (3) Each interleaving frame is mapped to a T2 frame, and each frame is divided into several TI blocks, as shown in Figure 6-15. Each TI block may use up to the full TI memory, thus increasing the maximum bit rate for a PLP.

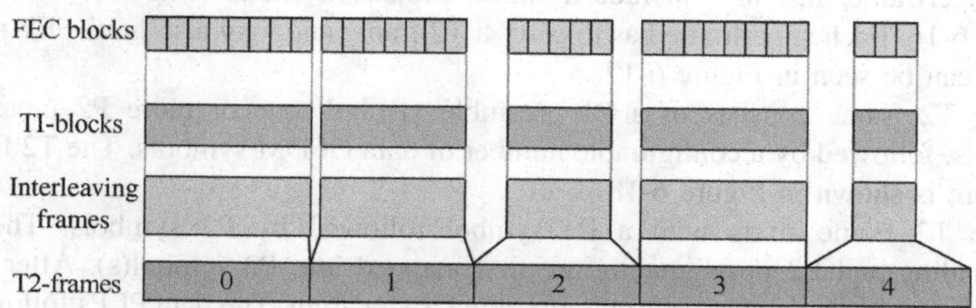

FIGURE 6-13 Time interleaving mode 1.

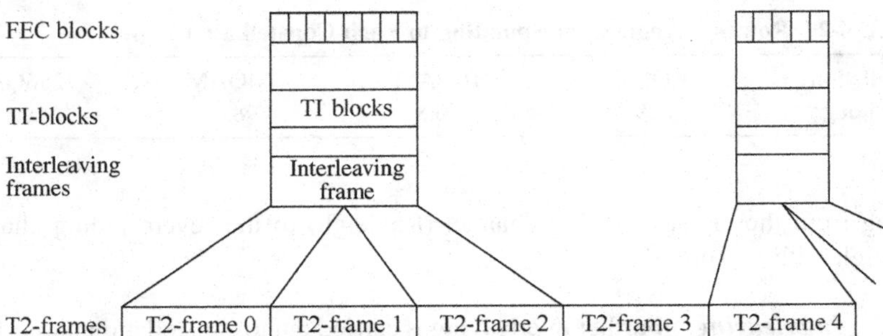

FIGURE 6-14 Time interleaving mode 2.

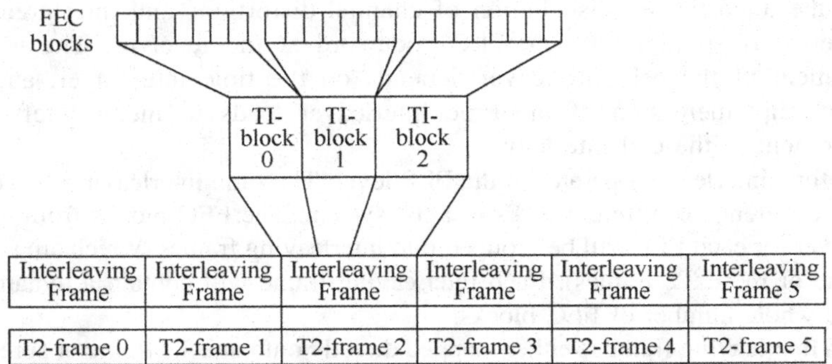

FIGURE 6-15 Time interleaving mode 3.

6.1.4 Frame Builder

The function of the frame builder is to assemble the cells produced by the time interleaver for each PLP and the cells of the modulated L1 signaling data into arrays of active OFDM cells corresponding to each of the OFDM symbols to form the overall frame structure. The frame builder operates according to the dynamic information produced by the scheduler and the configuration of the frame structure.

At the top level, the frame structure consists of the superframe, which is divided into multiple T2 frames, and these T2 frames are further divided into OFDM symbols. The superframe may also include a future expansion frame (FEF), as shown in Figure 6-16. Each superframe has to carry a T2 frame and may also have FEF parts, which can be seen in Figure 6-17.

The T2 frame consists of a P1 preamble symbol, one or more P2 preamble symbols, followed by a configurable number of data OFDM symbols. The T2 frame structure is shown in Figure 6-18.

The T2 frame starts with a P1 symbol followed by P2 symbols. The L1 presignaling and L2 postsignaling are first mapped into P2 symbol(s). After that, the common PLPs are mapped right after the L1 signaling. The data PLPs follow the common PLPs starting with type 1 PLPs. The type-2 PLPs follow the type-1 PLPs.

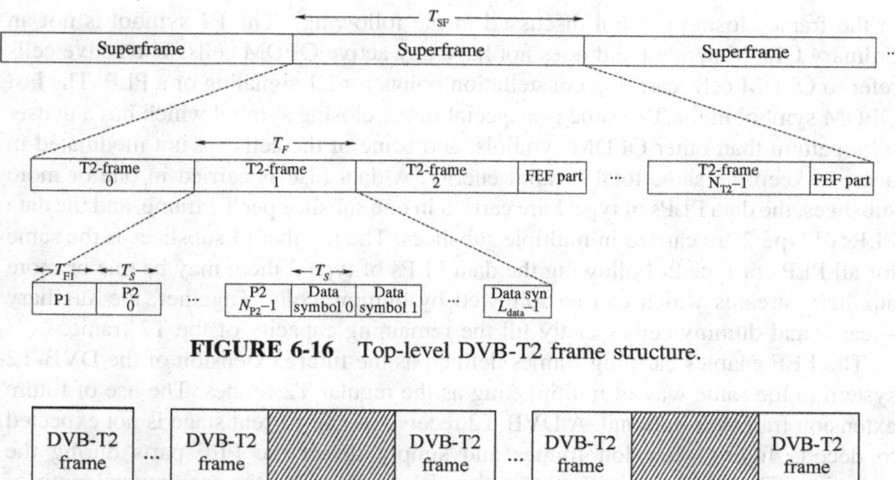

FIGURE 6-16 Top-level DVB-T2 frame structure.

FIGURE 6-17 Frame structure for each superframe.

The auxiliary stream or streams, if any, follow the type 2 PLPs, and this can be followed by dummy cells. Therefore, the PLPs, auxiliary streams, and dummy data cells will exactly fill the remaining cells in the frame.

The number of P2 symbols N_{P2} in the T2 frames is determined by the FFT size, whereas the number of data symbols L_{data} in the T2 frame is a configurable parameter signaled in the L1 presignaling. Therefore, the total number of OFDM symbols in a T2 frame (excluding P1 symbols) is given by $L_F = N_{P2} + L_{data}$. The duration of T2 frame is therefore given by

$$T_F = L_F \times T_s + T_{P1} \tag{6-1}$$

where T_s is the duration of a complete OFDM symbol, T_P is the duration of P1 symbol, and the maximum value for the frame duration T_F is 250 ms.

The frame builder will map the cells from both the time interleaver (for the PLPs) and the constellation mapper (for the L1 pre- and L1 postsignaling) onto the data cells of each OFDM symbol in each frame (the data cells are the cells in the OFDM symbols, not used for pilots or tone reservation, and may even be an unmodulated cell

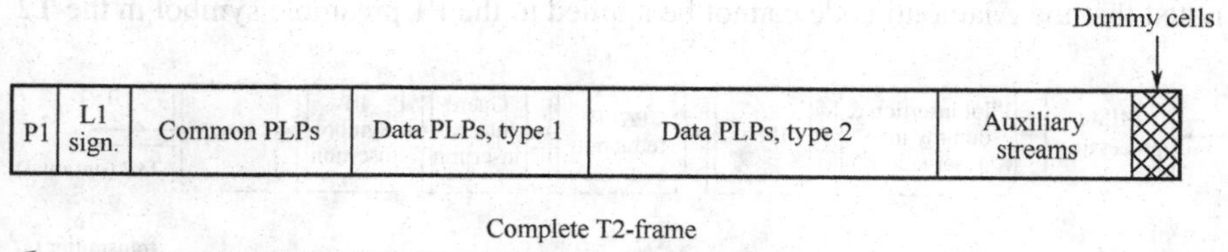

FIGURE 6-18 Frame structure of T2 frame.

in the frame closing symbol discussed in the following). The P1 symbol is not an ordinary OFDM symbol and does not have any active OFDM cells. The active cells refer to OFDM cells carrying constellation points for L1 signaling or a PLP. The last OFDM symbol in the T2 frame is a special frame closing symbol which has a denser pilot pattern than other OFDM symbols, and some of the cells are not modulated in order to keep the same total symbol energy. A data PLP is carried in one or more subslices, the data PLPs of type 1 are carried in one subslice per T2 frame, and the data PLPs of type 2 are carried in multiple subslices. The number of subslices is the same for all PLPs of type 2. Following the data PLPs of type 2 there may be one or more auxiliary streams which can be followed by dummy cells. Together, the auxiliary streams and dummy cells exactly fill the remaining capacity of the T2 frame.

The FEF enables carrying frames defined as the future extension of the DVB-T2 system in the same way of multiplexing as the regular T2 frames. The use of future extension frames is optional. A DVB-T2 receiver at the current stage is not expected to decode future extension frames and simply detect the FEF parts during the reception. The FEF part will start with a P1 symbol and the maximum length of FEF is 250 ms.

After the framing process is completed, the frequency interleaving will be done on the data cells of each OFDM symbol by mapping the data cells from the frame builder onto the N_{data} available data carriers.

6.1.5 OFDM Symbol Generation

The function of the OFDM generation module is to insert pilots into the cells by the frame builder for the OFDM modulation and generate the time-domain baseband data signal for the transmission. It then inserts the guard intervals and, if needed, applies PAPR reduction processing to create the completed T2 signal. Another optional initial stage, known as MISO processing, allows the initial frequency domain signal to be processed by a modified Alamouti encoding, which splits the T2 signal into two groups and then transmits at the same frequency in such a way that the two groups will not interfere with each other but instead help enhance the reception performance. The block diagram of the OFDM generation module is shown in Figure 6-19. The dashed lines in the figure indicate the second group of data generated by MISO processing.

6.1.5.1 MISO Processing The MISO processing in Figure 6-19 consists of taking the input data cells and producing two sets of data cells at the output which satisfy the modified Alamouti coding, each of which will be directed to a transmitter. It should be noted that the Alamouti code cannot be applied to the P1 preamble symbol in the T2

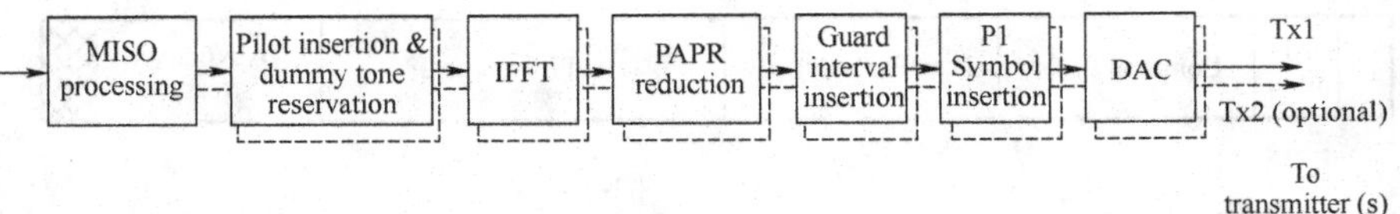

FIGURE 6-19 Block diagram of OFDM generation module.

frame. The MISO processing is applied to pairwise adjacent data cells after the frequency interleaver, so the storage required is minimal. Note that once the pilots have been inserted, the adjacent data cells might not be mapped to the adjacent carriers.

6.1.5.2 Pilot Insertion The pilots will be inserted into Alamouti encoded data cells. Various cells within the OFDM frame are modulated by the sequences/symbols whose values are exactly known to the receiver (called the pilots). Cells containing those symbols are usually transmitted at higher power level. These pilots help the receiver achieve the frame synchronization, frequency synchronization, timing synchronization, channel estimation, transmission mode identification, and phase noise tracking. The symbols transmitted by these cells are either scattered (scattered pilots) or continuous (continuous pilots?) and can be carried by the P2 symbol, the normal data symbol, or the frame closing symbol. The value of the pilot information is derived from a reference sequence, which can be derived from a symbol-level PRBS and frame-level PN sequence. The reference sequence is applied to all types of pilots for each symbol of the T2 frame, including scattered, continuous, edge, P2, and frame closing pilots. Table 6-3 shows the distribution of different types of pilots on different OFDM symbols. The scattered and continuous pilots are similar to those? in DVB-T. The newly added edge pilot is to apply the frequency-domain interpolation to the boundary of the spectrum (DVB-T2 uses spread spectrum to increase spectrum utilization). The edge pilot is modulated in the same way as the scattered pilot. The frame closing pilot will be inserted for the last special frame closing symbol of the T2 frame. This pilot is a certain combination of FFT size, guard intervals, and scattered pilot pattern. The P2 pilot is applied to the P2 symbol while the P1 symbol does not have dedicated P1 pilots.

The scattered pilots are applied to normal OFDM symbols other than P1, P2, and frame closing symbols. The principle of the scattered pilot is exactly the same as that of DVB-T. The biggest difference is that DVB-T2 supports various scattered pilot patterns, from PP1 to PP8, to meet a variety of applications in the multipath environment. To support MISO processing, the two transmitters will use different scattered pilots for insertion. Tables 6-4 and 6-5 list the effective combinations of scattered pilot patterns, FFT size, and guard intervals in SISO and MISO modes.

TABLE 6-3 Distribution of Various Types of Pilots in Each Type of Symbol (X = present)

	Pilot Type				
Symbol	Scattered	Continuous	Edge	P2 pilot	Frame closing
P1					
P2				X	
Data symbol	X	X	X		
Frame closing symbol			X		X

TABLE 6-4 Effective Combination of Scattered Pilot Pattern, FFT Size, and Guard Interval in SISO Mode

	Guard Interval						
FFT size	1/128	1/32	1/16	19/256	1/8	19/128	1/4
32 K	PP7	PP2, PP6	PP2, PP8, PP4	PP2, PP8, PP4	PP2, PP8	PP2, PP8	NA
16 K	PP7	PP7, PP4, PP6	PP2, PP8, PP4, PP5	PP2, PP8, PP4, PP5	PP2, PP3, PP8	PP2, PP3, PP8	PP1, PP8
8 K	PP7	PP7, PP4	PP8, PP4, PP5	PP8, PP4, PP5	PP2, PP3, PP8	PP2, PP3, PP8	PP1, PP8
4 K, 2 K	NA	PP7, PP4	PP4, PP5	PP4, PP5	PP2, PP3	PP2, PP3	PP1
1 K	NA	NA	PP4, PP5	PP4, PP5	PP2, PP3	PP2, PP3	PP1

The main difference between pilot patterns lies in different intervals of the pilots, i.e., the structure of the pilot patterns in both frequency and time domains. As an example, the patterns of scattered pilot PP3 in SISO and MISO modes are shown in Figures 6-20 and 6-21, respectively.

A number of continuous pilots will be inserted into each normal OFDM symbol. The number and location of continuous pilots are determined by the FFT size and the scattered pilot pattern. According to the FFT mode used in symbols, the locations of

TABLE 6-5 Effective Combination of Scattered Pilot Pattern, FFT Size, and Guard Interval in MISO Mode

	Guard Interval						
FFT size	1/128	1/32	1/16	19/256	1/8	19/128	1/4
32K	PP8, PP4, PP6	PP8, PP4	PP2, PP8	PP2, PP8	NA	NA	NA
16K	PP8, PP4, PP5	PP8, PP4, PP5	PP3, PP8	PP3, PP8	PP1, PP8	PP1, PP8	NA
8K	PP8, PP4, PP5	PP8, PP4, PP5	PP3, PP8	PP3, PP8	PP1, PP8	PP1, PP8	NA
4K, 2K	NA	PP4, PP5	PP3	PP3	PP1	PP1	NA
1K	NA	NA	PP3	PP3	PP1	PP1	NA

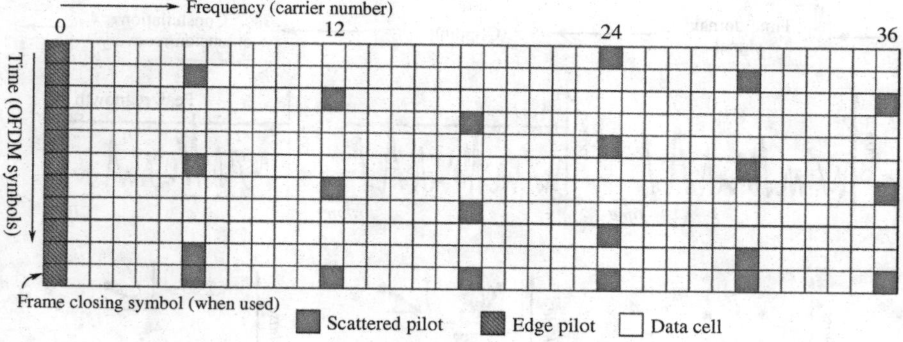

FIGURE 6-20 Scattered pilot pattern PP3 (SISO system).

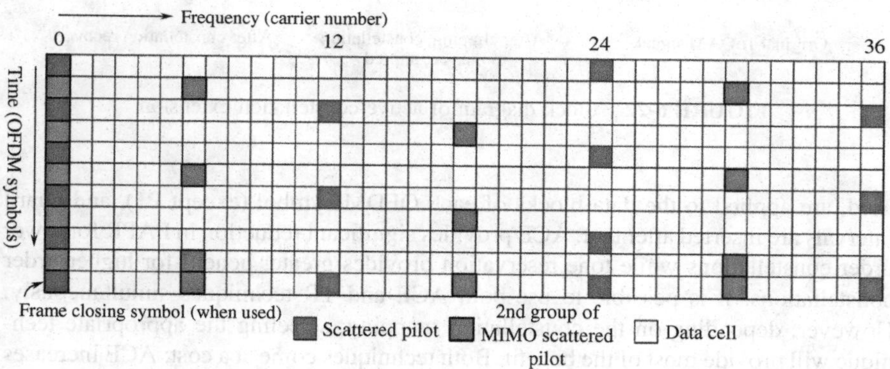

FIGURE 6-21 Scattered pilot pattern PP3 (MIMO system).

continuous pilots are taken from one or more continuous pilot (CP) groups. Table 6-6 shows CP groups in different FFT modes. The pilot locations belonging to the same CP group depend on the scattered pilot pattern in use.

6.1.5.3 PAPR Reduction DVB-T2 offers two PAPR reduction techniques: active constellation expansion (ACE) and tone reservation (TR). Both techniques, when

TABLE 6-6 Continuous Pilot Groups Used with Each FFT Size

FFT Size	CP Groups Used	K_{mod}
1 K	CP_1	1,632
2 K	CP_1, CP_2	1,632
4 K	CP_1, CP_2, CP_3	3,264
8 K	CP_1, CP_2, CP_3, CP_4	6,528
16 K	CP_1, CP_2, CP_3, CP_4, CP_5	13,056
32 K	CP_1, CP_2, CP_3, CP_4, CP_5, CP_6	NA

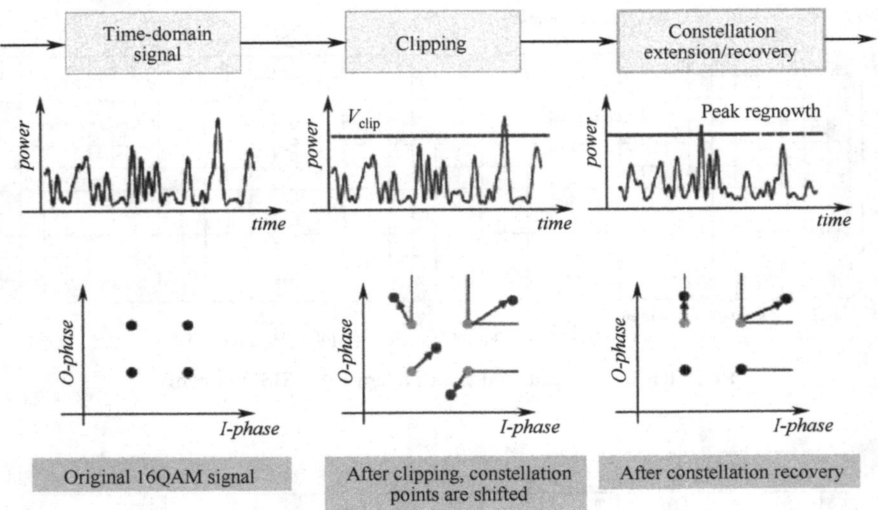

FIGURE 6-22 Block diagram of active constellation extension.

used, are applied to the data blocks of each OFDM symbol (except P1), and guard intervals are inserted after that. ACE provides significant reduction in PAPR for lower order constellations while tone reservation provides greater benefit for higher order constellations. It is possible to use both ACE and TR techniques simultaneously. However, depending on the constellation mapping, selecting the appropriate technique will provide most of the benefit. Both techniques come at a cost: ACE increases the noise level at the receiver while TR decreases the throughput. The major trade-off is the increased transmission power versus the loss in throughput. From the perspective of the receiver, ACE would reduce the SNR at the receiving end, while the reserved tone would not be supported by the traditional receiver.

The ACE algorithm modifies the power distribution of time-domain signal samples for an improvement in power efficiency of the power amplifiers or in out-of-band emission levels at the transmitter. The ACE technique cannot be used for modulating pilots or reserved subcarriers or after the constellation rotation. The basic principles are shown in Figure 6-22. For example, the time-domain waveform of a normal 16QAM signal is first obtained by IFFT, and then the waveform is clipped. The clipped time-domain signal is converted back into the frequency domain by FFT with the constellation point deviating from the original location, as shown in the clipped 16QAM signal constellation in Figure 6.22. For the constellation points at the edge, the space can be moved outside without increasing error rate, and expansion of the constellation points at the edge can be maintained while the inner constellation points have to be readjusted to the normal 16QAM locations. As some constellation point positions are extended, the PAPR might be lowered a certain extent when the signal is changed back to the time domain, again by IFFT. The above process can be repeatedly done to obtain a better PAPR performance. The extended constellations are fully defined from original data

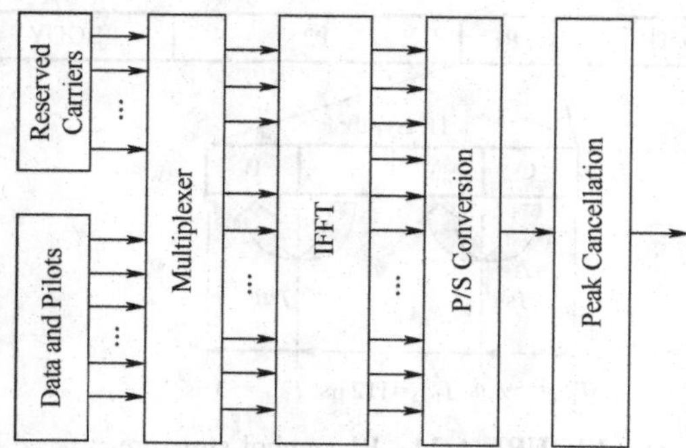

FIGURE 6-23 Block diagram of tone reservation technique.

cell constellations and the maximal extension value L, which is a parameter of the ACE algorithm. Better performance can be obtained by adjusting L along with other ACE parameters, such as clipping threshold V_{clip} and gain G.

The basic principle of tone reservation is that some carriers are reserved for PAPR reduction purposes. These reserved carriers do not carry any data or L1 pre- and postsignaling and are instead filled with complex values to help reduce the PAPR. The power of each reserved carrier cannot exceed 10 times the average power of the data carrier. Figure 6-23 shows the structure of the OFDM transmitter using tone reservation. The location of the reserved carrier is predetermined and peak cancellation is done to reduce PAPR using a predetermined signal after the IFFT.

6.1.5.4 Guard Interval Insertion Table 6-7 shows different guard interval fractions for different FFT sizes. The absolute duration of the guard interval is expressed as an integral multiple of the elementary period T.

The transmitted signal includes insertion of the guard interval and the guard interval should be inserted following PAPR reduction if the PAPR reduction algorithm is used.

6.1.5.5 P1 Symbol Insertion P1 is an OFDM symbol of 1 K long with two "guard interval–like" portions added. The total symbol lasts 224 μs in the 8-MHz system, comprising 112 μs, the duration of the useful part A of the symbol plus two modified

TABLE 6-7 Guard Interval Duration Expressed in Elementary Period *T*

	Guard Interval Fraction (Δ/T_u)						
FFT Size	1/128	1/32	1/16	19/256	1/8	19/128	1/4
32K	256T	1024T	2048T	2432T	4096T	4864T	NA
16K	128T	512T	1024T	1216T	2048T	2432T	4096T
8K	64T	256T	512T	608T	1024T	1216T	2048T
4K	NA	128T	256T	NA	512T	NA	1024T
2K	NA	64T	128T	NA	256T	NA	512T
1K	NA	NA	64T	NA	123T	NA	256T

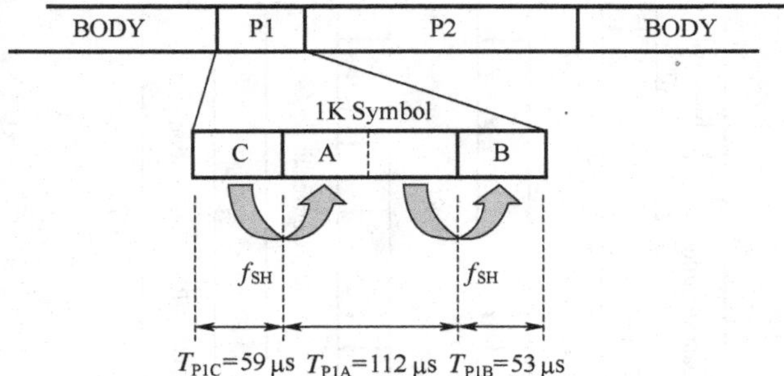

FIGURE 6-24 P1 symbol structure.

guard interval sections C and B of roughly 59 μs (542 samples) and 53 μs (482 samples); see Figure 6-24.

Sections C and B are derived by the cyclic extension and rotation of the useful part A. The 1 K OFDM modulation is applied to section A, including 853 useful carriers and the rest are virtual carriers. Out of the 853 useful carriers, only 384 are used and others are set to 0, as shown in Figure 6-25. The used carriers occupy approximately 6.83 MHz band from the middle of the 7.61 MHz signal bandwidth. The design is done such that even if a maximum offset of not more than 500 kHz is used, most of the used carriers in the P1 symbol are still within 7.61 MHz nominal bandwidth and the P1 symbol can be recovered with the receiver tuned to the nominal center frequency. The 384 active carriers are DBPSK modulated combined with scrambling for transmission, carrying 7-bit signaling.

The P1 symbol is designed to be very robust to ensure the symbol is decoded at the receiver even under fairly challenging conditions at minimal overhead. The P1 symbol is designed with the following characteristics:

1. The P1 symbol has strong anti-interference ability, which could be received and decoded under extremely adverse circumstances. The choice (for efficiency) of a short fixed-length P1 symbol means that intersymbol interference may occur but can be tolerated since the modulation and coding are designed to operate in

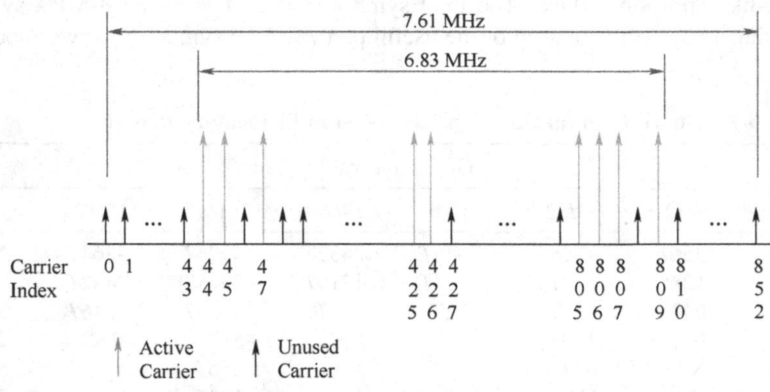

FIGURE 6-25 Active carriers of P1 symbol.

very low signal-to-noise ratio. (The length of portions C and B is not enough to absorb completely the response of the channel to the previous data symbol.) The presence of the two portions C and B at the beginning and end of the symbol improves robustness against both false detection and loss of detection which might otherwise occur in the presence of long-delayed echoes of the channel (even of opposite sign) or spurious signals (such as CW interference).

2. The P1 symbol can be received in a completely unknown channel. Due to its carrier distribution, the P1 symbol supports frequency offset of up to 500 kHz from the center frequency in the 8-MHz system, which can be correctly restored if the receiver is tuned to the nominal center frequency. The PAPR of the symbol has been optimized in order to make its reception better, ever if any AGC loop is not yet stable.
3. Offset correction capabilities: At initialization, the P1 symbol can be used to gain coarse time synchronization of the receiver as well as to detect (and correct) any frequency offset, for both fractional and whole-carrier shifts, from its nominal central bandwidth.
4. Robustness of signaling: The 7-bit signaling that is conveyed within the P1 is DBPSK modulated. The signaling is encoded by using a complementary set of sequences, and the orthogonality of the sequences improves robustness in terms of decoding the correct pattern. It has been ensured that this protection is enough to recover the signaling information even under negative values of SNR.

6.2 INTRODUCTION TO DTMB-A SYSTEM [4]

Digital Television Terrestrial Multimedia Broadcasting—Advanced (DTMB-A), another digital terrestrial television broadcasting system proposed by China, was submitted to ITU in 2013. DTMB-A can support high-definition TV, standard-definition TV, and data broadcasting services under indoor/outdoor and fixed/mobile reception conditions with improved spectrum efficiency and better system performance. DTMB-A can be used for large-area coverage within both multiple- and single-frequency networks. DTMB-A adopts multicarrier modulation and advanced forward error correction coding scheme and therefore can provide fast system synchronization, high receiving sensitivity, better performance against multipath effects, high spectrum efficiency, and flexibility for future extension.

6.2.1 System Architecture

The physical layer structure of DTMB-A is designed to provide one or multiple pipelines of the transmission channel for upper layer services. From the top, the DTMB-A physical layer channel contains a superframe comprising the synchronization channel, data channel, and control channel. The synchronization channel is used for initial synchronization and system parameters acquisition. The data channel contains S_{service} data, each of which utilizes the integral number of data frames, and each superframe contains F_{data} frames. The value of F_{data} could change according

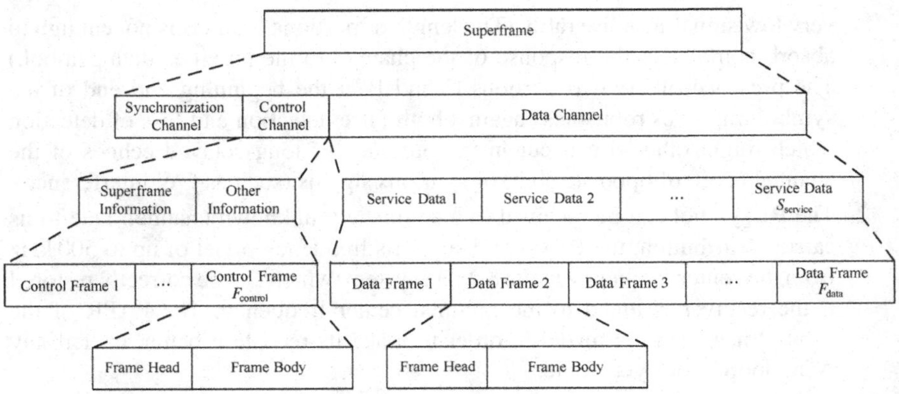

FIGURE 6-26 Superframe structure.

to the different system parameters. The control channel consists of F_{control} frames, and its main function is to carry the information of service configurations, system parameters required by demodulation and channel decoding, real-time information (short messages, wireless localization, etc.), and so on. The data frames and control frames adopt an identical frame structure, which is made up of a frame header and a frame body. Figure 6-26 shows the superframe structure of DTMB-A.

Figure 6-27 presents the block diagram of the DTMB-A transmitter, and each part of the transmitter will be detailed in the following:

The control frames within the control channel are generated by the following baseband processing: FEC channel coding at low code rates, bit interleaving, QPSK constellation mapping, symbol interleaving, IDFT, and frame header insertion.

The data frames within the data channel adopt independent channel coding and modulation schemes, and the parameters can be configured according to different system requirements in practice. The data frames are produced by the following baseband processing: FEC channel coding, bit interleaving, constellation mapping, symbol interleaving, IDFT, and frame header insertion.

Either identical or different lengths of the frame header and frame body can be adopted by control frames and data frames, wherein the frame header is composed of multicarrier pseudorandom noise (PN-MC) binary sequence in the frequency domain.

The control frames and the data frames corresponding to different service types are combined to generate the superframe, and then the baseband postprocessing and up conversion are used to finally generate the radio frequency (RF) signal for transmission.

6.2.2 Interface and Data Preprocessing

The input data interface of DTMB-A is fully compliant with the standard GB/T 17975.1, "Information Technology—Generic Coding of Moving Pictures and Associated Audio Information—Part 1: Systems."

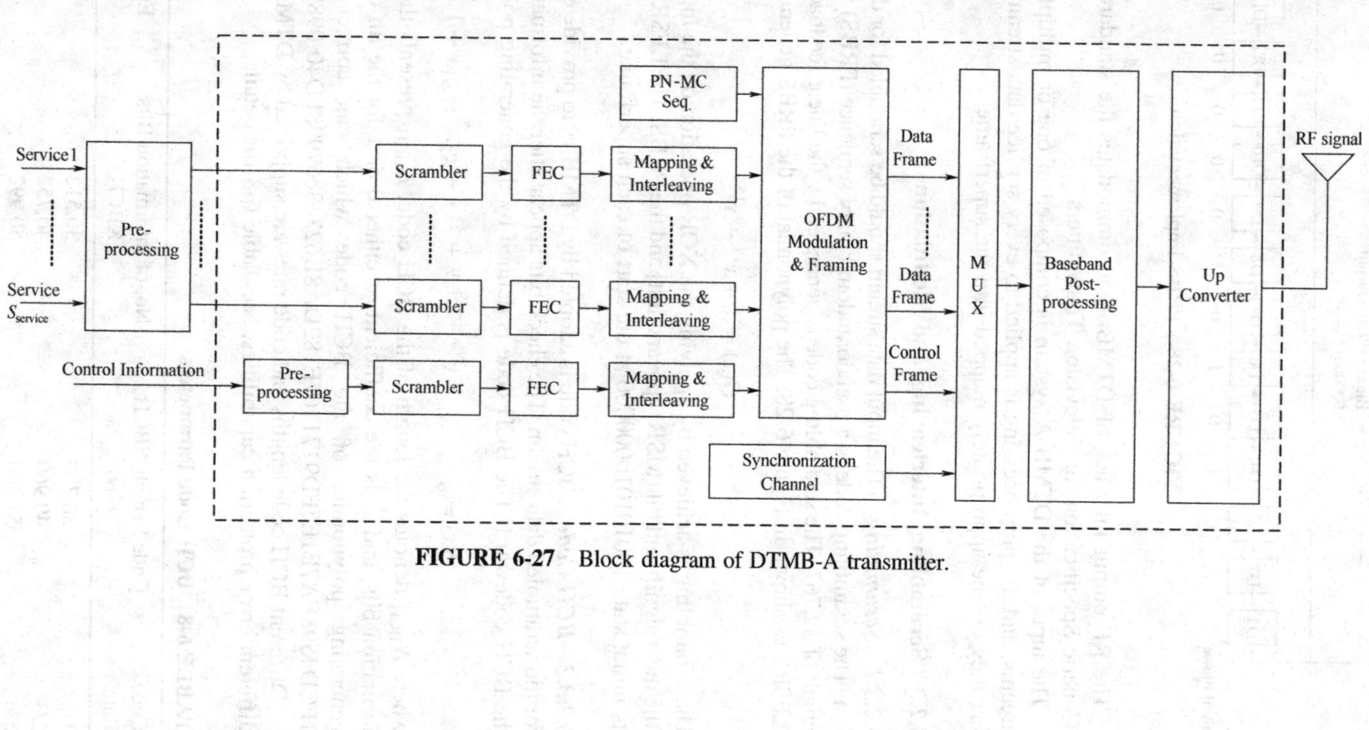

FIGURE 6-27 Block diagram of DTMB-A transmitter.

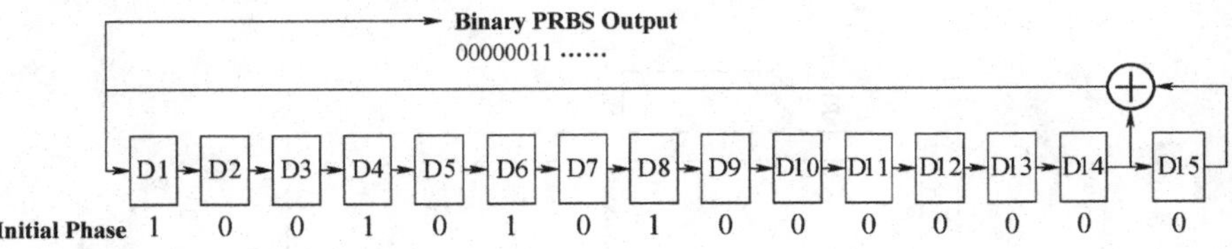

FIGURE 6-28 PRBS implementation.

The RF output interface of DTMB-A accommodates the standard SJ/T 10351, "Generic Specification for Television Transmitters."

The input of the DTMB-A system is composed of one or multiple service data streams, and the preprocessing is applied to every service data stream to make sure that they can be appropriately mapped into the superframe.

6.2.3 Scrambling, Interleaving, and Modulation

6.2.3.1 Scrambling The input data stream should be scrambled for data randomization. The scrambling code is a pseudorandom binary sequence (PRBS) with maximum length of 32,767. The scrambling code is generated by the linear feedback shift register (LFSR), as shown in Figure 6-28. The polynomial of the PRBS generator is

$$G(x) = 1 + x^{14} + x^{15} \tag{6-2}$$

The scrambling is achieved by applying the XOR operation on the input data stream [the most significant bit (MSB) appears first] and the PRBS. The LFSR will be reset to its initial state 100101010000000 at the start of each superframe.

6.2.3.2 BCH Code BCH code is adopted by DTMB-A to provide error protection for the scrambled data stream. The check bits appear after the information bits within the BCH codeword. The BCH code is defined by the generating polynomial

$$g(x) = g_{N_{\mathrm{BCH}}-K_{\mathrm{BCH}}}\, x^{N_{\mathrm{BCH}}-K_{\mathrm{BCH}}} + \cdots + g_2 x^2 + g_1 x + 1 \tag{6-3}$$

where N_{BCH} denotes the length of the BCH code, K_{BCH} presents the length of the formation bits, and g_i is the coefficient of either 0 or 1 for the ith order of x. The generating polynomial of the BCH code which can correct 13 errors is 1FCD4598EA7E1E5FD9171707E182DA8B62313c67998FD40 09895 in Hex.

Different BCH code lengths and code rates are supported by DTMB-A to provide different error protection capabilities; see Table 6-8 for details.

TABLE 6-8 BCH Code Parameters

Code Rate	Code Length NBCH (bits)	No. of Information Bits, KBCH	Error Correction (bits)
1/2	30,720	30,512	13
2/3	40,960	40,752	13
5/6	51,200	50,992	13

TABLE 6-9 LDPC Code Parameters

Code Rate	Code Length I	Information Bits	Code Length II	Information Bits
1/2	61,440	30,720	15,360	7,680
2/3	61,440	40,960	15,360	10,240
5/6	61,440	51,200	15,360	12,800

6.2.3.3 Forward Error Correction Low-density parity-check (LDPC) encoding will be applied to the BCH as shown in Table 6-9, where DTMB-A adopts the LDPC code with two different code lengths and three different code rates.

The code length can be either 61,440 bits or 15,360 bits, the number of information bits is determined by the code length as well as the code rate, and the check bits will be appended after the information bits, as shown in Figure 6-29 ($N_{\mathrm{BCH}} = K_{\mathrm{LDPC}}$ for code length 61,440, while $N_{\mathrm{BCH}} = 4K_{\mathrm{LDPC}}$ for code length 15,360).

DTMB-A adopts quasi-cyclic (QC) LDPC code whose generator matrix $\mathbf{G}_{\mathrm{qc}}$ and parity check matrix $\mathbf{H}_{\mathrm{qc}}$ are both composed of a circulant submatrix. The generator matrix $\mathbf{G}_{\mathrm{qc}}$ of the systematic LDPC code is uniquely determined by the parity-check matrix $\mathbf{H}_{\mathrm{qc}}$. The generator matrix $\mathbf{G}_{\mathrm{qc}}$ of the LDPC code is

$$\mathbf{G}_{\mathrm{qc}} = \begin{bmatrix} \mathbf{I} & \mathbf{O} & \cdots & \mathbf{O} & \mathbf{G}_{0,0} & \mathbf{G}_{0,1} & \cdots & \mathbf{G}_{0,n-k-1} \\ \mathbf{O} & \mathbf{I} & \cdots & \mathbf{O} & \mathbf{G}_{1,0} & \mathbf{G}_{0,1} & \cdots & \mathbf{G}_{1,n-k-1} \\ \vdots & \vdots & \ddots & \vdots & \vdots & \vdots & \mathbf{G}_{i,j} & \vdots \\ \mathbf{O} & \mathbf{O} & \cdots & \mathbf{I} & \mathbf{G}_{k-1,0} & \mathbf{G}_{k-1,1} & \cdots & \mathbf{G}_{k-1,n-k-1} \end{bmatrix} \quad (6\text{-}4)$$

where $\mathbf{I}$ is an identity matrix of size $b \times b$, $\mathbf{O}$ denotes a zero matrix of size $b \times b$, and $\mathbf{G}_{i,j}$ represents a circulant matrix of size $b \times b$, $0 \leq i \leq k-1$, and $0 \leq j \leq n-k-1$.

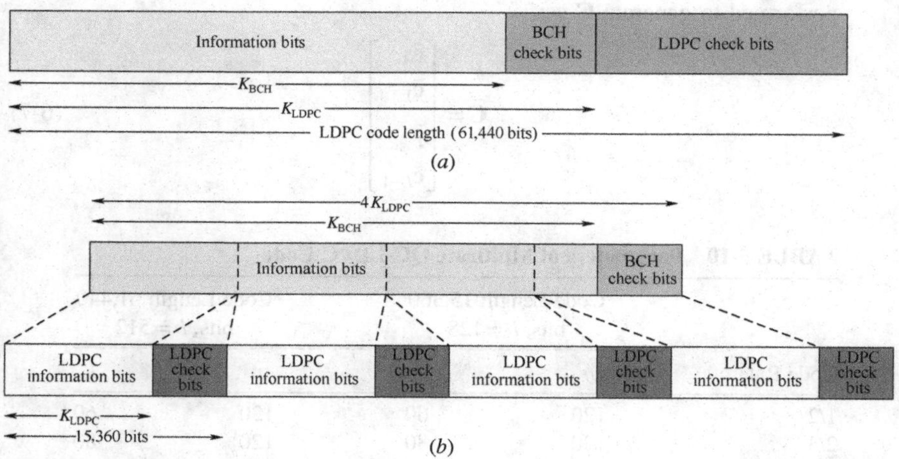

FIGURE 6-29 LDPC code format with lengths (*a*) of 61,440 and (*b*) 15,360 bits.

Similarly, the parity-check matrix $\mathbf{H}_{qc}$ can be written as

$$\mathbf{H}_{qc} = \begin{bmatrix} \mathbf{A}_{0,0} & \mathbf{A}_{0,1} & \cdots & \mathbf{A}_{0,n-1} \\ \mathbf{A}_{1,0} & \mathbf{A}_{1,1} & \cdots & \mathbf{A}_{1,n-1} \\ \vdots & \vdots & \ddots & \vdots \\ \mathbf{A}_{n-k-1,0} & \mathbf{A}_{n-k-1,1} & \cdots & \mathbf{A}_{n-k-1,n-1} \end{bmatrix} \quad (6\text{-}5)$$

where $\mathbf{A}_{i,j}$ $(0 \le i < n-k, 0 \le j < n)$ of size $b \times b$ is the (i,j)th submatrix of $\mathbf{H}_{qc}$. The specific values of n, k, b can be found in Table 6-10.

6.2.3.4 Bit Interleaving and Bit Permutation The bit stream after FEC will be mapped on the constellations to generate the symbol stream. To facilitate iterative demapping and decoding at the receiver, bit interleaving and bit permutation between FEC and constellation mapping are executed within the LDPC codeword.

1. *Bit Interleaving.* The LDPC codeword can be denoted by $\mathbf{c} = (c_0, c_1, \ldots, c_{N_{\text{LDPC-1}}})$, where N_{LDPC} is the LDPC code length, e.g., $N_{\text{LDPC}} = 15{,}360$ or $N_{\text{LDPC}} = 61{,}440$. The LDPC codeword $\mathbf{c}$ is rewritten row by row (row sizes are $b = 128$ for the LDPC code of length 15,360 and $b = 512$ for the LDPC code of length 61,440) to produce the following matrix $\mathbf{C}$ with $L = 120$ rows:

$$\mathbf{C} = \begin{bmatrix} \mathbf{c}_0 \\ \mathbf{c}_1 \\ \vdots \\ \mathbf{c}_{L-1} \end{bmatrix} \quad (6\text{-}6)$$

where $\mathbf{c}_l$ is the lth row of $\mathbf{C}$ and $0 \le l < L$. Then, the intrarow interleaving is performed to generate $\tilde{\mathbf{C}}$ as:

$$\tilde{\mathbf{C}} = \begin{bmatrix} \tilde{\mathbf{c}}_0 \\ \tilde{\mathbf{c}}_1 \\ \vdots \\ \tilde{\mathbf{c}}_{L-1} \end{bmatrix} \quad (6\text{-}7)$$

TABLE 6-10 Parameters of Multirate QC-LDPC Code

	Code Length 15,360 bits, $b = 128$		Code Length 61,440 bits, $b = 512$	
Code Rate	n	k	n	k
1/2	120	60	120	60
2/3	120	80	120	80
5/6	120	100	120	100

TABLE 6-11 Bit Permutation Pattern $\mathbf{p} = (p_0, p_1, \ldots, p_{m-1})$

Constellation/Code Rate	QPSK	16APSK	64APSK	256APSK
1/2	(0,1)	(2,0,1,3)	(2,0,5,3,1,4)	(5,4,0,1,6,2,3,7)
2/3	(0,1)	(1,2,3,0)	(0,2,3,4,5,1)	(1,0,2,4,5,3,6,7)
5/6	(0,1)	(0,1,2,3)	(4,0,1,2,3,5)	(2,3,0,1,4,6,5,7)

where $\tilde{\mathbf{c}}_l = \mathbf{c}_{120/m \times l_m + l_d}$, with $l_d = \lfloor l/m \rceil$ denoting the remainder of the Euclidean division of l by m, $l_d = \lfloor l/m \rceil$ the most positive integer no more than l/m, $m = \log_2 M$, with M presenting the constellation order. Finally, $\tilde{\mathbf{C}}$ is read out column by column to produce the bit-interleaved stream $\tilde{\mathbf{c}} = (\tilde{c}_0, \tilde{c}_1, \ldots, \tilde{c}_{N_{\text{LDPC}-1}})$.

2. *Bit Permutation.* The bit-interleaved stream $\tilde{\mathbf{c}}_0 = (\tilde{c}_0, \tilde{c}_1, \ldots, \tilde{c}_{N_{\text{LDPC}-1}})$ is equally divided into groups, whereby each m-bit group $\tilde{\mathbf{b}} = (\tilde{b}_0, \tilde{b}_1, \ldots, \tilde{b}_{m-1})$ corresponding to one data symbol is bit permutated to generate $\mathbf{b} = (b_0, b_1, \ldots, b_{m-1})$, where $b_{p_i} = \tilde{b}_i (0 \leq i < m)$, and $\mathbf{p} = (p_0, p_1, \ldots, p_{m-1})$ denotes the bit permutation pattern. The bit permutation pattern could vary with different LDPC codes and constellations, as shown in Table 6-11.

6.2.3.5 *Constellation Mapping* DTMB-A adopts the following constellations: QPSK, 16ASPK, 64APSK, and 256APSK. QPSK will be used by the frame body of the control frames and the synchronization frame as well as the frame headers. The frame bodies of the data frames can use QPSK, 16ASPK, 64APSK, and 256APSK. Each service data stream adopts fixed channel coding and a constellation mapping scheme.

When the constellation order is $M = 2^m$, the bit vector $\mathbf{b} = (b_0, b_1, \ldots, b_{m-1})$ in the form of "Left-MSB" corresponds to the decimal number (point index) $d = \sum_{l=0}^{m-1} b_l \times 2^{m-1-l}$, where $0 \leq d < M$. The constellation mapping rule can be presented by the vector $\mathbf{x} = (x_0, x_1, \ldots, x_{M-1})$ with length M, where $x_j, 0 \leq j < M$, denotes the jth constellation point in the complex plane and the bit vector $\mathbf{b}$ will be mapped onto x_d. The average power of the constellations will be normalized, e.g., $(1/M)\sum_{j=0}^{M-1} \|x_j\|^2 = 1$.

The APSK constellation mapping rules are shown in Figures 6-30 to 6-33, respectively. Note that QPSK can be regarded as a special APSK having only one circle. Those figures plot the decimal number d corresponding to the bit vector $\mathbf{b}$. Every circle of the APSK constellations has the identical M_C constellations points having the uniform phase distribution within the range $\lfloor 0, 2\pi)$ with the initial phase rotation of π/M_C.

1. *QPSK.* The QPSK constellation has only one circle with radius $\mathbf{r} = 1$ according to the requirement of power normalization.
2. *16APSK.* The 16APSK constellation has two circles with radii $\mathbf{r} = (r_0, r_1) = (0.586, 1.287), M_C = 8$.

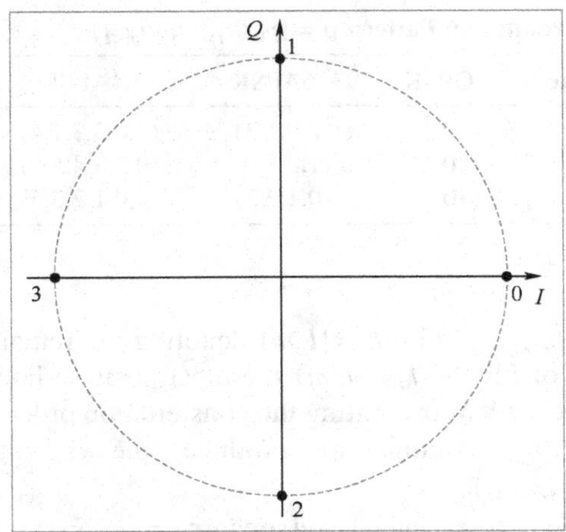

FIGURE 6-30 QPSK constellation mapping.

3. *64APSK*. The 64APSK constellation has four circles with radii $\mathbf{r} = (r_0, r_1, r_2, r_3) = (0.3610, 0.7221, 1.0590, 1.4922)$, $M_C = 16$.
4. *256APSK*. The 256APSK constellation has eight circles with the radii $\mathbf{r} = (r_0, r_1, \ldots, r_7) = (0.2639, 0.4750, 0.6333, 0.7916, 0.9499, 1.1346, 1.3457, 1.6360)$, $M_C = 32$.

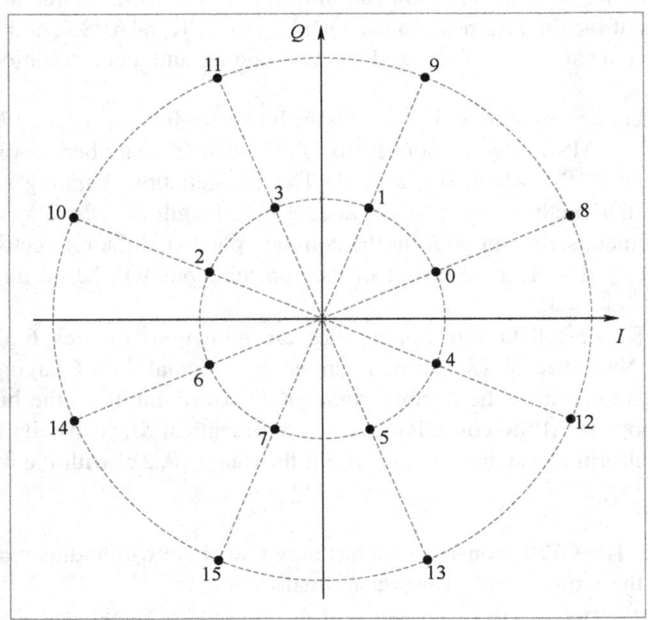

FIGURE 6-31 16APSK constellation mapping.

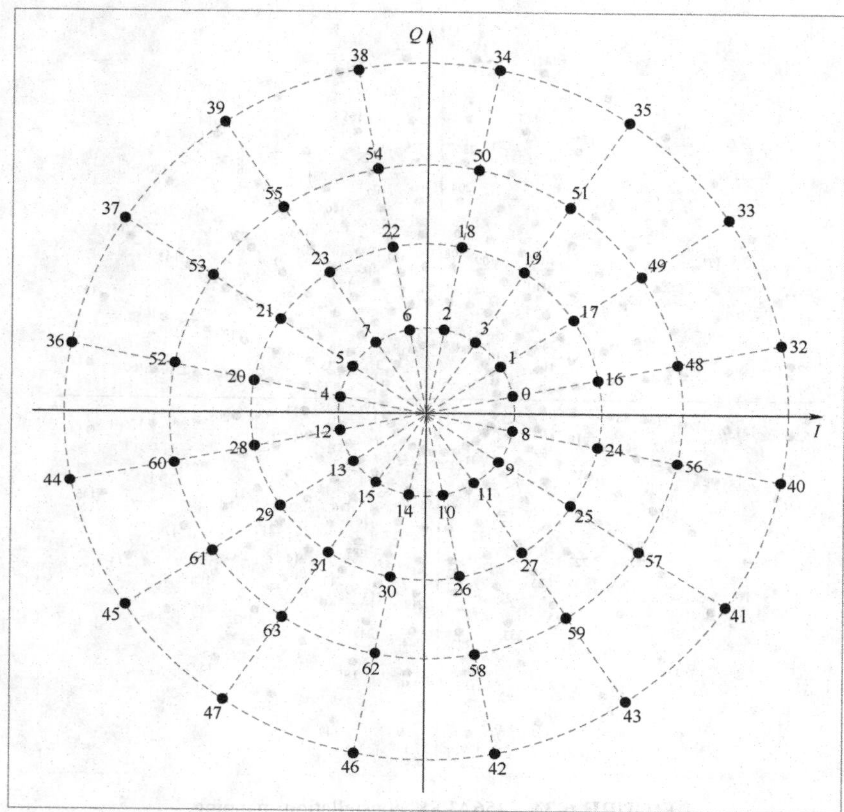

FIGURE 6-32 64APSK constellation mapping.

6.2.3.6 Symbol Interleaving To combat time- and frequency-selective fading channels, symbol interleaving with special cyclic-shifted rows will be applied to the symbol streams after constellation mapping. The symbol interleaver operates in the form of "write in row and read out column," but cyclic row shift will be applied to the matrix form data between the write and read operations. The symbol interleaver with M_S rows and N_S columns involves an integral number of OFDM symbols, i.e., $M_S \times N_S = T_S \times N$, where T_S is an integer.

1. *IQ Interleaving*. IQ interleaving only applies to QPSK constellation mapping. Every four adjacent symbols of the input symbol stream form a group of (u_0, u_1, u_2, u_3), whose Q components $(u_0^Q, u_1^Q, u_2^Q, u_3^Q)$ will be interleaved by $(v_0^Q = u_1^Q, v_1^Q = u_3^Q, v_2^Q = u_0^Q, v_3^Q = u_2^Q)$. Then, the interleaved Q components $(v_0^Q, v_1^Q, v_2^Q, v_3^Q)$ and the original I components $(u_0^I, u_1^I, u_2^I, u_3^I)$ will form the new symbol streams.
2. *Symbol Interleaving*. The symbol interleaving within each OFDM symbol consists of the following steps:

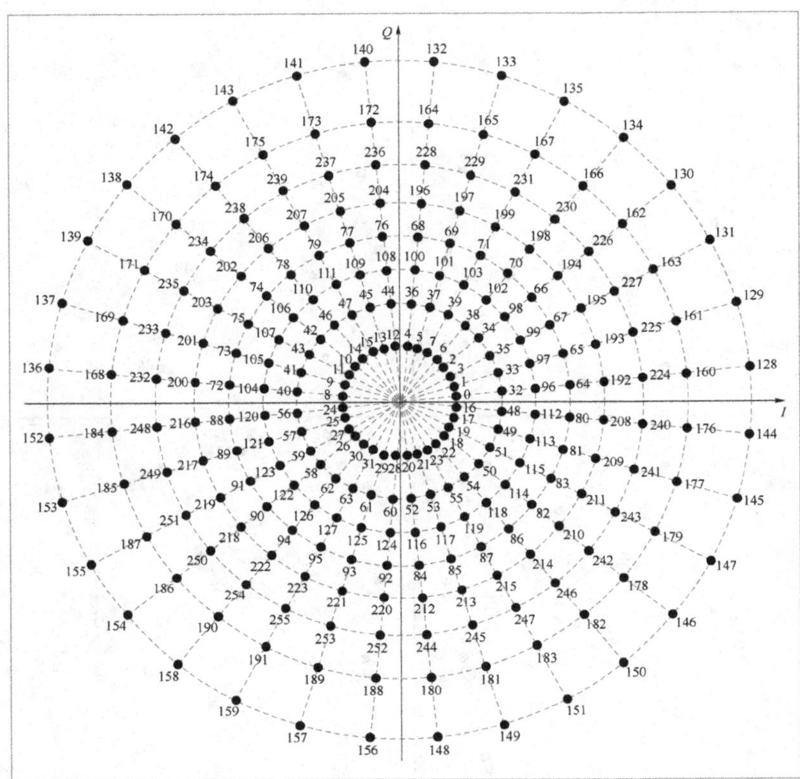

FIGURE 6-33 256APSK constellation mapping.

Step 1: *Write in Row.* The input symbol stream $\mathbf{x} = (x_0, x_1, \ldots, x_{M_s \times N_{s-1}})$ will be written row by row to produce the data $\mathbf{X} = \{X_{i,j}\}$ in the form of a matrix, where $X_{i,j}$ denotes the (i,j)th entry of $\mathbf{X}$, $0 \leq i < M_S$, $0 \leq j < N_S$, and $X_{i,j} = x_{i*N_{S+j}}$.

Step 2: *Cyclic Shift within Subblocks.* The matrix form data $\mathbf{X}$ will be divided into S subblocks $\mathbf{X} = [\mathbf{X}^{(0)}, \ldots, \mathbf{X}^{(S-1)}]$, where each subblock $\mathbf{X}^{(s)}$ has $G = N_S/S$ columns and $0 \leq s < S$. Cyclic shift will be executed within the subblock $\mathbf{X}^{(s)}$ with the down-shift value of f_s as

$$\mathbf{X}^{(s)} = \begin{bmatrix} \mathbf{x}_0^{(s)} \\ \mathbf{x}_1^{(s)} \\ \vdots \\ \mathbf{x}_{M_{S-1}}^{(s)} \end{bmatrix} \tag{6-8}$$

TABLE 6-12 Parameters of Symbol Interleaving

Subcarrier Number N	M_S	N_S	S	$f_s, 0 \le s < S$
4096	240	4096	16	$2s$
8192	240	4096	8	$4s$
32768	240	4096	2	$8s$

where $\mathbf{x}_i^{(s)}$ with G entries denotes the ith row of $\mathbf{X}^{(s)}$, $0 \le i < M_S$. Then, the f_s downward-shifted version $\tilde{\mathbf{X}}^{(s)}$ of $\mathbf{X}^{(s)}$ is

$$\tilde{\mathbf{X}}^{(s)} = \begin{bmatrix} \mathbf{x}_{M_s-f_s}^{(s)} \\ \mathbf{x}_{M_s-f_s+1}^{(s)} \\ \vdots \\ \mathbf{x}_{M_s-1}^{(s)} \\ \mathbf{x}_0 \\ \vdots \\ \mathbf{x}_{M_s-f_s-1}^{(s)} \end{bmatrix} \tag{6-9}$$

So the cyclically shifted matrix $\tilde{\mathbf{X}} = [\tilde{\mathbf{X}}^{(0)}, \ldots, \tilde{\mathbf{X}}^{(S-1)}]$ is formed, and the cyclic shift value $f_s, 0 \le s < S$ can be found in Table 6-12.

Step 3: *Read in Column.* The matrix $\tilde{\mathbf{X}}$ obtained in step 2 will be read out column by column to generate the finally interleaved symbol stream $\tilde{\mathbf{x}} = (\tilde{x}_0, \tilde{x}_1, \ldots, \tilde{x}_{M_S \times N_S - 1})$.

6.2.4 Superframe Structure

6.2.4.1 Definition of Superframe As indicated by Figure 6-26, the superframe is the basic unit of the physical layer channel of DTMB-A. All signals in the superframe adopt the unified symbol rate of 7.56 MSps or 70/9 MSps (extended mode), i.e., the system bandwidth is 7.56 MHz or 70/9 MHz (extended mode). The specific superframe structure is shown in Figure 6-34.

6.2.4.2 Synchronization Channel The synchronization channel is the preamble of each superframe used for fast signal acquisition, coarse timing synchronization, and carrier frequency offset estimation. The system transmission parameters can also be carried by the synchronization channel.

6.2.4.3 Control Channel The control channel comprises F_{control} frames directly after the synchronization channel, and it contains the superframe structure information, multiservice information, and other information. The control channel adopts the FFT size of 4096, the LDPC code length of 15,360, the LDPC code rate of 2/3 (BCH is not applied), QPSK modulation, and $K = 1024$ for the PN-MC sequence.

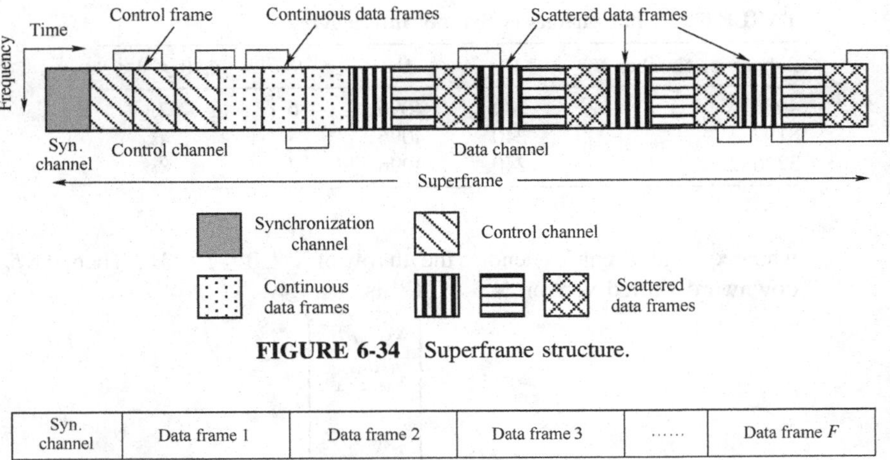

FIGURE 6-34 Superframe structure.

Syn. channel	Data frame 1	Data frame 2	Data frame 3	......	Data frame F

FIGURE 6-35 Superframe for single-service data.

In the control frame, the signaling data of 3072 bits are appended by 7168 bits of 0, and they are both encoded by LDPC code to obtain the data stream of 15,360 bits. After removing the inserted 7168 bits of 0, the 8192 encoded bits for transmission are obtained.

6.2.4.4 Data Channel The data channel is composed of data of multiple services. Each service data within one superframe contains an integral number of data frames, which can be either continuously or separately allocated within the superframe, as shown in Figure 6-34. The allocation scheme can be flexibly configured according to the system requirements of time-domain diversity and power saving in practice. Each service data adopts the symbol interleaving independently, and the number of subcarriers as well as the modulation schemes can be chosen flexibly.

6.2.4.5 Description of Superframe Structure The control frames and data frames are composed of signal frames, the basic units of the superframe. Each signal frame comprises the frame header and the frame body.

Particularly, when the single-service data are carried by one superframe, no control frame will be applied ($C = 0$), and the fixed FFT size will be used within one superframe; see Figure 6-35.

Each service data within one superframe can adopt different numbers of subcarriers, constellation modulations, channel coding schemes, etc. As shown in Figure 6-36, when no control frame ($F_{\text{control}} = 0$) is used for the multiservice

Syn. channel	Data frame 1	......	Data frame 15	Data frame 16	......	Data frame 60	Data frame 61	......	Data frame 120	Data Frame 121	......	Data Frame 135
	Service 1			Service 2			Service 3			Service 4		

FIGURE 6-36 Special superframe for multiservice transmission.

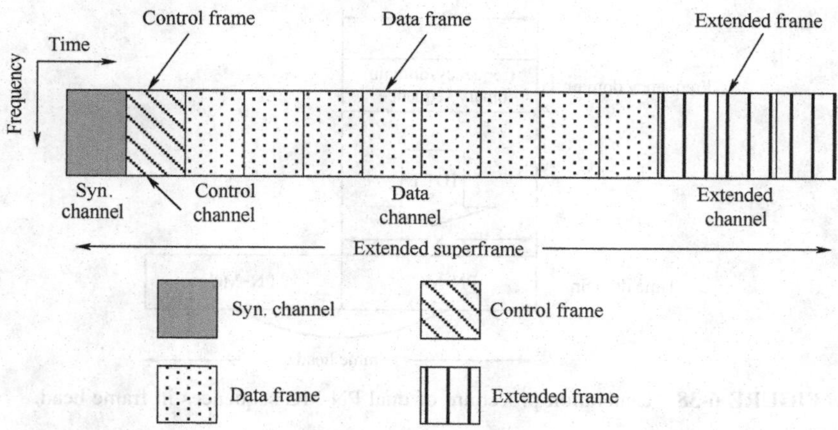

FIGURE 6-37 Signal frame structure.

transmission mode, up to four streams of service data can be supported by DTMB-A. Service 1 with subcarrier number 4096 occupies 15 continuous data frames, service 2 with subcarrier number 4096 occupies 45 continuous data frames, service 3 with subcarrier number 4096 occupies 60 continuous data frames, and service 4 with subcarrier number 32768 occupies 15 continuous data frames.

6.2.5 Signal Frame

6.2.5.1 Signal Frame Structure As shown in Figure 6-37, each signal frame is composed of the frame header and frame body. As the guard interval of the frame body, the frame header comprises two PN-MC sequences of length K, and it can be used for both synchronization and channel estimation. The frame body is generated by OFDM modulation.

In different application scenarios, the frame body length (the subcarrier number) N can be 4096, 8192, or 32768, and K can be in the range 1/4–1/256 of N; see Table 6-13 for details.

After the last signal frame of one superframe, a PN-MC sequence identical to the frame header of the last signal frame will be appended.

TABLE 6-13 Parameters for Signal Frame Structure

	PN-MC Sequence Length K					
Frame Body Length N	1/128	1/64	1/32	1/16	1/8	1/4
32,768	256	512	1024	N/A	N/A	N/A
8,192	N/A	N/A	256	512	1024	N/A
4,096	N/A	N/A	N/A	256	512	1024

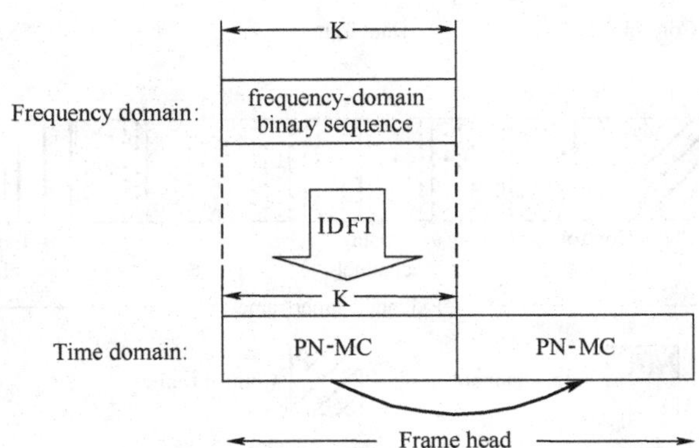

FIGURE 6-38 Generation procedure of dual PN-MC sequences in frame head.

6.2.5.2 PN-MC Sequence The PN-MC sequence in the frame header has the same average power as the frame body. The PN-MC sequence is the IDFT output of the frequency-domain binary sequence. The generation procedure of the dual PN-MC sequences in the frame header is shown in Figure 6-38.

6.2.5.3 Frame Body After symbol interleaving, N constellation symbols $\{X_k\}_{k=0}^{N-1}$ will be applied to IDFT to generate the time-domain OFDM signal (frame body) as

$$x_n = \frac{1}{\sqrt{N}} \sum_{k=0}^{N-1} X_k e^{j2\pi nk/N} \qquad n = 0, 1, \ldots, N-1 \tag{6-10}$$

which occupies the bandwidth of 7.56 MHz or 70/9 MHz (extended mode).

6.2.6 Synchronization Channel

6.2.6.1 Synchronization Channel Structure The synchronization channel of length 2048 is composed of an OFDM symbol of length 1024 and its cyclic extensions, as shown in Figure 6-39, where T_s denotes the symbol duration. The synchronization channel has the same average power as the frame body of the signal frame.

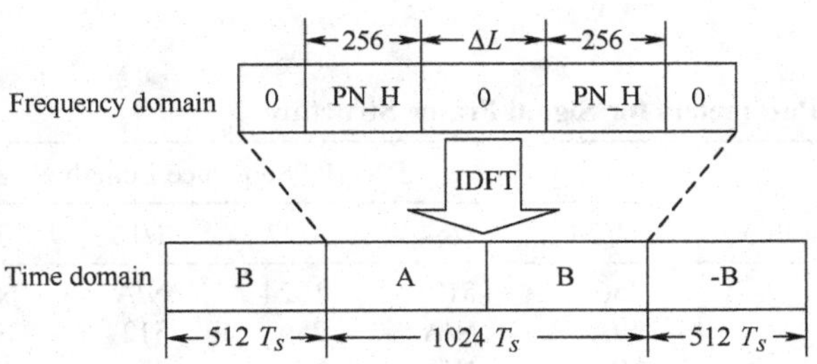

FIGURE 6-39 Synchronization channel structure.

The frequency-domain signal of the synchronization channel $\{Z_k\}_{k=0}^{1023}$ contains 1024 subcarriers, which is composed of a pair of length 256 PN_H sequences with distance ΔL and zeros on the remained subcarriers:

$$Z_k = \begin{cases} 0 & 0 \le k < 256 - \lceil \Delta L/2 \rfloor \\ \mathrm{PN\_H}_{k-256+\lceil \Delta L/2 \rfloor} & 256 - \lceil \Delta L/2 \rfloor \le k < 512 - \lceil \Delta L/2 \rfloor \\ 0 & 512 - \lceil \Delta L/2 \rfloor \le k < 512 + \lfloor \Delta L/2 \rceil \\ \mathrm{PN\_H}_{k-512-\lfloor \Delta/2 \rceil} & 512 + \lfloor \Delta L/2 \rceil \le k < 768 + \lfloor \Delta L/2 \rceil \\ 0 & 768 + \lfloor \Delta L/2 \rceil \le k < 1024 \end{cases} \quad (6\text{-}11)$$

where $\lceil \cdot \rfloor$ denotes round toward plus infinity and $\lfloor \cdot \rceil$ presents denotes round toward minus infinity. The PN_H sequence is the DBPSK modulated signal of the binary sequence PN_U given as

$$\mathrm{PN\_H}_i = \begin{cases} \mathrm{PN\_H}_{i-1} & \mathrm{PN\_U}_i = -1 \\ -\mathrm{PN\_H}_{i-1} & \mathrm{PN\_U}_i = 1 \end{cases}$$

where $0 \le i \le 255$. For $i = 0$, the initial bit is $\mathrm{PN\_H}_{-1} = 1$, which will not be transmitted.

The binary sequence PN_U is defined as

$$\begin{aligned} \mathrm{PN\_U} = \{&1,-1,1,-1,1,-1,-1,-1,1,-1,-1,-1,1,1,1,-1,-1,1,-1,-1,-1,1,-1,-1,1,1,-1,1,1,-1,-1,1,-1,\\ &1,-1,1,-1,-1,1,1,1,-1,1,1,-1,-1,-1,1,-1,1,-1,1,1,1,-1,-1,-1,1,-1,1,1,1,-1,1,1,-1,1,-1,-1,-1,\\ &-1,-1,-1,-1,1,1,1,1,1,1,-1,1,1,-1,-1,1,1,1,1,1,-1,-1,-1,1,1,-1,-1,-1,1,-1,-1,-1,-1,1,-1,\\ &-1,1,1,1,1,-1,-1,1,1,1,-1,-1,-1,1,1,1,-1,1,1,1,-1,1,1,1,1,1,-1,1,-1,1,1,-1,-1,-1,-1,1,-1,\\ &1,-1,-1,-1,-1,1,1,1,-1,-1,-1,-1,1,-1,-1,-1,1,-1,1,1,-1,-1,-1,1,1,1,1,1,-1,1,1,1,-1,-1,1,-1,1,\\ &-1,-1,-1,1,1,-1,1,1,-1,1,1,1,1,1,1,1,1,-1,-1,1,1,-1,-1,1,-1,-1,1,-1,1,1,-1,-1,1,1,-1,1,-1,\\ &1,1,-1,1,-1,1,1,1,1,1,1,-1,1,-1,-1,1,1,-1,1,-1,-1,1,1,1,1,1,-1,-1,1,-1,-1,1,1,1,-1,-1,1\} \end{aligned}$$

Applying the 1024-point IDFT to $\{Z_k\}_{k=0}^{1023}$, we obtain the time-domain signal $\{z\_n\}_{n=0}^{1023}$ as

$$z_n = \frac{1}{\sqrt{1024}} \sum_{k=0}^{1023} Z_k e^{j2\pi nk/1024} \qquad n = 0, 1, \ldots, 1023 \quad (6\text{-}12)$$

The time-domain OFDM signal $\{z_n\}_{n=0}^{1023}$ is equally divided into two parts A and B of the same length. Then, the cyclic structure (B A B –B) is generated by repeating B in front of the OFDM symbol and also repeating B after the OFDM symbol. Thus, the baseband synchronization signal $\{p_1(n)\}_{n=0}^{2047}$ is produced according to

$$p_1(n) = \begin{cases} z_B(n) & 0 \le n < 512 \\ z_A(n-512) & 512 \le n < 1024 \\ z_B(n-1024) & 1024 \le n < 1536 \\ -z_B(n-1536) & 1536 \le n < 2048 \end{cases} \quad (6\text{-}13)$$

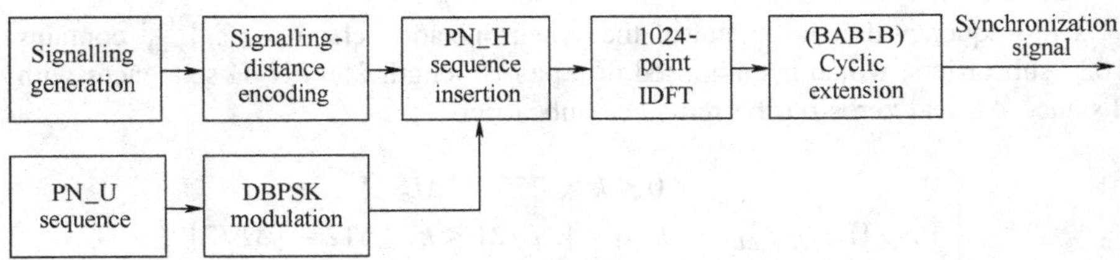

FIGURE 6-40 Overal generation procedure of superframe synchronization signal.

where $\{z_A(n)\}_{n=0}^{511}$ denotes the first half of $\{z_n\}_{n=0}^{1023}$, and $\{z_B(n)\}_{n=0}^{511}$ is the second half of $\{z_n\}_{n=0}^{1023}$.

The overall generation procedure of the synchronization signal is shown in Figure 6-40.

6.2.6.2 Signaling Format The system signaling is carried by the distance ΔL of two PN_H sequences within the range of [64,319], i.e., the 256 different values can carry 8-bit signaling. The binary-decimal encoding scheme is applied to associate the distance and the corresponding signaling according to

$$\Delta L - 64 = \sum_{i=0}^{7} s_i \times 2^i \tag{6-14}$$

Tables 6-14 to 6-18 show the specific definitions of signaling ($s_7 s_6 \cdots s_0$). Note that s_0 with the fixed value of 0 is the reserved bit.

TABLE 6-14 Definition of s_7

s_7	Definition
0	Single antenna
1	Two antennas

TABLE 6-15 Definition of s_6

s_6	Definition
0	Single service
1	Multiservice

TABLE 6-16 Definition of $s_5\, s_4$

$s_5\, s_4$	Definition
00	QPSK
01	16APSK
10	64APSK
11	256APSK

TABLE 6-17 Definition of s_3

s_3	Definition
0	15,360 LDPC
1	61,440 LDPC

TABLE 6-18 Definition of $s_2\, s_1$

$s_2\, s_1$	Definition
00	1/2 code rate
01	2/3 code rate
10	5/6 code rate
11	Reserved

6.2.7 Transmit Diversity

DTMB-A supports adaptive two-antenna transmit diversity. As shown in Figure 6-41, the signal frames transmitted by two transmit antennas are composed of space–frequency coded frame bodies and an orthogonal frame header. The frame header, which is identical to the PN-MC sequence used in the single-antenna scenario, will be alternatively used by two transmitted antennas. For simplicity, the frame header signal is denoted by **P** in this clause, and **0** denotes a zero sequence.

The frame headers adopted by two transmit antennas must also be orthogonal in the frequency domain (optional). As shown in Figure 6-42, the first antenna utilizes the

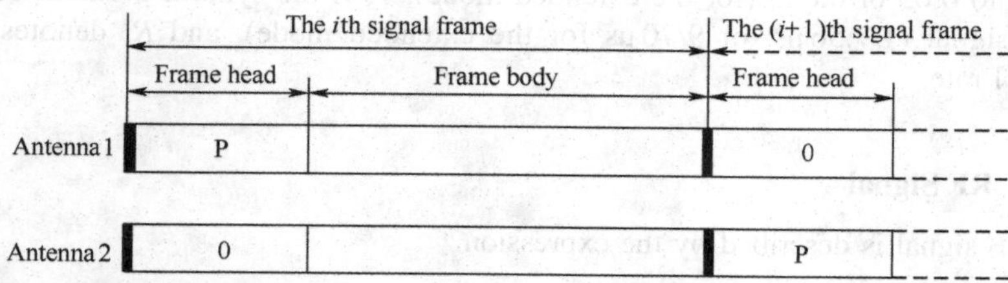

FIGURE 6-41 Signal frame structure for two-antenna transmit diversity.

FIGURE 6-42 Signal frame structure for two-antenna transmit diversity (optional).

same frame header as that in the single-antenna scenario, i.e., $\mathbf{P} = [P_1, P_2, P_3, P_4, \ldots]$, while the second antenna adopts $\mathbf{P}^{\perp} = [P_1^{\perp}, P_2^{\perp}, P_3^{\perp}, P_4^{\perp}, \ldots]$ that is orthogonal with $\mathbf{P} = [P_1, P_2, P_3, P_4, \ldots]$, where $P_1^{\perp} = P_1, P_2^{\perp} = -P_2, P_3^{\perp} = P_3, P_4^{\perp} = -P_4$, and so on.

The following space–frequency block coding (SFBC) scheme will be applied to the input symbol stream $[X_1, X_2, \ldots]$ to generate the frame bodies for two-antenna transmit diversity:

$$\mathbf{X} = \begin{bmatrix} X_1 & X_2 \\ X_2^* & -X_1^* \end{bmatrix} \tag{6-15}$$

where $*$ means complex conjugate.

6.2.8 Baseband Postprocessing

The baseband postprocessing utilizes the square-root raised-cosine (SRRC) filter for the baseband pulse shaping. The frequency response of the SRRC filter is

$$H(f) = \begin{cases} 1 & |f| \leq f_N(1-\alpha) \\ \left\{\dfrac{1}{2} + \dfrac{1}{2}\cos\left[\dfrac{\pi}{\alpha f_N}\left(\dfrac{|f| - f_N(1-\alpha)}{2}\right)\right]\right\}^{1/2} & f_N(1-\alpha) < |f| \leq f_N(1+\alpha) \\ 0 & |f| > f_N(1+\alpha) \end{cases} \tag{6-16}$$

where $f_N = 1/(2T_s) = R_s/2$ is the Nyquist frequency, α is the rolling-off factor and equals to 0.05 or 0.025 (for the extended mode), T_s is the symbol duration of the input signal (1/7.56 μs, or 9/70 μs for the extended mode), and R_s denotes the symbol rate.

6.2.9 RF Signal

The RF signal is described by the expression

$$s(t) = \Re\{\exp(j \cdot 2\pi f_c t) \times [h(t) \otimes x(t)]\} \tag{6-17}$$

where $s(t)$ denotes the RF signal, f_c is the carrier frequency, $h(t)$ presents the impulse response function of the SRRC filter, $x(t)$ is the superframe signal, $\Re\{\cdot\}$ means the real part of the argument, and $\otimes$ denotes linear convolution.

6.2.10 Baseband Signal Spectrum Characteristics and Spectrum Mask

The typical baseband signal spectrum after the SRRC filter is illustrated in Figure 6-43.

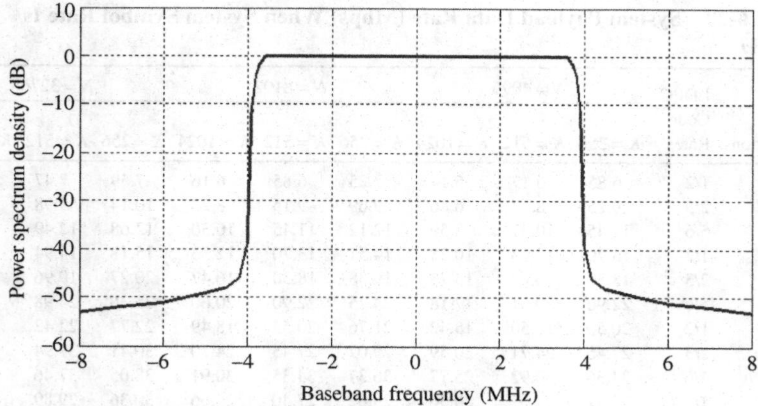

FIGURE 6-43 Typical DTMB-A baseband signal spectrum after SRRC filter.

6.2.11 System Payload Data Rate

DTMB-A provides payload data rates up to 50 Mbps with different signal frame lengths, LDPC code rates, and constellation mapping schemes when the system symbol rate is 7.56 MSps. Tables 6-19 and 6-20 list the payload data rates of different working modes of DTMB-A of different bandwidths.

TABLE 6-19 System Payload Data Rate (Mbps) When System Symbol Rate Is 7.56 MHz

Constellations	LDPC Code Rates	$N=4096$			$N=8192$			$N=32768$		
		$K=256$	$K=512$	$K=1024$	$K=256$	$K=512$	$K=1024$	$K=256$	$K=512$	$K=1024$
QPSK	1/2	6.66	5.99	5.00	7.05	6.66	5.99	7.38	7.26	7.05
	2/3	8.90	8.01	6.67	9.42	8.89	8.00	9.85	9.70	9.41
	5/6	11.13	10.02	8.35	11.78	11.13	10.01	12.33	12.14	11.78
16APSK	1/2	13.32	11.99	9.99	14.10	13.32	11.98	14.75	14.53	14.09
	2/3	17.79	16.01	13.34	18.84	17.79	16.01	19.70	19.40	18.82
	5/6	22.26	20.04	16.70	23.57	22.26	20.03	24.65	24.28	23.55
64APSK	1/2	19.98	17.98	14.99	21.15	19.98	17.98	22.13	21.79	21.14
	2/3	26.69	24.02	20.02	28.25	26.68	24.01	29.56	29.1	28.23
	5/6	33.39	30.05	25.05	35.35	33.39	30.04	36.98	36.41	35.33
256APSK	1/2	26.64	23.98	19.98	28.21	26.64	23.97	29.51	29.05	28.19
	2/3	35.58	32.03	26.69	37.67	35.58	32.01	39.41	38.80	37.64
	5/6	44.53	40.07	33.39	47.14	44.51	40.06	49.31	48.55	47.10

TABLE 6-20 System Payload Data Rate (Mbps) When System Symbol Rate Is 70/9 MHz

Constellations	LDPC Code Rates	$N=4096$			$N=8192$			$N=32768$		
		$K=256$	$K=512$	$K=1024$	$K=256$	$K=512$	$K=1024$	$K=256$	$K=512$	$K=1024$
QPSK	1/2	6.85	6.17	5.14	7.25	6.85	6.16	7.59	7.47	7.25
	2/3	9.15	8.24	6.86	9.69	9.15	8.23	10.14	9.98	9.68
	5/6	11.45	10.31	8.59	12.12	11.45	10.30	12.68	12.49	12.12
16APSK	1/2	13.70	12.33	10.28	14.51	13.70	12.33	15.18	14.94	14.50
	2/3	18.30	16.47	13.73	19.38	18.30	16.47	20.27	19.96	19.36
	5/6	22.90	20.61	17.18	24.25	22.90	20.60	25.36	24.98	24.23
64APSK	1/2	20.56	18.50	15.42	21.76	20.55	18.49	22.77	22.42	21.75
	2/3	27.46	24.71	20.59	29.07	27.45	24.70	30.41	29.94	29.05
	5/6	34.36	30.92	25.77	36.37	34.35	30.91	38.05	37.46	36.35
256APSK	1/2	27.41	24.67	20.56	29.02	27.40	24.66	30.36	29.89	29.00
	2/3	36.61	32.95	27.46	38.76	36.60	32.93	40.54	39.92	38.73
	5/6	45.81	41.23	34.36	48.50	45.80	41.21	50.73	49.95	48.46

6.3 SUMMARY

The DVB organization initiated the second generation of digital terrestrial television (DVB-T2) in 2008 to solve the shortcomings discovered during the implementation and promotion of the DVB-T standard. DVB-T2 used a variety of advanced technologies to increase system capacity and system transmission reliability compared with DVB-T. The advanced system of DTMB (DTMB-A) sponsored by the Chinese government has successfully demonstrated its capability of supporting a payload rate of around 40 Mbps for fixed reception and a payload rate of more than 20 Mbps for mobile reception. It also improved the signal-to-noise ratio (SNR) threshold in complex multipath channels by taking advantage of the latest technical breakthroughs. In this chapter, the frame structure, channel coding, and modulation techniques used in DVB-T2 and DTMB-A systems were introduced in detail to help readers obtain a comprehensive understanding.

REFERENCES

1. ETSI EN 302 755 V1.3.1, *Digital Video Broadcasting (DVB); Frame Structure Channel Coding and Modulation for a Second Generation Digital Terrestrial Television Broadcasting System (DVB-T2)*, Geneva: Digital Video Broadcasting (DVB), Apr. 2012.
2. DVB Document A133, *Implementation Guidelines for a Second Generation Digital Terrestrial Television Broadcasting System (DVB-T2)*, Geneva: Digital Video Broadcasting (DVB), Feb. 2009.
3. S. Alamouti, "A simple transmit diversity technique for wireless communications," *IEEE Journal on Selected Areas in Communications*, vol. 16, no. 8, pp. 1451–1458, Oct. 1998.
4. Z. Yang et al., "Technical review for chinese future DTTB system," the *IEEE 72nd Vehicular Technology Conference Fall (VTC 2010-Fall)*, pp.1–6, 6–9 Sept. 2010.

7

DESIGN AND IMPLEMENTATION OF DTV RECEIVER

7.1 INTRODUCTION

When studying digital transmission systems from an information theory perspective, the description can be simplified this way: The source sequence is mapped into the channel input symbol sequence $X = (x_1, \ldots, x_n, \ldots)$, and the corresponding channel output sequence at the receiver is $Y = (y_1, \cdots, y_n, \cdots)$. The transmitter sends $X = (x_1, \ldots, x_n, \ldots)$ by the continuous-time waveform $s(t, \mathbf{x})$, which means $X = (x_1, \cdots, x_n, \cdots)$ to $s(t, \mathbf{x})$ mapping is completed by the modulator. Therefore, in addition to $s(t, \mathbf{x})$, the distribution of the output sequence θ is determined by the parameter set $\theta = \{\theta_T, \theta_C\}$. Here, the subset θ_T is the transmitter parameter set, and subset θ_C is the channel parameter set for synchronization, including the phase θ and time delay ε. These parameters are usually unknown at the receiver, and the essential function of the receiver is to correctly recover the transmitted sequence based on the channel output sequence [1].

The digital receiver can be generally divided into the inner and outer receivers, as shown in Figure 7-1. First, the inner receiver with its functional blocks inside the dashed box is used to accurately estimate the unknown parameters $\theta = \{\theta_T, \theta_C\}$ from the received signals, and these estimated results are sent to the outer receiver to mimic the real channel conditions for the transmitted symbols for signal equalization. After removing the channel impact on the received symbols as much as possible by equalization, the outer receiver will perform the decoding to recover the transmitted sequence.

Digital Terrestrial Television Broadcasting: Technology and System, First Edition. Edited by Jian Song, Zhixing Yang, and Jun Wang.

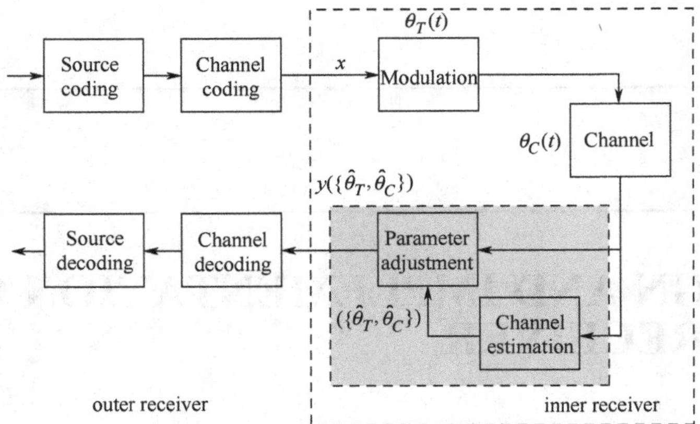

FIGURE 7-1 Generic model for digital receiver.

The principles of channel synchronization and channel estimation as well as equalization in the inner receiver for both single- and multicarrier systems are introduced in this chapter, and their applications for the DTMB system is then discussed with some examples.

The DTTB channels are allocated within VHF and UHF bands and characterized as the time-varying, noisy, and dispersive wireless channel, presenting a very harsh propagation environment for the DTTB signals. Distortions such as additive noise, cochannel interference, and multipath transmission will introduce not only the noise but also both the static and dynamic multipath interferences as well as the Doppler frequency shift, which would lead to intersymbol interference for the received signal. As DTV signals usually occupy several megahertz, it is difficult to remove all these negative impacts, and therefore, precise channel synchronization and estimation are needed to obtain the channel characteristics for good equalization results to achieve better receiving performance under such harsh transmission environments.

Channel synchronization plays a very important role at the inner receiver. In general, the synchronization process consists of timing recovery (also known as symbol synchronization) and carrier synchronization. Timing recovery includes both frame synchronization and sampling clock synchronization and can be briefly explained as follows. As the data sequence is usually segmented or packetized by a certain frame structure, this frame structure must be accurately identified at the receiver to correctly recover the transmitted data. This requires that the timing at the receiver is exactly the same as that of the data frame structure from the transmitter. Besides, the receiver must also "know" the starting time of each data symbol to correctly detect the symbol sequence, which requires that the sampling clock at the receiver synchronizes with the frequency and phase of the transmitted data sequence. The process to generate a local carrier is called carrier synchronization and demodulation can be divided into coherent and noncoherent demodulations according to whether a local coherent carrier is required. The antinoise performance of coherent demodulation is superior to that of noncoherent demodulation while coherent demodulation is more complicated as it requires the receiver to generate a local carrier with the same frequency and phase as that of the carrier at the transmitter.

According to whether data information is required, the synchronization algorithm can be divided into two categories: (1) decision directed (DD)/data aided (DA) and (2) non–data aided (NDA). The NDA algorithm is used by averaging all possible sequences when there is no known sequence, while the DA algorithm takes the advantage of the known sequence in the transmitted signal to facilitate the synchronization. According to where to extract the synchronization error signal, the synchronization algorithm can be divided into two categories: feedforward (FF) and feedback (FB). The FF estimation extracts the synchronization error signal before the synchronization recovery unit while the FB estimation extracts the error signal behind the synchronization recovery unit and then feeds back the calibrated signal to the units in front. As the FB structure has the ability of automatically tracking the slow-varying parameters, the FB estimator is also known as the error FB synchronizer. According to whether synchronization parameter estimation needs the information of other synchronization parameters, the synchronization algorithm can be divided into two categories: correlation and noncorrelation estimations. By name, no synchronization information of other parameters is required for the noncorrelation estimation algorithm, which allows independent synchronization between different parameters. According to the signal type used for synchronization, the synchronization algorithm can be divided into two categories: continuous-signal and burst-signal estimations. Continuous-signal estimation requires that the algorithm has the ability to track the timing variations over an extended period of time, while burst-signal estimation requires that the algorithm has quick capturing capability and the synchronization should be achieved within a relatively short period of time.

Channel estimation and equalization make up a very important module for the inner receiver other than channel synchronization. They estimate the channel characteristics the transmitted signals have experienced and equalize the channel impact by multiplying the frequency response inverse to that of the channel to the received signals, which helps cancel the intersymbol interference caused by the time-varying and multipath characteristics of the channel. That is, the time–frequency selective fading of the channel can be compensated for by the channel estimation and equalization.

Currently, there are many different channel estimation methods for wireless systems. According to which domain the channel compensation is performed, the channel equalization is mainly divided into time-domain equalization, frequency-domain equalization, and a combination of both. Time-domain equalization is the most commonly used method for the single-carrier system with the goal of making the time-domain impulse response satisfy the conditions of non-intersymbol interference at the sampling point. The frequency-domain equalization method is often applied to OFDM systems in the goal of making the overall transfer function (frequency response) meet the conditions of distortless transmission, i.e., correcting both the amplitude and phase distortion of the transmitted signals. According to the principles of the channel compensation algorithms, channel estimation can be mainly divided into two different types: least square (LS, also known as zero forcing) and minimum mean-square error (MMSE) criterion. In theory, the frequency-domain MMSE has the best theoretical performance but needs to know the statistical characteristics of the channel and therefore suffers from high implementation complexity. Several methods have been proposed to reduce the complexity of MMSE, including low-order LMMSE (low-rank linear MMSE) based on the best low-order theory, singular-value decomposition (SVD), and the transform-domain implementation (in either Fourier transform or wavelet transform domains). According to whether the channel estimation result is used for the feedback, the equalizer can be divided into linear and nonlinear equalizers. If the output is not used for feedback control, the equalizer is linear; otherwise, it is nonlinear, including the decision feedback equalizer (DFE), maximum-likelihood sequence equalizer (MLSE), etc. For ATSC, the channel equalization method for the 8-VSB single-carrier signal is DFE [10, 11]. Since the data sequence carried by the first data packet of each field within the ATSC data frame is used for training and the time interval between two adjacent training sequences is 24.2 ms, compensation for the rapidly changing multipath can only be achieved by the adaptive blind equalization method. To effectively eliminate the multipath interference, DFE generally requires a large number of filter taps, which greatly increases the complexity and cost of the receiver. Besides, DFE suffers from the problems of self-excitation and instability because of the infinite impulse response (IIR) structure. DVB-T and DVB-T2 are OFDM systems which can deal with the multipath effect effectively. The frequency-domain channel estimation methods generally used at the DVB-T receiver include the decision feedback frequency-domain estimation and pilot frequency-domain estimation [16, 17]. Based on the implementation method of channel compensation, channel estimation can be divided into data-aided, decision-directed, and blind estimation methods. In the DA method, the channel estimation is carried out by either pilots or training sequences which have been repeatedly transmitted with certain patterns since the channel is time varying. The time-varying, multipath fading channel can be characterized by a two-dimensional (time and frequency) matrix, so the distribution of either pilots or training sequences in both the time and frequency domains will mainly depend on the coherence time and coherence bandwidth of the channel. As long as the sampling frequency (or the sampling clock) in the time domain and the sampling interval in the frequency domain satisfy the Nyquist sampling criteria, the entire channel response can be obtained by interpolation. Unlike the DA method, the channel estimation of the DD method is carried out by using the recovered information data after a decision. For the blind estimation method, the channel estimation is performed with the completely unknown transmitted signal and the transmitter is not required to send any pilots or training sequences. This greatly improves the spectral efficiency of the system by eliminating the overhead from sending the known data. However, reliable estimation can be obtained only after enough data are received if the blind estimation method is used (which means longer delay and a larger data buffer), and the system performance will be quite sensitive to the time variation channel.

7.2 MATHEMATICAL PRINCIPLES

7.2.1 Channel Synchronization

In principle, all parameter estimation methods for channel synchronization use a maximum-likelihood (ML) algorithm. When the input data are equiprobable, the maximum posteriori probability (MAP) criterion is identical to the ML criterion. When the synchronization algorithm is derived in the following, the synchronization parameters (θ, ε) to be estimated are usually assumed to be unchanged for a certain of period time (within the time interval between two adjacent training sequences or two identical pilot patterns for DTTB systems). That is, although the synchronization parameters (θ, ε) are time varying, they change at a much slower pace so that they can be regarded as constant.

Figure 7-2 shows a typical baseband model for the carrier-modulated digital transmission system, where $H_c(\omega)$ is the frequency response for the linear channel, $n(t)$ is the additive white Gaussian noise (AWGN), and $F(\omega)$ is the frequency-domain transfer function of the prefilter. Suppose $s(t)$ is the linear modulated baseband signal of any type to be transmitted:

$$s(t) = \sum_n a_n g_T(t - nT - \varepsilon_0 T) \tag{7-1}$$

where $\{a_n\}$ is the baseband digital sequence in the complex plane, $g_T(t)$ is the impulse response of the pulse shaping filter, T is the data symbol period, and ε_0 is the fractional timing deviation. If the phase difference between the transmitter and receiver can be expressed as $\theta_0 = \theta_T(t) - \theta_R(t)$, the equivalent baseband signal at the receiver is

$$r_f(t) = s_f(t) + n(t) = \sum_n a_n g(t - nT - \varepsilon_0 T)e^{j\theta_0(t)} + n(t) \tag{7-2}$$

where $g(t)$ is the composite pulse waveform at the receiver and can be expressed as

$$g(t) = g_T(t) \otimes h_C(t) \otimes f(t) \tag{7-3}$$

$h_c(t)$ is the time-domain channel response, and $f(t)$ is the time-domain transfer function of the prefilter.

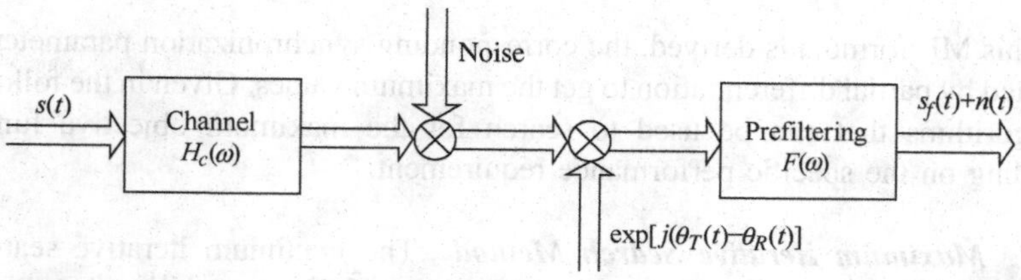

FIGURE 7-2 Generic linear baseband model of digital transmission system with carrier modulation.

The received baseband signal $r_f(t)$ enters into the synchronization module where synchronization parameters such as the carrier frequency offset and timing clock will be estimated. If we consider the detection of the sequence $\{a_n\}$ with N symbols, (7-2) can be further written as

$$r_f(t) = \sum_{n=0}^{N-1} a_n g(t - nT - \varepsilon_0 T) e^{j\theta_0(t)} + n(t) \tag{7-4}$$

Assume the sampling of $\{r_f(kT_s)\}$ from the $r_f(t)$ can provide sufficient statistics. Also assume it can be proved that the $|F(\omega)|^2$ of the prefilter is symmetric with respect to $T_s/2$, and the noise process is a complex white Gaussian process with power spectral density N_0. The match filter (MF) $g_{\mathrm{MF}}(nT + \varepsilon T - kT_s)$ at the receiver satisfies the Nyquist sampling theorem, namely,

$$h_{m,n} = \sum_k g(kT_s - mT) g_{\mathrm{MF}}(nT - kT_s) = \begin{cases} h_{0,0} & m = n \\ 0 & \text{otherwise} \end{cases} \tag{7-5}$$

where $h_{m,n}$ represents the equivalent channel response.

The likelihood function can be simplified as

$$L(r_f|a, \varepsilon, \theta) \propto \exp\left\{-\frac{1}{\sigma_n^2}\left[2\,\mathrm{Re}\left(\sum_{n=0}^{N-1} |h_{0,0}|^2 |a_n|^2 - 2a_n^* z_n(\varepsilon) e^{-j\theta}\right)\right]\right\} \tag{7-6}$$

where $z_n(\varepsilon) = z(nT + \varepsilon T)$ is the output of the matched filter, which can be expressed as

$$z_n(\varepsilon) = \sum_{k=-\infty}^{\infty} r_f(kT) g_{\mathrm{MF}}(nT + \varepsilon T - kT_s) \tag{7-7}$$

If N is sufficiently large, $\sum_n |a_n|^2 \to \sum E\left[|a_n|^2\right]$ based on the law of large numbers, (7-6) can be approximated as

$$L(r_f|a, \varepsilon, \theta) = \exp\left[-\frac{2}{\sigma_n^2}\mathrm{Re}\left(\sum_{n=0}^{N-1} a_n^* z_n(\varepsilon) e^{-j\theta}\right)\right] \tag{7-8}$$

After this ML formula is derived, the corresponding synchronization parameters are estimated by partial differentiation to get the maximum values. Given in the following are algorithms that can be used to search for the maximum objective function depending on the specific performance requirement.

*7.2.1.1 **Maximum Iterative Search Method*** The maximum iterative searching method is used to directly calculate the parameters $(\hat{\theta}, \hat{\varepsilon})$ to maximize the objective function by iteration with the given initial value. Suppose the estimation result $\hat{a}$ of the data sequence can be obtained or is known with the data sequence $\alpha = a_0$. Then the necessary but not sufficient condition for maximizing the value of the objective function will be

$$\begin{aligned}\frac{\partial}{\partial \theta} L(r_f | a, \theta, \varepsilon)|_{\hat{\theta},\hat{\varepsilon}} &= 0 \\ \frac{\partial}{\partial \varepsilon} L(r_f | a, \theta, \varepsilon)|_{\hat{\theta},\hat{\varepsilon}} &= 0\end{aligned} \tag{7-9}$$

As the objective function is the convex function of parameters (θ, ε), if the initial estimate is within the convergence region, one can find the values of θ and ε to satisfy (7-9) through iteration using the gradient method or maximum ascent method. The iterative search method is useful when the symbols are known in the training stage.

7.2.1.2 Error Feedback Method The error feedback method adjusts the synchronization parameters using the error signal from estimation, which is obtained by performing the partial differentiation on the following objective function and substituting the parameters with the latest estimation results $\hat{\theta}_n$, $\hat{\varepsilon}_n$:

$$\begin{aligned}&\frac{\partial}{\partial \varepsilon} L(\hat{a}, \theta = \hat{\theta}_n, \varepsilon = \hat{\varepsilon}_n) \\ &\frac{\partial}{\partial \theta} L(\hat{a}, \theta = \hat{\theta}_n, \varepsilon = \hat{\varepsilon}_n)\end{aligned} \tag{7-10}$$

The estimated error signal depends on the previous symbol α_n, which is assumed to be recovered correctly, and then it is used in the new estimation, that is,

$$\begin{aligned}\hat{\varepsilon}_{n+1} &= \hat{\varepsilon}_n + a_\varepsilon \frac{\partial}{\partial \varepsilon} L(a_n, \hat{\theta}_n, \hat{\varepsilon}_n) \\ \hat{\theta}_{n+1} &= \hat{\theta}_n + a_\theta \frac{\partial}{\partial \theta} L(a_n, \hat{\theta}_n, \hat{\varepsilon}_n)\end{aligned} \tag{7-11}$$

where the first-order discrete error feedback system is given with its loop bandwidth determined by a_ε and a_θ. A higher order tracking system can be achieved by an appropriate loop filter. When the error is small enough, the error feedback system will enter into the tracking operation mode. The process to change the error feedback system from the initial state to the tracking mode is called locking and is generally a nonlinear process.

It can be seen from the above introduction that maximum searching and error feedback methods need to get the error signal by performing partial differentiation on the objective function. However, there are essential differences: The maximum iterative searching algorithm converges to the final estimation value through iterations while the error feedback method can be done in real time. As a DTV signal is transmitted continuously, only an initial search process with less stringent time requirements is needed at the beginning. The error feedback structure can be implemented with reasonable complexity at the receiver for the tracking, and the synchronous module in the DTTB system generally adopts the error feedback structure.

7.2.2 Channel Estimation

7.2.2.1 Least Squares Criterion [2] The least square criterion, also known as the zero-forcing rule, is the most basic and simple channel estimation method. Assuming a linear discrete-time channel model with impulse response $\{h_k\}$ cascaded by a channel equalizer of infinite taps with impulse response $\{c_k\}$, one can have the equivalent filter

$$q_n = \sum_{j=-\infty}^{\infty} c_j h_{n-j} \tag{7-12}$$

For simplicity, q_0 is normalized, and the equalizer output at the sampling time k can be calculated by (7-12) as

$$\hat{I}_k = I_k + \sum_{n \neq k} I_n q_{k-n} + \sum_{j=-\infty}^{\infty} c_j n_{k-j} \tag{7-13}$$

where the first term is the desired data symbol, the second term is the intersymbol interference, and the third term is the combined noise.

The weighted coefficients of the equalizer tap are based on the equation

$$D = \sum_{\substack{k \neq -\infty \\ k \neq 0}}^{\infty} |q_k| = \sum_{\substack{k \neq -\infty \\ k \neq 0}}^{\infty} \left| \sum_{j=-\infty}^{\infty} c_j h_{k-j} \right| \tag{7-14}$$

If the intersymbol interference can be completely eliminated, the tap coefficient $D = 0$ should be

$$q_n = \sum_{j=-\infty}^{\infty} c_j h_{n-j} = \begin{cases} 1 & n = 0 \\ 0 & n \neq 0 \end{cases} \tag{7-15}$$

Since the interference value is forced to be zero when this algorithm is used, this is also known as a zero-forcing criterion. Performing the *Z*-transformation on (7-15), one has

$$Q(z) = C(z)H(z) = 1 \tag{7-16}$$

Or

$$C(z) = \frac{1}{H(z)} \tag{7-17}$$

If the intersymbol interference can be completely eliminated, a channel equalizer with frequency response the inverse of the channel response is required. The channel equalizer should have transfer function $C(z)$ inversely proportional to the frequency response of the channel model $H(z)$.

7.2.2.2 Minimum Mean-Square Error Criterion [2] The filter satisfying the minimum mean-square error criterion is a two-dimensional Wiener filter. The frequency-domain MMSE criterion is introduced below as an example, and the time-domain MMSE criterion based on the same principle will not be described in details.

Assume a linear discrete-time channel model has a frequency response $\{H_k\}$ and the transmitted data $\{X_k\}$ are in the frequency domain. The received data are

$$Y_k = X_k H_k + W_k \qquad 0 \le k < N \tag{7-18}$$

where W_k is the frequency-domain AWGN with variance σ_n^2. Using the MMSE criterion, the squared channel estimation error can be estimated by

$$J(N) = E|\varepsilon_k|^2 = E\left|H_k - \hat{H}_k\right|^2 \tag{7-19}$$

Namely,

$$J(N) = E|\varepsilon_k|^2 = \sum_{k=0}^{N-1} \frac{E(|W_k|^2)}{|X_k|^2} = \sigma_n^2 \sum_{k=0}^{N-1} \frac{1}{|X_k|^2} \tag{7-20}$$

The minimized squared channel estimation error is equivalent to

$$\min \sum_{k=0}^{N-1} \frac{1}{|X_k|^2} \quad \text{s.t.} \quad \sum\nolimits_{k=0}^{N-1} |X_k|^2 = \text{const} \tag{7-21}$$

which leads to the minimum error condition

$$|X_0| = |X_1| = \cdots = |X_{N-1}| \tag{7-22}$$

This means the frequency-domain pilots or the time-domain training sequence should have constant amplitude in the frequency domain.

In comparison:

1. The LS estimation algorithm is simple to implement, but the estimation accuracy is greatly affected by the AWGN and the intercarrier interference (ICI). In contrast, MMSE estimation algorithms usually have high implementation cost but good capability to suppress both ICI and AWGN.
2. The MMSE estimation algorithm is optimal in the sense of mean squares. Both the channel delay power spectrum (subcarrier frequency correlation) and the Doppler power spectrum (symbol–time correlation) are required. In practice, it is difficult to have the requested knowledge available at the receiver, and this

causes high implementation complexity. With the increased number of data points, the complexity increases exponentially. When coupled with the estimation error, the final performance of the MMSE estimator is slightly worse than theoretically predicted.

According to the sampling theorem, the signal can be accurately recovered from its samples if the sampling frequency satisfies the Nyquist frequency requirement. If the noise at this sampling point is additive, the original signal might be recovered using a Wiener filter in the sense of the minimum mean-square error. Consider the high complexity from this two-dimensional signal processing method. The practical way is to convert this two-dimensional signal into two one-dimensional signals, and then the one-dimensional Wiener filter can be applied. Since the complexity of the Wiener filter is still too high, a variety of simplified methods have been proposed and adopted in many applications. Although the performance is slightly worse than MMSE estimation, the computational complexity is greatly reduced. Detailed descriptions will be given below.

7.2.2.3 Simplified Multidimensional Linear Interpolation [2] Taking two-dimensional interpolation as an example, linear interpolation is carried out in both time and frequency domains. Two-dimensional linear interpolation can be decomposed into two independent interpolation processes: interpolation filter in the time direction followed by the interpolation filter in the frequency direction. The implementation process is shown in Figure 7-3.

Taking the DVB-T system as an example, the interval of the scattered pilot along the time axis is four OFDM symbols, that is, pilots repeatedly appear at the same subcarrier location after every three OFDM symbols, as shown in Figure 7-3*a*, and estimation along the time direction is given as

$$\hat{H}^{t}_{i,n+m} = \left(1 - \frac{m}{4}\right)\hat{H}^{p}_{i,n} + \frac{m}{4}\hat{H}^{p}_{i+4,n} \qquad 1 \le m \le 3 \tag{7-23}$$

where $\hat{H}^{t}_{i,n+m}$ is the channel estimation after interpolation along the time axis is completed. The interval of the pilots along the frequency axis changes to 3, $N_t/N_f = 12/4 = 3$. Then the estimation is given as

$$\hat{H}_{i,n+l} = \left(1 - \frac{l}{3}\right)\hat{H}^{t}_{i,n} + \frac{l}{3}\hat{H}^{t}_{i,n+3} \qquad 1 \le l \le 2 \tag{7-24}$$

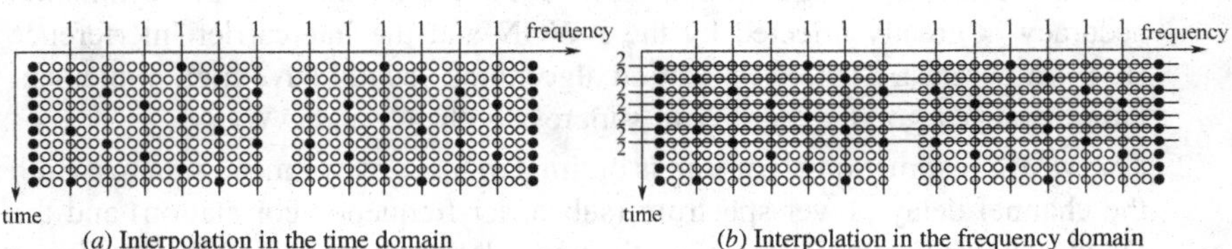

(*a*) Interpolation in the time domain (*b*) Interpolation in the frequency domain

FIGURE 7-3 Two-dimensional linear interpolation.

Once the interpolations in both the time and frequency directions are completed, the channel estimation at all points can be obtained, as shown in Figure 7-3*b*.

It is seen from the above description that, in principle, two-dimensional linear interpolation can correctly track the response of the channel, changing fast in the frequency direction, but its performance is very sensitive to the time-varying channel with large Doppler frequency shift. Meanwhile, multiple OFDM symbols are used for the estimation for better performance at the receiver, which means the frequency deviation and the common phase error among these OFDM symbols must be very small. In terms of the hardware, two-dimensional linear interpolation needs a lot of memory, which also increases the cost of the implementation. Theoretically speaking, higher order polynomial interpolation may estimate the channel response more accurately while its complexity may greatly increase. Therefore, there will be a trade-off between performance and complexity.

7.2.2.4 SVD-Based Channel Estimation [3] The simple interpolation algorithm may not be able to suppress both ICI and AWGN well and also suffers from interpolation error. If it will not provide satisfactory estimation performance, one can choose a method based on SVD channel estimation.

The SVD of the autocorrelation matrix R_{HH} for the channel transfer function can be expressed as

$$R_{HH} = U\Lambda U^H \tag{7-25}$$

where U is the unitary matrix containing the singular vector and Λ is the diagonal matrix containing the singular values $\lambda_1 \geq \lambda_2 \geq \cdots \geq \lambda_N$. The best p^{th}-order estimator is

$$H_p = U\begin{bmatrix} \Delta_p & 0 \\ 0 & 0 \end{bmatrix} U^H H_{\text{LS}} \tag{7-26}$$

where U^H represents the Hermitian transformation of U, $\Delta_p = \text{diag}[\delta_0, \delta_1, \delta_2, \ldots, \delta_{p-1}]$ is the $p \times p$ upper left submatrix of Δ, and Δ is defined as

$$\Delta = \Lambda\left(\Lambda + \frac{\beta}{\text{SNR}} I\right)^{-1} = \text{diag}\left(\frac{\lambda_1}{\lambda_1 + \beta/\text{SNR}}, \ldots, \frac{\lambda_N}{\lambda_N + \beta/\text{SNR}}\right) \tag{7-27}$$

where I is the identity matrix and β is a constant determined by the constellation of the modulation, which can be obtained using (7-31).

It can be seen from the above equations that U^H can be taken as the transformation matrix and the singular value λ_k of matrix R_{HH} is the channel energy of the kth transformation coefficient. Because U is a unitary matrix, U^H can be seen as the rotation of H_{LS}, and then the components of U^H are not relevant.

Figure 7-4 shows the block diagram of the low-order estimator of order p. First, the received data signal vector Y is multiplied by X^{-1}(X is the transmitted signal vector), and the LS estimation of H_{LS} can be obtained; then U^H is generated by rotating H_{LS}.

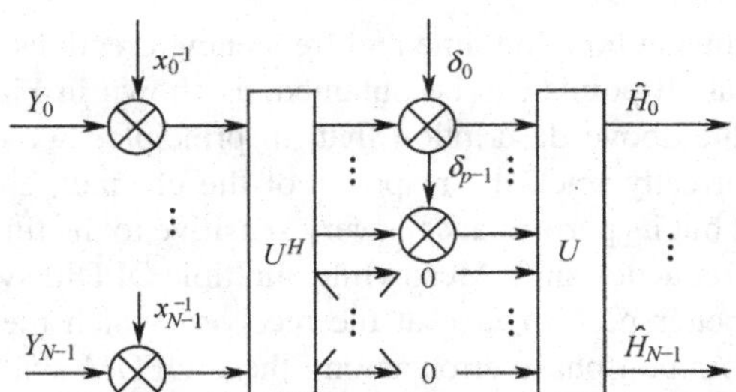

FIGURE 7-4 SVD-based pth-order channel estimation.

Next, the final channel estimations are calculated by (7-26). When passing through this low-order estimator, the results can be seen as a subspace of order p within the space of the final channel estimations by the LS estimator. If the dimension of the subspace is small and the channel features are well characterized by this subspace, an estimator with good performance and low complexity can be achieved. The channel performance at noninteger sampling points has also been proven to be satisfactory. Because the estimation is carried out only in the subspace of the channel, this method suffers from the errorfloor effect. To eliminate the error floor at a given SNR value, the order of the estimator must be large enough. However, an increase in the estimator's order means an increase in computational complexity, and there exists a trade-off. When designing an OFDM system, the length of the cyclic prefix should be larger than the maximum delay spread of the channel, and excellent estimation performance can be achieved if the order p is equal to the length of CP. Since the OFDM symbol length is usually much larger than that of CP, the computational complexity can be greatly reduced.

7.2.2.5 DFT-Based Channel Estimation [4] To reduce MMSE algorithm complexity, one can also use IDFT/DFT for channel estimation. Figure 7-5 is a diagram of the DFT-based channel estimation algorithm with the process described as follows. First, carry out the channel estimation for the LS algorithm in the time domain by IDFT and conduct a linear transformation. Then perform the filter, taking advantage

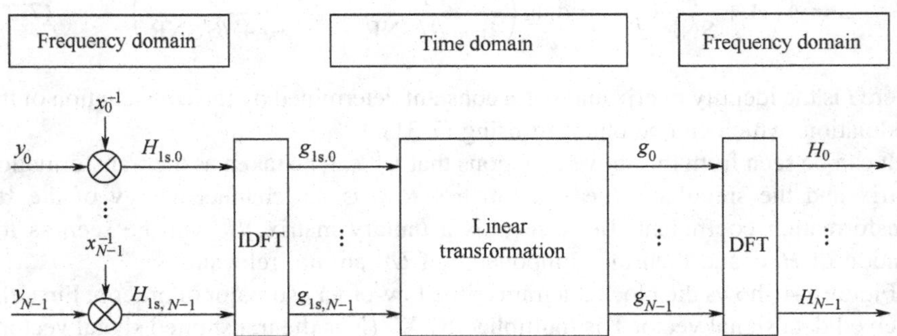

FIGURE 7-5 Diagram of DFT-based channel estimation algorithm.

that the energy of the time-domain channel impulse response is concentrated at a fewer sampling points. Finally output the estimation results in the frequency domain through DFT.

The LS estimation in an OFDM symbol can be expressed as

$$H_{\mathrm{LS}} = X^{-1}Y = \left[\frac{Y(0)}{X(0)}\frac{Y(1)}{X(1)}\cdots\frac{Y(N-1)}{X(N-1)}\right]^T \tag{7-28}$$

where N is the number of subcarriers and X and Y denote the vectors of the transmitted symbol sequence $\{X(n)\}$ and received symbol sequence $\{Y(n)\}$, respectively. The X for LS estimation can be either the pilots or training sequences inserted within each OFDM symbol and can also be the data after detection.

The linear minimum mean-square error (LMMSE) estimation may be obtained as

$$\begin{aligned} H_{\mathrm{LMMSE}} &= R_{HH}[R_{HH} + \sigma_n^2(XX^H)^{-1}]^{-1}H_{\mathrm{LS}} \\ R_{HH} &= E\{HH^H\} \end{aligned} \tag{7-29}$$

where R_{HH} is the autocorrelation matrix of the frequency-domain channel transfer function and σ_n^2 is the variance of AWGN. To further reduce the complexity of the MMSE algorithm, the expectation of $(XX^H)^{-1}$ (i.e., $E\{(XX^H)^{-1}\}$) can be used to replace $(XX^H)^{-1}$ and simulations confirm that the performance deterioration caused by the approximation is negligible. For equiprobable modulated symbols at the transmitter,

$$E\{(XX^H)^{-1}\} = E\{|1/x_k|^2\}I \tag{7-30}$$

The SNR is defined as $E\{|x_k|^2\}/\sigma_n^2$, and the following is obtained by the further simplification:

$$H_{\mathrm{LMMSE}} = WH_{\mathrm{LS}} \tag{7-31}$$

where

$$W = R_{HH}\left(R_{HH} + \frac{\beta}{\mathrm{SNR}}I\right)^{-1}$$

$$\beta = E\{|x_k|^2\}E\left\{|1/x_k|^2\right\}$$

where β is a constant determined by the constellation of the modulation. For example, the β of the 16QAM modulation constellation is 17/9. One can see that if the autocorrelation matrix and SNR are known in advance, W only needs to be calculated once. However, W needs to be calculated frequently in practice. The following can be obtained by performing IDFT to the channel transfer function H_{LS} derived from the

LS algorithm:

$$g_{\text{LS}} = \text{IDFT}(H_{\text{LS}}) \tag{7-32}$$

As the channel frequency response is usually oversampled when the DA and DD methods are used, the channel energy is mainly concentrated in a smaller portion of the equivalent channel impulse response vector g_{LS} while the noise energy is distributed throughout the whole range of the vector. The most straightforward method is to ignore those in g_{LS} with small SNR so that only the parameters with higher signal energy are used in the frequency domain via DFT. The complexity will be greatly reduced if this time-domain filtering is carried out. One can get $g = Qg_{\text{LS}}$ by performing the linear transform to get g_{LS} and then derive $H = \text{DFT}(g)$ by DFT.

It should be noted that methods based on DFT and SVD require frequency-domain statistics (R_{HH}) and SNR in advance yet these parameters are generally unknown in practice. The DFT algorithm is a simple and efficient interpolation algorithm. However, when the interpolation is carried out on a N-point, uniformly sampled sequence from an analog signal with negative values, the most accurate interpolation requires the original analog signal to not only be bandlimited (with its bandwidth less than the Nyquist frequency) but also have the discrete spectrum, which means that the distribution of the channel delay must have the same time interval as the sampling interval of the OFDM symbol. In practice, this requirement usually cannot be satisfied even though the channel impulse response is finite in length and its frequency response bandwidth is within the Nyquist frequency. The power distribution of the channel delay is not discrete but continuous, as shown in Figure 7-6. Therefore, simplifying the algorithm by setting some data points to zero will inevitably introduce errors due to the aliasing from the energy leakage during the DFT-based interpolation, which results in the error floor of the channel estimation.

The above two methods are based on frequency-domain correlation, and one can surely design the channel estimator using the time-domain correlation with a similar principle. The time-domain correlation matrix instead of R_{HH} can be introduced, and this can be figured out easily.

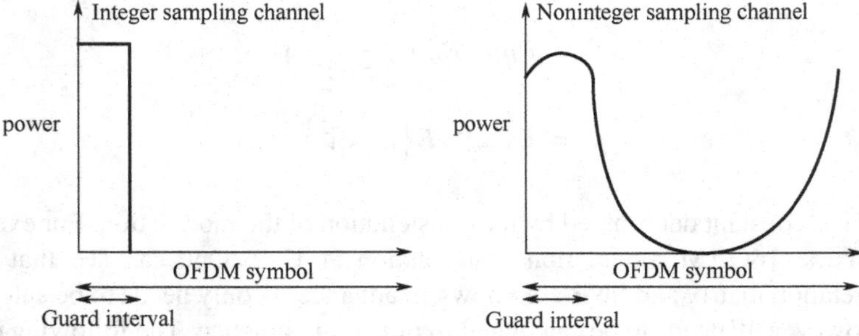

FIGURE 7-6 Integer sampling channel and noninteger sampling channel.

7.3 SINGLE-CARRIER SYSTEMS

7.3.1 Timing Synchronization [5,6]

In general, timing error includes both frame synchronization error and sampling clock error. The former refers to the offset εT between the transmitter and receiver assuming all other time delays have been successfully removed while the latter refers to the sampling clock T_s at the receiver that cannot be perfectly aligned to the clock T at the transmitter.

7.3.1.1 Typical NDA Timing Estimation Algorithms As previously introduced, when the objective function for the synchronization parameters (θ, ε) is obtained, the timing estimation independent of the data and the phase will first be derived, that is, obtaining the estimate on ε by removing the parameters a and θ. To eliminate the data correlation, equation (7-8) is multiplied by the symbol probability $P({}^i a)$, where ${}^i a$ denotes the ith symbol among all M symbols, and the overall possibilities of each symbol of these M symbols are then accumulated. Assume symbols are independent and equiprobable. Then the likelihood function will be

$$L(\theta, \varepsilon) = \prod_{n=0}^{N-1} \sum_{i=1}^{M} \exp\left[-\frac{2}{\sigma_n^2} \mathrm{Re}\left({}^i a_n^* z_n(\varepsilon) e^{-j\theta} \right) P({}^i a) \right] \tag{7-33}$$

There are different ways to solve (7-33). A typical method is to assume a uniformly distributed phase θ, and then one can have

$$\begin{aligned} L_2(a, \varepsilon) &\approx \prod_{n=0}^{N-1} \int_{-\pi}^{\pi} \exp\left[-\frac{2}{\sigma_n^2} |z_n(\varepsilon)| a_n^* \ \mathrm{Re}\left(e^{j(-\arg a_n - \theta + \arg z_n(\varepsilon))} \right) \right] d\theta \\ &= \prod_{n=0}^{N-1} I_0 \left(\frac{|z_n(\varepsilon) a_n^*|}{\sigma_n^2/2} \right) \end{aligned} \tag{7-34}$$

Since $|a_n|$ is constant, (7-34) can be applied to all systems with phase modulation (M-PSK). For M-QAM, the symbol-based average is needed to derive the NDA synchronization algorithm for the M-QAM signal as $|a_n|$ is no longer constant. The objective function can be further simplified by expanding the Bessel function. Because $I_0(x) \approx 1 + \frac{1}{2}x^2 \quad |x| \ll 1$, one can get

$$\begin{aligned} &\text{NDA}: \quad \hat{\varepsilon} = \arg \max_{\varepsilon} L_1(\varepsilon) \approx \arg \max_{\varepsilon} \sum_{n=0}^{N-1} |z_n(\varepsilon)|^2 \\ &\text{DA}: \quad \hat{\varepsilon} = \arg \max_{\varepsilon} L_2(a, \varepsilon) \approx \arg \max_{\varepsilon} \sum_{n=0}^{N-1} |z_n(\varepsilon)|^2 |a_n|^2 \end{aligned} \tag{7-35}$$

It can be seen that the algorithms of NDA and DA are the same for M-PSK. The logarithmic NDA auxiliary objective function will be created by the differential operation on the parameter ε, and the error feedback algorithm can be derived as

$$\frac{\partial L(\varepsilon)}{\partial \varepsilon} = \frac{\partial}{\partial \varepsilon}\sum_{n=0}^{N-1}|z\,(nT+\varepsilon T)|^2 = \sum_{n=0}^{N-1} 2\,\mathrm{Re}\,[z(nT+\varepsilon T)\dot{z}^*(nT+\varepsilon T)]$$
$$\dot{z}(nT+\varepsilon T) = \frac{\partial z(nT+\varepsilon T)}{\partial \varepsilon} \tag{7-36}$$

As the summation is carried out in the loop filter, let $\varepsilon = \hat{\varepsilon}$, and the error signal is derived as

$$x(nT) = \mathrm{Re}\,[z(nT+\hat{\varepsilon}T)\dot{z}^*(nT+\hat{\varepsilon}T)] \tag{7-37}$$

Finally, the approximate differential result is derived as

$$x(nT) = \mathrm{Re}\left\{z(nT+\hat{\varepsilon}T)\left[z^*\left(nT+\frac{T}{2}+\hat{\varepsilon}T\right) - z^*\left(nT-\frac{T}{2}+\hat{\varepsilon}T\right)\right]\right\} \tag{7-38}$$

In addition, according to the Gardner algorithm [7], two sampling values, independent of the carrier phase offset, are required to make the estimation on each data symbol: one near the decision point, the other in the middle of the adjacent two symbol decision points. Since this is done independent of carrier, timing adjustment can be completed before the carrier recovery, and the timing recovery loop and carrier recovery loop are therefore independent of each other. This facilitates the design of the demodulator. However, there is a problem with this algorithm: the high self-noise generally required to be suppressed using the loop filter.

Another commonly used NDA timing estimation algorithm is based on spectral estimation. Assume $|z_n(\varepsilon)|^2$ can be taken as the periodic signal and $\hat{\varepsilon}$ can be found without maximum searching. The unbiased estimate can be obtained from the first coefficient c_1 in the Fourier series expansion $|z_n(\varepsilon)|^2$,

$$\hat{\varepsilon} = -\frac{1}{2\pi}\arg\ c_1 \tag{7-39}$$

As the NDA algorithm is a unique and relatively simple estimation method for ε, it has been widely used in practice.

7.3.1.2 Typical DD/DA Timing Estimation Algorithm If DD/DA timing estimation is used, the estimated value of the parameter a should have been obtained or known. Substituting the estimated value into (7-8) gives

$$L(\hat{a},\theta,\varepsilon) = \exp\left[-\frac{2}{\sigma_n^2}\mathrm{Re}\left(\sum_{n=0}^{N-1}\hat{a}_n^* z_n(\varepsilon)e^{-j\theta}\right)\right] \tag{7-40}$$

As the phase θ may be unknown, one can consider the joint estimate of (θ, ε).

Define $\mu(\varepsilon) = \sum_{n=0}^{N-1} \hat{a}_n^* z(\varepsilon)$. The two-dimensional search for (θ, ε) can be changed to the one-dimensional search

$$\max_{\varepsilon,\theta} \text{ Re } \left[\mu(\varepsilon)e^{-j\theta}\right] = \max_{\varepsilon,\theta} |\mu(\varepsilon)| \text{Re } [e^{-j(\theta - \arg \mu(\varepsilon))}] \tag{7-41}$$

First, the timing estimate can be obtained by maximizing the absolute value of $\mu(\varepsilon)$,

$$\hat{\varepsilon} = \arg \max_{\varepsilon} |\mu(\varepsilon)| \tag{7-42}$$

where the second term Re $[e^{-j(\theta - \arg \mu(\varepsilon))}]$ in (7-41) reaches its maximum when $\theta = \arg\mu$ (ε), that is,

$$\hat{\theta} = \arg \mu(\hat{\varepsilon}) \tag{7-43}$$

One can use the timing error feedback system for timing recovery. Take the partial differentiation of the likelihood function with regard to the parameter ε:

$$\frac{\partial}{\partial \varepsilon} L(\hat{a}, \theta, \varepsilon) \propto -\frac{2}{\sigma_n^2} \text{Re} \left[\sum_{n=0}^{N-1} \hat{a}_n^* \frac{\partial}{\partial \varepsilon} z_n(nT + \varepsilon T) e^{-j\hat{\theta}}\right] \tag{7-44}$$

The error signal will be

$$x(nT) = \text{Re} \left[\sum_{n=0}^{N-1} \hat{a}_n^* \frac{\partial}{\partial \varepsilon} z_n(nT + \varepsilon T)\bigg|_{\varepsilon=\varepsilon'} e^{-j\hat{\theta}}\right] \tag{7-45}$$

The sequence differentiation can be done by letting the sequence $\dot{z}(nT + \varepsilon T)$ pass through the digital filter with impulse response $h_d(kT_s)$. The output of the filter will be the error signal, and this error signal will then be fed back for the timing adjustment.

The signal value at the best sampling point in the all-digital receiver is usually obtained by interpolation on the sampling sequence instead of from the sampling sequence directly. Therefore, the timing adjustment of the interpolation filter is based on the signal instead of the clock of the local oscillator. The continuous-time signal $r_f(t)$ at the receiver is with symbol period T. Since $r_f(t)$ is band limited, one can sample the input signal at the local fixed clock rate $1/T_s$ to get $r_f(mT_s)$. Because the symbol period T and the local sampling clock are independent and there is always a minor difference between two clocks, the ratio T/T_s in the system is generally an irrational number. The function of the interpolation filter $h_I(t)$ is to perform the interpolation on r_f (mT_s) and create a new sample data at time $nT_I + \varepsilon T_I$. Here, $T_I = T/M$ is the interpolation interval, M is an integer representing the oversampling ratio, and ε is the time delay relative to T_I. In DTTB systems, all timing loops generally use an error feedback structure. Only the interpolation control part of the error feedback system is described here. Since the ratio T/T_s is unknown for the actual loop, the loop is used to generate both the decimation coefficient m_k and the interpolated coefficient μ_k. The operation to generate the sequence $\{r_f(kT_s + \hat{\mu}_n T_s)\}$ out of the sequence $\{r_f(kT_s)\}$ is called interpolation. The sequence $\{r_f(kT_s + \hat{\mu}_n T_s)\}$ will be further processed only within the time slot $m_n\ T_s$. The operation is called decimation, which is easy to implement by simply discarding the unwanted interpolation points. The interpolation can be achieved through a variety of filters, including finite impulse response (FIR) filters, polynomial FIR filters, and Lagrange interpolation filters.

7.3.2 Carrier Synchronization [8, 9]

7.3.2.1 Carrier Phase Estimation

1. *DD Carrier Phase Estimation.* For the DD method, the objective function is derived by substituting a_n with the data decision value of $\hat{a}_n$ assuming that the timing synchronization has been completed and ε is known:

$$L(\hat{a}, \hat{\varepsilon}, \theta) = \mathrm{Re}\left[\sum_{n=0}^{N-1} \hat{a}_n^* z_n(\hat{\varepsilon}) e^{-j\theta}\right] \tag{7-46}$$

When the match filter output $z_n(\hat{\varepsilon})$ and decision data $\hat{a}_n$ satisfy certain phase relationships, the objective function reaches the maximal value, i.e.,

$$e^{j\hat{\theta}} = e^{j\arg \sum_n \hat{a}_n^* z_n(\hat{\varepsilon})} \tag{7-47}$$

It is easy to obtain the phase error signal after differentiation on θ when the feedback system is used. We now use the first-order digital phase-locked loop (PLL) as an example to illustrate how the phase error feedback system works with the assumption that there is no new noise generated with accurate timing as well as correct symbol decision, i.e., $a_n = \hat{a}_n$ (at this time, the DD mode is the same as the DA mode). In this case, the match filter output will be

$$z_n = a_n e^{j\theta_0} \tag{7-48}$$

and the error signal e_n is given as

$$e_n = \mathrm{Im}\left[|a_n|^2 e^{j(\theta_0 - \hat{\theta}_n)}\right] = |a_n|^2 \sin\ \psi_n \tag{7-49}$$

where $\psi_n = \theta_0 - \hat{\theta}_n$ is the phase error signal that will be further processed in the loop filter. The phase estimation value is then updated in an integrator based on the formula

$$\hat{\theta}_{n+1} = \hat{\theta}_n + k_1 e_n \tag{7-50}$$

Substituting the value into the above equation yields

$$\hat{\theta}_{n+1} = \hat{\theta}_n + k_1 \sin\ \psi_n \tag{7-51}$$

where k_1 is a loop constant and the whole feedback system is shown in Figure 7-7.

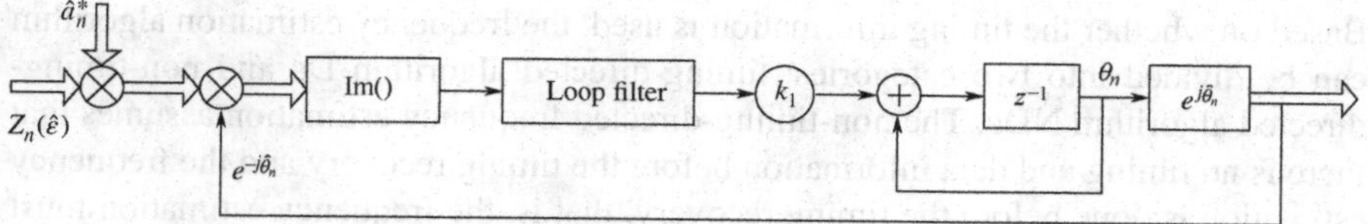

FIGURE 7-7 Phase error feedback system.

2. *NDA Carrier Phase Estimation.* Use of the NDA method would degrade the performance of the synchronization unit, and therefore the DD method is generally used for phase retrieval of the QAM signal at high SNR. When there is no reliable data estimate available such as in the low-SNR case, the NDA algorithm can be adopted for M-PSK. If implemented by the analog circuitry, the NDA method is almost the only one that can be used with low complexity and is not suitable for digital circuitry. The NDA algorithm can be briefly introduced as

$$F(|z_n(\hat{\varepsilon})|)e^{j \arg z_n(\hat{\varepsilon})M} \tag{7-52}$$

where $F(|x|)$ is an arbitrary function as shown in Figure 7-8 and the algorithm is called the Viterbi algorithm.

7.3.2.2 Carrier Frequency Estimation When there is frequency offset Ω, the linear model described above has to be modified and the transmission phase θ_T can be defined as $\theta_T = \Omega t + \theta$ with θ the fixed phase offset. The input signal $s(t)$ can then be written as $s(t)e^{j(\Omega t+\theta)}$. Assume that the channel frequency response $C(\omega)$ and prefilter frequency response $F(\omega)$ are flat within the frequency range $|\omega| \leq 2\pi B + |\Omega_{\max}|$ with B the single-sided bandwidth for $s(t)$ and $\Omega_{\max}$ the maximum frequency error. The signal $S_f(t,\Omega)$ can be written as

$$S_f(t, \Omega) = \sum_{n=0}^{N-1} a_n g(t - nT - \varepsilon T)e^{j(\Omega t+\theta)} \tag{7-53}$$

The matched filter output $z_n(\varepsilon, \Omega)$ is given as

$$z_n(\varepsilon, \Omega) = \sum_{k=-\infty}^{\infty} r_f(kT_s)e^{-j\Omega kT_s} g_{MF}(nT + \varepsilon T - kT_s) \tag{7-54}$$

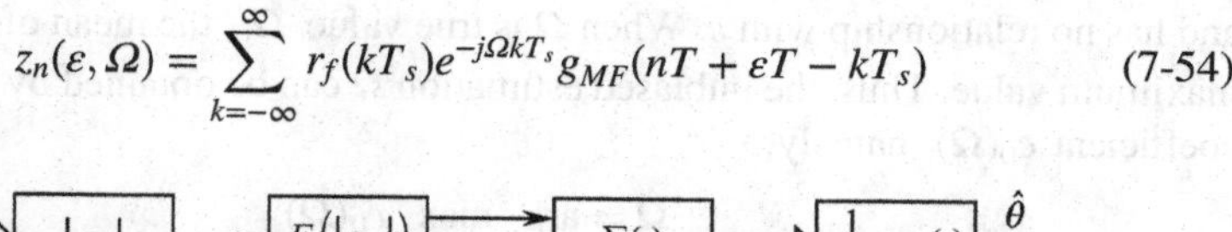

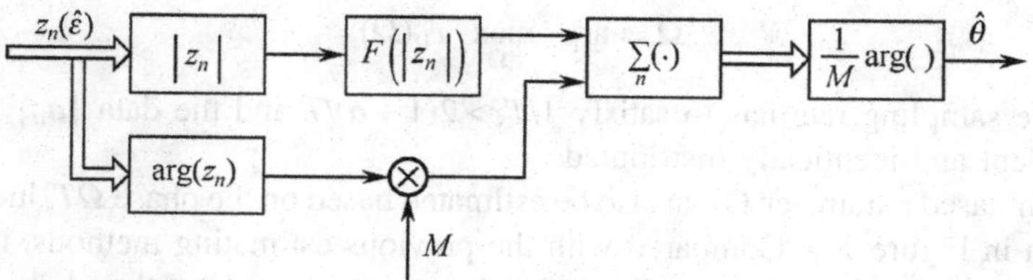

FIGURE 7-8 NDA phase retrieval algorithm proposed by Viterbi.

Based on whether the timing information is used, the frequency estimation algorithm can be divided into two categories: timing-directed algorithm Dε and non-timing-directed algorithm NDε. The non-timing-directed frequency estimation assumes that there is no timing and data information before the timing recovery and the frequency estimation is done before the timing recovery, that is, the frequency estimation must be independent of the estimation on other parameters. Without the timing information, the SNR from the match filter output is lower than that with the timing information, and the corresponding NDε algorithm is therefore vulnerable to AWGN. To solve this problem, a long average interval is used together with the two-stage approximation method, i.e., the frequency acquisition and frequency tracking stages. In this case, the frequency acquisition stage is designed to quickly acquire the rough frequency estimation within a large frequency range and at a short time while the tracking performance is not important. In the frequency tracking stage, the design goal is to optimize the tracking performance and a large capturing range is no longer needed.

The algorithms for the frequency tracking stage will be described below and share the requirement of correct timing, which requires that the timing recovery module operate properly within a certain frequency offset range. Similarly, the corresponding frequency feedback algorithm can be derived by differentiation on the likelihood function with respect to Ω.

1. *NDε Frequency Estimation.* Those algorithms are commonly used in the frequency acquisition stage without any timing information. That is, the receiver should estimate the frequency offset Ω first and then estimate other synchronization parameters after the compensation of Ω.

One can get the NDε algorithm by analyzing the spectrum of the Fourier expansion for the time-domain waveform $|z(lT+\varepsilon T)|^2$:

$$\sum_{l=-L}^{L} |z(lT + \varepsilon T, \Omega)|^2 = c_0 + 2\,\mathrm{Re}\,[c_1 e^{j2\pi\varepsilon}] + \sum_{|n|\geq 2} c_n e^{j2\pi n\varepsilon} \tag{7-55}$$

As the signal is band limited, only three terms of the Fourier coefficients, i.e., c_{-1}, c_0, c_1, have non zero mean for the further consideration while the summation of other terms ($\sum_{|n|\geq 2} c_n e^{j2\pi n\varepsilon}$) can be treated as disturbance. The value of c_0 is only related to Ω and has no relationship with ε. When Ω is true value Ω_0, the mean of c_0 will reach its maximum value. Thus, the unbiased estimation $\hat{\Omega}$ can be obtained by maximizing the coefficient $c_0(\Omega)$, namely,

$$\hat{\Omega} = \arg \max_{\Omega} \; c_0(\Omega) \tag{7-56}$$

where the sampling rate has to satisfy $1/T_s > 2(1+\alpha)/T$ and the data $\{a_n\}$ must be independent and identically distributed.

The unbiased parameter $\hat{\Omega}$ can also be estimated based on the phase ΩT_s increment, as shown in Figure 7-9. Compared with the previous estimating methods, the main difference is that the parameter Ω can be directly estimated by the phase rotation vector $e^{j\Omega t}$. The phase characterized by $\theta(kT_s) = \theta + \Omega kT_s$ has an increment of

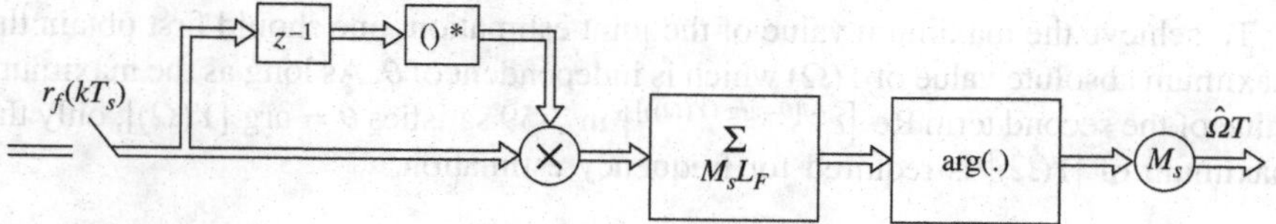

FIGURE 7-9 Frequency estimation from incremental phase.

$\Delta\theta(kT_s) = \Omega T_s$ between two sampling intervals, and the average of $\hat{\Omega}T$ can be replaced by the mathematical expectation of the negative phase $r_f(kT_s)r_f^*[(k-1)T_s]$, namely, $E[r_f(kT_s)r_f^*[(k-1)T_s]]$. As its magnitude is equal to Ω_0 T, an unbiased estimation of Ω can then be obtained by

$$\hat{\Omega}T_s = \arg\left[\sum_{L_s} r_f(kT_s)r_f^*[(k-1)T_s]\right] \tag{7-57}$$

2. *Dε Frequency Estimation.* The timing-directed algorithm Dε is used with the assumption that the timing synchronization has been established prior to the frequency synchronization. Since the frequency offset range is determined by not only the capturing range of the frequency estimation algorithm but also the timing recovery capability of the timing estimation algorithm under certain frequency offset, therefore the frequency offset for Dε should not too much. The algorithms introduced are generally applicable when the frequency offset satisfies $|\Omega T/2\pi| \leq 0.15$, yet this restriction is not very strict because these algorithms are generally used in the frequency tracking stage. If the frequency offset cannot meet this requirement, rough estimation on the frequency offset in the frequency acquisition stage can then be performed to reduce the frequency offset into the required range.

First, let's discuss the DD case with data symbols of length N noted as $\{a_{0,\ n}\}$ known. The ML algorithm requires the following joint estimation on $\{\Omega,\theta\}$:

$$\{\hat{\Omega},\hat{\theta}\} = \arg\max_{\Omega,\theta} \sum_{n=-(N-1)/2}^{(N-1)/2} a_{0,n}^* e^{-j\theta} e^{-jnT\Omega} z_n \tag{7-58}$$

The two-dimensional maximization problem for $\{\Omega,\theta\}$ in the above equation can be simplified to the one-dimensional maximization problem

$$\begin{aligned}\arg\max_{\Omega,\theta} \operatorname{Re}&\left[\sum_{n=-(N-1)/2}^{(N-1)/2} a_{0,n}^* e^{-j\theta} e^{-jnT\Omega} z_n\right] \\ &= \arg\max_{\Omega,\theta} |Y(\Omega)| \operatorname{Re}\left[e^{-j[\theta-\arg(Y(\Omega))]} z_n\right]\end{aligned} \tag{7-59}$$

where $Y(\Omega) = \sum_{n=-(N-1)/2}^{(N-1)/2} a_{0,n}^* e^{-jnT\Omega} z_n$.

To achieve the maximum value of the joint estimation, one should first obtain the maximum absolute value of $Y(\Omega)$ which is independent of θ. As long as the maximum value of the second term Re $\left[e^{-j[\theta-\arg(Y(\Omega))]}\right]$ in 7-59 satisfies $\hat{\theta} = \arg[Y(\hat{\Omega})]$, only the maximum of $|Y(\Omega)|$ is required for frequency estimation,

$$\hat{\Omega} = \arg \max_{\Omega} |Y(\Omega)| \tag{7-60}$$

The maximum value of $|Y(\Omega)|$ is equal to that of $Y(\Omega)Y^*(\Omega)$. Therefore, the sufficient condition obtaining this maximum value is that the partial differentiation of $|Y(\Omega)|$ relative Ω equals to zero, and one can get the approximation

$$\hat{\Omega}T = \arg\left[\sum_{n=-(N-1)/2}^{(N-1)/2} b_n \frac{a^*_{0,n+1}}{a^*_{0,n}} (z_{n+1} z^*_n)\right] \tag{7-61}$$

where $b_n = \frac{1}{2}[(\frac{N^2-1}{4}) - n(n+1)]$. The algorithm actually performs the weighted average on phase increments $\arg(z_{n+1} z^*_n)$ between two consecutive samples with the maximum phase increment caused by the frequency shift π. As even larger phase increment ($>\pi$) cannot be separated from the negative value acquired from $2\pi - \hat{\Omega}T$, the capture range is $|\Omega T|/2\pi < 1/2$.

7.3.3 Channel Estimation and Equalization

Time-domain channel estimation and equalization are often adopted in the single-carrier system.

7.3.3.1 Linear Equalizer [1] The most frequently used time-domain equalizer is the FIR transversal filter. Suppose that the channel has L multipaths. Then the input y_k to the equalizer can be expressed as

$$y_k = \sum_{j=0}^{L} h_j I_{k-j} + n_k \tag{7-62}$$

where $\{n_k\}$ is the AWGN sequence, $\{h_k\}$ is the channel impulse response tap coefficient, and $\{I_k\}$ is the input data sequence.

As shown in Figure 7-10, if the number of taps before and after the main path of the equalizer is N, one can get the estimate value on the kth symbol for the equalizer output as

$$\hat{I}_k = \sum_{j=-N}^{N} c_j y_{k-j} \tag{7-63}$$

where $\{c_j\}$ are the filter tap coefficients.

If the decision on the equalizer output $\hat{I}_k$ is different from transmitted data I_k, an error occurs. Whether the decision is correct largely depends on the filter tap

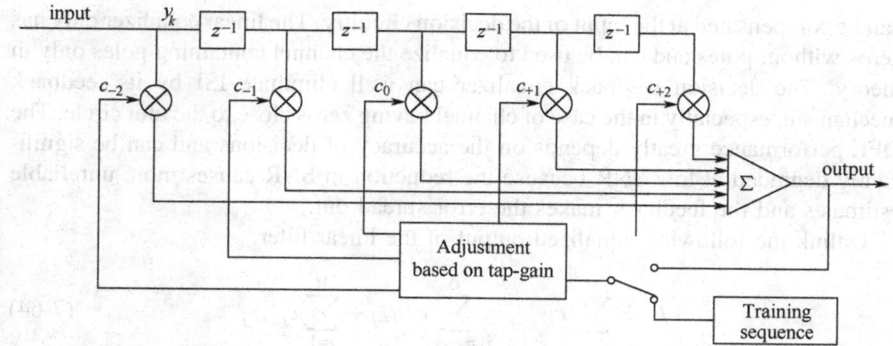

FIGURE 7-10 Linear equalizer with transversal filter structure.

coefficients $\{c_k\}$, and there is much literature focusing on optimal determination methods for filter tap coefficients. The most commonly used criteria for optimizing the equalizer coefficients are the above-mentioned least square criterion and the minimum mean-square error criterion.

7.3.3.2 Decision Feedback Equalizer [1] The linear FIR filter eliminates the intersymbol interference at the cost of SNR, and its capacity to handle the deep fading channel, especially the channel with null spectrum, is very limited. The decision feedback equalizers have been proved to be an effective solution to this problem.

The DFE equalizer is a typical nonlinear equalizer, and Figure 7-11 gives the block diagram of the operating principle of the DFE equalizer. The equalizer consists of two filters, a feedforward filter (FIR) and a feedback (IIR) filter, with the tap interval of both filters equal to the symbol interval T. The feedforward filter helps cancel the leading ISI (determined by the position of the reference tap) by converting this ISI into the lagging ISI; the feedback filter cancels the lagging ISI by using the output of decision circuitry as the input. Compared with the linear equalizer, the operating principle of the DFE equalizer is as follows: If a transmitted data symbol has been obtained, its interference on the following receiving symbols (ISI) can be calculated by the feedback filter, and its impact on the sampling point for the next data symbol

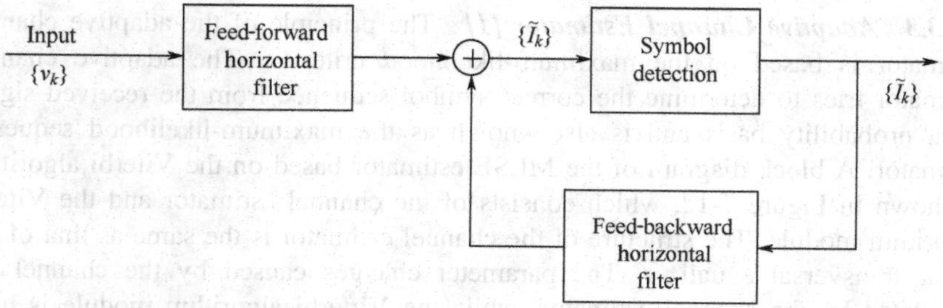

FIGURE 7-11 DFE equalizer structure.

can be compensated at the input of the decision circuitry. The linear equalizer only has zeros without poles and can be used to equalize the channel containing poles only in theory. The decision feedback equalizer can well eliminate ISI by its feedback mechanism, especially in the case of channel having zeros close to the unit circle. The DFE performance greatly depends on the accuracy of decisions and can be significantly degraded at low SNR because the reduction in SNR causes more unreliable estimates and the feedback makes the error spread out.

Unlink the following equalized output of the linear filter:

$$\hat{I}_k = \sum_{j=-N}^{M} c_j y_{k-j} = \sum_{j=-N}^{0} c_j y_{k-j} + \sum_{j=1}^{M} c_j y_{k-j} \tag{7-64}$$

The output of the decision feedback equalizer is expressed as

$$\hat{I}_k = \sum_{j=-N}^{0} c_j y_{k-j} + \sum_{j=1}^{M} b_j \hat{I}_{k-j} \tag{7-65}$$

The IIR filter in the feedback part calculates the impact on $\hat{I}_k$ using the M data symbols previously output from the decision circuitry, and the backward ISI interference can be completely eliminated without introducing any noise if there is no decision error, i.e., $\hat{I}_k = I_k$. As the noise only comes from the feedforward part (FIR) of the equalizer, the noise performance of the DFE can be greatly improved compared with the linear equalizer. In addition, the IIR filter in the feedback part does not produce a derivative multipath with the cancellation, and the length of the IIR is only determined by the length of the backward delay of ISI to be eliminated. Especially under strong channel fading, the performance improvement of the decision feedback equalizer over the linear equalizer becomes much more obvious. If decision error occurs, it will be fed back to the IIR for the next calculation and affect the several data symbols in the following. This may lead to new error decisions, introducing the error propagation problem, until the incorrect decision results in the symbol leaving the IIR. As in the case of linear equalization, the decision feedback equalizer calculates the tap coefficients based on the same LS and MMSE criteria.

7.3.3.3 Adaptive Channel Estimator [1] The principle of the adaptive channel estimator is based on the maximum-likelihood criterion. The adaptive channel estimator tries to determine the correct symbol sequence from the received signal on a probability basis and is also known as the maximum-likelihood sequence estimator. A block diagram of the MLSE estimator based on the Viterbi algorithm is shown in Figure 7-12, which consists of the channel estimator and the Viterbi algorithm module. The structure of the channel estimator is the same as that of the linear transversal equalizer. The parameter changes caused by the channel are calculated by the channel estimator, while the Virterbi algorithm module is used to generate the output by minimizing the errors between the received sequence and the

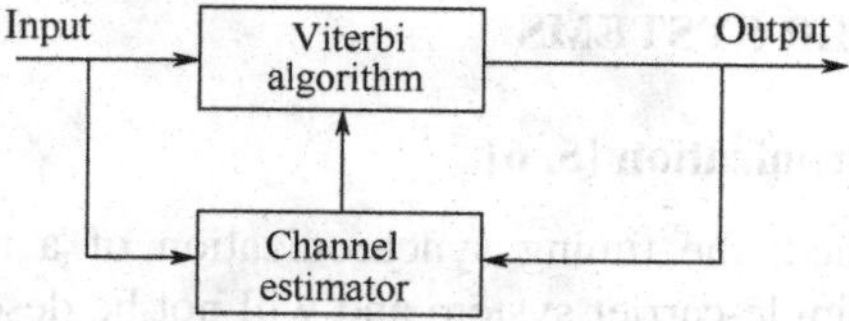

FIGURE 7-12 Maximum-likelihood sequence estimator based on Viterbi algorithm.

output of the channel estimator. A short, known sequence is used for the initial adjustment of the tap coefficients of the adaptive channel estimator before starting the equalization process, and the error signal can be directly obtained from the output of the decision circuitry during the tracking stage.

The MLSE estimator is optimal from an error probability point of view. However, the computational complexity of MLSE under the ISI channel will exponentially increase with channel length. For example, if the number of symbols is M and the number of symbols affected by the ISI is L, then $M\ (L+1)$ metrics have to be calculated for each newly received symbol by the Viterbi algorithm module, which cannot be implemented in practice. In contrast, the computational complexity of the linear filter is a linear function of channel length L, and as mentioned earlier, the computational complexity is quite limited. When there exists a deep fading in the channel, the linear equalizer usually introduces very high gain for the compensation, which inevitably causes the noise enhancement. So, it is highly preferable to use a nonlinear equalizer under this circumstance.

7.3.3.4 Adaptive Equalizer [1] When introducing the above equalization methods, it is assumed that the channel characteristics are known to the receiver, which is impractical, and the equalizer must be able to automatically track the channel change and make the adjustment in time, which is called adaptive equalization.

The adaptive equalizer usually consists of two operating states: training mode and tracking mode. In the training mode, the known sequence is utilized to train the equalizer, and the algorithm used should achieve rapid convergence and tap coefficients are therefore obtained. In the tracking mode, the equalizer generates the error signal directly from the decision output and adjusts the tap coefficients based on certain adaptive algorithms so that the optimal channel equalization can be achieved automatically. In practice, a specially designed training sequence is often inserted into the transmitted signals to facilitate the convergence and the tracking of the equalizer. Based on the zero forcing (ZF) or linear minimum mean-square error criteria discussed before, the adaptive algorithms can include ZF, linear least mean-square (LMS) error, recursive least squares (RLS), fast Kalman, etc., among which the LMS is one of the most commonly used adaptive equalization algorithms. To evaluate the performance of these algorithms, the major considerations include fast convergence properties, tracking capability for the fast time-varying channel, and computational complexity. Detailed introduction and performance analysis can be found in the references.

7.4 MULTICARRIER SYSTEMS

7.4.1 Timing Synchronization [5, 6]

As previously explained, the timing synchronization of a multicarrier system is similar to that of the single-carrier system and will not be described in detail in this section. The ISI in OFDM systems due to the timing offset will destroy the orthogonality among subcarriers and introduce ICI. For example, with $\varepsilon > 0$, the output of the inner receiver will be

$$Y_{i,n} = e^{j2\pi\frac{n}{N}\varepsilon}\frac{N-\varepsilon}{N}X_{i,n}H_{i,n} + n_{i,n} + n_{\varepsilon}(i,n) \tag{7-66}$$

where the ISI-introduced ICI can be treated as additional noise of $n_{\varepsilon}(i,n)$, and its power will be much smaller than AWGN $n_{i,n}$ if and only if the timing synchronization is very accurate.

7.4.2 Carrier Synchronization

For carrier frequency synchronization, the OFDM system is very sensitive to the carrier frequency offset due to its relatively small frequency spacing between two adjacent subcarriers. Therefore, very accurate carrier synchronization is required. The frequency offset in the OFDM system can be generally broken down into two parts relative to the subcarrier spacing: integer and fractional parts. The integer part is the frequency offset equal to an integer multiple of subcarrier spacing, and the fractional part is the frequency offset less than the subcarrier spacing. The integer part only introduces the cyclic shift of the subcarrier position and will not destroy the orthogonality between subcarriers while the fractional part will surely cause the ICI, leading to the decrease of SNR. Carrier synchronization for the OFDM system can be divided into two parts: fine synchronization using cyclic prefix in the time domain for the fractional part of the carrier offset estimation and coarse synchronization in the frequency domain using pilots for the integer part of the carrier offset estimation. The combination of the fractional carrier synchronization in the time domain and the integer carrier synchronization in the frequency domain could effectively eliminate the ICI caused by the fractional part of carrier offset to ensure the following coarse carrier synchronization will not be affected by ICI, and another round of the fine synchronization can be performed again using the pilots.

7.4.2.1 Carrier Synchronization Using the Pilot [12] We now first focus on the tracking stage and assume that the residual frequency offset in the tracking stage is far less than half of the subcarrier spacing of $1/T_{\text{sym}}$ (T_{sym} is the OFDM symbol period). If considering only one subchannel, the carrier synchronization is similar to that in the single-carrier system, and the ML method can be used, which converts the frequency estimation into the problem of estimating the frequency offset between the sample values (i.e., $Y_{i,n}$ and $Y_{i+1,\ n}$) of those two consecutive subcarriers.

In the presence of carrier offset Δf, the FFT output can be expressed as

$$Y_{i,n} = e^{-j\pi i \Delta f T_{sym}} \operatorname{sinc}(\pi \Delta f T_u) X_{i,n} H_{i,n} + n_{i,n} + n_{\Delta f}(i,n) \quad (7\text{-}67)$$

When the training symbol is transmitted, the complex conjugate of the training symbols can be used to eliminate the modulation impact on the phase of the data symbol,

$$\hat{f}\pi T_{sym} = \arg\left[\sum_{n=0}^{N}\left(Y_{i,n}(\hat{f}_{acq})Y^*_{i+1,n}(\hat{f}_{acq})\right)\left(X^*_{i+1,n}X_{i,n}\right)\right] \quad (7\text{-}68)$$

where $\hat{f}_{acq}$ denotes the frequency estimate from the capture stage and $X_{i,n}$ denotes the known symbol extracted from the training sequence. In most cases, especially in the frequency selective fading channel, there is no need to let the training sequence occupy the entire bandwidth. A number of subcarriers (i.e., N_p) evenly distributed within the system bandwidth can be used as the pilot and can be sent every D OFDM symbols in the time domain, as shown in Figure 7-13, where N_G denotes the number of protected subcarriers.

One can get the estimate

$$\hat{f}\pi T_{\text{sym}} = \frac{1}{D}\arg\left[\sum_{n=0}^{N_p}\left(Y_{i,p(n)}(\hat{f}_{\text{acq}})Y^*_{i+D,p(n)}(\hat{f}_{\text{acq}})\right)\left(X^*_{i+D,p(n)}X_{i,p(n)}\right)\right] \quad (7\text{-}69)$$

where $p(n)$ denotes the location of the nth pilot and $X^*_{i+D,p(n)}$ and $X_{i,p(n)}$ are the data transmitted at the same subcarrier position but in the $(i+D)$th OFDM symbol and ith OFDM symbol, respectively, which is similar to the frequency estimation with the interval of D symbols in the single-carrier system. The assumption for this arrangement is that the channel frequency response at $Y^*_{i+D,p(n)}$ and $Y_{i,p(n)}$ should remain the same. In other words, the channel will not change too fast. The accuracy for the tracking stage depends on $\hat{f}_{\text{acq}}$ in the capture stage, and the tracking performance is reliable only if the residual frequency deviation $\delta f = \Delta f - \hat{f}_{\text{acq}}$ satisfies $|\delta f\, T_{\text{sym}}| < 0.5$.

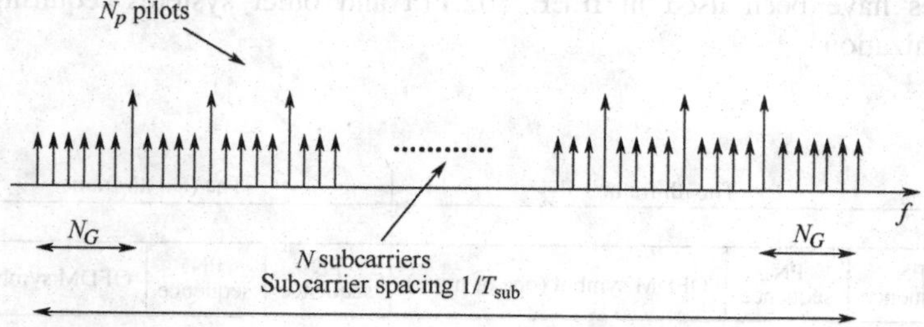

FIGURE 7-13 Diagram of pilot insertion.

The next stage is the capture stage. The same pilot sequence is used to avoid additional overhead with a goal of satisfying $|\delta f T_{\text{sym}}| < 0.5$. When residual frequency offset is zero, $Y_{Ii,n}$ reaches a maximum. Thus one can estimate the initial frequency by searching for the maximum amplitude of $\text{sinc}^2(\pi \,\Delta f\, T_u)$, that is, if $\hat{f}_{\text{acq}}$ is equal to Δf, the maximum can be achieved. One can get

$$\hat{f}\pi T_{\text{sym}} = \max_{f_{\text{trial}}} \left[\left| \sum_{n=0}^{N_p} \left(Y_{i,p(n)}(f_{\text{trial}}) Y^*_{i+D,p(n)}(f_{\text{trial}}) \right) (X^*_{i+D,p(n)} X_{i,p(n)}) \right| \right] \tag{7-70}$$

where f_{trial} is the initial frequency in the capture stage at the beginning and $Y_{I,\,p(n)}(f_{\text{trial}})$ is the output of the FFT operation after correction on f_{trial}. To avoid the maximum appearing at $f_{\text{trial}} - \Delta f = [p(j) - p(j-1)]/T_{\text{sym}}$ ($j = 1,2, \ldots, N_p$), the data symbols carried by the pilots should use a well-designed PN sequence: a fairly good auto correlation property that is easy to implement using shift registers. The capture time is directly proportional to the frequency step for the search, and capture time will be longer when the frequency offset is large, such as ±10 subcarrier spacing. For continuous transmission, only one capture process at the beginning of the transmission is needed.

7.4.2.2 Carrier Synchronization Using Time-Domain Training Sequence The carrier synchronization methods for single- and multicarrier systems are quite similar, the difference being that the synchronization is performed in the time domain for single-carrier systems and in the frequency domain for OFDM systems. Better estimation results can be achieved by using pilots with the number of the pilots less than the data to minimize the overhead. The impact of the frequency offset in OFDM systems will cause ICI, and that would degrade the performance of frequency estimation. Since the algorithm can only be used after accurate timing is achieved, the synchronization time tends to be very long. To solve this issue, the frame structure consists of training sequences and the OFDM symbol is used for the carrier synchronization in the time domain. Figure 7-14 gives an example of this structure with dual PN sequences.

The same algorithm as that for single-carrier systems can be applied to the synchronization estimation for OFDM systems by using either the current or adjacent signal frames. This method can be used to achieve faster synchronization, and similar structures have been used in IEEE 802.11a and other systems requiring rapid synchronization.

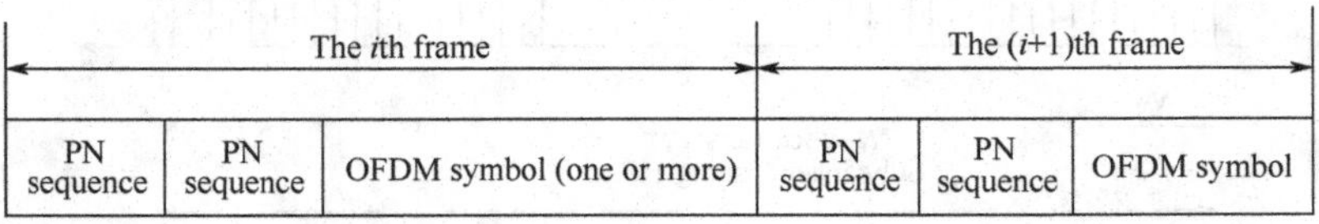

The *i*th frame			The (*i*+1)th frame		
PN sequence	PN sequence	OFDM symbol (one or more)	PN sequence	PN sequence	OFDM symbol

FIGURE 7-14 OFDM frame structure using time-domain dual PN sequence.

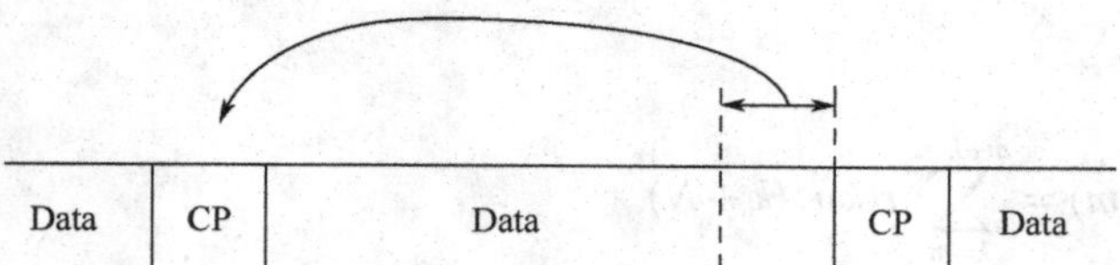

FIGURE 7-15 OFDM symbol structure with cyclic prefix.

7.4.2.3 Carrier Synchronization Using Cyclic Prefix [13] The intersymbol interference can be eliminated by inserting the cyclic prefix for OFDM systems, and one can use this cyclic property to achieve the synchronization. With this arrangement, the system throughput loss caused by the pilot insertion can be effectively reduced.

Consider a typical OFDM system with N subcarriers with cyclic prefix length L. Each OFDM symbol will have $(N+L)$ sampling data points, as shown in Figure 7-15.

Assuming the starting point of the OFDM symbol is θ and the carrier offset is Δf, the received signal is given as

$$r(k) = s(k-\theta)e^{j2\pi\Delta f k/N} + n(k) \tag{7-71}$$

Define two vectors

$$\begin{aligned} I &= \{\theta, \cdots, \theta+L-1\} \\ I' &= \{\theta+N, \cdots, \theta+N+L-1\} \end{aligned} \tag{7-72}$$

where the vector I is the cyclic prefix of the ith OFDM symbol, which has a strong correlation with the elements inside the vector I',

$$\forall k \in I \quad E[r(k)r^*(k+m)] = \begin{cases} \sigma_S^2 + \sigma_n^2 & m = 0 \\ \sigma_S^2 e^{-j2\pi\Delta f} & m = N \\ 0 & \text{otherwise} \end{cases} \tag{7-73}$$

where $\sigma_s^2 = E[|s(k)|^2]$ is the average signal energy and $\sigma_n^2 = E[|n(k)|^2]$ is the average AWGN power.

The logarithmic likelihood function $\Lambda(\theta, \Delta f)$ is defined as the logarithm of the probability density function:

$$\Lambda(\theta, \Delta f) = \log f(r|\theta, \Delta f) \tag{7-74}$$

Among all the $N+L$ sampling data points for the ith OFDM symbol, only the corresponding elements in the vectors I and I' are correlated, and the rest of the sampling data points are independent of each other. So one can get

$$\Lambda(\theta, \varepsilon) = |\gamma(\theta)| \cos\,[2\pi\;\Delta f + \angle\gamma(\theta)] - \rho\Phi(\theta) \tag{7-75}$$

where

$$\gamma(m) = \sum_{k=m}^{m+L-1} r(k)r^*(k+N) \tag{7-76}$$

$$\Phi(m) = \frac{1}{2}\sum_{k=m}^{m+L-1} |r(k)|^2 + |r(k+N)|^2 \tag{7-77}$$

$$\rho = \left|\frac{E\{r(k)r^*(k+N)\}}{\sqrt{E\{|r(k)|^2\}E\{|r(k+N)|^2\}}}\right| = \frac{\sigma_s^2}{\sigma_s^2+\sigma_n^2} = \frac{\text{SNR}}{\text{SNR}+1} \tag{7-78}$$

Here, ρ denotes the correlation coefficient between $r(k)$ and $r(k+N)$ and $\angle\gamma(\theta)$ denotes the phase of $\gamma(\theta)$. The likelihood algorithm can estimate both the beginning of the frame and the carrier frequency offset simultaneously. The maximization for the likelihood function $\Lambda(\theta, \Delta f)$ is done in two steps,

$$\max_{(\theta,\Delta f)} \Lambda(\theta,\Delta f) = \max_{\theta} \max_{\Delta f} \Lambda(\theta,\Delta f) = \max_{\theta} \Lambda(\theta,\Delta\hat{f}_{\text{ML}}(\theta)) \tag{7-79}$$

To maximize $\Lambda(\theta, \Delta f)$ in terms of Δf, the cosine term in (7-75) should be 1, that is,

$$\Delta\hat{f}_{\text{ML}}(\theta) = -\frac{1}{2\pi}\angle\gamma(\theta) + n \tag{7-80}$$

In most cases, the carrier frequency offset should be in a smaller range, that is, $n=0$. After the coarse synchronization, the residual frequency offset can meet the condition $|\Delta f| \leq 0.5$. Let the cosine term be 1. Then

$$\Lambda(\theta,\Delta f) = |\gamma(\theta)| - \rho\Phi(\theta) \tag{7-81}$$

Equation (7-81) is only related to θ, and the maximum-likelihood estimate *of* θ can be obtained by

$$\hat{\theta}_{\text{ML}} = \arg \max_{\theta} [|\gamma(\theta)| - \rho\Phi(\theta)] \tag{7-82}$$

$$\Delta\hat{f}_{\text{ML}} = -\frac{1}{2\pi}\angle\gamma(\hat{\theta}_{\text{ML}}) \tag{7-83}$$

The description above only provides open-loop structure of the maximum-likelihood estimation. One can also use a closed-loop structure with the maximum-likelihood estimation signals $\hat{\theta}_{\text{ML}}$ and $\Delta\hat{f}_{\text{ML}}$ being fed back to the PLL. If the assumption that the estimated parameters remain constant within a certain time is valid, the accumulation by PLL can significantly improve the performance of the maximum-likelihood estimation. The performance of the above maximum-likelihood estimation method can be analyzed using the Monte Carlo method. It is simple to implement the maximum-likelihood algorithm with a cyclic prefix, and this can be done without pilots/training sequences to reduce overhead. However, frequency estimation and frame synchronization for this algorithm without pilots is relatively poor, especially in a multipath environment. Therefore, in practice, carrier synchronization is generally achieved by using either pilots in the frequency domain or training sequences the in time domain.

7.4.3 Channel Estimation/Equalization for OFDM System

7.4.3.1 Channel Estimation by Frequency Domain Pilots [14] Most frequency-domain channel estimation algorithms for OFDM systems are based on pilot estimation. A DTTB channel is a time-varying and frequency-selective fading channel, and the DTTB signal is often segmented by a frame structure. As the change of the DTTB channel is relatively slow, one can usually assume the channel within the frame period is unchanged so that DFT can be performed under this linear and time-invariant assumption, which greatly simplifies the implementation complexity of the channel estimator.

1. *Pilot Pattern.* The two most important parameters for pilot pattern selection are the maximum Doppler frequency F_d and maximum multipath delay $\tau_{\max}$. The pilot symbols are transmitted at the specific location of the time–frequency grid, which can be taken as the two-dimensional sampling on the channel transfer function $H(f, t)$, and the samples must be close enough to meet the Nyquist sampling theorem so that distortion can be avoided. Therefore, the pilot density is determined by the Nyquist sampling theorem. Suppose the interval of OFDM symbols is N_t along the time direction and the subcarrier spacing is N_f along the frequency direction. Then

$$N_t = \frac{1}{2F_d T_{\text{sym}}} \qquad N_f = \frac{1}{\tau_{\max}\,\Delta F_c} \tag{7-84}$$

Again $T_{\text{sym}} = T_g + T_u$ is the OFDM symbol period with T_g the guard interval and T_u the duration of the OFDM data block and $\Delta F_c = 1/T_u$ is the subcarrier spacing.

To ensure robust channel estimation performance, the worst-case channel condition with both maximum Doppler frequency and channel delay should be considered. Meanwhile, more sampling is required for better estimation on the fading channel. The pilot symbols should be placed close enough to track the changes of the channel in both time and frequency domains with increased overhead. A compromise between data throughput and channel estimation performance should be made, depending on the application requirements and the transmission environment. In general, it is suggested that the number of pilots be twice that specified by the sampling theorem along both the time and frequency directions.

The commonly used pilot patterns are given in Figure 7-16. Figure 7-16*a* shows a comb-type pilot pattern: The pilot signals are evenly distributed in each OFDM symbol, and this pattern is sensitive to frequency selectivity. Figure 7-16*b* shows

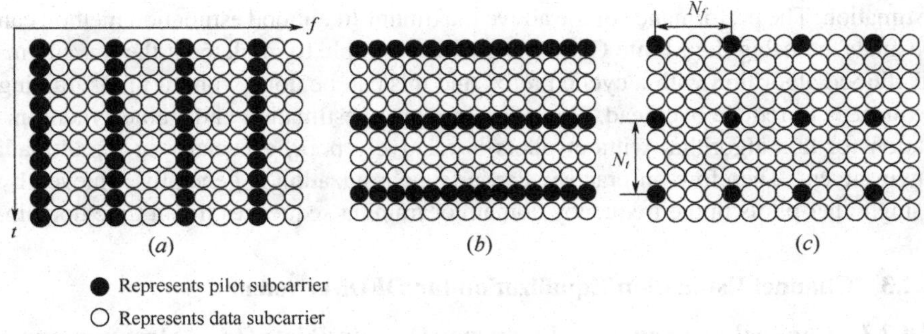

FIGURE 7-16 Three commonly used pilot patterns.

periodic transmitted pilots with the entire OFDM symbol used for pilots, and this pattern is suitable for slow-fading channels. Figure 7-16*c* shows the interpolation along both time and frequency directions with less pilots than either Figure 7-16*a* or 7-16*b*. The frequency response for other subcarriers to be estimated in the figure can be obtained by interpolation using neighboring pilots. As shown in Figure 7-16*b*, after the pilot symbol is sent, there will be no pilots in the OFDM symbols that follow, which requires the channel to be quasi-static or even unchanged within a certain time, i.e., the channel is quasi-stationary or a slow-fading channel. This pattern is suitable for an unchanged channel and can also be applicable for wireless local area networks (WLANs). The WLAN data are transmitted in the burst mode and usually contain a number of OFDM symbols with the first few as pilot symbols. Because the channel is slowly faded, the channel estimation result obtained from the first few symbols is still valid when OFDM data symbols pass through the channel. The pattern in Figure 7-16*c* has higher effectiveness than the other two while the complexity for processing is higher, and it has been widely used for broadcasting services such as the DVB and DAB.

2. *Pilot-Based Estimation Method.* The example of the scattered pilot for DVB-T systems is used to illustrate the pilot-based estimation method with the pilot pattern derived from Figure 7-16*c* for better channel estimation performance, and the power of the pilot signal is about 3 dB higher than that of the average power of the data.

With the assumption that the guard interval is longer than the maximum channel delay, the data symbol received on the *n*th subcarrier of the *i*th OFDM symbol will be

$$Y_{i,n} = X_{i,n} H_{i,n} + n_{i,n} \tag{7-85}$$

where $X_{i,n}$ is the data symbol from the transmitter, $H_{i,n}$ is the channel frequency response, and $n_{i,n}$ is the AWGN. The detailed estimation algorithm based on two-dimensional pilots can generally be divided into the following two steps:

Step 1: The channel frequency response at the pilot position can be obtained by the LS estimation algorithm. The data value $Y^p_{i,n}$ of the pilot is extracted from

$Y_{i,n}$, and one can get the channel estimation at the position of the pilot directly as the data value carried by the pilot is known:

$$\hat{H}^p_{i,n} = \frac{Y^p_{i,n}}{X^p_{i,n}} = H^p_{i,n} + \frac{n_{i,n}}{X^p_{i,n}} \quad (7\text{-}86)$$

Step 2: Two-dimensional interpolation is performed based on LS estimates after the channel estimation $\hat{H}^p_{i,n}$ at the pilot location is obtained, and the Wiener filter in the following is used:

$$\hat{H}_{i,n} = \sum w_{i',n',i,n} \hat{H}^p_{i',n'} \quad (7\text{-}87)$$

where $w_{i',n',i,n}$ is the weighting factor for the interpolation. The complete channel response for all subcarriers is obtained after interpolation filtering along both time and frequency directions.

7.4.3.2 Frequency Domain Channel Estimation for Training Sequence [15] In addition to the frequency-domain pilot, the synchronization can be achieved by using the special training sequence between OFDM symbols, such as TDS-OFDM technology in the DTMB system. Unlike the C-OFDM, the time-domain PN sequence instead of the frequency-domain pilot signal is utilized in the TDS-OFDM system, and the receiver can use the PN sequence for channel estimation by acquiring the time-domain impulse response of the channel first and then getting the frequency-domain response through DFT, which is described in detail in the following.

Assuming the training sequence is $\{c(k)\}$ ($k = 1,2, \ldots, K$) at the transmitter, the received time domain signal is

$$r(k) = c(k)^* h + n(k) = \sum_{l=0}^{L-1} c(k-l)h(l) + n(k) \qquad k = 1, 2, \ldots, K \quad (7\text{-}88)$$

where K is the length of the training sequence, L is the channel impulse response length, $n(k)$ is AWGN, and $*$ is the convolution operation. The matrix form is $r = Ch + n$, where, $r = [r(1), r(2), \ldots, r(K)]^{\mathrm{T}}$ denotes the column vector of the received training symbols, $h = [h(1), h(2), \ldots, h(L)]^{\mathrm{T}}$ denotes the column vector of the channel impulse response, $n = [n(1), n(2), \ldots, n(K)]^{\mathrm{T}}$ denotes the column vector of AWGN, and $\boldsymbol{C}$ is the $K \times K$ matrix generated by cyclic shifting the training sequence $\{c(k)\}$:

$$C = \begin{bmatrix} c(1) & c(K) & c(K-1) & \cdots & c(2) \\ c(2) & c(1) & c(K) & \cdots & c(3) \\ \vdots & \vdots & \vdots & & \vdots \\ c(K-1) & c(K-2) & c(K-3) & \cdots & c(K) \\ c(K) & c(K-1) & c(K-2) & \cdots & c(1) \end{bmatrix} \quad (7\text{-}89)$$

For simplicity, the PN sequence with good correlation properties is generally chosen as $\{c(k)\}$, and the normalized correlation function $\rho(n)$ can be expressed as

$$\rho(n) = \frac{1}{K}\sum_{k=0}^{K-1} c(n-k)^* c(k) \approx \begin{cases} 1 & n = k \\ 0 & \text{otherwise} \end{cases} \tag{7-90}$$

where $C^H C \approx KI$ and H as the superscript is the Hermitian operator.

As the original training sequence of $\{c(k)\}$ of the transmitter is known to the receiver, channel estimation for the DTMB receiver can generally be divided into the following three steps:

Step 1: The time-domain channel impulse response estimation can be obtained by

$$\hat{h} = \frac{1}{K}C^H r = \frac{1}{K}C^H Ch + \frac{1}{K}C^H n \approx h + \frac{1}{K}C^H n \tag{7-91}$$

Step 2: Disregard those small values from the coarse estimation of $\hat{h}$ based on a certain threshold as they are generally unreliable in the presence of noise and multipath. The threshold is determined by the antinoise and the sensitivity requirements from the applications.

Step 3: Obtain the channel frequency-domain response by performing DFT on the time-domain channel estimation from the step 2:

$$\hat{H} = \text{DFT}(\hat{h}) \tag{7-92}$$

7.5 INTRODUCTION TO DTMB INNER RECEIVER [18–20]

Functional block diagrams for the DTMB transmitter and receiver are shown in Figure 7-17. From the transmitter, the transport stream will form the basic data frame

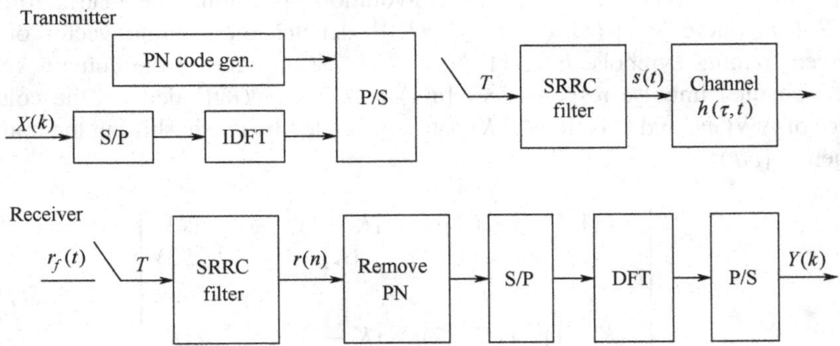

FIGURE 7-17 General structure of transmitter and receiver for DTMB system.

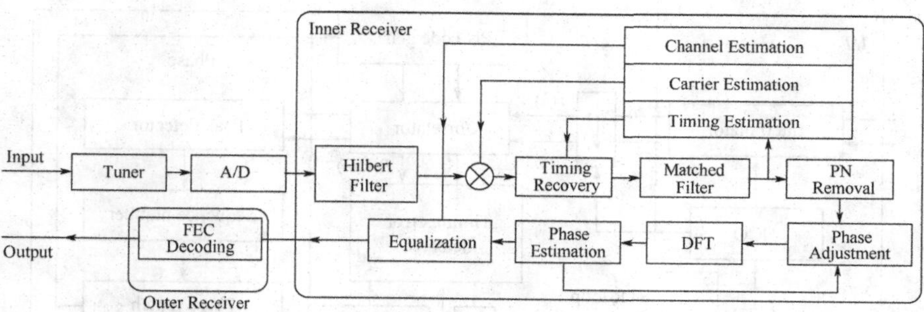

FIGURE 7-18 Detailed structure of DTMB receiver.

after scrambling, FEC, and constellation mapping and then become the interleaving frame through the time-domain interleaving. This time-interweaved frame will be further processed and then multiplexed with the TPS signal to form the OFDM data block (also known as the frame body). After IDFT, the time-domain frame body signal can be obtained and will be further multiplexed with the corresponding PN sequence (known as the frame header) to form the signal frame (which is the basic building block of the hierarchical frame structure for the DTMB system). It is converted into the baseband output signal via the post-baseband processing block with pulse shaping. Finally, the baseband signal is upconverted to the RF signal by the quadrature upconversion for transmission.

The reverse processing will be carried out at the receiver assuming perfect timing and carrier synchronization can be achieved with accurate channel estimation as well as good equalization performance.

Figure 7-18 shows the detailed structure of the DTMB receiver, including both inner and outer receivers. The DTMB inner receiver mainly focuses on the frame synchronization, timing recovery, carrier recovery, channel estimation, equalization, etc. As these can be done with the help of the time-domain PN sequence at the receiver, it can achieve fast synchronization and accurate channel estimation within a relatively short time as all these can be done in the time domain. It has the merits of high spectrum efficiency (as the PN sequence with the same length as that of the cyclic prefix can serve as both a training sequence and a guard interval), fast synchronization and accurate channel estimation, etc., which help improve the overall system performance. The DTMB outer receiver mainly includes time-domain de-interleaving, QAM demapping, LDPC and BCH decoding, and data descrambling.

The different modules of the inner receiver will be introduced below.

7.5.1 Frame Synchronization

The timing recovery process is generally divided into two stages: coarse data symbol synchronization and fine symbol synchronization. As the DTTB system continuously transmits the data stream, its fine symbol synchronization algorithm is usually the one

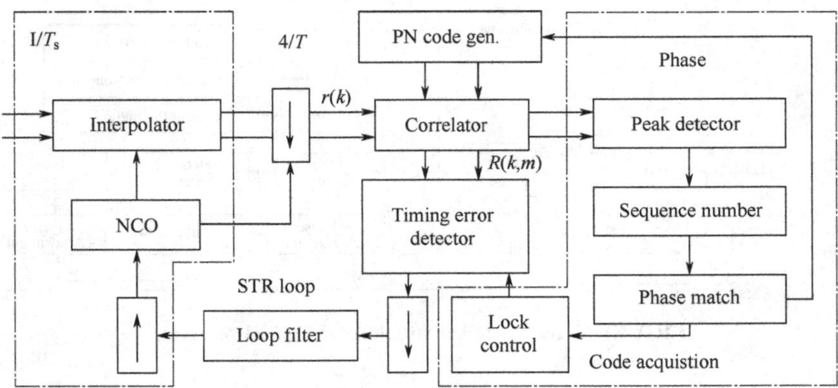

FIGURE 7-19 Block diagram of STR system.

with feedback structure to achieve better tracking performance. The main function of the coarse symbol synchronization of the DTMB system is to find the appropriate phase for the local PN sequence to make sure that the local PN sequence has the same phase with the transmitted PN sequence. In the tracking stage where the symbol timing recovery (STR) by the closed-loop tracking technology is used, the phase error between two sequences must be further reduced so that this phase match state (with very high accuracy) can be maintained in the presence of various interferences. The major interference is from sampling frequency drift, which will lead to the ICI and timing error drift for OFDM systems and further deterioration in frame synchronization performance. Figure 7-19 shows the block diagram achieving the frame synchronization, from which it can be seen that the frame synchronization can be mainly divided into the following steps:

1. *Finding Correlated Peak by Sliding Correlation.* The autocorrelation function of the PN sequence will reach its peak value K (equal to its period), and this is a good indicator of whether the phase of both sequences (the PN sequence at the transmitter and the local PN sequence) is aligned. To reduce the number of parallel correlators, the sliding search method can be used, which requires only one correlator with length K. Using the sliding correlation technique could greatly reduce the capture time with only one correlator, which also reduces implementation complexity. All possible phases for the PN sequence are thoroughly searched within the time period KT (T stands for data symbol duration).
2. *Phase Matching.* In the DTMB system, the PN sequence for each signal frame within one superframe is predetermined to ensure the unique phase shift among adjacent signal frames. With this arrangement, one signal frame in a superframe can be fully identified by ΔPhase when received, and thus the phases of the PN sequences in the subsequent signal frames can be accurately predicted in advance.

Due to noise or other interference, the correlation peak detector may generate the wrong decision, which causes false alarm and misdetection with certain probability. Especially when a misdetection occurs, the search for the phase of the local PN sequence will stop, and the incorrect synchronization signal will make the system work in a nonsynchronous state and it is impossible for the receiver to operate properly. To prevent this from happening, one needs to use the phase matching circuit for verification when the detection is confirmed: Even after the correlation peak for the second signal frame is obtained, the comparison between the correlation peak of the third signal frame and the threshold will help verify whether synchronization is actually achieved. If the synchronization has been successfully achieved and verified, the STR tracking circuit will start to work and only the update of the local PN sequence phase is required to maintain the synchronization. The frame synchronization signal will be used to extract the timing error of the STR loop to estimate the timing error signal greater than $\pm T_s$. To reduce the impact of false alarm in the synchronous state, the system is considered to have loss of synchronization only when the correlation peak cannot be found for three consecutive signal frames. If the synchronization loss is confirmed, the circuit will reenter the capture stage.

The STR loop includes two main functionalities: to extract estimation from the timing errors and to use the filtered estimates to drive the numerically controlled oscillator (NCO) to complete the sampling clock adjustment in the time domain. The synchronization between the received and the transmitted data symbols can then be achieved by a linear interpolation filter. To ensure the accuracy of the timing synchronization and the linear interpolation, four times oversampling is performed at STR. The loop improves the accuracy of the timing synchronization through the correlation operation even for the multipath channel.

7.5.2 Carrier Synchronization

At the DTMB receiver, the carrier synchronization is completed in the time domain using the received PN sequence of the OFDM signal, and a low-complexity time-domain carrier estimation algorithm is proposed. There usually exists a relatively large frequency offset between the transmitter and receiver when the receiver is turned on. The algorithms adopted here should ensure a large enough capture range for frequency estimation with good estimation accuracy.

The frequency estimation can also be divided into three stages for the DTMB inner receiver. If Δf is more than a certain threshold at the beginning of timing synchronization, coarse frequency estimation (CFE) needs to be done first to capture the frequency and reduce its offset below that threshold before timing synchronization. As CFE is performed before the timing synchronization with no timing information available, the CFE algorithm is NDε based. In (7-55) $z(k)z^*(k-l)$ can be replaced by $[r(k)r^*(k-l)]^P$, where P is the number of constellation points of the modulation. Since the PN sequence is modulated by BPSK, $P=2$. The disadvantages for the CFE algorithm are that the capture range becomes $1/P$ of the original and the variance of the estimation becomes larger due to the introduction of the Pth power. After the CFE stage, the noncoherent AFC is performed before the coherent AFC. The difference between coherent and noncoherent AFC is that the timing information can be used for the coherent one while timing information is unavailable for noncoherent AFC. The timing synchronization loop and fine frequency estimation loop turn on at the same time after CFE, but the timing loop needs several more signal frames to achieve the very accurate synchronization and complete the timing recovery. Since the timing information during this period is not accurate enough, noncoherent AFC without a requirement on the timing information can be used for synchronization, and that is why noncoherent AFC is done first. Both analysis and simulation results show that this method is close to the Cramer–Rao bound (CRB) at low computational complexity and short capture time on frequency when the SNR is high.

7.5.3 Channel Estimation and Equalization

The channel estimation and equalization methods discussed here are based on the correlation algorithm for the time-domain synchronous PN sequence. The estimation on the channel impulse response is done by the correlation in the time domain and the channel equalization is achieved in the frequency domain. With this arrangement, the channel estimation will be less affected by AWGN and the time variation of the channel and is relatively simple to achieve.

The channel estimation methods for the DTMB receiver can be divided into three categories: are time-domain correlation, iterative channel estimation with PN sequence impact removal, and frequency-domain decision feedback methods. The PN sequence frame header and data frame body are separated first at the receiver, and one of the three methods will be used for channel estimation. In general, the PN sequence-based time-domain correlation method is suitable when the multipath interference is limited within a short time; iterative channel estimation with PN sequence impact removal is suitable when the multipath delay is fairly long; and the frequency-domain decision feedback method is suitable when the time-invariant effect of the channel is strong. After the channel frequency response estimate $\hat{H}(n,k)$ is available, DFT is performed on the data portion of the signal frame to get the frequency-domain data $Y(n, k)$, and the data signal of $Z(n, k)$ after channel equalization can then be obtained by $Z(n, k) = Y(n, k)/\hat{H}(n,k)$.

As the DTTB channel is usually time varying, the PN sequence-based time-domain correlation and iterative channel estimation with PN sequence impact removal methods will be further introduced.

7.5.3.1 PN Sequence-Based Time-Domain Correlation Method Without considering the interference of the data (from the previous OFDM data block), the received PN sequence $r(k)$ can be expressed as

$$r(k) = \sum_{l=0}^{L-1} c(k-1)h_c(l) + n(k) \tag{7-93}$$

where $h_c(l)$ is the time-domain channel impulse response, $n(k)$ is AWGN, and again $c(k)$ is the PN sequence where its normalized correlation function$\rho(n)$ can be expressed as

$$\rho(n) = \frac{1}{K}\sum_{k=0}^{K-1} c(n-k)^* c(k) \approx \begin{cases} 1 & n = k \\ 0 & \text{otherwise} \end{cases} \tag{7-94}$$

where n, k denote the serial numbers and K is the length of the PN sequence.

The coarse estimation for the channel impulse response in the time domain can be obtained through the time-domain correlation

$$\begin{aligned} \hat{h}_{\mathrm{tc}}(n) &= \frac{1}{K}\sum_{k=0}^{K-1} c(n-k)^* r(k) \\ &= h(n) + \frac{1}{K}\sum_{k=0}^{K-1} c(k)^* n(k) \\ &= h_c(n) + n_c(n) \qquad n \in [0, K-1] \end{aligned} \tag{7-95}$$

where $h_c(n)$ is the channel impulse response and $n_c(n)$ is the linearly combined AWGN.

From the coarse estimation result $\hat{h}_{\mathrm{tc}}(n)$, estimates with very small value below a given threshold are discarded. Channel echoes ahead of the main path are called preechos while those behind the main path are called postechos. DTMB has defined the preamble and postamble in its frame header, which are the cyclic extensions of the PN sequence and are used to protect the PN sequence for the more accurate estimation on the time-domain channel impulse response as long as the channel length of the preecho and postecho are less than that of the preamble and postamble, respectively. The inner receiver should choose appropriate preecho length L_{pre} and postecho length L_{post} so that most channel energy (the correlation output) concentrates within $[k' - L_{\mathrm{pre}} + 1, k' + L_{\mathrm{post}}]$ and k' corresponds to the correlation peak position, i.e., the locked main path position.

Using the main path position k' as the reference, the correlation output data within $[k' - L_{\mathrm{pre}} + 1, k' - 1]$ is selected as the preecho part $\hat{h}_{\mathrm{tc,pre}}$, and the correlation output data within $[k' + 1, k' + L_{\mathrm{post}}]$ is selected as the post-echo part $\hat{h}_{\mathrm{tc,post}}$. Then $\hat{h}_{\mathrm{tc,post}}$ and $\hat{h}_{\mathrm{tc,pre}}$ are combined by to form a sequence with length N after zero padding:

$$\hat{h}_{\mathrm{tc},N}(n) = \begin{cases} \hat{h}_{\mathrm{tc,post}} & 0 < \ n \le L_{\mathrm{post}} \\ 0 & L_{\mathrm{post}} < \ n < N - L_{\mathrm{pre}} \\ \hat{h}_{\mathrm{tc,pre}} & N - L_{\mathrm{pre}} \le \ n < N \end{cases} \tag{7-96}$$

The resulting $\hat{h}_{\mathrm{tc},N}(n)$ satisfies the cyclic characteristics of the DFT, and the frequency response estimate $\hat{H}(n,k)$ for each OFDM subcarrier can be obtained by N-point DFT.

7.5.3.2 *Iterative Channel Estimation with PN Sequence Impact Removal Method*

After passing through the multipath channel, the mutual interference between the OFDM data block and the PN sequence occurs. If the multipath delay is longer than the cycle length of the PN sequence (which is equivalent to the sum of the length of preamble and postamble), the errors caused by the data interference would become inevitable for the time-domain correlation channel estimation method, and this will seriously affect the system performance. Therefore, the iterative channel estimation and the PN sequence impact removal method for channel estimation should be adopted.

The *i*th signal frame at the transmitter consists of two parts without any overlap: PN sequence $\{c_{i,k}\}_{k=0}^{M-1}$ and data sequence $\{s_{i,k}\}_{k=0}^{N-1}$, as shown in Figures 7-20*a*, *b*. Due to the multipath effect of the channel, these two parts become overlapped at the receiver with $\{y_{i,k}\}_{k=0}^{M+L-1}$ representing the linear convolution result of the PN sequence and the channel impulse response and $\{x_{i,k}\}_{k=0}^{N+L-1}$ representing the linear convolution result of the OFDM data block $\{s_{i,k}\}_{k=0}^{N-1}$ and the channel impulse response, as shown in Figures 7-20*c*, *d*.

The iterative cancellation method is adopted to separate $\{x_{i,k}\}_{k=0}^{N+L-1}$ and $\{y_{i,k}\}_{k=0}^{M+L-1}$, which includes the following steps:

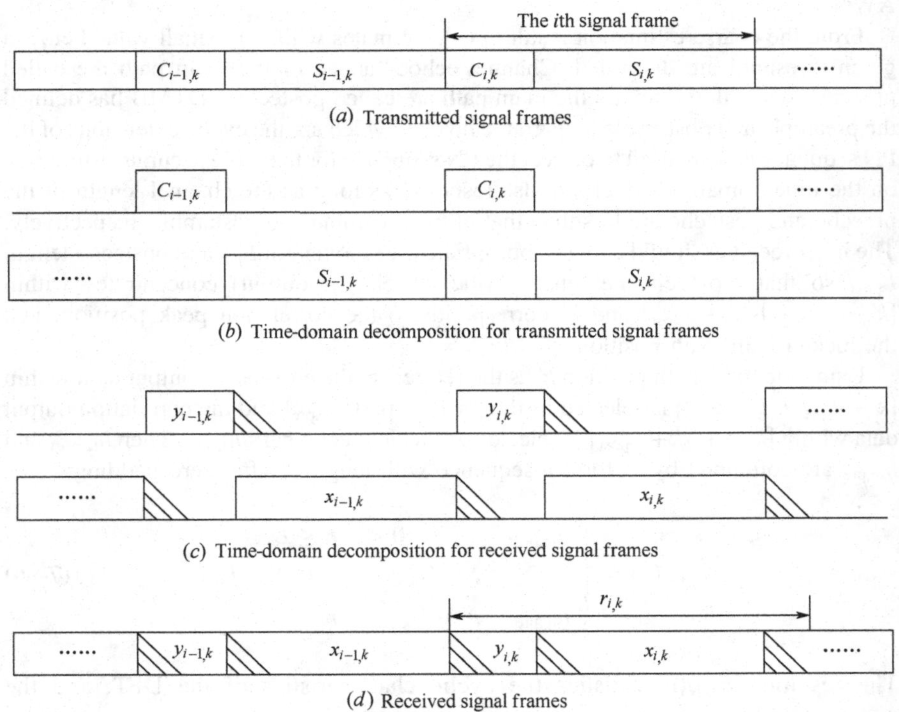

FIGURE 7-20 Signal decomposition at transmitter and receiver after multipath channel.

Step 1: The channel impulse response $\left\{\hat{h}_{i,l}^{\text{iter}=0}\right\}_{l=0}^{L-1}$ of the ith frame can be obtained by the linear interpolation based on the channel impulse responses $\{\hat{h}_{i-2,l}\}_{l=0}^{L-1}$ and $\{\hat{h}_{i-1,l}\}_{l=0}^{L-1}$ for the corresponding known $(i-2)$th and $(i-1)$th signal frames. The iteration number is set to 0.

Step 2: The channel impulse response $\left\{\hat{h}_{i+1,l}^{\text{iter}=I}\right\}_{l=0}^{L-1}$ of the $(i+1)$th signal frame can be obtained by the linear interpolation using $\{\hat{h}_{i-1,l}\}_{l=0}^{L-1}$ and $\left\{\hat{h}_{i,l}^{\text{iter}=I}\right\}_{l=0}^{L-1}$ during this Ith iteration.

Step 3: After the synchronization is achieved at the receiver, the PN sequences $\{c_{i,k}\}_{k=0}^{M}$ and $\{c_{i+1,k}\}_{k=0}^{M}$ for the ith and $(i+1)$th signal frames are known, and the linear convolution between $\{c_{i,k}\}_{k=0}^{M}/\{c_{i+1,k}\}_{k=0}^{M}$ and channel impulse response can be calculated individually with the corresponding results of $\left\{\hat{y}_{i,k}^{\text{iter}=I}\right\}_{k=0}^{M+L-1}/\left\{\hat{y}_{i+1,k}^{\text{iter}=I}\right\}_{k=0}^{M+L-1}$, respectively.

Step 4: The estimate on the linear convolution $\left\{\hat{x}_{i,k}^{\text{iter}=I}\right\}_{k=0}^{M+N-1}$ between the ith signal frame and the channel impulse response can be achieved by removing the impact of both $\left\{\hat{y}_{i,k}^{\text{iter}=I}\right\}_{k=0}^{M+L-1}$ and $\{\hat{y}_{i+1,k}\}_{k=0}^{M+L-1}$ from the received sampling data $\{r_{i,k}\}_{k=0}^{M+N-1}$ in the ith signal frame:

$$\hat{x}_{i,k}^{\text{iter}=I} = \begin{cases} r_{i,k+M} - \hat{y}_{i,k+M}^{\text{iter}=I} & 0 \le \quad k < 1 \\ r_{i,k+M} & L-1 \le \quad k < N \\ r_{i,k+M} - \hat{y}_{i+1,k-N}^{\text{iter}=I} & N \le \quad k < M+N \end{cases} \tag{7-97}$$

Step 5: The resulting $\left\{\hat{x}_{i,k}^{\text{iter}=I}\right\}_{k=0}^{M+N-1}$ is equivalent to the zero-padded OFDM (ZP-OFDM) signal, and ZP-OFDM equalization algorithms can be applied.

Step 6: The iteration will stop if the preset iteration number J is reached; $\left\{\hat{x}_{i,k}^{\text{iter}=J}\right\}_{k=0}^{M+N-1}$ and $\left\{\hat{h}_{i,l}^{\text{iter}=J}\right\}_{l=0}^{L-1}$ are the final estimates of $\{x_{i,k}\}_{k=0}^{N+L-1}$ and $\{h_{i,k}\}_{k=0}^{L-1}$, respectively. After the decision is made for $\{\hat{x}_{i,k}\}_{k=0}^{N+L-1}$, the $(i+1)$th signal frame can be processed using the same method.

Step 7: If J is not reached, the time-domain filtering and decision feedback method are applied to $\left\{\hat{x}_{i,k}^{\text{iter}=I}\right\}_{k=0}^{M+N-1}$ for the residual ISI removal and the noise suppression, and $\left\{z_{i,k}^{\text{iter}=I}\right\}_{k=0}^{M+N-1}$ will be obtained.

Step 8: Reconstruct the $\left\{\hat{y}_{i,k}^{\text{iter}=I+1}\right\}_{k=0}^{M+L-1}$ by

$$\hat{y}_{i,k}^{\text{iter}=I+1} = \begin{cases} r_{i,k} - \hat{x}_{i-1,k+N}^{\text{iter}=J} & 0 \le \quad k < L-1 \\ r_{i,k} & L-1 \le \quad k < M \\ r_{i,k} - z_{i,k-M}^{\text{iter}=I} & M \le \quad k < M+L-1 \end{cases} \tag{7-98}$$

Step 9: More accurate channel estimation $\left\{\hat{h}_{i,l}^{\text{iter}=I+1}\right\}_{l=0}^{L-1}$ will be generated from $\left\{\hat{y}_{i,k}^{\text{iter}=I+1}\right\}_{k=0}^{M+L-1}$ after one iteration. Increase I by 1 and return to step 2 for another iteration process.

With the above functional description, the channel estimation is updated every signal frame, and this could be jointly performed in both the time and frequency domains as follows:

1. The initial estimate of the channel is given by cross-correlation of the locally generated PN sequence at the receiver and the received signal containing the PN sequence (together with the noise and other interferences) in the time domain.
2. The frequency-domain estimation algorithm is then used in the iterative calculation. After step 9 of the above process, the N_1-point DFT is performed on both $\left\{\hat{y}_{i,k}^{\text{iter}=I+1}\right\}_{k=0}^{M+L-1}$ and $\{c_{i,k}\}_{k=0}^{M-1}$ (if the sequence length is less than N_1, then zero pad the length to N_1), and the channel estimation $\left\{\hat{h}_{i,k}^{\text{iter}=I+1}\right\}_{k=0}^{N_1-1}$ will be obtained as

$$\hat{h}_{i,k}^{\text{iter}=I+1} = \text{IDFT}\left\{\frac{Y_{i,k}^{\text{iter}=I+1}}{C_{i,k}}\right\} \quad 0 \le k < N_1 - 1 \tag{7-99}$$

Set all the elements with the index $k \ge L$ within $\left\{\hat{h}_{i,k}^{\text{iter}=I+1}\right\}_{k=0}^{N_1-1}$ to zero and use the newly generated $\left\{\hat{h}_{i,k}^{\text{iter}=I+1}\right\}_{k=0}^{L-1}$ for the next iteration.

If $\left\{\hat{x}_{i,k}^{\text{iter}=I}\right\}_{k=0}^{M+N-1}$ still suffers from the noise and residual ISI after step 7 with the impact of the PN sequence on the OFDM data block removed, further filtering can be done as follows:

1. Calculate channel estimation $\left\{\tilde{h}_{i,k}^{\text{iter}=I}\right\}_{k=0}^{L-1}$ of the ith signal frame. The average of $\left\{\hat{h}_{i,l}^{\text{iter}=I}\right\}_{l=0}^{L-1}$ and $\left\{\hat{h}_{i+1,l}^{\text{iter}=I}\right\}_{l=0}^{L-1}$ can be used for simplicity, that is,

$$\tilde{h}_{i,k}^{\text{iter}=I} = \left(\hat{h}_{i,k}^{\text{iter}=I} + \hat{h}_{i+1,k}^{\text{iter}=I}\right)/2 \tag{7-100}$$

2. After the frequency-domain equalization, one can get

$$\hat{s}_{i,k}^{\text{iter}=I} = \text{IDFT}\left\{\frac{\text{DFT}\left(\hat{x}_{i,k}^{\text{iter}=I}\right)}{\text{DFT}\left(\tilde{h}_{i,k}^{\text{iter}=I}\right)}\right\} \quad 0 \le k < N_2 \tag{7-101}$$

where $N_2 \ge N$.

3. By setting the elements with the index $k \geq N$ to zero, one can get $\left\{\hat{s}_{i,k}^{\text{iter}=I}\right\}_{k=0}^{N-1}$.
4. To further eliminate the noise impact, the equalized data sequence $\left\{\hat{s}_{i,k}^{\text{iter}=I}\right\}_{k=0}^{N-1}$ can be transformed into the frequency domain with an N-point DFT to get the frequency decision of $\left\{\hat{S}_{i,k}^{\text{iter}=I}\right\}_{k=0}^{N-1}$, and then it can be transformed back to the time domain to $\left\{\tilde{s}_{i,k}^{\text{iter}=I}\right\}_{k=0}^{N-1}$.
5. The final filter output $\left\{z_{i,k}^{\text{iter}=I}\right\}_{k=0}^{M+N-1}$ is the convolution of $\left\{\tilde{s}_{i,k}^{\text{iter}=I}\right\}_{k=0}^{N-1}$ and $\left\{\tilde{h}_{i,l}^{\text{iter}=I}\right\}_{l=0}^{L-1}$, which can be calculated by N_2-point DFT.

Through the above iterative method, more accurate channel estimation can be achieved. It has been widely used for DTMB receiver design.

7.6 SUMMARY

This chapter introduced the inner receiver structure of the DTV system. Based on mathematical fundamentals, a detailed analysis of timing synchronization, carrier synchronization, channel estimation, and equalization for both single- and multi-carrier systems were presented. Moreover, to facilitate the understanding of the inner receiver, the DTMB inner receiver structure was introduced in details.

REFERENCES

1. H. Meyr, M. Moeneclaey, and S. A. Fechtel, *Digital Communication Receivers: Synchronization and Channel Estimation*, New York: Wiley, 1997.
2. J. G. Proakis, *Digital Communications*, 5th ed., Columbus, OH: McGraw-Hill, 2007.
3. O. Edfors, M. Sandell, J.-J. van de Beek, S. K. Wilson, and P. O. Borjesson, "OFDM channel estimation by singular value decomposition," *IEEE Transactions on Communications*, vol. 46, no. 7, pp. 931–939, July 1998.
4. O. Edfors, M. Sandell, J.-J. van de Beek, S. K. Wilson, and P. O. Borjesson, "Analysis of DFT-based channel estimators for OFDM," *Wireless Personal Communications*, vol. 12, no. 1, pp. 55–70, 2000.
5. F. M. Gardner, "A BPSK/QPSK timing-error detector for sampled receivers," *IEEE Transactions on Communications*, vol. 34, no. 5, pp. 423–429, May 1986.
6. Y. Deng, Z. Yang, and X. Guo, "A simple feedforward timing offsets estimation algorithm for BPSK/QPSK," in *The 2004 Joint Conference of the 10th Asia-Pacific Conference on Communications, 2004 and the 5th International Symposium on Multi-Dimensional Mobile Communications Proceedings*, vol. 1, Piscataway NJ: IEEE, Sept. 2004, pp. 214–217.
7. W. Leng, Y. Zhang, and Z. Yang,"A modified Gardner detector for multilevel PAM/QAM system," in *Proc. ICCCAS 2008: IEEE International Conference on Communications, Circuits and Systems*, Piscataway NJ: IEEE, 2008, pp. 891–895.

8. K. Peng, A. Xu, and Z. Yang, "Optimal correlation based frequency estimator with maximal estimation range," in *Proc. ICCCAS 2008: IEEE International Conference on Communications, Circuits and Systems*, Piscataway NJ: IEEE, 2008, pp. 259–263.
9. Q. Liu, Z. Yang, J. Song, and C. Pan,"A novel QAM joint frequency-phase carrier recovery method," in *Proc. ICACT 2006: The 8th IEEE International Conference Advanced Communication Technology*, Piscataway NJ: IEEE, 2006, pp. 1617–1621.
10. Advanced Television Systems Committee, *A/54A Guide to Use of the ATSC Digital Television Standard, with Corrigendum No. 1*, Washington, DC: ATSC, 2006.
11. J. Whitake, *Television Receivers: Digital Video for DTV, Cable, and Satellite*, New York: McGraw-Hill, 2001.
12. F. Classen and H. Meyr, "Frequency synchronization algorithms for OFDM systems suitable for communications over frequency selective fading channels," in *Proc. 1994 IEEE 44th Vehicular Technology Conference*, vol. 3, Piscataway NJ: IEEE, 1994, pp. 1655–1659.
13. J.-J. van de Beek, M. Sandell, and P. O. Borjesson, "ML estimation of time and frequency offset in OFDM systems," *IEEE Transactions on Signal Processing*, vol. 45, no. 7, pp. 1800–1805, July 1997.
14. Y. Li, "Pilot-symbol-aided channel estimation for OFDM in wireless systems," *IEEE Transactions on Vehicular Technology*, vol. 49, no. 4, pp. 1207–1215, July 2000.
15. L. Házy, *Initial Channel Estimation and Frame Synchronization in OFDM Systems for Frequency Selective Channels*, Ottawa: Carleton University, 1997.
16. P. Combelles et al., "A receiver architecture conforming to the OFDM based digital video broadcasting standard for terrestrial transmission (DVB-T)," in *Proc. 1998 IEEE International Conference on Communications*, vol. 2, pp. 780–785, June 1998.
17. P. Renzo, *Advanced OFDM Systems for Terrestrial Multimedia Links*, Lausanne, Switzerland: EPFL, 2005.
18. J. Song, Z. Yang, K. Gong, C. Pan, J. Wang, and Y. Wu, "Technical review on Chinese digital terrestrial television broadcasting standard and measurements on some working modes," *IEEE Transactions on Broadcasting*, vol. 53, no. 1, pp. 1–7, Mar. 2007.
19. J. Wang, Z. Yang, C. Pan, M. Han, and L. Yang, "A combined code acquisition and symbol timing recovery method for TDS-OFDM," *IEEE Transactions on Broadcasting*, vol. 49, no. 3, pp. 304–308, Sept. 2003.
20. J. Wang, *"Studies on synchronization and channel estimation algorithms for digital TV terrestrial broadcasting,"* Ph.D. thesis, Tsinghua University, 2005.

8

NETWORK PLANNING FOR DTTB SYSTEMS

8.1 INTRODUCTION

In the early stages of deploying DTTB services, a multifrequency network (MFN) was generally adopted due to the construction limitation and planning complexity. With the ever-increasing demand on spectrum resources and technology advances, RF reuse for DTV broadcasting has become popular and necessary to ensure reliable and efficient coverage for a large region, resulting in the creation of single-frequency networks (SFNs).

In addition, the study of diversity technologies has enabled the increase in system capacity and reduction in system bit error rate (BER), which also play a very importance role in network planning. Transmit diversity and/or receive diversity further improve the transmission capacity of the system and have been widely used due to their good antifading capability and ability to increase system capacity.

In this chapter we will discuss the issues of coverage and network planning of DTV terrestrial broadcasting networks and give a detailed introduction of SFNs. The characteristics and implementation of the diversity technology will also be introduced.

Digital Terrestrial Television Broadcasting: Technology and System, First Edition. Edited by Jian Song, Zhixing Yang, and Jun Wang.

8.2 BASIC CONCEPTS

8.2.1 Carrier-to-Noise Ratio

The carrier-to-noise ratio (C/N or CNR), which refers to the power ratio of signal to noise at the receiver and generally expressed in decibels, is an important parameter to evaluate signal quality. The thresholds of BER and C/N at the receiver jointly determine whether the output signal suffers from noticeable distortion, i.e., whether the output image can be correctly received. Generally, for different modulation and coding schemes, C/R values guaranteed for correct reception are different.

8.2.2 Minimal Field Strength

The coverage quality of signals can be evaluated by the field strength of the received signal. The minimal field strength of the received signal must satisfy the voltage requirement of correct reception with such factors as the noise figure of the tuner, the C/N for quasi-error-free (QEF) reception (defined as less than one uncorrected error per hour at the input of the MPEG-2 decoder) required by the system for different modulation parameters, channel bandwidth, and noise power. The calculation is as follows: First, calculate the noise input power P_n in dBW of the receiver by the noise figure F in dB and channel bandwidth B in Hz:

$$P_n = F + 10\log(kT_{\varphi}B) \tag{8-1}$$

where $k = 1.38 \times 10^{-23}\ J/K$ is Boltzmann's constant and $T_{\varphi} = 290$ K is the typical temperature in the absolute scale. Then, one can calculate the minimum signal power $P_{s,\min}$ in dBW required for correct reception at the terminal:

$$P_{s,\min} = P_n + \frac{C}{N} \tag{8-2}$$

Converting the minimum signal power to the minimum input voltage in dBV yields

$$V_{s,\min} = P_{s,\min} + 120 + 10\log Z \tag{8-3}$$

where Z is the receiver input impedance in Ω (the typical value is 75 Ω).

Table 8-1 compares the minimum field strengths required for correct reception for an analog PAL system and the DVB-T system with 64QAM as an example. In this table, the position coverage of the analog signal is 50% and that defined for the digital signal is higher.

TABLE 8-1 Comparison of Minimum Field Strengths Required by Analog and Digital Systems

Signal Type	Position Coverage Ratio (%)	Minimum Field Strengths Required for Fixed Reception (V/m)
Analog (PAL)	50	64
DVB-T	70	46
DVB-T	95	55
DVB-T	97	57
DVB-T	99	61

In general, analog TV signal power is mainly concentrated around the video carrier; therefore, the power of the analog TV signal can be measured by simply measuring the peak voltage of the video carrier. However, the energy of the DTV system signal is usually uniformly distributed within the channel, so it is required to measure the signal strength over the entire bandwidth.

8.2.3 Cliff Effect

It is well known that analog TV may provide smooth transition coverage at the service area boundary, and reception performance gradually degrades when the receiver moves outward of the coverage area. When the transmitter cannot provide the field strength required for the signal reception without or with little error outside the broadcasting area, there will be significant signal attenuation, which causes image quality degradation, and eventually the TV program becomes unwatchable. For DTV services, things are slightly different. Although the digital signal strength suffers from attenuation similar to that of analog TV services, the reception failure mechanism is quite different. The DTV system generally adopts a very powerful error correction code, which can correct most of the errors to meet the BER threshold requirement. Therefore, users will not notice any errors even when the signal suffers from very severe field strength attenuation. However, once the BER threshold cannot be satisfied, reception failure occurs, and the program "suddenly" becomes unwatchable. This is the inherent "cliff effect" of the DTV system, and the satisfactory coverage area difference should be considered when planning the transition from analog to digital TV service [1].

8.2.4 Location Coverage Probability

Location coverage probability refers to the percentage of positions having the required field strength among all the testing points in a given area. Due to the cliff effect of the DTV system, the location coverage probability standard used for

DTV system is usually different from that of the analog TV system. In analog broadcasting, a measured location coverage probability of 50% is acceptable in a service area. But for a digital signal, a much higher position coverage percentage is required to ensure sufficient receiving quality. According to the location coverage probability, the small area coverage can be defined as the following two classes: (1) good, meaning the location coverage probability for fixed and mobile reception can reach 95 and 99%, respectively, and (2) acceptable, i.e., in a small area, the location coverage probability for fixed and mobile reception can reach 70 and 90%, respectively.

For any position, if the received signal has a C/N meeting the required threshold and the received minimum signal strength is also satisfied, this position is considered successfully covered. The criterion used here assumes that the receiving position does not introduce any attenuation such as that caused by the superposition of the direct path signal and reflected signals. If the position can receive a signal satisfying the required field strength 99% of the time, the position is considered sufficiently covered; if the position coverage requirement for transmitter in a small area can be satisfied 99% of the time, this small area is considered adequately covered. For a small area, the terminology "coverage" means at least 70% of positions can receive enough field strength 99% of the time.

8.2.5 Protection Ratio

Protection ratio refers to the minimum power ratio of a signal to the interference at the measurement receiver. The reference power of the DTV signal is the average signal power measured within the entire system bandwidth, and the reference power of the analog TV signal is the root-mean-square power of the RF signal measured at the peak synchronizing signal. Under certain conditions, if the reception quality of the signal is specified, the protection ratio must also be determined. For a DTV system, the receiving quality is usually represented by BER. Taking the DVB-T system as an example, a BER of 2×10^{-4} after Viterbi decoding but before RS decoding is usually used to represent good receiving quality. Interference can be classified as follows:

1. *Cochannel Interference.* The cochannel interference occurs when more than one transmitter sends the signal using the same broadcasting channel, which is inevitable in DTTB networks. Measures should be taken to prevent or minimize the impact of cochannel interference in the network.
2. *Adjacent-Channel Interference.* A TV signal cannot be perfectly bandwidth limited due to imperfect filter characteristics, and there are always sidelobes outside the channel bandwidth. Adjacent-channel interference refers to interference from adjacent channels that leaks into the current channel and cannot be completely removed by the receiver filter. The situation becomes even worse when the transmitters operating in adjacent channels have much higher power. For example, in digital and analog simulcast systems, if the DTV signal and high-power analog TV signal are allocated to channels next to each other, adjacent-channel interference will become nonnegligible.

3. *Remote-Transmitter Interference.* In the large-area coverage of SFNs, there are many transmitters which transmit the same signals at the same time, and those signals may arrive at one location with different time delays from different paths. Among all these signals, one is considered the local signal while others are considered remote signals. The remote-transmitter interference happens in SFNs when the relative delay between the local signal and remote signal(s) exceeds the guard interval length of the system and remote signal(s) will become interference for the local signal. Such interference must be considered under specific meteorological and geographical conditions.

8.3 ANALOG AND DIGITAL TV BROADCASTING

8.3.1 Comparison between Analog and Digital Transmissions

For wireless broadcasting channels, reflection from buildings or any other obstacles or multipath fading is inevitable, and this causes waveform distortion on the received signal. In addition, interferences such as cochannel interference or adjacent-channel interference will also cause degradation in the quality of the received signal. These two factors introduce errors at the receiver. For analog TV signal transmission, distortion due to nonperfect channel and system nonlinearity always exists, while in DTTB systems, these errors might be corrected by FEC, and the original information bit sequence can be correctly restored. In this case, image quality at the receiver will be primarily determined by video compression and demultiplex schemes.

It is obvious that the spectrum efficiency of the DTV system is generally much higher than the analog system: A DTTB system can support several TV programs in one RF channel, which can only carry one analog TV program.

8.3.2 Frequency Planning for Terrestrial Broadcasting

Almost in every country around the world, bands I, II, IV, and V are used for TV broadcasting. Among them, band I can realize wide-range coverage due to its low frequency and long wavelength. However, it is not recommended by ITU because this band easily suffers from cochannel interference and is too narrow to accommodate many programs. The Stockholm agreement concluded in 1961 recommended a series of methods for spectrum allocation, e.g., four UHF channels for the nationwide terrestrial television broadcasting coverage, arrangement that ensures minimum interference to neighboring countries.

In the UHF band, signal transmission can only be realized within the coverage slightly larger than the line-of-sight range due to its poor diffraction property. This characteristic, however, makes the coverage model easily established after enough geographical data are collected [1]. When broadcasting network planning is performed, the position of the transmitter is determined based on certain criteria, such as the largest possible population coverage, the minimum cochannel interference, environmental factors, and economic factors.

The coverage area can also be influenced by the curvature of Earth. For example, with a transmitter using an antenna with height h_1 (m) above sea level, its coverage is in a circular shape with radius d (km). Assuming that the height of the receiving antenna is h_2 above sea level yields

$$d = 3.57\sqrt{h_1 + h_2} \text{ km} \tag{8-4}$$

Of course, the radius of the coverage area may be more than this predicted value if climate conditions permit.

Because the receiving antenna gain is high enough only within the narrow frequency range of UHF, the RF allocation for TV channels in the same area must be well planned such that all the allocated channels in this area will work within the same narrow band at high receiving antenna gain. Restrictions for the 8-MHz-bandwidth systems are provided below.

1. The oscillation frequency of a local oscillator (LO) of 38.9 MHz in the TV receiver is the summation of the intermediate frequency (IF) and the channel frequency to be received:

$$\begin{aligned} \text{LO} &= \text{receiving frequency} + \text{IF} \\ \text{IF} &= 38.9\,\text{MHz} \end{aligned} \tag{8-5}$$

2. After the LO signal mixes with IF signal, the image frequency occurs in the channel with the channel index of the current channel number plus 10. This is a first-image channel and cannot be used.
3. Repeat step 2. The second image channel will be generated, which cannot be used either.
4. All broadcasters do not want to use upper and lower adjacent-frequency channels to avoid the potential mutual interferences.

8.3.3 Simulcast of Digital and Analog TV

For historical and economic reasons, analog TV receivers will exist for a certain time and in large quantities. To promote the DTV under the current situation, terrestrial simulcast is commonly used, which is achieved by using the Taboo channels to broadcast the same analog programs using DTV broadcasting technologies within the current terrestrial broadcasting band.

According to the frequency planning for DTV broadcasting with fixed transmit power, the lower the required SNR threshold is, the larger the coverage area will be if there is no parameter change. The more effective the methods used to guard against cochannel interference are, the larger the coverage area will be. Since the simulcast system requires that each program appears in digital and analog channels simultaneously, the following requirements on DTTB systems must be satisfied:

1. Use the same RF bandwidth as that of the existing analog TV system.
2. Use a better channel coding scheme with lower SNR threshold for larger coverage area of DTV broadcasting network at low cost.
3. Provide the capability for successful reception under various types of interferences such as cochannel interference and adjacent-channel interference. Also, the interference from the DTTB system to analog TV services should be less than the mutual interference among analog TV systems.

8.3.4 Frequency Utilization of Terrestrial Broadcasting

Simulcast has been widely used to ensure graceful transition from analog TV to digital TV broadcasting. This method, however, requires more bandwidth. To save the precious spectrum resource, it is highly recommended to reallocate all the existing channels to ensure an efficient utilization of the terrestrial TV broadcasting spectrum.

For an analog TV receiver, good receiving quality can be guaranteed only when the C/N is very high while perfect receiving quality can be achieved with a C/N of 15 dB for DTTB systems at the spectrum efficiency of 4 bits/s/Hz. At the same transmitter power, the coverage area can be quite different. It is, therefore, necessary to reevaluate the cochannel interference in the simulcast mode.

Reference [1] gives an analysis of the relationship of the coverage area, C/N threshold, and protection ratio of D/U using U.S. spectrum setting and planning in the UHF band as an example.

Analog TV requires a much larger reuse distance due to its poor anti-interference ability, while this distance is greatly reduced in a DTTB system due to its strong anti-interference ability. Therefore, frequency planning for the simulcast mode will surely be different from the analog TV age and should be reconsidered.

Assume channel planning in the UHF band for the analog TV network shown in Figure 8-1 is a cell structure consisting of seven frequency groups (labeled F1–F7) and the reuse distance for the same frequency is at least three times as much as the spacing of the two adjacent transmitters.

When simulcast needs to be supported, another DTV broadcasting station with a new frequency DF1 (digital frequency 1) can be ideally set up at the joint location of

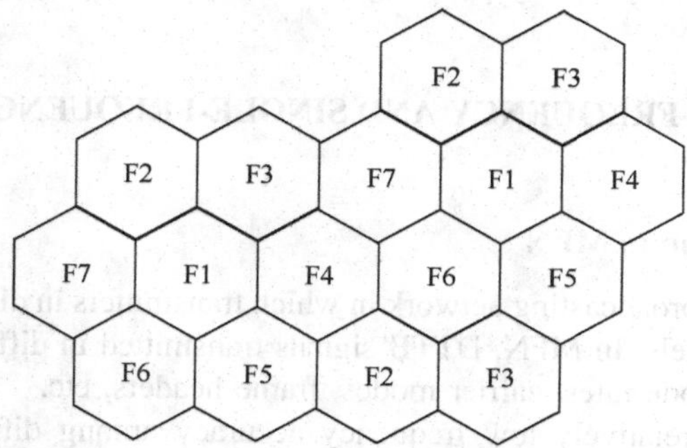

FIGURE 8-1 Example of frequency planning on analog TV network for UHF band.

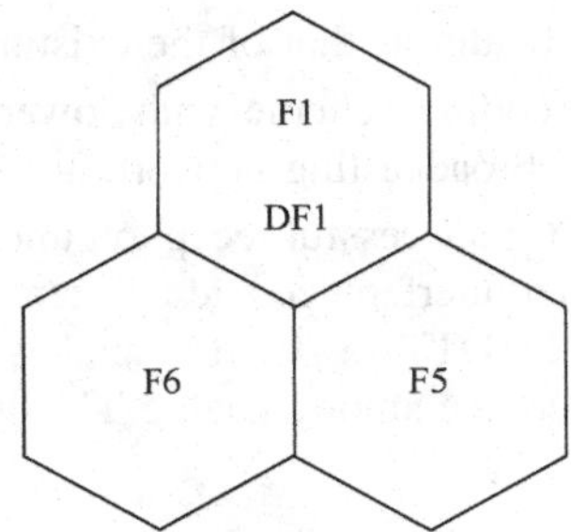

FIGURE 8-2 Frequency allocation when simulcast is applied.

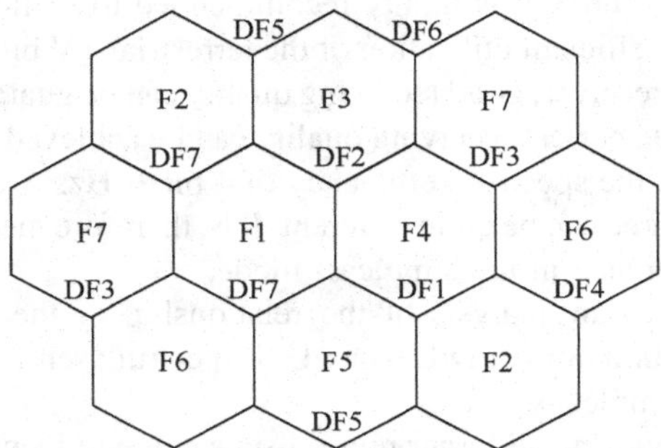

FIGURE 8-3 Frequency planning for simulcast.

F1, F5, and F6 in Figure 8-1, and this is illustrated in Figure 8-2. In this way, the cell structure of seven frequencies will have a new cell structure, and the ideal allocation can be achieved as shown in Figure 8-3.

The analog TV signals generally introduce serious out-of-band leakage, causing strong interference on digital channels. Some countries have decided to offset the center frequency of DTV signals to reduce the impact from the analog signals when doing DTTB system frequency planning.

8.4 MULTIPLE-FREQUENCY AND SINGLE-FREQUENCY NETWORKS

8.4.1 Introduction to MFN

MFN refers to the broadcasting network in which transmitters in different regions use different RF channels. In MFN, DTTB signals transmitted in different regions may choose different code rates, carrier modes, frame headers, etc.

MFN needs a relatively low frequency accuracy among different transmitters (5×10^{-5} is acceptable), and it is unnecessary to coordinate the output delay of different transmitters. Therefore, network operation is more flexible, and network construction is simpler. In this case, the frequency planning and network construction of MFNs are relatively straightforward and flexible, and it is generally easier to design receivers because there is no interference from the propagation delay spread from different transmitters working at the same frequency (to be more precise, the distance between transmitters at the same frequency is far enough, and the effect becomes negligible). Since RF channels for neighboring transmitters are different, no cochannel interference exists and therefore no coordination is needed for the frequency planning. With simple network topology and almost no echo, network deployment is very much simplified, and there is no need for synchronization for transmitters within the same region.

However, MFNs require much more spectrum than SFNs provide the same amount of programs, which is the price one has to pay for the simplicity. For the boundary area covered by different transmitters, the quality of the received signal is relatively poor and that is why it is very difficult to achieve highly reliable coverage by MFNs. To overcome this problem, regional SFNs may be applied to a specific service area under an overall MFN architecture when needed [2].

8.4.2 Introduction to SFN

SFN refers to the broadcasting network with a number of transmitting stations sending the same RF signals at the same frequency in the same time to achieve a reliable coverage in bigger service areas than that of the MFN [3]. With limited spectrum resources, frequency coordination among neighboring countries or regions faces great limitation on both channel allocation and large coverage realization when the MFN is used. The introduction of the SFN will help overcome this limitation.

The most obvious benefit from the SFN is the improvement of spectrum efficiency. An MFN needs to use different RF channels to transmit the same program to users in different regions by different transmitters, and the spectrum efficiency is quite low. A frequency reuse factor measures how efficiently each system uses the precious spectrum resources. Generally, a large frequency reuse number means low spectrum utilization efficiency, and the reuse factor for the SFN is 1! The example shown in Figure 8-4 gives that the reuse factor of the MFN is 7.

The fundamental challenge of the SFN is how to handle the time delay spread from different transmitters. The situation is quite complicated in the regional SFN, as shown in Figure 8-5. The same signal coming from different transmitters at the same time will arrive at the receiver with different delays from the different paths, which is called an artificial multipath. Multicarrier modulation such as OFDM technology can solve this problem very efficiently by introducing the guard interval (GI) to the OFDM data block. If the GI is shorter than the maximum delay spread of this artificial multipath effect, the signal will suffer from multipath delay. If it is longer, the signal can be enhanced as the effective signal power increases. As a result, the OFDM system can easily handle the artificial multipath problem and support SFN utilization. In fact, the ATSC system, which is a single-carrier system, also supports SFN. However, the equalizer design will be complicated to support deployment of a large

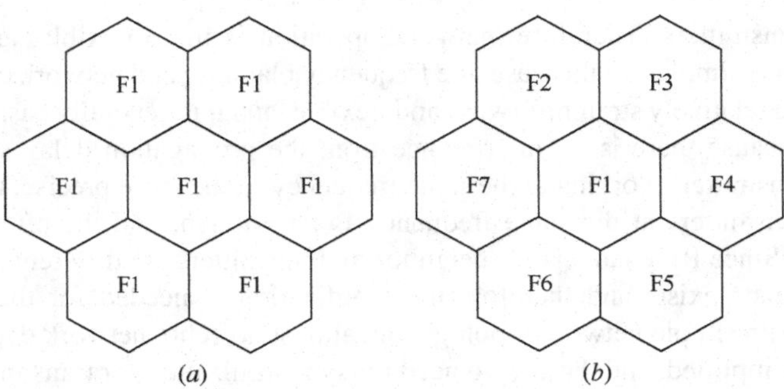

FIGURE 8-4 Comparison of frequency reuse factor for (*a*) SFN and (*b*) MFN.

SFN coverage (i.e., the maximum delay spread is too large in this application scenario, and the number of taps of the equalizer will be very large).

Another requirement for SFN utilization is very accurate synchronization among all transmitters in the coverage area because, in general, SFN operation requires all transmitters to send the same program using the same frequency at the same time.

8.4.3 Classification of SFNs

In this section, we will mainly focus on the important SFNs as follows:

1. *Nationwide SFN.* It usually covers a very large region, even the whole country, and has the features of high transmit power for many transmitters and long-distance transmission.
2. *Regional SFN.* It only has one or several high-power transmitters, capable of transmitting the signal over a long distance.

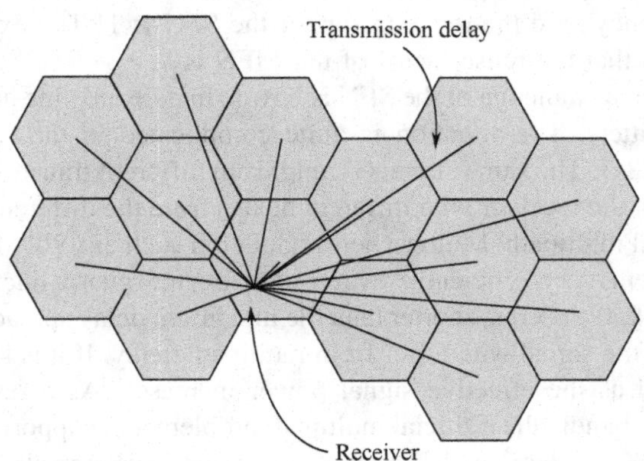

FIGURE 8-5 Illustration of echo delays in SFN.

3. *Local SFN.* It has one high-power transmitter in the main station at one channel, while a group of local transmitters in SFN close to the main high-power transmitter operate at another channel.

In practical applications of SFN, one type of transmitter, known as a "gap-filler," is mainly deployed in certain smaller areas to compensate for the signal power loss either from the long propagation distance or from the severe attenuation. Considering economic factors, the synchronization requirement on these low-power gap fillers is not very strict

8.4.4 Interference Analysis of SFN

Self-interference occurs when the distance between transmitters exceeds a certain range so that the maximum delay spread for the signal from transmitters in SFNs is longer than the time protection range of the DTTB signal. In this case, those signals with much longer time delay spread (usually have lower amplitude) will be treated as noise or interference at the receiver, and this reduces the SNR of the desired signal. In general, for SFN design, the maximum delay spread due to the different physical locations of the transmitters should not be longer than the protection range.

A receiver located within the coverage area of the transmitter (noted as T_{x1}) may receive signals from neighboring transmitters (such as T_{x2}). In this case, the delay spread will be determined by the difference in distance between the two propagation paths from transmitters to the receiver instead of the distance between these two transmitters. Let the distance from transmitter T_{x1} to the receiver be $d_1(km)$ and the distance from the transmitter T_{x2} to the receiver be $d_2(km)$, and the transmitting path distance difference be $D = d_2 - d_1(km)$. If $d_2 > d_1$, then the delay spread is

$$T = \frac{D \times 10}{3 \times 10^8}(\text{s})$$

After the parameters of GI length of the SFN are determined, the maximum path difference permitted in the SFN is also determined. If the actual path difference exceeds this value, those signals arriving at the receiver with longer delay than this threshold will be taken as interference. Such interference usually is weaker than the desired signal and has the characteristics of noise. A fixed receiving antenna pointing to the desired transmitter helps reduce the interfering signal and hence improves the SNR at the receiver. It is generally considered that the signal from neighboring transmitters will not lead to destructive results at the receiver under most receiving conditions.

An experimental SFN usually involves two or three transmitters operating at the same frequency, and the time delay of the two signals reaching the reference receiver can be changed if the transmitters are set to send the signal from different starting times. The signal amplitude may also be changed to test the performance of the SFN at extreme conditions. Usually, performance of the SFN should be tested under the condition of two signals with the same magnitude but different arrival times, called the 0-dB echo condition. The performance should be tested for maximum time delay equal to GI. The extreme condition of a 0-dB echo seldom happens in reality, but this experiment is helpful to determine the appropriate GI of OFDM signals.

The received signal power might double when the signals sent by two transmitters arrive at the receiver at the same amplitude (0-dB echo condition) and in phase. Due to the so-called artificial multipath effect, the C/N threshold for successful reception may also increase, i.e., a higher C/N is required to resist the interference of the 0-dB echo at the receiver. Therefore, the extreme condition of the 0-dB echo can be accommodated by the SFN, but generally it may not bring the network gain due to the increases of required C/N.

8.4.5 Synchronization in SFN

SFN operation requires a stable timing reference signal for all transmitters in the network, and the accuracy of the reference signal should be better than 1 μs. One method is to use the signal at the rate of 1 pulse/s provided by a GPS receiver as reference.

8.4.5.1 Time Synchronization Using MPEG-2 as example, a time tag needs to be inserted into the MPEG.-2 transport stream to ensure the synchronization of all the transmitters within the SFN. Furthermore, the transmission time of different transmitters in the SFN might be different because of the transmission and processing delay in the network. It is very important to compensate for this time difference when constructing SFNs. These time delays may be calculated beforehand, and the compensation is done by putting the maximal delay value in the time tag. This maximal delay value gives the time (usually less than 100 μs) that can be delayed for the signal transmission, and it is usually longer than the time delay to reach the most distant transmitter in the network. All the transmitters other than the most distant one should wait for this maximal delay time to ensure the signal to be transmitted can reach the most distant transmitter before all the transmitters in the SFN can transmit the same signal at the same time. With this arrangement, the network planner can easily compensate and balance the distributed delay in the SFN, and the timing synchronization is guaranteed. The DTTB modulator only needs to buffer the MPEG-2 transport stream with the preset amount of time to realize network synchronization.

In the DVB-T system, the synchronization is usually realized through a megaframe. The transmitted signal is organized in the frame groups, and each consists of 68 OFDM frames. Four frame groups constitute one superframe. The output of the SFN adapter should be MPEG-2 TS, where the individual packets are organized in groups to form a megaframe. Each megaframe consists of n packets, where n is an integer number which depends on the number of RS packets per superframe in the DVB-T. In the 8K mode n = number of RS packets per superframe × 2. In the 2K mode n = number of RS packets per superframe × 8. Each megaframe contains exactly one megaframe initialization packet (MIP), which includes a time stamp indicating when the megaframe will be sent. This time stamp will be compared with the absolute time with the reference frequency coming from the local GPS receiver. Based on the

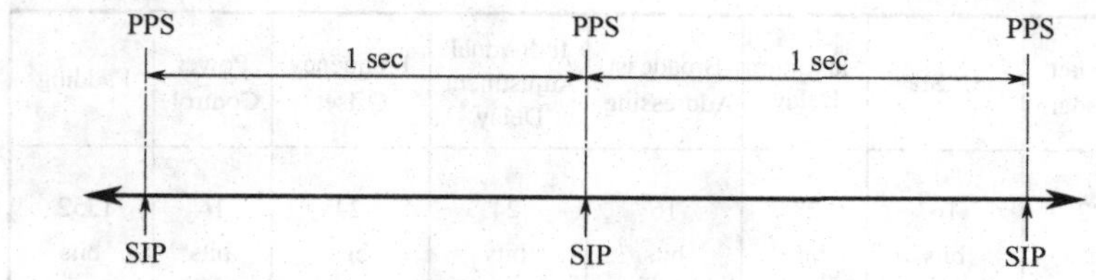

FIGURE 8-6 Illustration of SIP insertion.

above, the exact starting time for the signal broadcasting is determined. All the transmitters will be synchronized in a second.

In the DTMB system, the second frame initial packet (SIP) is used for synchronization with its function and corresponding synchronization method similar to that of MIP. The SFN adapter of the DTMB system inserts one SIP packet into the TS every second using 1 pps (pulse per second) signal of GPS receiver as reference, which is shown in Figure 8-6.

In SFNs, each transmitter obtains the maximal delay time T_{delay_max} and the network transmission delay time $T_{delay\_transmitted}$ by detecting the received SIP packet in the transport stream. Then the additional time delay T_{delay_add} of the exciter will be by $T_{delay\_add} = \mathrm{T}_{delay\_max} - T_{delay\_transmitted}$. Where the maximal delay time T_{delay_max} refers to the time delay for TS to be transmitted by each transmitter relative to the 1 pps signal of the GPS receiver and the transmission delay time $T_{delay\_transmitted}$ refers to the actual propagation delay caused by the distribution network, the additional delay time T_{delay_add} is the time delay each transmitter needs to wait before transmitting the signal. As the signal of the GPS receiver is used as reference, T_{delay_max} that can be handled by the transmitter in the DTMB SFN system will be 0.9999999s. The SIP processing is shown in Figure 8-7.

The SIP format is the same as that of the MPEG-2 TS packet, which consists of a 4-byte packet header and 184 data bytes. The structure with important bit definitions of SIP is given in Figure 8-8, where the fields are defined as follows:

1. Packet header: 32 bits long, with its definition satisfying GB/T 17975.1-2000, and the PID is 0x0015.

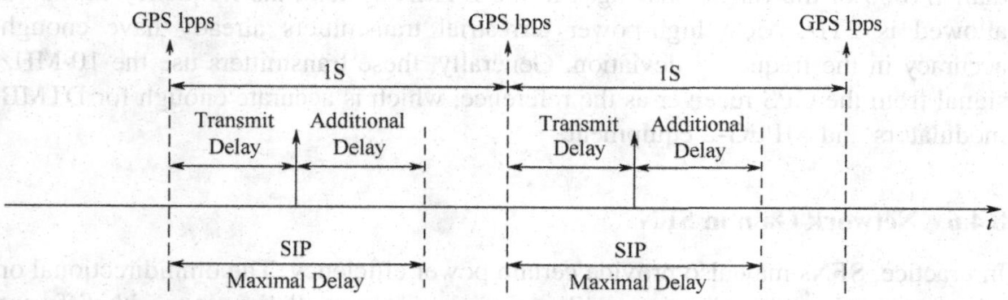

FIGURE 8-7 Illustration of SIP insertion for DTMB system.

Packet Header	SI_SIP	Maximum Delay	Broadcast Addressing	Individual Adjustment Delay	Frequency Offset	Power Control	Padding
32 bits	16 bits	24 bits	16 bits	24 bits	24 bits	16 bits	1352 bits

FIGURE 8-8 Structure and important bit definition for SIP.

2. SI_SIP: consisting of 16 bits (from P0–P15) with the definition given in Table 8-2.
   1. Maximal delay: 24 bits long with the setting range of 0x000000-0x98967F, in unit of 100 ns.
   2. Transmitter addressing: 16 bits long, used to find transmitters in the SFN. Addressable range is 0x0000-0xFFFF, where 0x0000 means to find all the transmitters in the SFN.
   3. Independent delay adjustment: 24 bit longs. Based on the maximal delay time, the independent delay adjustment is applied to the specific transmitter to ensure all the signals transmitted by each transmitter in the SFN satisfy a certain delay relationship.
   4. Frequency offset setting: 24 bits long. It is used to set the frequency offset value for the specific transmitter. Its unit is 1 Hz and the range is -8388608-8388607.
   5. Power control: 16 bit longs, used to control the transmission power of the specific transmitter. The most significant bit is the power control switch with bit 1 indicating "on" and 0 indicating "off." The rest of the 15 bits give the power control value in unit of 0.1 dBm with range [0, 32767] × 0.1 dBm.
   6. Padding: 169 bytes long and all are 0xFF.

8.4.5.2 Frequency Synchronization Frequency synchronization requires that the working frequency of each transmitter in the single-frequency network must be identical. In practice, the frequency deviation of the transmitter is required to be less than 1/1000 of the carrier spacing. For the DTMB system, the frequency deviation allowed is 2 Hz. Most high-power terrestrial transmitters already have enough accuracy in the frequency deviation. Generally, these transmitters use the 10-MHz signal from the GPS receiver as the reference, which is accurate enough for DTMB modulators and MPEG-2 equipment.

8.4.6 Network Gain in SFN

In practice, SFNs may also provide certain power efficiency. The omnidirectional or whip antenna in real applications will receive signals from all directions with different

TABLE 8-2 SIP Definition

P0, P1, Frame Header Mode	P2, Number of Carriers	P3–P5, Mapping Mode	P6, P7, Code Rate	P8, Interleaving Mode	P9, Double Pilot	P10, PN Phase	P11–P15, None
00:PN420 01:PN595 10:PN945 11: Reserved	0:$C=1$ 1:$C=3780$	000:4QAM-NR 001:4QAM 010:16QAM 011:32QAM 100:64QAM 101–111: Reserved	00:0.4 01:0.6 10:0.8 11: Reserved	0:240 1:720	0: Without 1: With	0: Without rotation 1: With Rotation	Reserved

amplitudes and delays. Such a composite signal at the receiver, in theory, includes all signal components of transmitters in the SFN. In this case, if one signal suffers from the severe fading introduced by its propagation path, the receiver can still use signals from other transmitters through different paths to ensure successful reception. From this perspective, the use of SFNs may help reduce the power of the transmitter while maintaining the good reliability for the receiving signal, and this is considered the network gain provided by the SFN.

Considering the transition from analog to digital TV, it is usually required to maximally utilize the existing infrastructure and equipment to minimize construction cost and time. As equidistant transmitter allocation is not common practice and a directional antenna is commonly used in the network design and planning, it is therefore very difficult for network operators to design, optimize, and implement SFNs under the above-mentioned constraints. If existing facilities must be fully utilized, the network gain may only be achievable near the boundaries of areas originally covered by the individual transmitters.

8.4.7 Application of SFN

Taking the DTMB standard as an example, the basic structure of an SFN system consists of an SFN adapter at the transmitter and a synchronization system at the receiver, as shown in Figure 8-9 [4,5].

The GI length of the OFDM system determines the maximal delay time of the multipath signals to be handled by the receiver and therefore limits the maximum distance between the transmitters in an SFN. The DVB-T 8K mode and ISDB-T systems use a large number of subcarriers (that is, narrower subcarrier spacing and longer OFDM symbol duration), and a relatively larger GI is good to support large coverage of a region or a country by SFNs. The mode with smaller GI is used to provide the coverage of a local SFN. In a DVB-T system, the relationship between the maximal delay spread and transmitter distance is given in Table 8-3. It can be seen from the table that the duration of GI in the 8K mode is four times as much as that in 2K mode, indicating the allowable maximal distance between transmitters in the 8K mode is exactly four times as much as that in 2K mode [4].

Accordingly, with three different guard interval lengths supported by DTMB, the relationship between the maximum delay spread and the maximum transmitter distance is shown in the Table 8-4.

8.5 TRANSMISSION SYSTEM OF DTTB

The terrestrial television transmission system mainly consists of a transmitter, filters, an antenna, and its feeder cables while the transmitter mainly consists of the exciter, power amplifier, and cooling and corresponding control system.

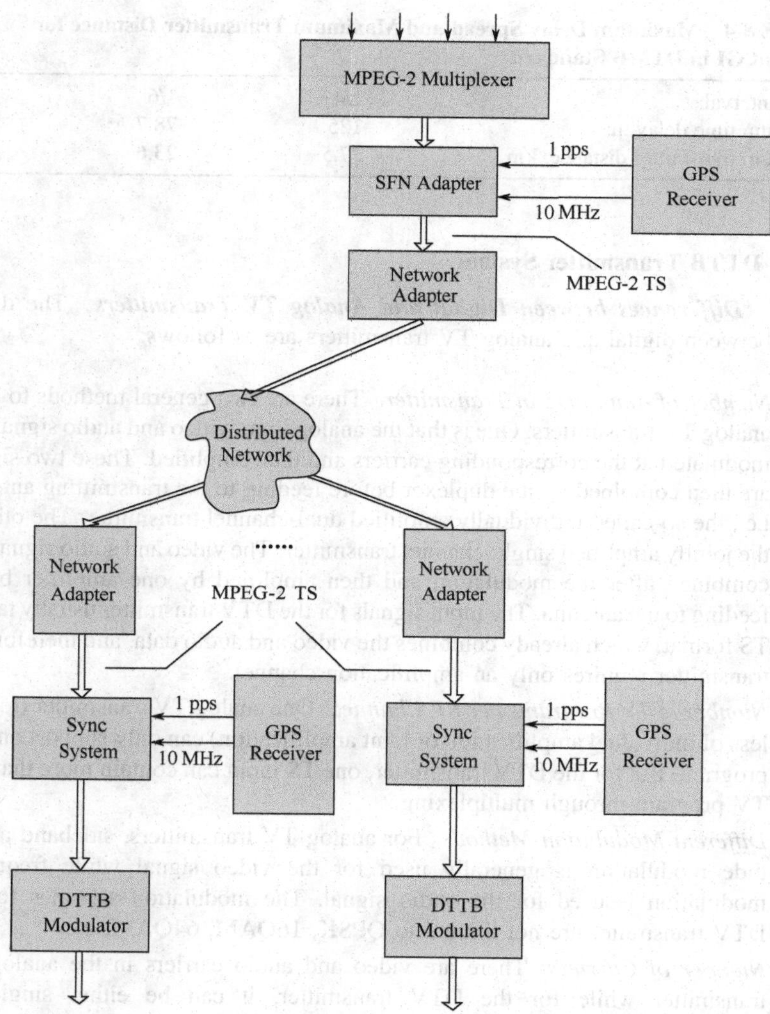

FIGURE 8-9 Basic structure of single-frequency network system for DTMB system.

TABLE 8-3 Maximum Delay Spread and Transmitter Distance in Different GI in DVB-T Standard

Guard Interval	1/4	1/8	1/16	1/32
2k Maximum delay time, μs	56	28	14	7
2k Maximum transmitter distance, km	16.8	8.4	4.2	2.1
8k Maximum time delay, μs	224	112	56	28
8k Maximum transmitter distance, km	67.2	33.6	17.8	8.4

TABLE 8-4 Maximum Delay Spread and Maximum Transmitter Distance for Different GI in DTMB Standard

Guard interval	1/4	1/6	1/9
Maximum time delay, μs	125	78.7	55.6
Maximum transmitter distance, km	37.5	23.6	16.7

8.5.1 DTTB Transmitter System

8.5.1.1 Differences between Digital and Analog TV Transmitters The differences between digital and analog TV transmitters are as follows:

1. *Number of Amplifiers in Transmitter.* There are two general methods to build analog TV transmitters. One is that the analog input video and audio signals are modulated at the corresponding carriers and then amplified. These two signals are then combined by the duplexer before feeding to the transmitting antenna, i.e., the so-called individually amplified dual-channel transmitter. The other is the jointly amplified single-channel transmitter: The video and audio signals are combined after the modulation and then amplified by one amplifier before feeding to the antenna. The input signals for the DTV transmitter usually takes a TS format, which already combines the video and audio data, and therefore the transmitter requires only an amplification channel.
2. *Number of TV Programs per RF Channel.* One analog TV transmitter (regardless of individual amplification or joint amplification) can only support one TV program. But for the DTV transmitter, one TS input can contain more than one TV program through multiplexing.
3. *Different Modulation Methods.* For analog TV transmitters, sideband amplitude modulation is generally used for the video signal while frequency modulation is used for the audio signal. The modulation schemes for the DTV transmitter are not limited to QPSK, 16QAM, 64QAM, etc.
4. *Number of Carriers.* There are video and audio carriers in the analog TV transmitter while for the DTV transmitter, it can be either single- or multicarrier.
5. *Different Definitions of Output Power.* The nominal power P_{syn} of the analog TV transmitter is the power measured at the peak of the synchronization signal while the nominal power of the DTV transmitter is the average signal power P_{RMS}. In addition, the receiving threshold of the DTV signal is much lower than that of the analog TV signal, and the transmit power of the DTV transmitter can be 10 dB lower than that of the analog TV transmitter to maintain the same coverage. The typical power level of the DTV transmitter in large cities is within the range of 1–3 kW for SFN operation.
6. *Difference in Peak-to-Average Power Ratio (PAPR).* For the analog TV transmitter, there are two PAPR values. One refers to the ratio of P_{syn} to the average signal power and is measured when the transmitter sends a black

field signal. The PAPR is around 1.68 (2.25 dB). The other refers to the normal analog broadcast condition. When the image carrier sends 0 dB power, the sound carrier sends −10 dB, and the chrominance subcarrier sends −16 dB power (all relative to P_{syn}), the PAPR is 2.2 (3.4 dB). The PAPR of the DTV signal is much higher than that of the analog TV signal. The test results show that the PAPR of DVB-T is about 9.5 dB. The PAPR for the DTMB multicarrier is about 9.72 dB for the multicarrier mode and 7.01 dB for the single-carrier mode with dual pilots. These numbers are under the condition of 16QAM modulation at the probability of 99.99%.

7. *Different Technical Requirements.* The performance of analog TV transmitters is evaluated by the differential gain, differential phase, luminance nonlinearity, intermodulation distortion (IMD), etc. While the performance of DTV transmitters is evaluated by modulation error ratio (MER), the shoulder distance, and so on.
8. *Different Unwanted Radiation Spectrum.* The unwanted out-of-band radiation of analog TV transmitters has discrete spectrum, mainly including the second- and the third-order harmonics, while that of the DTV transmitter has continuous spectrum and harmonics.

8.5.1.2 Structure of DTTB Transmitter The input TS for the DTV transmitter contains the video and audio information, and the channel coding will be applied to it. After all the baseband processing, the baseband digital signal is then modulated to the desired carrier(s), filtered, and sent to the antenna. The functional blocks include the exciter, which carries out all the baseband processing on the signal and frequency up conversion, power amplifier, power divider and combiner, filters, control systems, power supply system, cooling system, feedback loop, etc. It is usually assembled in a standard chassis, and the maximum transmission power can be up to 4 kW and weighing about 600 kg. The DTV transmitter is smaller than its analog counterpart. Figure 8-10 is the block diagram of the DTV transmitter.

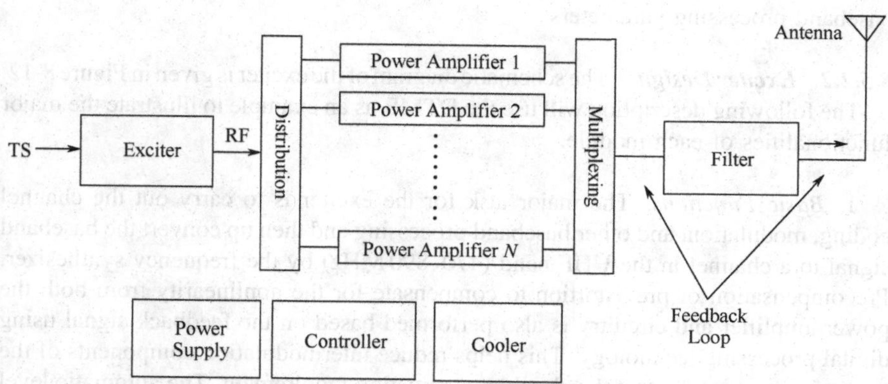

FIGURE 8-10 Block diagram of DTV transmitter.

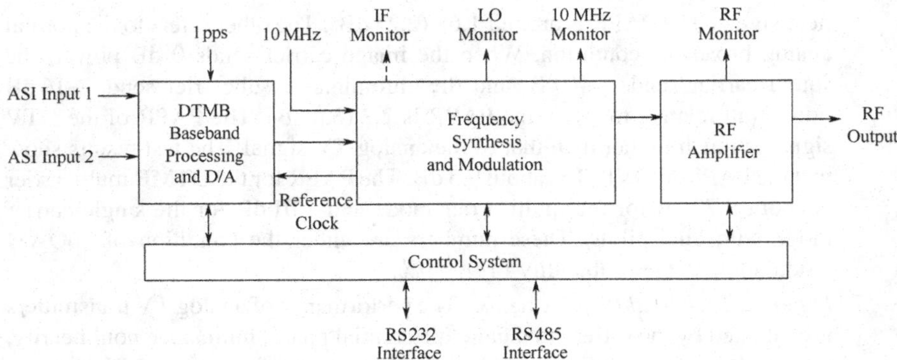

FIGURE 8-11 Block diagram of DTMB exciter.

8.5.2 DTTB Exciter

The exciter is the core part of the DTV transmitter, and its main function is to modulate the signal on the carrier in accordance with technical standards. To reduce the distortion from the circuitry nonlinearity, the exciter also has calibration functionalities, such as predistortion circuitry.

8.5.2.1 Structure of DTTB Exciter The DTTB exciter mainly consists of a baseband processor, D/A, frequency combiner and up converter, RF output amplifier, and the monitoring system. For different DTTB standards, the major difference lies in the baseband signal processing modules of the exciter. A block diagram of the DTMB exciter is shown in Figure 8-11.

The input signal of the exciter uses asynchronous serial interface (ASI), after channel coding, modulation, other baseband processing, and up conversion. The modules of the frequency combiner, up converter, and RF output amplifier are similar to that of the analog TV exciter, and the operation mode of the exciter depends on the baseband processing parameters.

8.5.2.2 Exciter Design The schematic diagram of the exciter is given in Figure 8-12. The following description will use the DTMB as an example to illustrate the major functionalities of each module.

1. *Basic Functions.* The major task for the exciter is to carry out the channel coding, modulation, and other baseband processing and then up convert the baseband signal to a channel in the UHF band (470–890 MHz) by the frequency synthesizer. Precompensation or predistortion to compensate for the nonlinearity from both the power amplifier and circuitry is also performed based on the feedback signal using digital processing technology. This helps reduce intermodulation components of the transmitter output to meet the shoulder attenuation requirement. The automatic level control (ALC) function can automatically control the power within the predefined

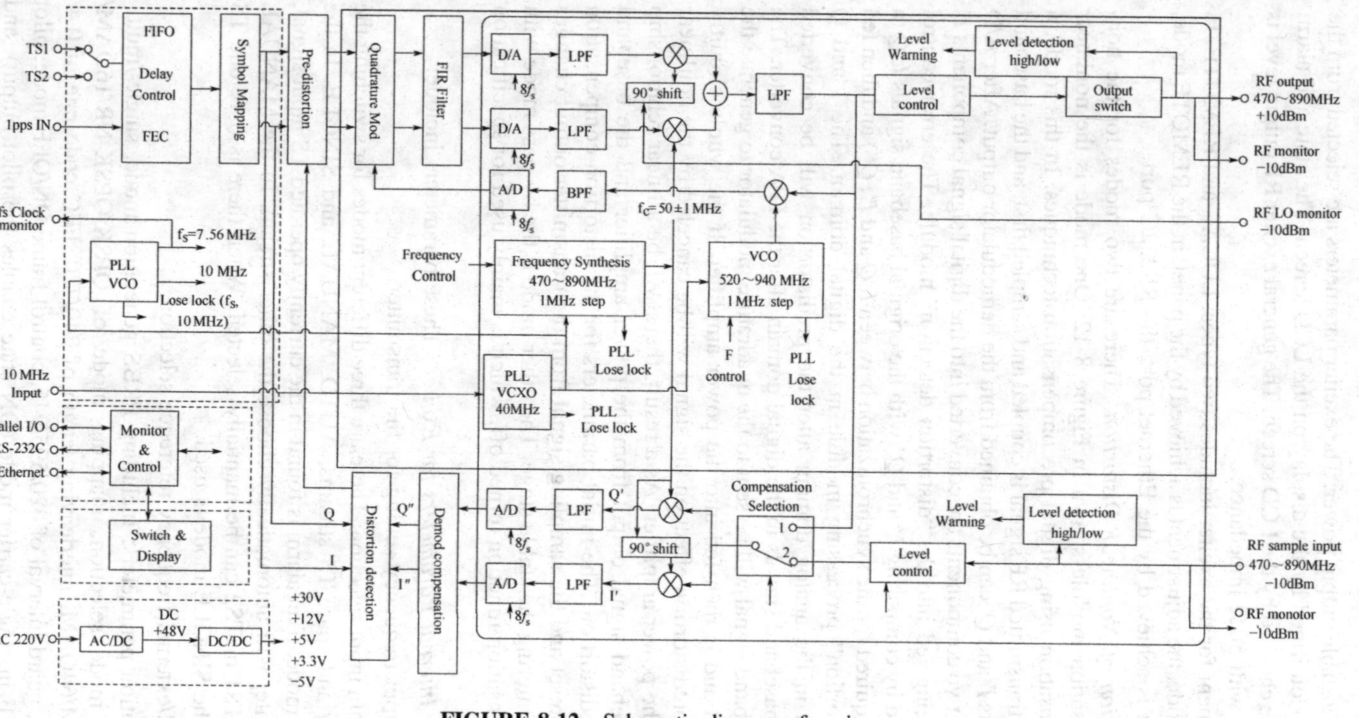

FIGURE 8-12 Schematic diagram of exciter.

range to maintain a stable output power. The exciter parameters are selected from the control panel and confirmed by the display on the LCD screen. The state and alarm records can also be checked by the LCD screen. The general exciter RF output level is $+10\,\text{dBm} \pm 1\,\text{dB}$, with $50\,\Omega$ impedance.

The adjustment for the exciter includes two types: LOCAL and REMOTE. In the LOCAL mode, the adjustment is achieved by the panel; in the REMOTE mode, the adjustment is achieved by the Ethernet port, the RS232C port.

2. *Compensation of Nonlinear Distortion.* There are two modes for the nonlinearity compensation module shown in Figure 8-12. One mode is the nonlinear distortion compensation using digital precompensation techniques. In this mode, a percentage of the transmitted RF signal is coupled and sampled first, and the baseband signal components I' and Q' can be obtained from the demodulator output. After A/D conversion, these two components are converted into the digital signal components I'' and Q'' before being fed into the "distortion detection" module. The error control signal is generated by comparing I'' and Q'' with the original baseband signals I, Q (a certain delay is required for the synchronization between I, Q and I'', Q'') and then fed into the "predistortion" processing module in the digital domain. The I and Q baseband signals in the analog domain after the predistortion will be converted into the I and Q baseband signals in the digital domain after the D/A converter. The predistorted baseband signal will be sent to the quadrature modulator to generate the RF analog signal and is then fed into the power amplifier. If the whole circuitry works well, the predistortion effect on the signal will be canceled by the nonlinear distortion from the power amplifier. As a result, there will be a linear relationship between the baseband signal output from the power amplifier and the baseband signal before predistortion. The initial parameters for the distortion compensation module can be preset and the sampling signal from the transmitter output changes very little when entering the stable stage. The other mode is the by-pass mode with distortion compensation function turned off, which is mainly used for specification testing.

3. *Adjustable Internal Parameters for Exciter.* These parameters include:

   1. The output frequency setting for the transmitter.
   2. TS stream input selection: There are three different modes for switching the primary and spare TS streams, AUTO, MANUAL, and SINGLE. In the AUTO mode, if an alarm is found in the currently operated TS input signal, the exciter will automatically choose the other stream. In the MANUAL mode, TS1 or TS2 can be manually selected. When there is only one TS input, the SINGLE mode is used.
   3. Internal/external frequency reference selection.
   4. Modulation parameters: Including PRBS generation mode, single-/multi-carrier mode selection, mapping mode of QPSK/QPSK-NR/16QAM/32QAM/64QAM, interleaving length of 240/720, FEC code rate of 0.4/0.6/0.8, guard interval of 420/595/945, control frame ON/OFF mode, pilot ON/OFF in single-carrier mode (ON state enables the pilot output), and so on.

5. Single-frequency network mode selection: ON/OFF (selecting ON for single-frequency network mode), setting the SIP to valid, setting the delay compensation (delay offset).

4. *Alarm Display and Monitoring.* The exciter must have complete alarm and monitoring functions. The alarm menu must include abnormal RF output, abnormal input TS stream, abnormal power supply, abnormal temperature, abnormal cooling fan, abnormal RF-AGC in distortion compensation, 1-pps abnormalities, abnormal external 10-MHz reference signal input, abnormal 10-MHz PLL of the frequency combiner unit, abnormal local oscillator PLL of frequency combiner unit, abnormal distortion compensation, and abnormal sampling clock PLL.

8.5.3 Power Amplifier

The power amplifier is used to amplify the modulated RF signal to the desired power level before the signal is sent to the antenna. The main space of the transmitter chassis is occupied by the power amplifier. Most power amplifiers used for the VHFband in the early stage are planar tetrodes, and the single-tube output power is generally 10 kW. The majority of klystrons and IOTs (inductive output tubes) are used in the UHF band with the single-tube output power up to 30–40 kW. The supply voltage of the vacuum tube is generally from thousands of volts to 20,000 V. Due to the high voltage applied, it has the shortcomings of low life expectancy, large volume, and high maintenance risk. The situation changed when the all-solid-state transmitters appeared in the 1990s.

Due to continuous technical breakthroughs for solid-state components and devices in the past decades, single-tube output power has significantly increased (with dissipated power of 300 W). With the advantages of low operating voltage (28–50 V) and long life expectancy, most transmitters now use solid-state devices such as transistors and FETs for the entire TV broadcasting band. Because of its high amplification gain and superior stability, the MOSFET (metal–oxide–semiconductor FET) is becoming the most popular choice.

The power amplifier of the DTV transmitter usually consists of a three-stage amplifier. The first and second stages are the preamplifiers, which generate enough power to drive the final stage of the power amplifier, consisting of several parallel amplification branches. In general, the amplifier for the preamplification stage operates in the class A mode, and the final stage amplifier operates in class AB mode.

The exciter output is sent to the power amplifier after passing through the divider. The average RF input power of this unit is around 0.5 mW. The signal after amplification will be nearly 500 W if the gain is 50–60 dB. The power amplifier usually operates in parallel such that even if a power amplifier fails, other power amplifiers can still operate well to ensure continuous service operation. If this situation happens, the transmitter output power will be lower than the desired value, and the compensation function for the exciter will then be stopped.

The power amplifier operating at high current will generate and then radiate a huge amount of heat to the environment and cooling must be applied. The methods for cooling the power amplifier can be generally divided into two classes: air cooling and liquid cooling. Air-cooling power amplitude has a large number of heat sinks with cooling achieved through the exhaust and forced air supply systems for the transmitter. With a simple structure, it tends to become quite dirty yet is difficult to clean and usually suffers from loud noise. The liquid-cooling power amplifier is equipped with an internal coolant circulating pipe connected to the main pipe of the cooling system (snap-fit conduit connection). There is no need to shut down the liquid-cooling transmitter and the unit of circular cooling when doing the replacement. Other advantages include no liquid leakage, not easily stained by the environment, and has clean and quiet operation. The disadvantage is that the system structure is complicated and operation cost is high.

The power amplifier and its power supply can be either integrated into one module or put separately. The DC power module is supplied by three-phase 380 V AC.

In addition, the power amplifiers of the DTV transmitter are required to have good consistency in parameters to ensure minimum intermodulation distortion and improve overall efficiency.

8.5.4 Multiplexer

The function of the multiplexer, also known as a channel combiner, is to combine the output signals from multiple transmitters working at different frequencies and transmit these signals by one antenna. This must be done interference free for signals from various transmitters. With the number of antennas on the transmit towers being quite limited, most DTV signals must be broadcast by sharing the antenna with existing analog TV signals and the multiplexer has become one of the most commonly used devices in DTTB systems.

The most basic multiplexer is the two-channel combiner, also known as a duplexer. The multiplexer can be obtained by the combination of duplexers based on certain rules. The structure of multiplexers commonly used includes star type, constant-impedance type (bridge type), and the delay line type (phase sensitive).

8.5.4.1 Star-Type Multiplexer The star-type multiplexer consists of the filters, multiport contact points, and associated connecting lines. All the transmitters are connected to the star-type multiplexer after passing through the filter and eventually are connected to the antenna. The filter includes the bandpass filter and bandstop (notch) filter, so the star-type multiplexer can be bandpass type or notch type.

Figure 8-13*a* gives the schematic diagram of the bandpass-type star diplexer, where S is the star contact. Transmitter 1 operating at frequency F1 is connected

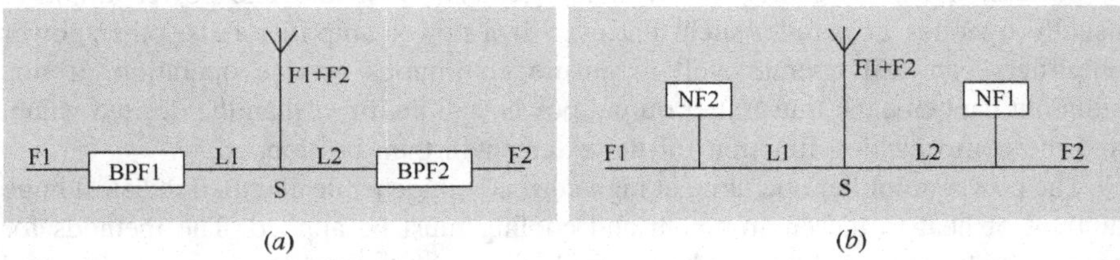

FIGURE 8-13 Schematic diagram of star-type duplexer of (*a*) bandpass and (*b*) notch.

to S through the bandpass filter BPF1 and connecting line L1 while transmitter 2 operating at frequency F2 is connected to S through the bandpass filter BPF2 and connecting line L2. To ensure isolation between these two transmitters, the impedance to frequency F2 should be infinity (i.e., open circuit) if looked from the point S to the left, and the impedance to frequency F1 should also be infinity (i.e., open circuit) if looked from the point S to the right. This requirement can be satisfied by adjusting the line lengths of L1 and L2. Similarly, the lengths of the connecting lines must be optimized to ensure the signal isolation among different transmitters when there are more than two transmitters connected to the multiplexer. Connecting three transmitters to the star multiplexer as an example, for any branch X, the impedance from any other branch should be open to this channel, which is not easy to realize. In addition, the bandwidth of the star contact has certain restrictions, and the channel spacing should be as large as possible to ensure enough isolation among all the channels. This means the number of combined channels cannot be too large, e.g., four within the UHF band if those channels are relatively close to each other.

The principles of the notch-type star multiplexer are similar to that of the bandpass type, and the difference is that the number of notch filters in the connecting line of each channel should increase when the number of channels increases. The notch-type multiplexer is generally used for combining two channels.

8.5.4.2 Constant-Impedance Multiplexer The constant-impedance multiplexer can also be divided into two types: constant-impedance bandpass type (CIB) and constant-impedance notch type (CIN).

Figure 8-14*a* is a schematic diagram of the constant-impedance bandpass duplexer, consisting of two 3-dB directional couplers (D1, D2), two bandpass filters (B1, B2), one balanced load, and the connection feeder. The input signal F1 arrives at the 3-dB directional coupler D1 from port 1 and two signals with half the power of the original output from ports 2 and 4, with the signal from port 2 having the same phase as port 1 input and the signal from port 4 lagging by 90° compared with the port 1 input. These

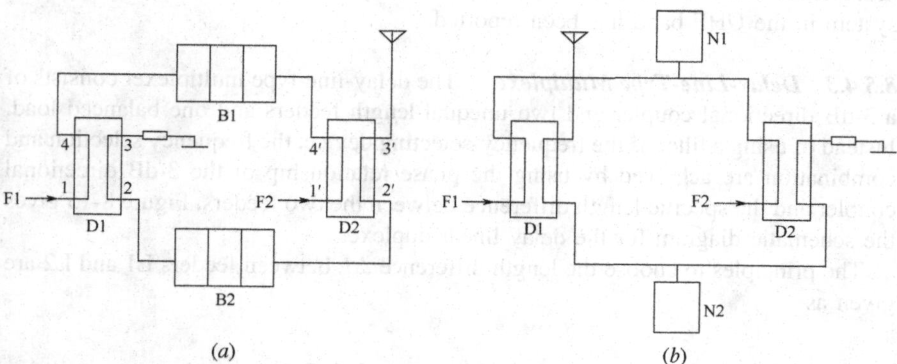

FIGURE 8-14 Schematic diagram of constant-impedance duplexer with type of (*a*) bandpass and (*b*) notch.

two signals then reach ports 2′ and 4′ of D2 through the bandpass filters B1 and B2. As B1 and B2 have the same working band of F1 and port 4′ has the same phase as port 3′ while port 2′ port lags by 90° relative to port 3′, the two in-phase signals are combined and output from port 3′ while there is no signal from port 1′.

The input signal F2 arrives at the 3-dB directional coupler D2 from port 1′ and two signals with half the power of the original output from ports 2′ and 4′, and they are totally reflected back to D2 from the bandpass filter. These two signals are combined and output from port 3′ while there is no signal from port 1′ due to the specific phase relationship.

As port 1 of D1 is only used to input signal F1, it is called a narrowband input port. Since port 1′ of D2 can be used to input all the signals other than F1, it is called a broadband input port.

The 3-dB directional couplers can provide more than 30 dB isolation in general, so the isolation from the narrowband port to the broadband port of the diplexer shown in Figure 8-14*a* will be greater than 30 dB. The isolation from the broadband port to the narrowband port is the isolation provided by the 3-dB directional coupler and the out-of-band attenuation of the bandpass filter, and it is up to 60 dB when the duplexer is applied to the nonadjacent channel case. To further improve the isolation from the narrowband port to the broadband port, an additional bandpass filter can be used at the broadband port.

Under the condition of the perfect symmetric system structure, the 3-dB coupler helps ensure the minimal reflection of this type of multiplexer at both narrowband and broadband ports within a broad frequency band. This constant-impedance characteristic is the reason it is named a constant-impedance multiplexer and helps absorb the clutter outside the transmitter band.

Figure 8-14*b* is a schematic diagram of a constant-impedance notch duplexer with its structure and operating principles the same as that of the bandpass counterpart except that the filter is changed to bandstop.

The duplexer shown in Figure 8-14 is also known as the bridge unit with its output connected to the broadband port of the next unit. With several units connected, a multiplexer will be obtained. The use of a multiplexer with 10 ports for a DTTB system in the UHF band has been reported.

8.5.4.3 Delay-Line-Type Multiplexer The delay-line-type multiplexer consists of a 3-dB directional coupler and two unequal-length feeders and one balanced load. Instead of using a filter as the frequency-selecting device, the frequency selection and combination are achieved by using the phase relationship of the 3-dB directional coupler and the specific length difference between the two feeders. Figure 8-15 gives the schematic diagram for the delay linear duplexer.

The principles to choose the length difference ΔL between feeders L1 and L2 are given as

$$\begin{aligned} \Delta L &= (2m+1)\frac{\lambda_1}{2} \quad & m &= 0, 1, 2, \ldots \\ \Delta L &= n\lambda_2 \quad & n &= 1, 2, 3, \ldots \end{aligned} \tag{8-6}$$

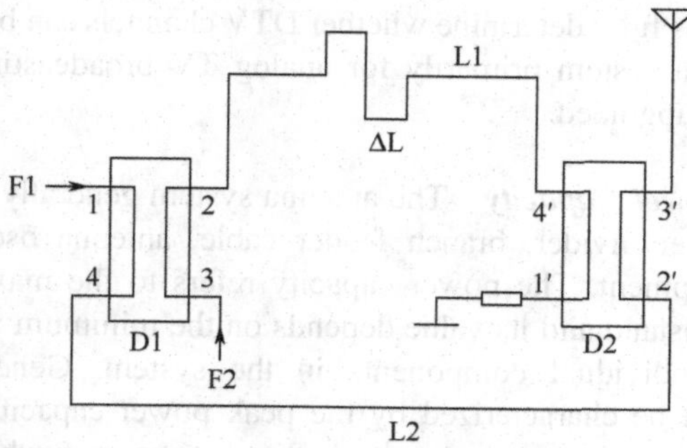

FIGURE 8-15 Schematic diagram of delay line duplexer.

When ΔL satisfies the above relationship, the phase difference $(2m+1)\pi$ will be produced for signal F1 with wavelength λ_1 after passing through feeders L1 and L2, and the phase difference of $2n\pi$ will be produced for signal F2 with wavelength λ_2 after passing through L1 and L2.

The F1 input comes from port 1 of the 3-dB directional coupler D1, and two signals with half the power each can be obtained and then output from ports 2 and 4 with port 2 output having the same phase as that of the port 1input and port 4 output lagging by 90° compared to the port 1 input. These two signals reach ports 2′ and 4′ of D2 via two feeders. Due to the existence of ΔL, the signal from port 2′ is lagged by 90° the signal from port 4′. Two signals are then combined by D2 and fed to the antenna from port 1′. The F2 input comes from port 3 of D1, and output from ports 2 and 4 with port 2 output lagged by 90° compared with the port 4 output. The signals reach D2 via two feeders; as ΔL does not produce a phase difference relative to F2, the signal to port 2′ is still lagged by 90° the signal to port 4′, and the two signals are combined and fed to the antenna.

The delay-line-type multiplexer has the features of simple structure, fewer tunable components, stable performance, and low cost as it does not contain any cavity filter. However, the isolation is poor without the filter, so the generally required channel spacing when operated in the UHF band is at least three channels.

The delay line duplexers can also be concatenated in series to form a multiplexer. But the more duplexers in series, the narrower its passband is, and the more difficult it is to optimize the length ΔL.

8.5.5 Transmitting Antenna

The antenna is considered the passive device in the transmission system. In general, there is no essential distinction between antenna systems for digital and analog TV. Currently, most TV transmitting stations share antennas broadcasting both DTV and analog channels. However, special considerations for certain technical specifications of DTV broadcasting are still required for its unique characteristics, and therefore requirements for the equipment are different from analog TV broadcasting. These special specifications fully determine whether DTV channels can be handily added to the existing antenna system primarily for analog TV broadcasting or parts of the system need to be upgraded.

8.5.5.1 System Power Capacity The antenna system generally includes the main feeder cables, power divider, branch feeder cable, antenna oscillator, and other auxiliary feed equipment. The power capacity refers to the maximum power that the system can withstand and its value depends on the minimum value of the power capacity for the individual components in the system. Generally, the system power capacity can be characterized by the peak power capacity and the average power capacity. The peak power capacity is the signal power when there is voltage breakdown of the device dielectric, and the average power capacity is the power that the system can withstand before the heating limit is reached. The two parameters must be considered together when evaluating the power capacity of the system.

The power of the analog TV signal refers to the average signal power of the synchronization within a single RF period, also known as the rated power. The average image signal power changes with the content and will not provide much useful meaning, while the power during the pulse synchronization is a constant, and the rated power is the peak power for the analog TV signal. In contrast, the average signal power of DTV is a constant while the instantaneous peak power does not provide meaningful distribution characteristics. Therefore, the power of the DTV signal is characterized by the average power. The DTV signal has a much higher peak-to-average power ratio even with low probability, and its impact on the antenna and feeding systems cannot be ignored.

8.5.5.2 Calculation of Combined Signal Power for Multiple Channels that Share One Antenna When using the multiplexer for frequency combining, the total signal power received by the antenna system must be correctly calculated, including the phase relationship of the multichannel signals. In practice, the value of the combined signal power can be calculated by the following according to the conditions of in-phase superposition to ensure maximum possible safety:

$$P = \left(\sqrt{P_1} + \sqrt{P_2} + \sqrt{P_3} + \cdots\right)^2 \tag{8-7}$$

where P is peak power of the combined signals with individual signal powers P_1, P_2, P_3, ...

In 8-7, it is assumed that all the individual signals are superposed in phase, and the resulting value is the maximum possible signal power. In fact, the phase relationship between signals is more complicated and depends on the frequencies, fully in-phase superimposition is almost impossible, so the actual total power is less than this value. Generally, 8-7 is used to calculate the maximum peak power instead of the average power. The result of 8-7 is quite conservative and system power exceeding this value will never happen in practice. However, this assessment methodology may seem too conservative from the power efficiency point of view, because it is an extremely low probability event and the probability of the multichannel signal having in-phase superimposition is even lower. In practice, the probability of the combined signal reaching its peak power can be obtained by using statistical methods. This helps determine the value of the system power capacity that meets safety requirements and is not too conservative. For example, if the beta distribution model is used for the calculation, the resulting peak power capacity of the combined signal is 2–4 dB less than that of the direct superposition method. The reduction of the required system power capacity greatly lowers the cost of equipment selection and system upgrade.

Regarding the system margin, around 40% of the margin will be needed in the system design when considering factors such as standing wave ratio (SWR), impedance compensation, and component aging on the system power capacity.

8.5.5.3 Bandwidth The antenna of the existing analog TV system is generally designed for single-channel operation, and therefore its technical specifications cannot meet the requirements when additional channels are included. If new channels are introduced to the system, system bandwidth assessment must be done and replacement for those components without sufficient bandwidth is needed. In most cases, the components limiting the bandwidth are the power splitter, coaxial adapters, and tuning elements. The length of the branch feeder cable may also lead to a poor radiation pattern as it is not optimized for the newly introduced channels, and replacement should be done based on the optimization for both the original and newly introduced channels.

8.5.5.4 Matching of Antenna System For analog TV broadcasting, the picture quality will suffer from ghosting effects if the RF signal reflection is too high due to poor matching of the transmitting antenna system. Furthermore, since the transmit power is often relatively high, the high-voltage standing wave formed by the excessive reflection may even damage the devices. Therefore, the requirements on antenna system matching for analog TV signal transmission are generally high, especially for high-power transmitting stations, and the SWR of antenna systems is often required to be below 1.1.

Compared with analog TV broadcasting, the signal power of the DTV transmitter is generally low and there will be no ghost effect due to anti-multipath interference and error correction capabilities of the DTV system. Therefore, the matching requirement of the antenna system dedicated to DTV signal transmission can be properly relaxed. The generally accepted specification is that the system SWR is less than 1.2.

8.6 SIGNAL RECEPTION OF DTTB

8.6.1 Main Impact Factors of Physical DTTB Channel

The impact factors for successful DTTB signal reception include the following:

1. *Fresnel Zone.* The Fresnel zone, named after physicist Augustin-Jean Fresnelis, is one of a (theoretically infinite) number of concentric ellipsoids which

define volumes in the radiation pattern of a circular aperture. It comes from diffraction by the circular aperture. The cross section of the first (innermost) Fresnel zone is circular while subsequent Fresnel zones are doughnut shape in cross section (concentric with the first). To maximize signal strength at the receiver, the effect of out-of-phase signals must be minimized by removing obstacles from the RF LOS. The strongest signals are on the direct path between transmitter and receiver and always lie in the first Fresnel zone.

2. *Reflection.* The RF signal may be reflected by buildings and other obstacles, especially underground tunnels. Due to the phase difference between the main path and the reflected signals, the quality of the received signal is generally degraded.
3. *Refraction.* When the RF signal is transmitted through one medium to another, refraction occurs, and this is caused by the air temperature change, atmospheric layer change, and air density change. In general, the refractive index of the air is proportional to the air density.
4. *Thermocline and Inversion Layer.* The air temperature of different atmospheric layers is different during the day, which makes the RF signal transmission characteristics different. The characteristics will change in a few minutes and the largest impact is from the thermocline and the inversion layers. A thermocline layer is created in places with a large-area water body (such as seas, lakes). In winter, the temperature in the city is usually higher than its surrounding area, and a high-temperature air layer is created in the middle of the down. As a result, a fading of nearly 20 dB might be introduced by the thermocline layer.

8.6.2 Fixed Reception

Fixed reception refers to receiving the DTTB signal with a directional antenna on the roof. The gain of the directional antenna helps provide better signal strength at the receiver than that of the omnidirectional whip antenna. The antenna height for the fixed reception is generally 10 m, and such a reception condition is generally the case in rural areas [6].

For the fixed reception scenario, the channel model is relatively simple since multipath fading is not too critical and fairly good reception performance can be achieved.

8.6.3 Portable Reception

Portable reception needs a small antenna directly connected to the receiver. Some antennas, such as retractable whip antennas or more sophisticated antennas, can even be installed inside the receivers. The portable reception is usually further defined into two types: A and B [6].

Under the class A portable reception scenario, the receiver is placed either outdoors with the antenna height no less than 1.5 m above the ground or on upper floors indoor with the field strength similar to that of the outdoor. The loss of signal strength as a result of penetrating through the building is called building penetration loss. The building penetration loss is quite large if the house is on the ground level and decreases with building height.

Under the class B portable reception scenario, the receiver is indoors and near ground level but there are windows of no less than 1.5 m high above the ground in the room where the receiver is located. In this case, the building penetration loss on the received signal is quite large, and the required field strength is higher than that of the class A portable reception scenario.

The portable reception only allows for the use of small antennas, and this limits the antenna gain to support successful reception, especially under the class B portable reception condition since the field strength will be greatly attenuated by the penetration loss.

8.6.4 Mobile Reception

Mobile reception refers to the scenario in which the transmitter and receiver have relative movement, and usually a vehicle-borne omnidirectional whip antenna is used. Mobile reception generally requires much higher field strength for successful reception, and modulating schemes such as OFDM that help eliminate multipath fading are more favorable for this type of application. In addition, hierarchical modulation technology can also be applied.

The channel model for the mobile reception application is generally considered a Rayleigh channel combined with the Doppler frequency shift from the movement of the mobile terminals (depending on signal frequency and terminal moving speed). A higher system margin for mobile reception is hence required during the network planning to provide satisfactory mobile reception performance.

8.7 DIVERSITY TECHNIQUES

The SNR of the received signal is usually quite unstable in the frequency- and time-selective channel, which is inevitable when a signal passes through the very sophisticated wireless environment. The interruption of the video stream is unacceptable for DTTB services when it is caused by the unsuccessful data reception under very low SNR. The diversity technique is an effective method to overcome this SNR instability and ensure good reception performance.

If the signal is transmitted through several uncorrelated path, the fading impact can be effectively minimized since the probability of several independent paths simultaneously in the deep-fading mode is fairly low, and video stream interruption will be significantly eliminated because of the reasonably stable SNR of the received signal.

8.7.1 Various Diversity Schemes

In DTTB and other telecommunication systems, diversity scheme refers to the signal reliability improvement using two or more paths with different characteristics.

Diversity plays an important role against fading, cochannel interference, and error bursts. The commonly used diversity technologies include time diversity, frequency diversity, space diversity, or a combination of them [7]. Among these techniques, time diversity and frequency diversity can somehow be taken as equivalent to the repetition code and interleaving technique in the channel coding. The costs paid for improvement of the system BER performance are transmission efficiency reduction, additional implementation cost, etc., and a trade-off must be made during network design. Space diversity uses multiple antennas to send and/or receive signals from multiple antennas (to form multiple independent paths) to overcome the deep-fading impact and ensure reliable signal reception. Currently, DVB-T and ISDB-T adopt space diversity reception due to its advantage of supporting reliable reception without sacrificing transmission efficiency.

Time diversity can be realized by sending the same data in different time slots (hoping that the data experience different fading conditions) to provide better reception performance. The minimum time interval for sending two identical information data must be longer than the channel coherence time to ensure the independent fading. In general, FEC and interleaving techniques are commonly used in DTTB systems to provide the time diversity. The redundancy by FEC can be considered as the duplication of the information bit, and the interleaving technique is also used to ensure the separation interval is longer than the channel coherence time. Inevitably, this technology will cause reduction in transmission efficiency, additional delay, and other implementation costs. Under the fast-fading environment, the channel coherence time is short and interleaving depth can be shallow, and the time diversity technique can be adopted with a small delay. However, when the channel is static, time diversity techniques cannot effectively provide improvement in system performance.

The frequency diversity gain can be realized by sending the same data at different frequencies, and redundancy is usually introduced. Similarly, the frequency diversity gain can be achieved only when the coherent bandwidth of the channel is narrow.

Space diversity (antenna diversity) is achieved by having multiple antennas at the transmitter end and/or the receiver. When the antennas are separated by a certain distance, the RF channels between the antennas can be considered independent from each other, and space diversity gain can therefore be achieved. Unlike the time and frequency diversities, space diversity neither reduces spectrum efficiency nor relies on the channel characteristics such as the channel coherence time and coherence bandwidth and is therefore widely used.

Space diversity can be further classified as transmit diversity and receive diversity and can be used either individually or jointly. Transmit diversity can be achieved by the space–time or space–frequency coding techniques, and various signal combination methods can be used at the receiver. A detailed description is given in the following.

8.7.2 Design Principles of Diversity Schemes for DTTB System

Considering the number of receivers to be covered by the TV stations, the additional cost from the transmit side will not significantly increase overall system cost, and the transmit diversity technology is more appropriate to further improve the antifading capability of DTTB systems while receiving diversity also plays an irreplaceable role. Many research results on transmit diversity have been published which require appropriate signal vector allocation among all transmit antennas to maximize the diversity order [8–10]. The transmit diversity scheme is also regarded as another layer of "inner code," called diversity code after the channel code. For DTTB systems, selection of the transmit diversity scheme should follow the following design principles:

1. *Highest Possible Transmission Efficiency.* The diversity scheme should ensure the highest possible system throughput by minimizing the redundancy introduced.
2. *Largest Possible Diversity Gain.* The diversity scheme should be optimized to improve the system performance under various fading channels while maintaining the system throughput.
3. *Keeping Transmitting Systems as Unchanged as Possible.* The diversity scheme can be treated as an alternative "accessory" of the existing DTTB transmitting system so that operators can determine whether the diversity technology is needed based on the overall practical considerations on the coverage requirement, channel characteristics, etc. This helps provide flexible network planning and minimizes the cost of the transmit station.
4. *Minimizing Whole Implementation Complexity.* Increasing the number of transmit antennas will certainly increase the complexity of the transmit system and will inevitably increase the functional blocks at the receiver, such as channel estimation and multibranch signal combination. But efforts must be made to reduce this change and lower the additional complexity without affecting the normal operation of the existing receivers.
5. *"Soft Failure" Support.* This is originally the advantage of the receiving diversity, but it is also expected to be a feature of the transmit diversity technique. For the transmit diversity, supporting "soft failure" means that the allocation of signal vectors among the transmitting antennas should satisfy the following criterion: When some transmitting branches fail to function normally, all the remaining branches must still work well to ensure normal reception of the system except for the loss of the average received signal power and the diversity gain reduction. From this viewpoint, the transmit diversity technique also helps improve the overall system reliability.

8.7.3 Transmit Diversity Technique

Among all the transmit diversity schemes, the space–time block code (STBC) has drawn significant attention due to its simplicity and effectiveness [8–10]. Figure 8-16 shows the block diagram of the transmit diversity scheme based on STBC.

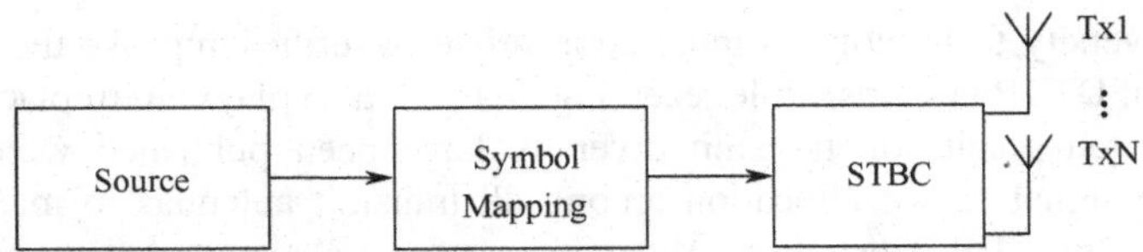

FIGURE 8-16 Block diagram of transmit diversity.

8.7.3.1 STBC Based on Orthogonal Structure Space–time block coding is used to encode the signal in two dimensions of space and time to allow different antennas to send different signals or the redundant information within different time slots to achieve the diversity. The results shows that using orthogonal STBC enables full transmit diversity. Constructing an orthogonal space–time code matrix may provide full transmit diversity with a number of transmit antennas n_T, while decoding can be easily done at the receiver by utilizing a simple maximum-likelihood decoding algorithm with linear signal processing [9].

In general, the matrix of space–time block codes can be defined as $\mathbf{X}_{n_T \times p}$ (also simplified as $\mathbf{X}$ in this chapter), where n_T again denotes the number of transmit antennas and p denotes the time periods to transmit a group of coded data symbols. The n_T parallel symbol sequences with a length of p are generated according to transmission matrix $\mathbf{X}_{n_T \times p}$, and these symbol sequences are then transmitted simultaneously by n_T transmitting antennas within this time length.

The transmission matrix $\mathbf{X}$ consists of k modulated data symbols, their complex conjugates $x_1^*, x_2^*, \ldots, x_k^*$, and their linear combination. To achieve full transmit diversity, $\mathbf{X}$ is constructed based on the orthogonal principle

$$\mathbf{X} \cdot \mathbf{X^H} = c\left(|x_1|^2 + |x_2|^2 + \cdots + |x_k|^2\right)\mathbf{I_{n_T}} \tag{8-8}$$

where c is a constant, $\mathbf{X^H}$ is the Hermitian transpose of $\mathbf{X}$, and $\mathbf{I_{n_T}}$ is a unit matrix of $n_T \times n_T$. The symbol $\mathbf{X}_{i,j}$ denotes the signal to be transmitted by the ith transmitting antenna and in the jth period within the p transmission periods.

Based on the orthogonal design for STBC, all the rows of transmission matrix $\mathbf{X}$ are mutually orthogonal, i.e.,

$$x_i x_j = \sum_{t=1}^{p} x_{i,t} x_{j,t}^* = 0 \quad i \neq j \quad i,j \in \{1,2,\ldots,n_T\} \tag{8-9}$$

where $x_i x_j$ denotes the inner product of the sequences x_i and x_j. The orthogonality enables the full transmit diversity by a specific number of transmitting antennas. With this design, the receiver can use the simple maximum-likelihood decoding algorithm with linear signal processing for decoding.

The rate of STBC is defined as the ratio of the number of input symbols to the number of space–time coded symbols transmitted by each antenna, i.e., $R = k/p$, $R \leq 1$.

For the STBC with constellations of complex signals, there exists the orthogonal square matrix which can provide full transmit diversity and full transmission rate only if $n_T = 2$, namely the Alamouti scheme [11].

8.7.3.2 Alamouti Space–Time Block Code The well-known Alamouti STBC coding method is given in the following. During the first transmission time period, antennas 1 and 2 transmit signals x_1 and x_2, respectively; during the second transmission period, antenna 1 transmits $-x_2^*$, while antenna 2 transmits x_1^*,

$$\mathbf{X} = \begin{bmatrix} x_1 & -x_2^* \\ x_2 & x_1^* \end{bmatrix} \tag{8-10}$$

The coding matrix satisfies the orthogonal property

$$\mathbf{X} \cdot \mathbf{X^H} = \begin{bmatrix} |x_1|^2 + |x_2|^2 & 0 \\ 0 & |x_1|^2 + |x_2|^2 \end{bmatrix} = \left(|x_1|^2 + |x_2|^2\right)\mathbf{I_2} \tag{8-11}$$

Assuming there is only one receiving antenna, the receiver block diagram for the transmit system using the Alamouti scheme is shown in Figure 8-17. The channel impulse responses from the first and second transmitting antennas to the receiver at time t are denoted by $h_1(t)$ and $h_2(t)$, respectively.

Assuming that the channel impulse responses remain unchanged during two consecutive symbol transmission periods, i.e., $h_1(t) = h_1(t+T)$, $h_2(t) = h_2(t+T)$ (T is the symbol period), one can get

$$\begin{aligned} r_1 &= h_1 x_1 + h_2 x_2 + n_1 \\ r_2 &= -h_1 x_2^* + h_2 x_1^* + n_2 \end{aligned} \tag{8-12}$$

where r_1 and r_2 denote the received signals at times t and $t+T$, respectively, assuming n_1 and n_2 independent complex Gaussian noises with power spectral density N_0.

If the channel impulse responses of h_1 and h_2 can be accurately estimated at the receiver and all the constellation points have equal probability, the maximum-

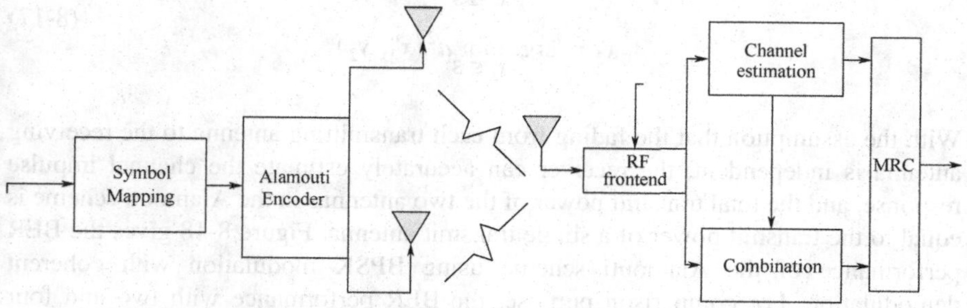

FIGURE 8-17 Block diagram of Alamouti STBC system.

likelihood decoding algorithm is equivalent to the minimum-distance decoding algorithm. The decoder can choose a pair of symbols (x_1', x_2') from all the possible constellation point combinations with the minimum distance:

$$(x_1', x_2') = \arg \min_{(x_1', x_2') \in \mathbf{C}} \left[\left(|h_1|^2 + |h_2|^2 - 1\right) \left(|x_1'|^2 + |x_2'|^2 \right) + |x_1' - y_1|^2 + |x_2' - y_2|^2 \right] \tag{8-13}$$

where x_1' and x_2' are the decisions on the data symbols based on the received signals and channel impulse responses and **C** is the set of all possible constellation symbol. The two statistics variables, i.e., y_1 and y_2, are given as

$$\begin{aligned} y_1 &= h_1^* r_1 + h_2 r_2^* \\ y_2 &= h_2^* r_1 + h_1 r_2^* \end{aligned} \tag{8-14}$$

Substituting 8-12 into 8-14, one has

$$\begin{aligned} y_1 &= \left(|h_1|^2 + |h_2|^2\right) x_1 + h_1^* n_1 + h_2 n_2^* \\ y_2 &= \left(|h_1|^2 + |h_2|^2\right) x_2 - h_1 n_2^* + h_2^* n_1 \end{aligned} \tag{8-15}$$

It can be seen from 8-15 that the maximum-likelihood decoding can be divided into two separate steps based on the known channel impulse responses h_1 and h_2 since y_i is only a function of x_i $(i = 1, 2)$:

$$\begin{aligned} x_1' &= \arg \min_{x_1' \in \mathbf{S}} \left[\left(|h_1|^2 + |h_2|^2 - 1\right) |x_1'|^2 + d^2(x_1', y_1) \right] \\ x_2' &= \arg \min_{x_2' \in \mathbf{S}} \left[\left(|h_1|^2 + |h_2|^2 - 1\right) |x_2'|^2 + d^2(x_2', y_2) \right] \end{aligned} \tag{8-16}$$

where **S** is the set of constellation symbols of x_i. For M-PSK modulation, 8-16 can be further simplified to the following with known h_1 and h_2:

$$\begin{aligned} x_1' &= \arg \min_{x_1' \in \mathbf{S}} d^2(x_1', y_1) \\ x_2' &= \arg \min_{x_2' \in \mathbf{S}} d^2(x_2', y_2) \end{aligned} \tag{8-17}$$

With the assumption that the fading from each transmitting antenna to the receiving antenna is independent, the receiver can accurately estimate the channel impulse response, and the total transmit power of the two antennas in the Alamouti scheme is equal to the transmit power of a single transmit antenna, Figure 8-18 gives the BER performance of the Alamouti scheme using BPSK modulation with coherent demodulation. For comparison purpose, the BER performance with two and four receiving antennas (with maximum ratio combining) are also shown in the figure.

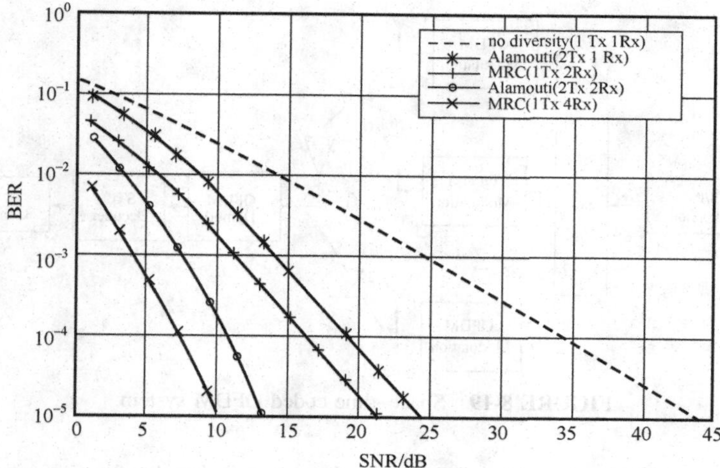

FIGURE 8-18 BER performance of Alamouti transmit diversity scheme in slow fading channel.

It can be seen from the figure that the slope of the BER curves for the Alamouti scheme with two transmit antennas and one receiving antenna is identical to that of the receiving diversity scheme with two receiving antennas with maximum-ratio combining (MRC). Because the transmitted power of each transmit antenna in the Alamouti scheme is half of the transmitted power for the single transmit antenna, the BER curve for the Alamouti scheme is 3 dB to the right of that for the receiving diversity scheme. A similar observation can be made from the figure for the performance of the Alamouti scheme with two receiving antennas versus the performance of the receive diversity scheme with four antennas. Since the slope of the BER curve in the figure fully reflects the diversity order, it can be summarized that the Alamouti scheme with two transmit antennas and n_R receiving antennas will provide the same diversity gain (diversity order) as the receive diversity scheme with one transmit antenna and $2n_R$ receiving antennae using MRC. As mentioned above, the Alamouti scheme provides full-rate orthogonal coding.

In OFDM systems, there are two methods of combining space coding with OFDM system, i.e., space–time coded [12] and space–frequency coded [13] OFDM systems. The space–time coded OFDM system means that every two consecutive OFDM symbols are block coded according to the Alamouti transmission scheme, or each symbol x_1 or x_2 in the transmission matrix denotes one OFDM symbol, as shown in Figure 8-19. The signals transmitted by the first transmitter during two consecutive time periods $[nT, (n+1)T]$ are $-\mathbf{X}^*(\mathbf{n+1})$ and $\mathbf{X}(\mathbf{n})$, and the signals transmitted by the second transmitter at two consecutive time periods are $\mathbf{X}^*(\mathbf{n})$ and $\mathbf{X}(\mathbf{n+1})$.

The space–frequency coded OFDM system means that the symbols on two adjacent subcarriers in one OFDM symbol are encoded based on the transmission matrix of the Alamouti scheme. The block diagram of a space–frequency coded

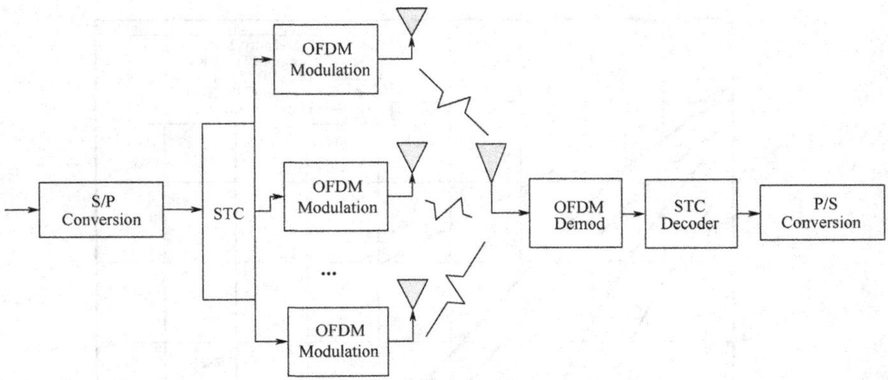

FIGURE 8-19 Space–time coded OFDM system.

OFDM system is shown in Figure 8-20 with the transmit signals of two transmitters as follows:

$$\begin{aligned} \mathbf{X}_1 &= [X(0), X(1) \cdots X(2k), X(2k+1) \cdots X(N-2), X(N-1)] \\ \mathbf{X}_2 &= [X^*(1), -X^*(0) \cdots X^*(2k+1), -X^*(2k) \cdots X^*(N-1), -X^*(N-2)] \end{aligned} \tag{8-18}$$

where N is the number of subcarriers in the OFDM signal and $0 \le k \le N/2 - 1$.

Performance of the space–time coded OFDM signal will deteriorate rapidly in the fast-fading channel, and eventually there will be no diversity gain. This is because the space–time coded OFDM scheme requires the channel to remain unchanged during two consecutive OFDM symbols, and this apparently cannot be satisfied in the fast-fading channel. The space–frequency coded OFDM scheme, however, exhibits excellent performance in the fast-fading channel.

Analysis of the conventional Alamouti scheme always needs the channel coefficients $h_1(t)$ and $h_2(t)$ to remain unchanged during two consecutive OFDM symbol

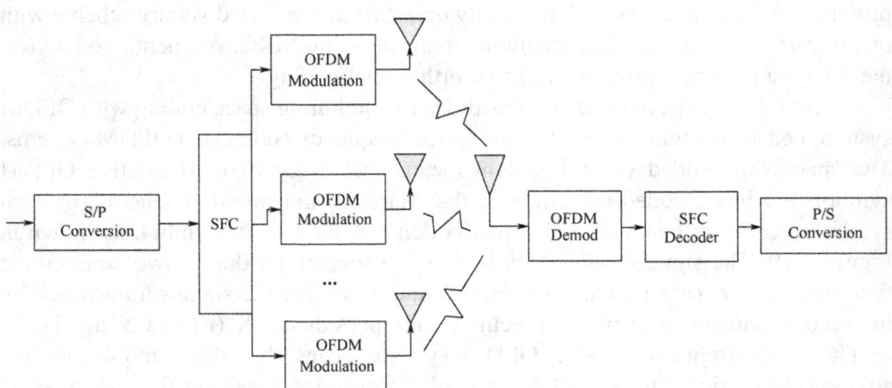

FIGURE 8-20 Space–frequency coded OFDM system.

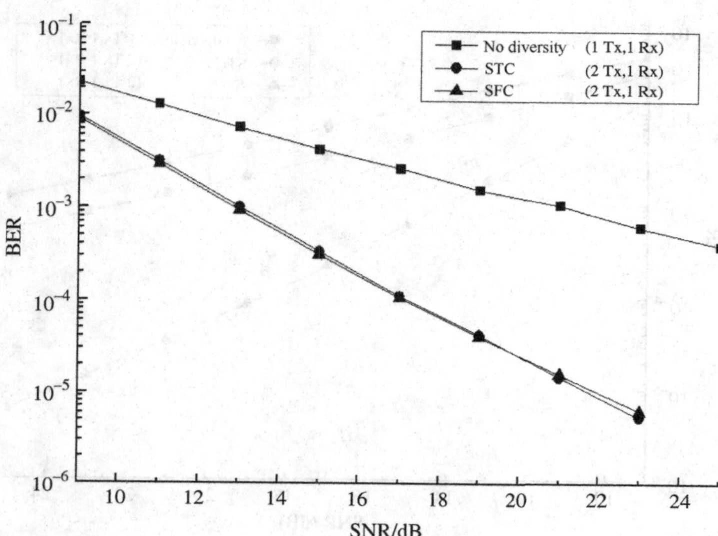

FIGURE 8-21 Performance comparison between space–time and space–frequency OFDM systems under slow-fading channel with Doppler frequency shift f_d of 20 Hz.

periods for the correct decoding at the receiver, namely, $h_1(t) = h_1(t + T)$, $h_2(t) = h_2(t + T)$, with T the symbol period. Therefore, for the space–time coded OFDM system, one needs to assume that the channel remains unchanged during the two consecutive OFDM symbols, while for the space–frequency coded OFDM systems, one needs to assume that the channel conditions are the same for the two adjacent subcarriers within the same OFDM symbol. Obviously, for the fast-fading channel, the channel assumption for the space–frequency coded OFDM system is more easily satisfied than that for the space–time coded OFDM system, and that is why its reception outperforms the space–time coded OFDM system. For the uncoded system with QPSK modulation, Figures 8-21 and 8-22 compare the performance of the space–time coded and space–frequency coded OFDM systems under slow-fading (maximum Doppler frequency shift $f_d = 20$ Hz) and fast-fading (maximum Doppler frequency shift $f_d = 100$ Hz) channels, respectively.

For the space–time coded OFDM system, only one IDFT operation on the first OFDM symbol is needed. The second OFDM symbol is obtained by a complex conjugate operation and rearranging the data of the first OFDM symbol, which greatly reduces implementation cost. Implementation cost for the space–frequency coded OFDM system can also be reduced in a similar way [16].

8.7.4 Receiving Diversity

Receiving diversity can be obtained by receiving the independently faded signals from multiple antennas at the receiver and combining all of them based on a certain algorithm. In general, signal combination can lead to a higher diversity gain yet at

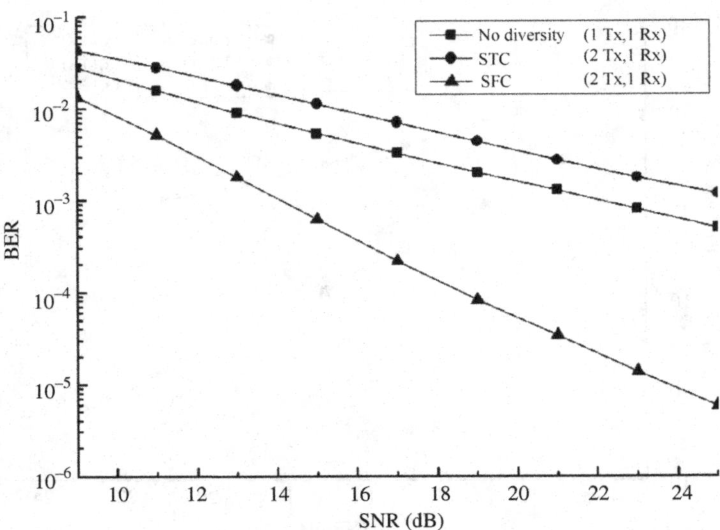

FIGURE 8-22 Performance comparison between space–time and space–frequency OFDM systems under fast-fading channel with Doppler frequency shift f_d of 100 Hz.

higher implementation cost because the signal combining requires complete channel state information (CSI) and multiple RF processing modules. Thus, the antenna selection method greatly reduces system complexity and implementation cost at the expense of small diversity gain.

8.7.4.1 Signal Combining Techniques The commonly used signal combining methods include maximum-ratio combining (MRC) and equal-gain combining (EGC), and they are introduced in the following.

1. *Maximum-Ratio Combining.* Maximum-ratio combining is a commonly used linear combining method with the output signal obtained by adding the various input signals by the corresponding weighting factors. The weighting factor is determined by the CSI, as shown in Figure 8-23. This method has been shown achieve the maximum output SNR which is equivalent to the addition of instantaneous SNR of individual signals, and diversity gain can therefore be achieved.

Suppose the number of antennas at the receiver is N_R, the signal passes through the N_R independent identically distributed (i.i.d.) Rayleigh fading channels, and the MRC method is used. Let $\overline{\gamma}_k = E[\gamma_k]$ be the E_b/N_0 in the kth channel, where E_b/N_0 of each channel is assumed equal for the i.i.d. Rayleigh fading channels, i.e., $\overline{\gamma}_k = \overline{\gamma} (k = 1, 2, \ldots, N_R)$. In this case, the average BER can be calculated by

$$P_b(e) = \left[\frac{1}{2}\left(1-\sqrt{\frac{\overline{\gamma}}{1+\overline{\gamma}}}\right)\right]^{N_R} \sum_{k=0}^{N_R-1} \binom{N_R-1+k}{k} \left[\frac{1}{2}\left(1+\sqrt{\frac{\overline{\gamma}}{1+\overline{\gamma}}}\right)\right]^{k} \tag{8-19}$$

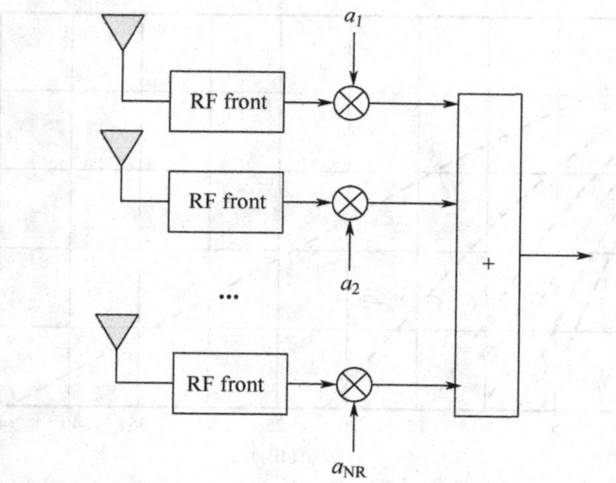

FIGURE 8-23 Schematic diagram of maximum ratio combining method.

where the factorial operation is

$$\binom{n}{m} = \frac{n!}{m!(n-m)!}$$

If $\overline{\gamma}$ is high, the formula can be further approximated as

$$P_b(e) = \left(\frac{1}{4\overline{\gamma}}\right)^{N_R} \binom{2N_R - 1}{N_R} \tag{8-20}$$

Figure 8-24 shows the BER curves versus N_R. BER performance with receiving diversity is significantly higher than the system without diversity ($N_R = 1$), and BER is inversely proportional to the (N_R)th power of $\overline{\gamma}$. At BER of 10^{-4}, diversity gain with $N_R = 2, 3, 4, 5, 6$ compared with that of no diversity is 17, 23, 26, 28, and 29.6 dB, respectively.

2. *Equal-Gain Combining*. As a suboptimal but quite simple linear combination method, EGC only requires the phase information of the channel, which eliminates the amplitude estimation of each channel. The equal-gain addition of all the received signals is done after each signal is equalized in phase and the implementation complexity of EGC is greatly reduced at the expense of diversity gain.

8.7.4.2 Antenna Selection Techniques Although signal combining helps achieve higher diversity gain, the size, cost, and complexity of the receiver significantly increase because the receiver requires multiple antennas, RF front ends, and accurate CSI. If the size of the RF front-end processing module is difficult to shrink, it will be unacceptable for mobile and portable devices, and therefore the trade-off between

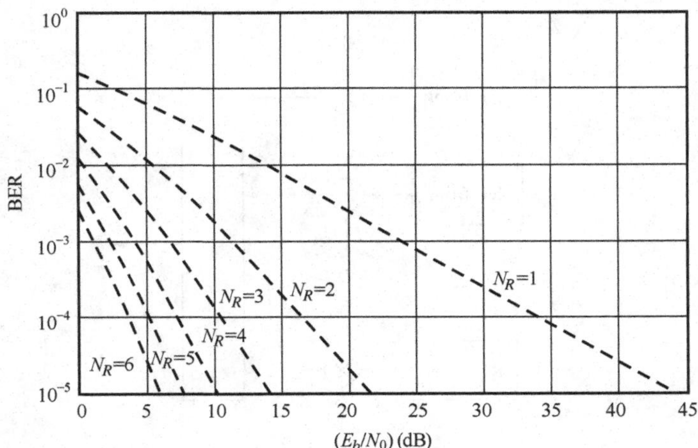

FIGURE 8-24 BER performance of coherent demodulation BPSK system under Rayleigh fading channel using MRC receive diversity.

receiver complexity and diversity gain must be considered. A simple way is to select the link with the best signal quality among all the received signals, which is why the antenna selection method is proposed, with significant reduction in implementation complexity but at the cost of diversity gain.

The antenna selection method analyzes a number of symbols of the received signals and finally selects the receiving antenna based on certain criteria. With different selection strategies and different frequencies of the switching, different system performance can be obtained at different overhead. Usually, there are two categories in antenna selection. One estimates the CSI of each channel for the selection while the other presets the signal quality threshold and chooses the channel with signal quality above this threshold for the reception. When the signal quality of the current path is below this threshold, other paths will be estimated for selection purposes. When the channel fading changes very drastically, the antenna selection scheme helps guarantee successful operation. As for DTTB systems, error-free reception can be realized if CNR of the received signal is always above a certain threshold with the help of the antenna selection technique.

Commonly used antenna selection methods include the maximum log-likelihood ratio (LLR), maximum SNR selection, maximum power/amplitude selection, and switching selection. Their major features can be summarized as follows:

1. *Maximum Log-Likelihood Ratio Selection.* The maximum LLR method selects the channel with the maximum log-likelihood ratio and can achieve very high diversity gain as the best antenna selection method [14]. Unfortunately, the instantaneous CSI and the statistical CSI are prerequisites in the LLR method, which makes the overhead reduction very difficult.

Taking the BPSK system as an example, if the channel gain of the ith channel is h_i, the received signal of the ith antenna, y_i, will be

$$y_i = h_i x + n_i \tag{8-21}$$

where n_i is the AWGN. The log-likelihood ratio Λ_i of the input symbol x is

$$\Lambda_i = \ln \frac{P(x = H_0 | h_i, y_i)}{P(x = H_1 | h_i, y_i)} \tag{8-22}$$

Here, the symbol Λ_i is the decision variable with its absolute value giving the reliability of a hard decision. Studies showed that BER decreases with the increase of $|\Lambda_i|$, so one can simply select the channel with maximum $|\Lambda_i|$ for the best system performance. The mathematical relationship between the log-likelihood ratio and the BER is given as

$$P_{e,i} = \frac{1}{1 + e^{|\Lambda_i|}} \tag{8-23}$$

2. *Maximum SNR Selection.* The maximum SNR selection method is to evaluate the SNR of each received signal from different antennas and then select the channel with the maximum SNR. Only the SNR instead of all the CSI is estimated in this method.

Compared with the optimal signal combining methods, the SNR selection method has the advantages of requiring only a single RF front-end processing module and performing the CSI estimation for the selected channel at the expense of only a certain diversity gain.

The disadvantage of the SNR selection method is that considerable resources are taken and a time delay is introduced by the SNR estimation.

3. *Maximum Power/Amplitude Selection.* The maximum power/amplitude selection is derived from the SNR selection method with the assumption that the noise powers of received signals from the antennas are approximately identical. In this case, the higher the power of the received signal, the larger the signal SNR will be. This method selects the channel with the maximum power by comparing the power of all the received signals from different antennas. This method requires no CSI, is easy and flexible to implement, and considerable diversity gain can be obtained with almost no additional overhead. The system schematic is given in Figure 8-25.

Amplitude-based antenna selection is increasingly used to further simplify system design and reduce overhead. In this method, the channel with the maximum amplitude is selected for reception by the envelope detection of all received signals [15].

4. *Switching Selection.* Due to the characteristics of the DTTB system, successful reception can be achieved as long as the CNR of the received signal is above a certain threshold, and this can be used as the criterion for antenna selection and switching. In fact, even if the channel with the received signal at higher CNR can be identified by

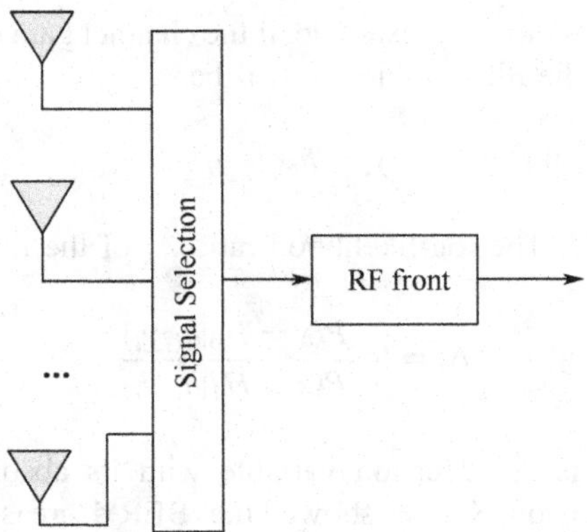

FIGURE 8-25 Schematic diagram of power selection method.

more sophisticated algorithms, the reception performance cannot be improved further. The steps for switching selection are as follows: The CNR threshold is determined based on the FEC code and the receiver demodulation performance, and successful reception is guaranteed if the CNR of the received signal from one channel is above this threshold. The channel is then selected. If the CNR of the received signal from the selected channel is below this threshold at a certain time, other channels are scanned to find a channel with a CNR of the received signal above that threshold. This method is known as the scan switching method, with the system block diagram shown in Figure 8-26.

Since the output signal is not an instantaneous optimal signal, the method cannot yield the same gain as other selection methods. However, it does not need to continuously carry out selection operations, and system complexity is greatly reduced. Since this method does not require accurate CSI, it is applicable to both coherent reception and noncoherent reception.

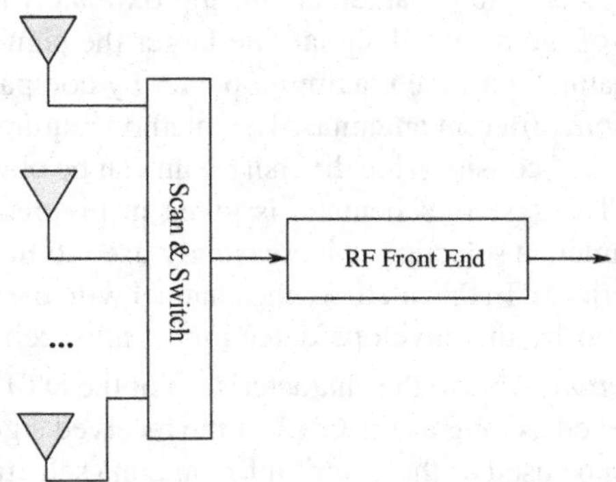

FIGURE 8-26 Schematic diagram of scan switching.

Antenna selection techniques should consider the compromise between diversity gain and implementation cost. Due to restrictions of terminal size, cost, power consumption, complexity, delay, etc., MRC cannot always be utilized in practice, which is why antenna selection techniques are gaining momentum even though the antenna selection method will not achieve the same diversity gain as that by MRC.

8.8 SUMMARY

This chapter has introduced the concepts of coverage and network planning of DTTB systems. Due to the different signal processing schemes in the transmission, first the network planning for analog TV and DTV broadcasting was compared. This was followed by a detailed discussion of SFNs with the focus on the synchronization technique in SFNs. Then the general DTTB transmission system and the different signal reception scenarios were introduced, respectively. Finally, a brief theoretical analysis and implementation of the diversity technology for the DTTB system were presented.

REFERENCES

1. S. O'Leary, *Understanding Digital Terrestrial Broadcasting*, Boston and London: Artech House, 2000.
2. A. Mattsson, "Single frequency networks in DTV," *IEEE Transactions on Broadcasting*, vol.51, no. 4, pp. 413–422, Dec. 2005.
3. A. Ligeti, "Single frequency network planning for DAB/DVB with individual data services," Nordic Radio Symposium, 1998.
4. ETSI 300 744 V1.2.1, *Digital Broadcasting Systems for Television, Sound and Data Services, Framing Structure, Channel Coding and Modulation for Digital Terrestrial Television*, Sophia-Antipolis, France: ETSI, 1999.
5. ETSI TS 101 191, *DVB Mega-Frame for Single Frequency Network (SFN) Synchronization*, Sophia-Antipolis, France: ETSI, Apr. 1997.
6. ETSI TR 101 190, *Implementation Guidelines for DVB Terrestrial Services: Transmission Aspects*, Sophia-Antipolis, France: ETSI, Dec. 1997.
7. B. Vucetic and J. Yuan, *Space-Time Coding*, Hoboken, NJ: Wiley, 2003.
8. G. Faria, *Mobile DVB-T Using Antenna Diversity Receivers*, 2001, http://www.teamcast.com/data/upload/files/BIBFILE_FILE_J5RoETZ.pdf.
9. V. Tarokh, H. Jafarkhani, and A. R. Calderbank, "Space-time block codes from orthogonal designs," *IEEE Transactions on Information Theory*, vol. 45, no. 5, pp. 1456–1467, July 1999.
10. V. Tarokh, H. Jafarkhani, and A. R. Calderbank, "Space-time block coding for wireless communications: Performance results," *IEEE Journal on Selected Areas in Communications*, vol. 17, no. 3, pp. 451–460, Mar. 1999.

11. S. Alamouti, "A simple transmit diversity technique for wireless communications," *IEEE Journal on Selected Areas in Communications*, vol. 16, no. 8, pp. 1451–1458, Oct. 1998.
12. K. Lee and D. Williams, "A space-time coded transmitter diversity technique for frequency selective fading channels." in *Proc. IEEE Sensor Array and Multichannel Signal Processing Workshop*, 2000.
13. K. F. Lee and D. B. Williams, "A space-frequency transmitter diversity technique for OFDM systems," *IEEE Global Telecommunications Conference, 2000. GLOBECOM '00*, vol. 3, 2000, pp. 1473–1477.
14. S. W. Kim and E. Y. Kim, "Optimum receive antenna selection minimizing error probability," *Wireless Communications Network Conference*, vol. 1, Mar. 2003.
15. W. Li and N. C. Beaulieu, "Effects of channel-estimation errors on receiver selection—Combining schemes for Alamouti MIMO systems with BPSK," *IEEE Transactions on Communications*, vol. 54, no. 1, pp. 169–178, Jan. 2006.
16. J. Wang, Z. Yang, C. Pan, and J. L. Yang, "Design of space-time-frequency transmitter diversity scheme for TDS-OFDM system," *IEEE Transactions on Consumer Electronics*, vol. 51, no. 3, pp. 759–764, Aug. 2005.

9

PERFORMANCE MEASUREMENT ON DTTB SYSTEMS

9.1 INTRODUCTION

The main content of this chapter is to provide the measurement procedure and compare the performance of ATSC [1], DVB-T [2], ISDB-T [3], and DTMB transmission standards under different interference conditions.

The chapter will provide a general description of the system-level performance test, including the physical meaning of the test content, test methodologies, and test requirements.

9.2 MEASUREMENT DESCRIPTION [4,5]

To obtain the comprehensive performance of a DTTB system, rigorous laboratory tests and open-circuit tests should be carried out not only for the transmitters but also for the receiver [6–8]. As the signal to be transmitted is the information bit stream for DTV transmission systems, the measurement methodology is very different from existing analog TV broadcasting systems. The measurement items for the DTV broadcasting system include:

1. The bit error rate (BER), the sensitivity at a given BER, to carrier-to-noise ratio (C/N) at a given BER, anti-multipath interference, anti-impulse interference, anti-single-frequency interference, and the constellation analysis at the receiver

Digital Terrestrial Television Broadcasting: Technology and System, First Edition. Edited by Jian Song, Zhixing Yang, and Jun Wang.

2. Actual output power, local oscillator (LO) phase noise, spectrum flatness, and modulation error ratio (MER) at the transmitter
3. The definitions of these parameters and their measurement techniques will be described below.

9.2.1 BER Measurement and Decision Threshold

The BER test is a very important measurement for the DTTB system and is one of the most common metrics of coverage and signal quality. It can be used to measure the signal quality at the receiver, to evaluate transmitter performance, and to determine the transmitter power backoff. The BER curve is usually expressed as a function of C/N, which is the most commonly used indicator of system performance.

The BER test is usually carried out as follows: The pseudorandom binary sequence (PRBS) is first packetized into the MPEG-2 TS structure and then transmitted through the channel to the standard receiver. The BER will be measured on the received signal and is used to evaluate the performance of the entire transmission system. As the measured BER curve as a function of C/N is very steep around the decision threshold of C/N due to the use of FEC, the working status of the system will quickly change from error free to an unacceptable error level; even the change of C/N is within 1 dB. The phenomenon whereby the BER changes drastically near the threshold under a very small change of C/N is called the cliff effect.

During system measurement, two criteria are usually used for the fixed reception:

1. *Objective Failure Criterion.* The reception is considered as a failure if the BER at the receiver (after FEC decoding) is more than 3×10^{-6} with an observation time of 1 min.
2. *Subjective Failure Criterion.* Refers to the subjective failure point (SFP) method defined in ITU-R BT.1368-11. It corresponds to the picture quality where, in three successive observation periods of 20 s, more than one error is visible in the picture for each 20 s.

In Recommendation ITU-R BT.1368, the criterion for mobile reception failure is the erroneous seconds ratio (ESR) of 5% and packet error ratio (PER) of 10^{-4}.

For ATSC, the criterion for system BER performance measurement is 3×10^{-6} at the input of the MPEG-2 decoder. For both DVB-T and ISDB-T, the BER can be measured between the inner and outer codes, and the criterion for system BER performance measurement is 2×10^{-4} before Reed–Solomon decoding and 1×10^{-11} at the input of the MPEG-2 decoder. For DTMB, the criterion for system BER performance measurement is at the output of the BCH decoder with BER of 3×10^{-6}.

The evaluation criteria for analog signal quality should be in accordance with Recommendation ITU-R BT.655. For image evaluation, the quality of the corresponding three-score quality corresponds to a subjective feeling of some nuisance, and four-score quality corresponds to the perceptible damage without nuisance; for an FM received signal, three-score quality corresponds to a signal–to-noise ratio (S/N) of 40 dB for a demodulated audio signal and four-score quality corresponds to S/N of 49 dB.

9.2.1.1 Gaussian Noise Test This performance test is intended to test the ability of the DTTB receiver to resist AWGN.

In ITU-R BT.2035, *Guidelines and Techniques for the Evaluation of DTTB Systems* [9], the input signal power of the receiver can be roughly divided into four levels: very strong signal (−15 dBm), strong signal (−28 dBm), moderate signal (−53 dBm), and weak signal (−68 dBm). During this test, the threshold can be obtained by adjusting the noise level, and the performance of different systems under different working modes can be evaluated by comparing the C/N when the threshold is reached. The visual threshold or objective threshold can be used for testing.

9.2.1.2 Dynamic Range of Input Signal The dynamic range of the input signal reflects the receiver's capability to handle either the very strong or the very weak input signal. Larger dynamic range generally means good design or implementation as this usually ensures that the receiver can work well under a wide range of working conditions. The maximum input signal level is generally no higher than −10 dBm considering the characteristics of terrestrial broadcasting, so it is generally expected that the minimum received signal level should be as low as possible. If the transmit power is fixed, the lower the minimum received signal level is, the wider the coverage of the same transmitter. For a terrestrial broadcasting system, the minimum input level can be obtained as

$$P_{s,\min} = \text{C/N} + \text{NF} - 105.2 \tag{9-1}$$

where $P_{s,\min}$ is the minimum received signal input power (in dBm), NF is the receiver noise figure (in dB), and C/N (in dB) is for the RF signal.

For a given operating mode, the minimum input signal level is directly determined by the NF. Using a low-noise amplifier as the first-stage amplifier could significantly improve the minimum received signal level of the receiver. The receiver must be able to work properly under the same four signal levels as specified in ITU-R BT.2035.

9.2.1.3 Static Multipath Interference Unlike the channel of either satellite broadcasting or cable broadcasting, which is a more Gaussian-like channel, the terrestrial TV broadcasting signal usually suffers from the multipath effect, which, in general, can be divided into Ricean and Rayleigh. The system performance under AWGN is often determined by the selection of FEC and system noise, independent of other system parameters. Even if the performance of the two systems under AWGN is roughly the same, there might be significant performance discrepancy between these two systems under multipath channels. To assess the performance of the DTV terrestrial broadcasting system, which usually operates under very sophisticated multipath environments, it is absolutely necessary to evaluate its performance under the multipath channel. The system decision threshold is reached by adjusting the interference level (i.e., S/N) during the test, and the performance difference can be obtained by comparing the interference level or S/N value. The input signal for this test is set to be the moderate signal (−53 dBm). It should be noted that the signal level refers to the sum of the signal power from all the paths in the presence of multipath, including the following two scenarios:

Single-Echo Scenario. It is mainly used to test the length of the guard interval and the system performance degradation if this echo is within the guard interval. The system performance can also be tested by setting the preecho (the echo ahead of the main-path signal) during the test or introducing the phase rotation to the echo signal. Generally, the time delay of the single-echo signal should not exceed the length of the guard interval during the test, and it is recommended that the phase rotation is between 0 and 5 Hz.

Multiecho Scenario. There should be no less than four echoes in this scenario, and the phase rotation can also be applied to each echo. The multipath interference will be introduced in detail in the following sections of this chapter.

9.2.1.4 Dynamic Multipath Interference The test for the dynamic multipath interference is intended to evaluate the performance of the DTTB receiver or system under the mobility. Since the channel is now dynamic, the test reflects the mobile reception performance of the receiver. For dynamic multipath tests, the channel of the urban model of GSM as well as UMTS and the DVB-T model can be considered. However, all these channel models can only characterize one particular situation and far from reflect the actual mobile reception environment. It is therefore necessary to carry out the open-air test to obtain the comprehensive performance of the system in terms of the mobile reception. For the definition of dynamic multipath echoes, refer to the multipath model section.

9.2.1.5 Cochannel Interference The test of cochannel interference is to check the performance of the DTTB receiver or system in the presence of analog or digital cochannel broadcast (per the discussion in Chapter 8, both desired, denoted by D and unwanted, denoted by U, signals are at the same frequency). This helps determine how far in distance the same frequency can be reused.

1. *Analog-to-Digital Interference.* The system performance is measured by obtaining the power ratio D/U when the decision threshold is reached. It is recommended that the analog TV signal use the color bar signal of 75% at the intermediate-frequency (IF) modulator. When measuring the power of the analog RF signal, the input signal of the IF modulator is a black field signal. The level of the interfered digital signal is generally recommended to be −53 dBm.
2. *Digital-to-Digital Interference.* The system performance is measured by obtaining the power ratio D/U when the decision threshold is reached. Due to the nature of the DTTB signal spectrum, there is no special requirement on the working mode of the interfering digital signals. Again, the level of the interfered digital signal is generally recommended to be −53 dBm.

9.2.1.6 Adjacent-Channel Interference The test of adjacent-channel interference is intended to check the performance of the DTTB receiver or system in the presence of analog or digital adjacent-channel broadcast (i.e., both desired, denoted by D, and unwanted, denoted by U, signals adjacent to the channel). This reflects that at the same transmitting station the maximum allowable transmitting power of the digital broadcasting system can be used in the adjacent channel and the interference of the analog signal in the adjacent channel on the digital signal.

1. *Analog to Digital.* The system performance is measured by obtaining the power ratio D/U when the decision threshold is reached. It is recommended that the analog TV signal use the color bar signal of 75% at the intermediate-frequency modulator. When measuring the power of the analog RF signal, the input signal of the IF modulator is a black field signal. If the analog signal is in the lower adjacent channel next to the digital signal, the input sound signal should be as large as possible. The level of the interfered digital signal is generally recommended to be −60 dBm.
2. *Digital to Digital.* The system performance is measured by obtaining the power ratio D/U when the decision threshold is reached. Due to the nature of the DTTB signal spectrum, there is no special requirement on the working mode of the interfering digital signals. Again, the level of the interfered digital signal is generally recommended to be −60 dBm.

9.2.1.7 Impulse Noise Interference This test is mainly for evaluating the performance of the DTTB system to resist the high-intensity impulse interference within a short duration, usually coming from the operation as well as on/off switches of home appliances such as the blender, hair dryer, and refrigerator. Demodulator chips from different vendors may have the same C/N threshold under AWGN but quite different capability to handle impulse interferences with different pulse widths. Under the same working mode and multipath condition, the demodulator is considered to have better performance if it can handle the impulse interference with longer pulse width.

There are two ways to generate the impulse noise, as shown in Figure 9-1. It is highly recommended to use the second method for the impulse noise generation.

9.2.1.8 Phase Noise In theory, the duration of an ideal pulse signal with fixed frequency of 1 MHz as an example should last exactly 1 μs, and the signal phase will

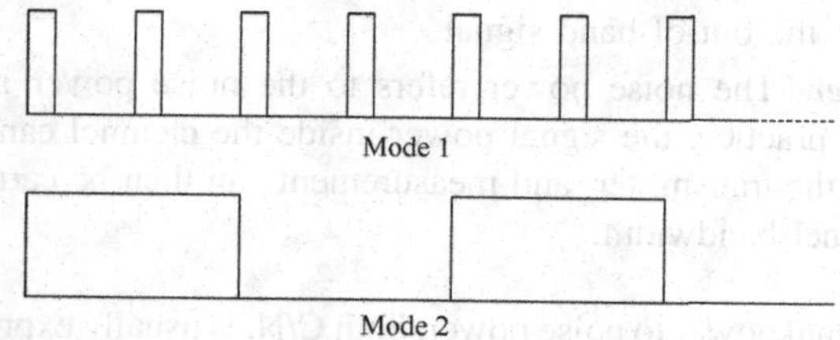

FIGURE 9-1 Two methods for impulse noise generation.

change exactly to the opposite every 500 ns, known as the edge flip-over. However, there always exists the small deviation of the signal's period in practice, which leads to the uncertainty of the arrival time of the next edge flip-over. This uncertainty is called the phase noise and is a concept in the frequency domain.

Without phase noise, the power of the oscillator oscillating at frequency f_0 should be concentrated exactly at frequency $f = f_0$ and there will be no power at any other frequencies. However, with the phase noise, power is no longer concentrated but is spread out instead, which results in the pop-up of the sidebands. Phase noise is another way to measure the signal timing changes and represent this change in the frequency domain and is usually characterized in units of dBc/Hz with a given offset frequency. Here, dBc is the ratio of the power (in dB) at this frequency to the total signal power. The phase noise of an oscillator for a given frequency offset is defined as the ratio of the signal power within 1 Hz bandwidth at this frequency offset to the total signal power.

In the DTTB system, the LO signals of the modulator and LO inside the tuner of the receiver will surely be affected by the phase noise, and its impact on system performance should be carefully considered in the design. The trade-off must be made as an unrealistic requirement on the phase noise performance will increase the cost of devices and the whole system.

9.2.2 C/N Measurement

C/N is an important parameter to measure DTTB system performance. In general, the lower the C/N required for a given BER at the receiver, the better the overall system performance will be.

To obtain CNR, first both the signal (carrier) power and noise power (average power) must be measured:

1. *Signal Power Measurement.* The signal power here refers to the RF signal power of the DTTB system at the receiver, defined as the average signal level measured by thermal power sensors. It should be pointed out that signal power measurements for analog and digital systems are quite different. The power of the analog TV signal is obtained by means the sync peak power while power measurement of the DTV signal is carried out through the whole channel bandwidth. To make sure that only the power of the signal in this channel is measured, out-of-band power should be removed by a channel filter at the transmitter output. If a spectrum analyzer is used to measure the signal power inside this channel, one can choose the measurement bandwidth appropriately to eliminate the out-of-band signal.
2. *Noise Power.* The noise power refers to the noise power inside the signal channel. In practice, the signal power inside the channel can be removed by turning off the transmitter, and measurement can then be carried out over the entire channel bandwidth.

The ratio of signal power to noise power, both C/N, is usually expressed in decibels for convenience.

If the spectrum analyzer is used to measure the in-band power, the RBW (resolution bandwidth) setting should be automatic for the analyzer.

9.2.3 Input Signal Level to the Receiver

As specified by ITU-R BT.2035, if the measurements for the DTTB receiver are to be performed by introducing AWGN, the power of the input digital signal is generally set to the moderate level (−53 dBm). Otherwise, the power of the input digital signal should generally be set to the weak level (−60 dBm).

9.2.4 Interface Parameters

The interface parameters are as follows:

TS Stream Interface. ASI or SPI interface according to IEC 60728-9.

Input/Output Impedance. The impedance of the RF test instrument is 50 Ω (N-type connector), the impedance of the video test instrument is 75 Ω (BNC connector), the impedance of the video cable is 75 Ω, and the impedance of the RF cable is 50 Ω.

As the impedances of the test instrument and receiver are not identical, impedance match and conversion are therefore required.

9.2.5 Multipath Models

The multipath model is an effective way to objectively describe the channel, which gives the amplitude variation, phase change, and Doppler frequency shift if applicable at each echo with different delays. Three multipath models are widely used in the test as shown in Tables 9-1 to 9-3, respectively. They are used to represent the multipath channel characteristics in the reception scenarios of non–line of sight, line of sight, and mobility.

9.2.6 Laboratory Test

Depending on where the tests on DTTB systems are conducted, the system performance test can be divided into laboratory tests and field tests. The laboratory tests are carried out on the closed-loop RF test platform built in the laboratory so that the tests on the DTTB transmission systems and equipment are both strict and accurate and can be reproduced. The results of laboratory tests could mimic the actual operation status of DTTB systems and equipment to a large extent. The closed-loop test is the basis for the on-site open-air test, and its results can provide a really important guide for field test data analysis.

In addition to the completeness of the test plan and the accuracy of the testing instruments/equipment, a good testing environment is an important prerequisite for reliable and accurate laboratory tests. The laboratory must meet the following

TABLE 9-1 Rayleigh Channel Model

Path	Amplitude (dB)	Delay (μs)
Main path	−7.8	0.518650
Echo 1	−24.8	1.003019
Echo 2	−15.0	5.422091
Echo 3	−10.4	2.751772
Echo 4	−11.7	0.602895
Echo 5	−24.2	1.016585
Echo 6	−16.5	0.143556
Echo 7	−25.8	0.153832
Echo 8	−14.7	3.324886
Echo 9	−7.9	1.935570
Echo 10	−10.6	0.429948
Echo 11	−9.1	3.228872
Echo 12	−11.6	0.848831
Echo 13	−12.9	0.073883
Echo 14	−15.3	0.203952
Echo 15	−16.5	0.194207
Echo 16	−12.4	0.924450
Echo 17	−18.7	1.381320
Echo 18	−13.1	0.640512
Echo 19	−11.7	1.368671

TABLE 9-2 Ricean Channel Model

Path	Amplitude (dB)	Delay (μs)
Main path	0	0
Echo 1	−19.2	0.518650
Echo 2	−36.2	1.003019
Echo 3	−26.4	5.422091
Echo 4	−21.8	2.751772
Echo 5	−23.1	0.602895
Echo 6	−35.6	1.016585
Echo 7	−27.9	0.143556
Echo 8	−26.1	3.324886
Echo 9	−19.3	1.935570
Echo 10	−22.0	0.429948
Echo 11	−20.5	3.228872
Echo 12	−23.0	0.848831
Echo 13	−24.3	0.073883
Echo 14	−26.7	0.203952
Echo 15	−27.9	0.194207
Echo 16	−23.8	0.924450
Echo 17	−30.1	1.381320
Echo 18	−24.5	0.640512
Echo 19	−23.1	1.368671

TABLE 9-3 Dynamic Multipath Model

Path	Amplitude (dB)	Delay (μs)	Doppler Mode
Echo 1	−3	0	Ricean
Echo 2	0	0.2	Ricean
Echo 3	−2	0.5	Ricean
Echo 4	−6	1.6	Ricean
Echo 5	−8	2.3	Ricean
Echo 6	−10	5	Ricean

requirements: It should be an excellent electromagnetic compatible environment, the test platform should have good grounding, and the AC power supply should be very stable. To reduce and eventually eliminate the impact of external factors, it is preferable to carry out the test in a metal-shielded room.

The main items of a laboratory test are the interference characteristics of the system, anti-interference capability, transmission performances, and reception performances, which include the following:

1. *System Performance Test in AWGN Channel.* This is to evaluate the noise level of the receiver, demodulation performance, and anti-inference capability by testing receiver noise characteristics, receiving signal level, BER, etc. Specific tests are:
   - Carrier-to-noise ratio threshold
   - Minimum received signal level
2. *System Performance in Multipath Channel.* The anti-interference ability of the DTTB system is tested by simulating various interference environments. Specific tests are:
   - Anti-noise performance in the static multipath channel
   - Performance in the static multipath channels with long echo delay
   - Antinoise performance in the dynamic multipath channel
   - Anti-Doppler frequency shift capability in the dynamic multipath channel
   - Performance of anti-single-echo capability in the dynamic channel
3. *Anti-Interference Performance and Electromagnetic Compatibility.* The anti-interference ability of DTTB system is tested by simulating various interference environments. Specific tests are:
   - Anti-impulse interference performance
   - Anti-single-frequency interference performance
   - Anti-cochannel and adjacent-channel analog interference performance
   - Anti-cochannel and adjacent-channel digital interference performance
   - Impact to the cochannel, adjacent channel, and image channel analog system
4. *Signal Characteristics Test.* The transmission efficiency and the possibility of interference against other systems are analyzed by testing the emission

spectrum, peak-to-average-power ratio (PAPR), and spurious outputs of the transmitter. Specific tests are:

- Antiphase noise capability
- Nominal data throughput test
- Peak-to-average-power ratio
- Effective bandwidth and roll-off factor of the pulse shape filter
- Transmitter spectrum characteristics

9.3 LABORATORY TEST PLAN USING DTMB SYSTEM AS EXAMPLE

In the following, the DTMB system is used as an example of a laboratory test for illustration purposes. The principles for the test on other standards are the same.

9.3.1 Laboratory Test Platform

A diagram of the laboratory test platform is shown in Figure 9-2.

In the figure, the RF interference generator, phase noise generator, and adjustable attenuator can either be individual devices or an integrated single device.

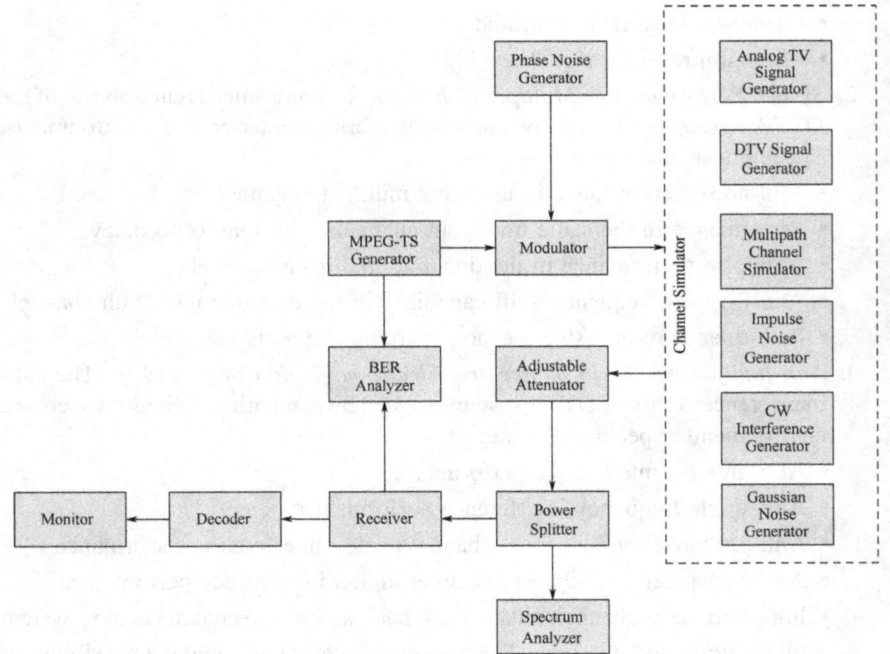

FIGURE 9-2 Schematic diagram of laboratory test platform.

9.3.2 Interface Setup of Test Platform

1. The transmitter under test (including the upconverter) should be able to output the RF signal with impedance 50 Ω. The output power should be adjusted according to the requirements of the RF channel simulator;
2. For the receiver, the input impedance of the tuner is generally 75 Ω, and the impedance match device should be inserted between the power splitter and the receiver to ensure accurate testing.
3. The input signal power for the DTTB receiver is generally set to a moderate level (−50 to −53 dBm) when testing, and the receiver input can be further reduced to −60 dBm when the protection ratio is tested.
4. For all the testing items, according to the test conditions or the related frequency, either VHF/UHF channels are selected upon request for the testing.

Table 9-4 provides a list of the commonly used instruments (with models only for reference purposes) for the laboratory test platform. Instruments from other companies with the same functionality and performance can also be used to complete the test.

9.3.3 C/N Threshold under Gaussian Channel

Test Purpose This test assesses the tolerance level of the DTTB transmission system to AWGN, which is the most fundamental characteristic reflecting system performance.

The C/N threshold in this case reflects the FEC performance of the system with the background thermal noise of the circuitry. The smaller the C/N threshold is (i.e., the less C/N required for quasi-error-free reception), the better the reception performance in the open-air testing environment will be, and this means better system performance.

The C/N threshold is defined as the power ratio (in dB) when the receiver under test reaches the failure criterion for the fixed reception.

TABLE 9-4 Testing Instruments for Laboratory Test Platform

No.	Names	Manufacturer	Model
1	Transport stream generator	Tektronix	MTX100
2	Low-noise spectrum analyzer	Agilent	N9020
3	PAL modulator	R&S	SFM 50
4	Digital TV test receiver	R&S	EFA
5	Noise generator	NOISE COM INC	UFX99CA
6	Channel simulator (multipath interference generator)	Elektrobit	Propsim FE
7	Digital transmission analyzer	EIDEN	7706A-002
8	Digital television signal generator	R&S	SFU

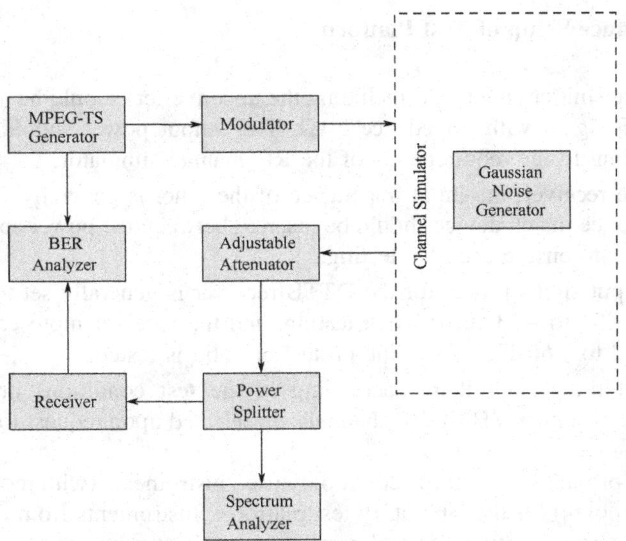

FIGURE 9-3 Experimental setup for C/N test.

The objective failure criterion for the fixed reception is used for the test, and the C/N threshold is calculated by the measurement results of both the received signal level and the noise level within the system bandwidth. The measurement bandwidth is 7.56 MHz and the signal power is set to −53 dBm.

Test Description

- The experimental setup for the C/N test is shown in Figure 9-3.
- Set the channel of the "multipath channel simulator" in the block diagram of the test platform in Figure 9-2 to the AWGN channel (or Gaussian channel) and the interference just comes from the RF noise.
- Set the attenuation of the noise to the maximum value of 127.9 dB (With the attenuation this high, it is equivalent to turning off the noise generator or modulator in the following. Some devices have no functionality to shut down its RF output, and it takes time to turn it off and then on again. That's why it is arranged this way. Then change the signal attenuation so that the average power (with the average done over 50 measurements) in the bandwidth of RF center frequency ±3.78 MHz from the spectrum analyzer is around −53 dBm, denoting this measured value as C (dBm).
- The noise attenuation is reduced until reception fails.
- The noise attenuation is increased in steps of 0.1 dB until the failure criterion is met.
- Set the attenuation of the signal to the maximum value of 127.9 dB and the average noise level (with the average done over 50 measurements) in the

TABLE 9-5 C/N Threshold of Preferred Working Modes for DTMB Receiver

Working Mode	Carrier Mode	FEC Code Rate	Constellation	Frame Header	Interleaving Depth	Payload Data Rate (Mbps)	C/N Threshold (dB)
1	$C=3780$	0.4	16QAM	PN = 945	720	9.626	8.0
2	$C=1$	0.8	4QAM	PN = 595	720	10.396	6.0
3	$C=3780$	0.6	16QAM	PN = 945	720	14.438	10.7
4	$C=1$	0.8	16QAM	PN = 595	720	20.791	12.6
5	$C=3780$	0.8	16QAM	PN = 420	720	21.658	13.2
6	$C=3780$	0.6	64QAM	PN = 420	720	24.365	15.7
7	$C=1$	0.8	32QAM	PN = 595	720	25.989	16.6

bandwidth of RF center frequency ±3.78 MHz from the spectrum analyzer, denoted as N (dBm).

- The C/N threshold can then be obtained.

DTMB Standard Requirement According to the DTV receiver industry standard of China (The official name of this standard is General Specification for the Digital Terrestrial Television Receivers but will be named Chinese digital TV receiver standard for simplicity in this chapter), the C/N threshold for the typical working modes should meet the requirements in Table 9-5.

9.3.4 Minimum Reception Level in Gaussian Channel

Test Purpose This test evaluates the receiving sensitivity of the DTTB receiver.

The minimum signal level required for the DTTB receiver is called the receiving sensitivity, which is jointly determined by the system NF, the thermal noise of the circuit, and the C/N threshold. The noise figure depends on the amplifier of the tuner at the receiver and is independent of system design. The front end with lower NF can significantly reduce the required minimum signal level, improving receiving sensitivity.

The receiving sensitivity plays an important role in DTTB network planning, which reflects not only the NF but also the system stability in the presence of self-interference, intermodulation products, etc. The minimum receiving signal level is closely related to the RF subsystem performance of the receiver. For two receiver designs with identical C/N, the minimum receiving signal level may be quite different.

The objective failure criterion for the fixed reception is used for the test and the receiver input signal level is measured in the bandwidth of 7.56 MHz.

Test Description

- The experimental setup for the minimum signal level test is shown in Figure 9-4.
- The "multipath simulator" in the test platform block diagram is set to transparently passthrough (the signal will transparently pass through the RF multipath channel emulator to the receiver, without suffering from any interference).

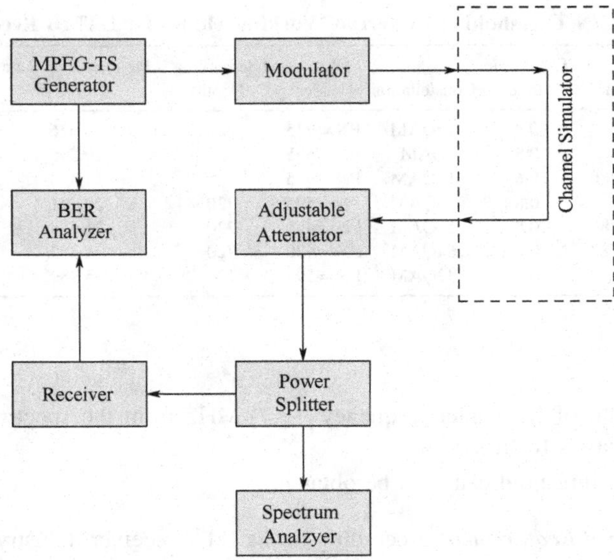

FIGURE 9-4 Experimental setup for both maximum and minimum signal level tests.

- The attenuation of the signal attenuator is adjusted by the appropriate step so that the minimum signal power in decibels of the receiver under test meets the objective criterion for successful reception, and this is the minimum received signal level.
- The average signal level measured by the spectrum analyzer (average power in the RF band within the center frequency of ±3.78 MHz and the number of tests for the average is 50) is denoted by P in dBm.

Reference Test Results According to the Chinese digital TV receiver standard, the minimum signal level in the Gaussian channel for the typical working modes should meet the values in Table 9-6.

TABLE 9-6 Minimum Reception Level in Gaussian Channel

Working Mode	Carrier Mode	FEC Code Rate	Constellation	Frame Header	Interleaving Depth	Payload Bit Rate (Mbps)	Minimum Input Signal Level (dBm) VHF	UHF
1	$C=3780$	0.4	16QAM	PN = 945	720	9.626	−92	−90
2	$C=1$	0.8	4QAM	PN = 595	720	10.396	−93	−91
3	$C=3780$	0.6	16QAM	PN = 945	720	14.438	−89	−87
4	$C=1$	0.8	16QAM	PN = 595	720	20.791	−87	−85
5	$C=3780$	0.8	16QAM	PN = 420	720	21.658	−86	−84
6	$C=3780$	0.6	64QAM	PN = 420	720	24.365	−84	−82
7	C_1	0.8	32QAM	PN = 595	720	25.989	−84	−82

9.3.5 Maximum Reception Level

Test Purpose This item evaluates the resilience of the receiver under the signal saturation. This parameter is closely related to the antisaturation property of the RF subsystem and the frequency selectivity. The AGC range of the tuner has direct impact on the testing result. The large reading of the input level usually indicates a better capability of the receiver when handling the maximum input level.

The objective failure criterion for the fixed reception is selected for the test, and the receiver input signal level is measured in the bandwidth of 7.56 MHz.

Test Description

- The experimental setup for the maximum signal level test is also shown in Figure 9-4.
- The "multipath simulator" in the test platform block diagram is set to transparently pass through (the signal will transparently pass through the RF multipath channel emulator to the receiver, without suffering from any interference).
- Adjust the receiver attenuator to increase the input signal power until the receiver under test meets the objective criterion for successful reception; the signal power in dBm at this time is the maximum received signal level, which can be recorded. If the error does not occur when the received input signal level reaches −10 dBm, record the maximum receiving level as −10 dBm. Do not continue to increase the receiver input signal level, which may cause damage to the tuner.
- The average signal level measured by the spectrum analyzer (average power in the RF band within the center frequency of ±3.78 MHz and the number of tests for the average is 50) is denoted by P in dBm.

Reference Test Results According to the Chinese digital TV receiver standard, the maximum signal level in the Gaussian channel for typical working modes should meet the values in Table 9-7.

TABLE 9-7 Maximum Receiving Levels

Mode	Carrier Mode	FEC Code Rate	Constellation	Frame Header	Interleaving Depth	Payload Bit Rate (Mbps)	Maximum Input Signal Level (dBm)	
							VHF	UHF
1	$C=3780$	0.4	16QAM	PN=945	720	9.626	−10	−10
2	$C=1$	0.8	4QAM	PN=595	720	10.396	−10	−10
3	$C=3780$	0.6	16QAM	PN=945	720	14.438	−10	−10
4	$C=1$	0.8	16QAM	PN=595	720	20.791	−10	−10
5	$C=3780$	0.8	16QAM	PN=420	720	21.658	−10	−10
6	$C=3780$	0.6	64QAM	PN=420	720	24.365	−10	−10
7	$C=1$	0.8	32QAM	PN=595	720	25.989	−10	−10

9.3.6 C/N Threshold in Ricean Channel

Test Purpose The Ricean channel is characterized by a line-of-sight main path, which represents the reception scenario of using the fixed roof antenna. This test result reflects the channel estimation and equalization performance of the receiver under the Ricean channel condition for fixed reception. The receiver demodulator chips designed by different vendors can have the C/N threshold at AWGN and the same sensitivity, but their C/N thresholds in multipath channel conditions may be quite different. For the same working mode and multipath condition, the lower the C/N threshold, the better performance the system has.

The objective failure criterion for the fixed reception is selected for the test. The C/N threshold is calculated by measuring the input signal level and noise level within the measurement bandwidth of 7.56 MHz and the signal level is −53 dBm.

Test Description

- The experimental setup for the C/N test in the Ricean channel is shown in Figure 9-5.
- The "multipath simulator" in the block diagram of the test platform is set to the static Ricean multipath channel with AWGN.

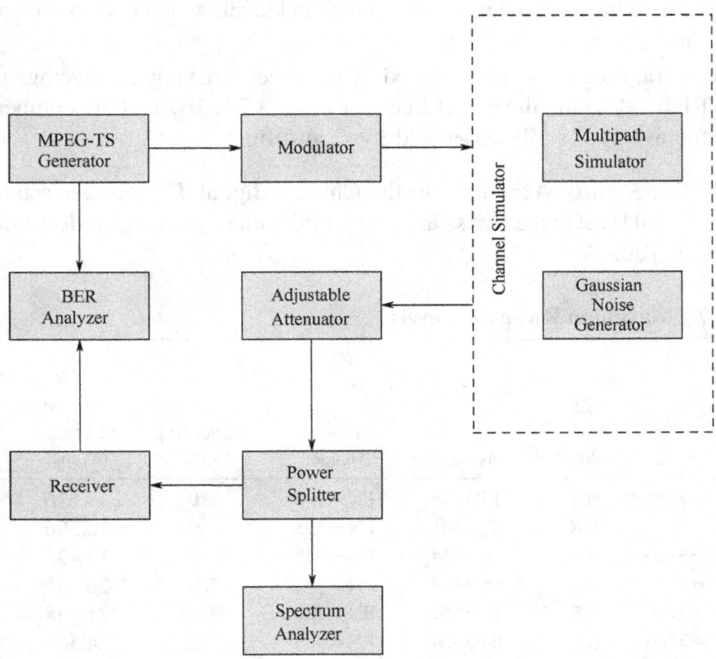

FIGURE 9-5 Experimental setup for C/N test under different channel conditions.

TABLE 9-8 C/N Threshold in Ricean Channel

Working Mode	Carrier Mode	FEC Code Rate	Constellation	Frame Header	Interleaving Depth	Payload Bit Rate (Mbps)	C/N Threshold (dB)
1	$C=3780$	0.4	16QAM	PN=945	720	9.626	8.7
2	$C=1$	0.8	4QAM	PN=595	720	10.396	6.5
3	$C=3780$	0.6	16QAM	PN=945	720	14.438	11.2
4	$C=1$	0.8	16QAM	PN=595	720	20.791	13.3
5	$C=3780$	0.8	16QAM	PN=420	720	21.658	14.0
6	$C=3780$	0.6	64QAM	PN=420	720	24.365	16.6
7	$C=1$	0.8	32QAM	PN=595	720	25.989	17.3

- Set the attenuation of the noise to the maximum value of 127.9 dB. The attenuation of the signal attenuator is adjusted so that the average power within the bandwidth ±3.78 MHz of the RF center frequency measured by the spectrum analyzer is around −53 dBm and the number of tests for the average is 50, denoted by C in dBm.
- The noise attenuation is reduced until reception fails.
- The noise attenuation is increased in steps of 0.1 dB until the failure criterion is met.
- Set the attenuation of the signal to the maximum of 127.9 dB; the average noise level measured by the spectrum analyzer (with bandwidth ±3.78 MHz of RF center frequency, the number of tests for the average is 50) is denoted by N in dBm.
- The C/N threshold is then obtained.

Reference Test Results According to the Chinese digital TV receiver standard, the C/N threshold in the Ricean channel for typical working modes should meet the values in Table 9-8.

9.3.7 C/N Threshold in Rayleigh Channel

Test Purpose The characteristic of the Rayleigh channel is that there is no obvious main path, which represents the reception scenario of the omnidirectional antenna located in a lower position. This test result reflects the channel estimation and equalization performance for the receiver under the Rayleigh channel condition for fixed reception. For the same working mode and multipath condition, the lower the C/N threshold, the better performance the system has.

The objective failure criterion for the fixed reception is selected for the test. The C/N threshold is calculated by measuring the input signal level and noise level within the bandwidth of 7.56 MHz and the signal level is −53 dBm.

Test Description

- The experimental setup for the C/N test for a Rayleigh channel is shown in Figure 9-5.

TABLE 9-9 C/N Threshold in Rayleigh Channel

Working Mode	Carrier Mode	FEC Code Rate	Constellation	Frame Header	Interleaving Depth	Payload Bit Rate (Mbps)	C/N Threshold (dB)
1	$C=3780$	0.4	16QAM	PN = 945	720	9.626	10.5
2	$C=1$	0.8	4QAM	PN = 595	720	10.396	9.5
3	$C=3780$	0.6	16QAM	PN = 945	720	14.438	14.0
4	$C=1$	0.8	16QAM	PN = 595	720	20.791	18.5
5	$C=3780$	0.8	16QAM	PN = 420	720	21.658	18.5
6	$C=3780$	0.6	64QAM	PN = 420	720	24.365	19.4
7	$C=1$	0.8	32QAM	PN = 595	720	25.989	22.4

- The "multipath simulator" in the block diagram of the test platform is set to the static Rayleigh multipath channel with AWGN.
- Set the attenuation of the noise to the maximum value of 127.9 dB. The attenuation of the signal attenuator is adjusted so that the average power within the bandwidth ±3.78 MHz of the RF center frequency measured by the spectrum analyzer is around −53 dBm and the number of tests for the average is 50, denoted by C in dBm.
- The noise attenuation is reduced until reception fails.
- The noise attenuation is increased in steps of 0.1 dB until the failure criterion is met.
- Set the attenuation of the signal to the maximum of 127.9 dB; the average noise level measured by the spectrum analyzer (with bandwidth ±3.78 MHz of RF center frequency, the number of tests for the average is 50) is denoted by N in dBm.
- The C/N threshold is then obtained.

Reference Test Results According to the Chinese digital TV receiver standard, the C/N threshold in a Rayleigh channel for typical working modes should meet the values in Table 9-9.

9.3.8 Maximum Doppler Frequency Shift in Dynamic Multipath Channel

Test Purpose This test evaluates the performance of the digital terrestrial television transmission system under dynamic multipath conditions. The measured C/N+3 dB is used to determine the minimum signal level required for mobile reception.

Dynamic multipath is characterized by the presence of Doppler frequency shift in the multipath channel, which reflects the mobile reception scenario of the omnidirectional antenna located at a lower position for the mobile terminal. This test result reflects the channel estimation and equalization performance of the receiver in multipath transmission conditions where there is no main path for mobile reception. This test result reflects the channel estimation, equalization, and tracking performance of the receiver. For the same mode and multipath conditions, the higher the Doppler frequency shift can be handled under the dynamic multipath condition, the better mobile reception performance the system has.

The objective failure criterion for fixed reception is selected for the test, and the maximum Doppler frequency shift for the received input signal is measured at the signal level of −53 dBm. Unlike other testing items, the test time for this test is 300 s.

Test Description

- The experimental setup for the maximum Doppler frequency shift test is shown in Figure 9-5.
- The "multipath simulator" in the block diagram of the test platform is set to the dynamic multipath channel with AWGN.
- Set the attenuation on the noise to the maximum value of 127.9 dB. The attenuation of the signal attenuator is adjusted so that the average power within bandwidth ±3.78 MHz of the RF center frequency measured by the spectrum analyzer is around −53 dBm and the number of tests for the average is 50, denoted by C in dBm.
- Set the Doppler frequency shift to 70 Hz, which is equivalent to the moving speed of 99.2 km/h at central frequency of 762 MHz.
- The noise attenuation is reduced until reception fails. Set the attenuation of the signal to the maximum of 127.9 dB and measure the average noise level by the spectrum analyzer (with bandwidth ±3.78 MHz of RF center frequency, the number of tests is 50). After the noise level denoted by N in dBm is obtained, calculate C/N as (C/N) min.
- Increase the noise attenuation, making C/N = (C/N) min + 3 dB. Increase the Doppler frequency shift until the reception fails. Record the Doppler frequency shift at this time as the maximum Doppler frequency.

Reference Test Results According to the Chinese digital TV receiver standard, the upper limit of maximum Doppler frequency shift in the dynamic multipath channel for some typical working modes should meet the values in Table 9-10.

TABLE 9-10 Mobility Performance in Dynamic Multipath Channel

Working Mode	Carrier Mode	FEC Code Rate	Constellation	Frame Header	Interleaving Depth	Payload Bit Rate (Mbps)	C/N_{min} (dB) ($f_d = 70$ Hz)	Maximum Doppler Frequency Shift (Hz)
1	$C = 3780$	0.4	16QAM	PN = 945	720	9.626	12.0	130
2	$C = 1$	0.8	4QAM	PN = 595	720	10.396	13.5	120
3	$C = 3780$	0.6	16QAM	PN = 945	720	14.438	17.0	115

9.3.9 Maximum Delay Spread in Two-Path Channel with 0-dB Echo

Test Purpose This test measures the capability of the digital terrestrial TV transmission system versus the 0-dB echo (the echo has the same amplitude as that of the main path, and with relatively long time delay, the time difference between the main path and this echo is still within the length of the guard interval), reflecting the ability of receivers to handle maximum delay spread, regardless of whether it is under the very harsh reception condition or in the SFN condition. In general, the data to be measured should not exceed the guard interval size. For the same working mode and multipath condition, the larger the maximum delay that can be handled, the better performance the system has.

This test result reflects the channel estimation and equalization performance of the receiver in the static two-path channel. The performance has a close relationship with the choice of guard interval and the implementation algorithm of the receiver.

The objective failure criterion for the fixed reception is selected for the test, and the 0-dB echo at the receiver is set to the signal level of −53 dBm.

Test Description

- The experimental setup for the maximum delay spread in a two-path channel with 0-dB echo is shown in Figure 9-6.
- The "multipath simulator" in the block diagram of the test platform is set to the two-path model (RF multipath channel emulator is set to two paths without any interference).

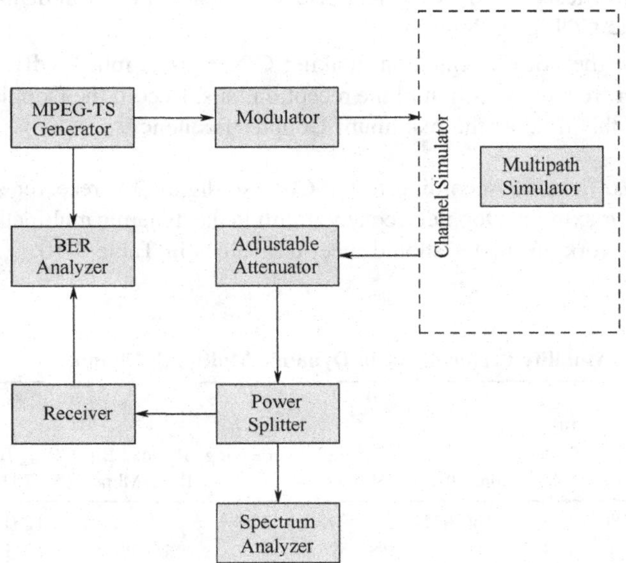

FIGURE 9-6 Experimental setup for 0-dB echo performance test.

TABLE 9-11 Maximum Delay of 0-dB Echo in Two-Path Channel

Working Mode	Carrier Mode	FEC Code Rate	Constellation	Frame Header	Interleaving Depth	Payload Bit Rate (Mbps)	Maximum Delay of 0-dB (μs)
1	$C=3780$	0.4	16QAM	PN=945	720	9.626	110
2	$C=1$	0.8	4QAM	PN=595	720	10.396	60
3	$C=3780$	0.6	16QAM	PN=945	720	14.438	110
4	$C=1$	0.8	16QAM	PN=595	720	20.791	60
5	$C=3780$	0.8	16QAM	PN=420	720	21.658	50
6	$C=3780$	0.6	64QAM	PN=420	720	24.365	50
7	$C=1$	0.8	32QAM	PN=595	720	25.989	60

- The echo intensity ratio for this given two-path channel is set to 0 dB.
- The attenuation on the noise is set to the maximum value of 127.9 dB. The attenuation of the signal attenuator is adjusted so that the average power within the bandwidth ±3.78 MHz of the RF center frequency measured by the spectrum analyzer is around −53 dBm and the number of tests for the average is 50, denoted by C in dBm.
- Set the main path delay to 0 μs. Adjust the delay of the echo signal; record the maximum echo delay (μs) when the receiver under test still satisfies the objective criterion. If the normal reception is possible even when 999 μs is reached, record it as ">999 μs."

Reference Test Results According to the Chinese digital TV receiver standard, the maximum 0-dB echo for the typical working modes should meet the values in Table 9-11.

9.3.10 C/N Threshold in Two-Path Channel with 0-dB Echo

Test Purpose This test result reflects the channel estimation and equalization performance of the receiver in the static two-path channel. The performance is closely related to the implementation algorithm of the demodulator.

This test measures the C/N threshold of the digital terrestrial TV transmission system under the 0-dB echo (with time delay within the guard interval) channel condition. The smaller the measured value, the less the C/N required for the normal reception in 0-dB echo channel. It is obvious that the measured value will not be less than the C/N threshold under AWGN. For the same working mode and multipath conditions, the lower the required C/N threshold, the better performance the system has.

The objective failure criterion for fixed reception is selected for the test, and the C/N threshold under the 0-dB echo signal at the receiver input is measured at the signal level of −53 dBm.

TABLE 9-12 C/N Thresholds in Two-Path Channel with 0-dB Echo Channel

Working Mode	Carrier Mode	FEC Code Rate	Constellation	Frame Header	Interleaving Depth	Payload Bit Rate (Mbps)	C/N (dB)
1	$C=3780$	0.4	16QAM	PN=945	720	9.626	11
2	$C=1$	0.8	4QAM	PN=595	720	10.396	11
3	$C=3780$	0.6	16QAM	PN=945	720	14.438	15
4	$C=1$	0.8	16QAM	PN=595	720	20.791	20.5
5	$C=3780$	0.8	16QAM	PN=420	720	21.658	20.5
6	$C=3780$	0.6	64QAM	PN=420	720	24.365	20.5
7	$C=1$	0.8	32QAM	PN=595	720	25.989	24.5

Test Description

- The experimental setup for the C/N test in the 0-dB echo channel is shown in Figure 9-5.
- The "multipath simulator" in the block diagram of the test platform is set to the two-path model (RF multipath channel emulator is set to two paths without noise and interference).
- Set the time delay of the main path to zero, and the time delay of the echo is 30 μs.
- The attenuation on the noise is set to the maximum value of 127.9 dB. The attenuation of the signal attenuator is adjusted so that the average power within the bandwidth ±3.78 MHz of the RF center frequency measured by the spectrum analyzer is around −53 dBm and the number of tests for the average is 50, denoted by C in dBm.
- The noise attenuation is reduced in steps of 0.1 dB until reception fails. Set the attenuation of the signal to the maximum of 127.9 dB and measure the average noise level by the spectrum analyzer (with bandwidth ±3.78 MHz of RF center frequency, the number of tests is 50). After the noise level, denoted by N in dBm, is obtained, calculate C/N as (C/N) min.

Reference Test Results According to the Chinese digital TV receiver standard, the C/N threshold in the 0-dB echo channel for typical working modes should meet the values in the Table 9-12.

9.3.11 Maximum Pulse Width of Impulse Noise Interference

Test Purpose This test evaluates the performance of the digital terrestrial television transmission system against the high-intensity impulse interference with short duration.

The performance is mainly determined by the time-domain interleaving design of the DTTB system. For the same operating mode and multipath conditions, the larger

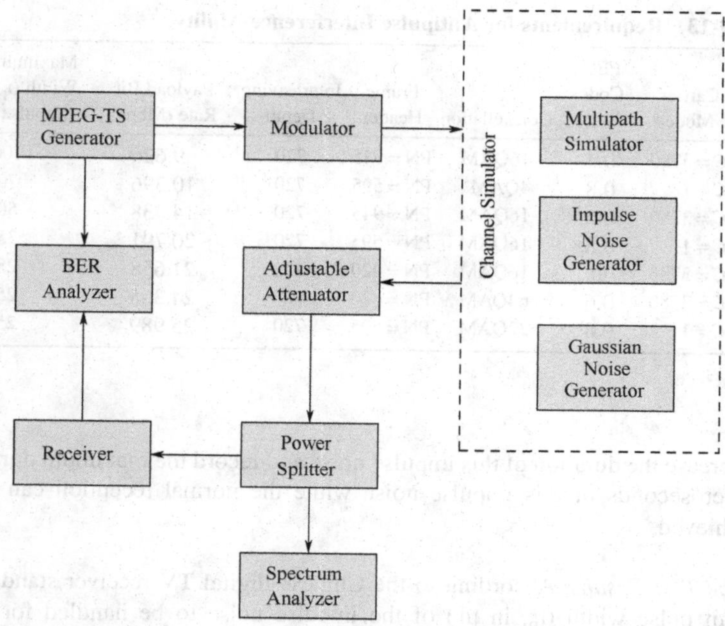

FIGURE 9-7 Experimental setup for maximum pulse width of impulse noise test.

the time duration of the impulse noise interference the system can handle, the better the system performance will be.

The objective failure criterion for the fixed reception is selected for the test, and the receiver performance of the maximum pulse width of the impulse noise is measured at the signal level of −53 dBm.

Test Description

- The experimental setup for the maximum pulse width of the impulse noise is shown in Figure 9-7.
- The "multipath simulator" in the test platform block diagram is set to the Gaussian channel (the signal will transparently pass through the RF multipath channel emulator to the receiver and the only inference comes from the RF circuitry.)
- The attenuation on the noise is set to the maximum value of 127.9 dB. The attenuation of the signal attenuator is adjusted so that the average power within the bandwidth ±3.78 MHz of the RF center frequency measured by the spectrum analyzer is around −53 dBm and the number of tests for the average is 50, denoted by C in dBm.
- Impulse noise is represented by the on/off of the high-power AWGN with repetition frequency of 100 Hz. The C/N during the "on" state is −3 dB.

TABLE 9-13 Requirements for Antipulse Interference Ability

Working Mode	Carrier Mode	FEC Code Rate	Constellation	Frame Header	Interleaving Depth	Payload Bit Rate (Mbps)	Maximum Pulse Width (t_p, μs) of Impulse Noise
1	$C=3780$	0.4	16QAM	PN = 945	720	9.626	100
2	$C=1$	0.8	4QAM	PN = 595	720	10.396	70
3	$C=3780$	0.6	16QAM	PN = 945	720	14.438	50
4	$C=1$	0.8	16QAM	PN = 595	720	20.791	35
5	$C=3780$	0.8	16QAM	PN = 420	720	21.658	25
6	$C=3780$	0.6	64QAM	PN = 420	720	24.365	25
7	$C=1$	0.8	32QAM	PN = 595	720	25.989	25

- Increase the duration of this impulse noise and record the maximum duration in microseconds of this impulse noise while the normal reception can still be achieved.

Reference Test Results According to the Chinese digital TV receiver standard, the maximum pulse width (t_p, in μs) of the impulse noise to be handled for typical working modes should meet the requirements in Table 9-13.

9.3.12 C/I Measurement with Cochannel and Adjacent-Channel Analog TV Signal Interference

Test Purpose This test evaluates the performance of the DTTB system against both cochannel and adjacent-channel interference from the analog terrestrial television broadcasting system.

During the switch-off period, it is usually required that both analog and digital TVs coexist well for a certain period of time. This is because there is very little or even no additional spectrum resource that can be allocated for DTV services before the analog TV service is shut down. The DTV services can be deployed in channels only if they cause very limited interference to existing analog TV services. During the transition from analog to digital TV, one of the most important issues to be addressed is the interference of DTV services on existing analog TV services.

The test is used to measure the protection rate C/I (i.e., the power ratio between the carrier and interference from either cochannel or adjacent-channel analog TV services), which reflects the capability of the DTTB system against interference from the cochannel and either upper or lower adjacent-channel analog TV systems under Gaussian channel conditions. This C/I is the power ratio of the useful signal to the interference signal. The smaller the measured value, the better system performance in the presence of cochannel and adjacent-channel interference of analog TV services will be.

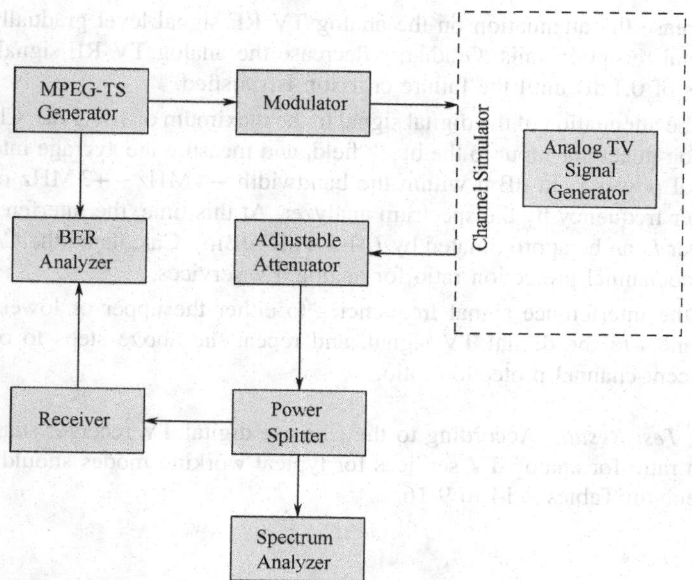

FIGURE 9-8 Experimental setup for protection ratio test for analog TV services.

The objective failure criterion for the fixed reception is selected for the test. The interference signal level at the receiver input is measured without any additional AWGN. The measurement bandwidth is 7.56 MHz and signal level is −60 dBm.

It is recommended that the analog TV signal uses the color bar signal of 75% at the intermediate-frequency modulator. When measuring the power of the analog RF signal, the input signal of the intermediate-frequency modulator is a black field signal.

Test Description

- The experimental setup for the protection ratio test is shown in Figure 9-8.
- Transmit the DTV signal exactly on the same channel of the existing analog TV service, the DTV signal is serving as the interference.
- The "multipath simulator" in the block diagram of the test platform is set to the cochannel analog TV signal (fixed 75% of the color bar signal), and the input video signal of the analog generator is the black field signal when measuring the analog RF signal power.
- Set the attenuation of the RF output of the analog TV signal generator to the maximum of 127.9 dB.
- The attenuation of the signal attenuator is adjusted so that the average digital signal power within the bandwidth ±3.78 MHz of the RF center frequency measured by the spectrum analyzer is around −60 dBm and the number of tests for the average is 50; record this as C in dBm.

- Decrease the attenuation on the analog TV RF signal level gradually so that normal reception fails. Gradually decrease the analog TV RF signal level in steps of 0.1 dB until the failure criterion is satisfied.
- Set the attenuation of the digital signal to the maximum of 127.9 dB. Change the analog generator input to the black field, and measure the average interference signal power I_1 in dBm within the bandwidth −4 MHz ~ +3 MHz of the RF center frequency by the spectrum analyzer. At this time, the interference peak power I can be approximated by $I_1 + 2.4$ dB (dBm). Calculated the C/I (dB) as the cochannel protection ratio for analog TV services.
- Set the interference signal frequencies to either the upper or lower adjacent channels to the digital TV signal, and repeat the above steps to obtain the adjacent-channel protection ratio.

Reference Test Results According to the Chinese digital TV receiver standard, the protection ratio for analog TV services for typical working modes should meet the requirements in Tables 9-14 to 9-16.

TABLE 9-14 Protection Ratio Requirements for Cochannel Analog TV Interference

Working Mode	Carrier Mode	FEC Code Rate	Constellation	Frame Header	Interleaving Depth	Payload Bit Rate (Mbps)	Protection Ratio *C/I* (dB)
1	$C = 3780$	0.4	16QAM	PN = 945	720	9.626	−5
2	$C = 1$	0.8	4QAM	PN = 595	720	10.396	−7
3	$C = 3780$	0.6	16QAM	PN = 945	720	14.438	−3
4	$C = 1$	0.8	16QAM	PN = 595	720	20.791	2
5	$C = 3780$	0.8	16QAM	PN = 420	720	21.658	4
6	$C = 3780$	0.6	64QAM	PN = 420	720	24.365	3
7	$C = 1$	0.8	32QAM	PN = 595	720	25.989	5

TABLE 9-15 Protection Ratio Requirements for Lower Adjacent Channel Analog TV Interference

Working Mode	Carrier Mode	FEC Code Rate	Constellation	Frame Header	Interleaving Depth	Payload Bit Rate (Mbps)	Protection Ratio *C/I* (dB)
1	$C = 3780$	0.4	16QAM	PN = 945	720	9.626	−46
2	$C = 1$	0.8	4QAM	PN = 595	720	10.396	−44
3	$C = 3780$	0.6	16QAM	PN = 945	720	14.438	−45
4	$C = 1$	0.8	16QAM	PN = 595	720	20.791	−41
5	$C = 3780$	0.8	16QAM	PN = 420	720	21.658	−42
6	$C = 3780$	0.6	64QAM	PN = 420	720	24.365	−41
7	$C = 1$	0.8	32QAM	PN = 595	720	25.989	−40

TABLE 9-16 Protection Ratio Requirements for Upper Adjacent Channel Analog TV Interference

Working Mode	Carrier Mode	FEC Code Rate	Constellation	Frame Header	Interleaving Depth	Payload Bit Rate (Mbps)	Protection Ratio *C/I* (dB)
1	$C = 3780$	0.4	16QAM	PN = 945	720	9.626	−46
2	$C = 1$	0.8	4QAM	PN = 595	720	10.396	−44
3	$C = 3780$	0.6	16QAM	PN = 945	720	14.438	−45
4	$C = 1$	0.8	16QAM	PN = 595	720	20.791	−41
5	$C = 3780$	0.8	16QAM	PN = 420	720	21.658	−42
6	$C = 3780$	0.6	64QAM	PN = 420	720	24.365	−41
7	$C = 1$	0.8	32QAM	PN = 595	720	25.989	−40

9.3.13 C/I Measurement with Cochannel and Adjacent-Channel DTV Signal Interference

Test Purpose This test evaluates the performance of the DTTB system against cochannel and adjacent-channel interference from DTV services.

Due to the limited frequency resources, adjacent channels have to be used for DTV service in many countries. This test measures the protection ratio C/I (i.e., power ratio between the carrier and interference from either cochannel or adjacent-channel DTV services) of the DTTB system, which reflects the capability of the DTTB system against the interference from the cochannel and either upper or lower adjacent-channel DTV service under Gaussian channel conditions. The smaller the measured value, the better system performance in the presence of cochannel and adjacent-channel interference of DTV services will be.

The objective failure criterion for the fixed reception is selected for the test. The interference signal level at the receiver input is measured without any additional AWGN. The measurement bandwidth is 7.56 MHz and signal level is −60 dBm.

Test Description

- The experimental setup for the protection ratio test is shown in Figure 9-9.
- Transmit the DTV signal exactly on the same channel of another DTV service, the first DTV signal is serving as the interference.
- Set the "multipath simulator" in the block diagram of the test platform to the cochannel DTV signal and turn off the interference DTV signal at this moment.
- Adjust the attenuation of signal attenuator so that the average power within the bandwidth ±3.78 MHz of the RF center frequency measured by the spectrum analyzer is around −60 dBm and the number of tests for the average is 50; record this as C in dBm.
- Turn on the RF output of the DTV signal serving as the interference and adjust its RF signal level so that normal reception fails. Gradually decrease the

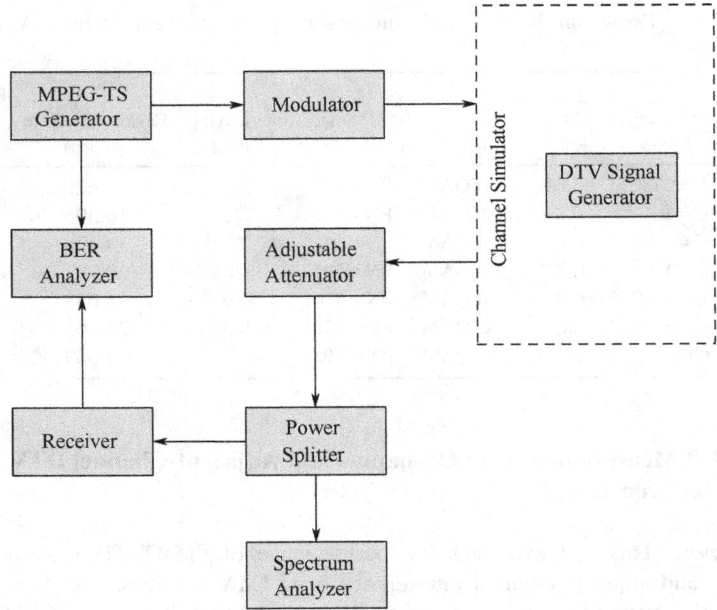

FIGURE 9-9 Experimental setup for protection ratio test for DTV services.

interfering DTV signal level in steps of 0.1 dB until the failure criterion is satisfied.

- Set the attenuation of the digital signal to the maximum of 127.9 dB. Measure the average interference signal power I in dBm within the bandwidth ±3.78 MHz of the RF center frequency by the spectrum analyzer. Calculated the C/I in dB as the cochannel protection ratio for digital services.
- Set the interference signal frequencies to either the upper or lower adjacent channels to the digital TV signal and repeat the above steps to obtain the adjacent-channel protection ratio.

Reference Test Results According to the Chinese digital TV receiver standard, the protection ratio for DTV services for typical working modes should meet the requirements in Tables 9-17 and 9-18.

9.3.14 C/I Measurement with Single-Tone Interference

Test Purpose This test evaluates the performance of the DTTB system against single-tone interference.

In the wireless environment, there are usually a variety of narrowband interferences from different systems, such as mobile telecommunications, FM broadcasting,

TABLE 9-17 Protection Ratio Requirements for Cochannel DTV Interference

Working Mode	Carrier Mode	FEC Code Rate	Constellation	Frame Header	Interleaving Depth	Payload Bit Rate (Mbps)	Protection ratio C/I (dB)
1	$C=3780$	0.4	16QAM	PN=945	720	9.626	8.5
2	$C=1$	0.8	4QAM	PN=595	720	10.396	7.0
3	$C=3780$	0.6	16QAM	PN=945	720	14.438	11.0
4	$C=1$	0.8	16QAM	PN=595	720	20.791	13.5
5	$C=3780$	0.8	16QAM	PN=420	720	21.658	13.5
6	$C=3780$	0.6	64QAM	PN=420	720	24.365	16.0
7	$C=1$	0.8	32QAM	PN=595	720	25.989	17.0

TABLE 9-18 Protection Ratio Requirements for Adjacent-Channel (Upper or Lower) DTV Interference

Working Mode	Carrier Mode	FEC Code Rate	Constellation	Frame Header	Interleaving Depth	Payload Bit Rate (Mbps)	Protection Ratio C/I (dB)
1	$C=3780$	0.4	16QAM	PN=945	720	9.626	−40
2	$C=1$	0.8	4QAM	PN=595	720	10.396	−41
3	$C=3780$	0.6	16QAM	PN=945	720	14.438	−38
4	$C=1$	0.8	16QAM	PN=595	720	20.791	−36
5	$C=3780$	0.8	16QAM	PN=420	720	21.658	−36
6	$C=3780$	0.6	64QAM	PN=420	720	24.365	−35
7	$C=1$	0.8	32QAM	PN=595	720	25.989	−34

and walkie-talkies. These interferences are often characterized by large power with very narrow bandwidth.

The test measures the capability of the digital terrestrial television transmission system versus single-tone interference under Gaussian channel conditions. The C/I in this case is the power ratio of the useful signal to the interference signal. The smaller the measured value, the better the system performance will be in the presence of single-tone interference.

The objective failure criterion for the fixed reception is selected for the test. The interference signal level at the receiver input is measured without any additional AWGN. The measurement bandwidth is 7.56 MHz and signal level is −53 dBm.

Test Description

- The experimental setup for the C/I with the single-tone interference test is shown in Figure 9-10.

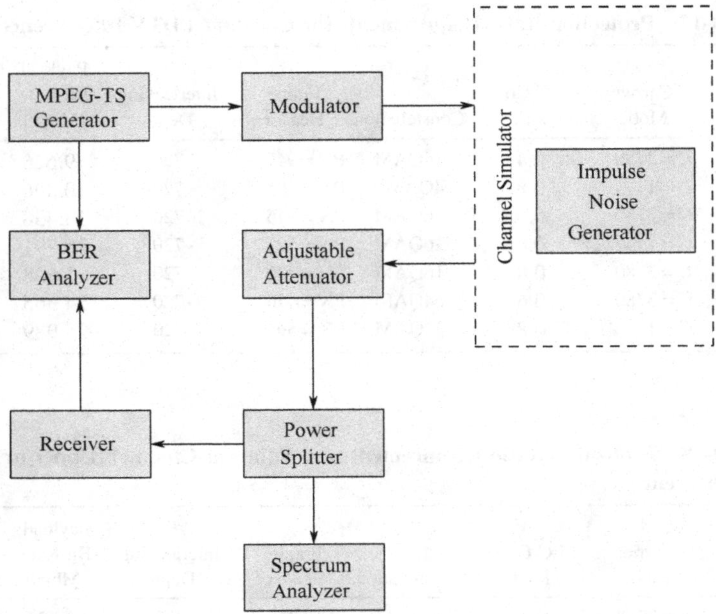

FIGURE 9-10 Experimental setup for protection ratio test under single-tone interference.

- Set the interference source of the "multipath channel simulator" in the block diagram of the test platform to an additive single-frequency signal, and then set the attenuation of the interference signal to the maximum of 127.9 dB.
- The attenuation of the signal attenuator is adjusted so that the average power within the bandwidth ±3.78 MHz of the RF center frequency measured by the spectrum analyzer is around −53 dBm and the number of tests for the average is 50; record this as C in dBm.
- The frequency offset of the interference signal to the channel center frequency is set to 0 MHz.
- Turn on the RF output of the interference signal and adjust its signal level so that normal reception fails. Gradually decrease the interfering DTV signal level in steps of 0.1 dB until the failure criterion is satisfied.
- Set the attenuation of the useful signal to the maximum of 127.9 dB and measure the average interference power I in dBm within the bandwidth ±3.78 MHz of the RF center frequency by the spectrum analyzer. Calculated the C/I in dB as the protection ratio.
- Adjust the frequency offset of the single-tone interference signal from the channel center frequency by ±1 MHz, ±2 MHz, ±3.5 MHz, and ±5 MHz. Repeat the above steps and get the protection ratio.

Reference Test Results There is no explicit requirement on anti-single-tone interference performance in the Chinese digital TV receiver standard.

9.3.15 Antiphase Noise Measurement

Test Purpose This test evaluates the performance of the DTTB system against the phase noise interference. The lower the measured value, the better antiphase noise capability the system has.

The received DTV RF signal is in either the VHF or UHF band and must be down converted by a local oscillator at the receiver to the low-IF band and baseband so that the channel demodulator chip can handle it. The phase noise of this local oscillator will introduce performance loss such as intersymbol interference and C/N degradation.

The objective failure criterion for the fixed reception is selected for the test. The interference signal level at the receiver input is measured without any additional AWGN. The measurement bandwidth is 7.56 MHz and signal level is −53 dBm.

Test Description

- The experimental setup for the antiphase noise capability test is shown in Figure 9-11.
- Connect the local oscillator input of the modulator with an external phase noise generator and set the output power and the frequency of the signal.

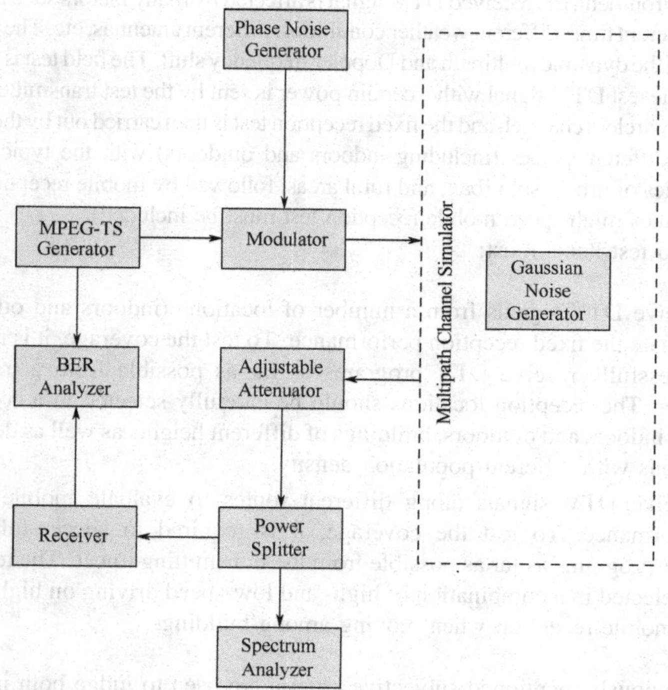

FIGURE 9-11 Experimental setup for C/N test in presence of phase noise.

- Connect the signal generator to the spectrum analyzer so that the phase noise to the local oscillator to be measured by the spectrum analyzer meets the specified requirements.
- The attenuation of the signal attenuator is adjusted so that the average power within the bandwidth ±3.78 MHz of the RF center frequency measured by the spectrum analyzer is around −53 dBm and the number of tests for the average is 50; record this as C in dBm.
- Gradually decrease the interfering DTV signal level in steps of 0.1 dB until the failure criterion is satisfied. Set the signal attenuation to the maximum and measure the average interference signal power N in dBm within the bandwidth ±3.78 MHz of the RF center frequency by the spectrum analyzer. The number of tests for the average is 50. Calculated the C/N in dB in the presence of phase noise.

Reference Test Results There is no explicit requirement on antiphase noise interference performance in the Chinese digital TV receiver standard.

9.4 FIELD TEST PLAN

9.4.1 Field Test

In a real environment, the received DTV signal is affected by many factors such as different terrains, different time, different weather conditions, different antennas, etc. The signal also suffers from the dynamic multipath and Doppler frequency shift. The field test is performed this way: The test DTV signal with a certain power is sent by the test transmitter, it passes through the wireless channel, and the fixed reception test is then carried out by the reference receiver at different places (including indoors and outdoors) with the typical channel characteristics of urban, suburban, and rural areas, followed by mobile reception tests on different routes (high-speed mobile reception test must be included).

The field test items must:

1. Receive DTV signals from a number of locations (indoors and outdoors) to evaluate the fixed reception performance. To test the coverage, it is required to successfully receive DTV programs as far as possible from a transmitting tower. The reception locations should be carefully selected in a combination with indoors and outdoors, buildings of different heights as well as density, and regions with different population density.
2. Receive DTV signals along different routes to evaluate mobile reception performance. To test the coverage, it is required to successfully receive DTV programs as far as possible from the transmitting tower. The tests should be selected in a combination of high- and low-speed driving on highways, and for mobile reception when moving among buildings.

The previously-mentioned subjective criteria are used to judge both images and sound quality for the field test, while the field strength and spectrum of the received

DTV signal should be recorded. The margin at the receiver needs to be checked for the fixed-point test while information such as location and speed (or even error seconds if appropriate) needs to be recorded for the mobile test.

The field tests are specifically defined in ITU BT.2035.

9.4.2 Objectives of Field Test

The main objectives of the field test include:

1. To confirm the impact of environmental changes on DTV system performance
2. To measure the service area and coverage
3. To collect the data for further improvement of DTV system performance
4. To evaluate the performance of DTV system in different working modes
5. To evaluate the performance of different DTV systems

Based on test results from different organizations and in different areas as well as time, the DTTB system performance under different conditions can be well understood. The benefits of conducting field tests include but are not limited to the following:

1. Performance comparison among different DTV transmission systems
2. Performance comparison between digital and analog TV transmission systems
3. Comparison of various transmitters as well as among receivers
4. Performance comparison of receivers of different generations or different manufacturers
5. Performance comparison of receivers under different environments
6. Service area and coverage measurement
7. Data collection for future improvement of DTV systems
8. Performance of DTV systems with different working modes

9.4.2.1 Coverage Test This measures the actual field strength from a transmitter at different locations. In general, there are two main purposes for the coverage test:

1. Confirm the transmitting antenna is functioning properly
2. Provide reference data for frequency planning and coverage prediction within the network

The coverage test is carried out by placing a standard dipole receiving antenna at a height of 9.1 m from the ground. This method is widely used for coverage measurement, the radiation pattern of the transmitting antenna, and other important data for frequency planning. In practice, the measurement can also be made using the Yagi receiving antenna with a correction factor. The coverage test can be carried out in different ways, such as along different angles from the transmitter and in grids and/or sectors. The huge amount data from this test are very helpful to analyze and understand the coverage situation of this transmitter. If the test is on a daily basis, whether the transmitting system normally operates or whether the signal inside the coverage area is interfered can also be monitored.

9.4.2.2 Service Test The service test is used to confirm whether the DTV program can be received successfully. Unlike the coverage test, the service test is more concentrated on whether the signal can be received normally at the test locations. If possible, the measurement of the C/N, link margin, and BER is required. Due to the existence of the various external interferences, the service test is much more complicated than the coverage test and may not necessarily be repeatable. In general, the coverage test reflects the performance of the transmitter, antenna, cable connector, and other equipment at the transmit side, while the service test is more concerned with the reception performance and the link margin. Due to the adoption of SFNs, multipath effect, interferences, etc., there always exists the case that the receiver cannot receive the signal successfully while the signal field strength is good from the coverage test. The service test is, therefore, more directly related to end users. Most of the field tests for the DTV system are the service tests.

9.4.2.3 Channel Parameter Recording As the RF DTV signal from the transmitter has to pass through the wireless channel, the channel parameters used to characterize the channel are very important, including the field strength variation, impulse noise, in-band interference, and multipath, interference. Measuring channel parameters usually takes a lot of time (not only the data recording time but also the time to extract the channel parameters because not all the parameters can be taken directly from the measurement data) as well as storage space, but it plays a fairly important role in the analysis of nonsatisfactory receiving performance. If possible, the data reflecting the channel parameters should be recorded as much as possible, which will help to further optimize and improve the system.

9.4.2.4 Working Modes Reception for the DTTB system includes the five receiving modes defined in ITU-R BT 2035. When conducting the test, one of these five modes is usually selected (see Table 9-19).

The five receiving modes are defined as follows:

1. *Fixed Reception.* It is generally defined as reception by a location-fixed antenna or receiver. The antenna may be either a roof-mounted antenna or a fixed indoor antenna.

TABLE 9-19 Receiving Modes

Moving mode	Outdoor	Indoor
Fixed	Fixed	Fixed
Low-speed moving	Pedestrian	Portable
High-speed moving	Mobile	Hand held

2. *Pedestrian Reception.* It is defined as the reception by a receiver moving at very low speed of no more than 5 km/h (i.e., walking speed). This reflects the reception scenario that the receiver is moving at the walking speed.
3. *Mobile Reception.* It is defined as the reception by a receiver moving at high speed, faster than 5 km/h. This reflects the reception scenario that the receiver is inside a vehicle moving much faster than the walking speed.
4. *Portable Reception.* It is defined as reception by a device with built-in receiving module of the DTV signal that remains stationary during the reception.
5. *Hand-Held Reception.* It is defined as the reception by a built-in, low-gain antenna of the hand-held device that moves during the reception. This represents the worst-case reception scenario.

9.4.3 Testing Signal

9.4.3.1 In-Service Program Signal This test will use a program signal broadcast by a transmitting station directly or a repetitive video sequence for picture quality evaluation. If the test video sequence is used, one should be very careful to ensure the seamless program switching. Using the real-time program source is also a good option. The test person can determine the service quality by observing the deterioration of either image or sound or the mosaic in the picture in the field test. This test can only be made by using the decision threshold (TOV), and the test results under the same conditions might be different because of the human factor of the testing personnel. However, the test is very simple, it is easy to conduct, and there is no need to get authorization. Most importantly, it can be carried out without affecting the existing broadcasting services.

9.4.3.2 Out-of-Service Program Signal The nonservice signal is used for the objective test. The data stream used in the test is the PN sequence with length of $2^{23} - 1$ as the DTTB modulator input. The exact BER performance can be obtained by measuring the BER at the receiver. Using the PN sequence can provide BER information, but it is generally not used as the test method for TV stations because this test requires the use of specialized test instruments and normal program broadcasting has to be interrupted.

9.4.4 Antenna

9.4.4.1 Antenna for Coverage Test The receiving antenna for the coverage test must be calibrated with respect to a dipole antenna at the height of 9.1 m from the ground before the test starts, and this antenna should be clearly described in the test report. In general, the receiving antenna should point toward the direction of the transmitting tower with the same polarization. However, during the service test, the best receiving direction may not be directly toward the transmitting tower; therefore, the antenna direction should also be recorded in the test results.

9.4.4.2 Service Test Antenna The receiving antennas for the service test may be either professional (high-end) or consumer products depending on the test purpose. The antenna for the service test should be easy to install and move and can be pointed with either maximal signal strength or optimal reception angle toward to the transmitter tower. Receiving antennas for the service test generally include the following:

1. The receiving antenna for the fixed outdoor test should be installed at the height of 9.1 m above the ground. If the orientation is nonoptimal, it must be documented in the test report.
2. The fixed-located, indoor receiving antenna is usually the consumer product. It should be mounted about 1.5 m above the floor and calibrated for gain and radiation pattern with respect to a dipole. If the orientation is nonoptimal, it must be documented in the test report. It should also be noted that the performance and characteristics of the antenna may significantly change in the indoor environment. It is therefore required that both the setup and personnel position remain unchanged during measurement to ensure a controlled test environment.
3. Antennas for portable reception are also consumer products that could be designed as omnidirectional or directional. They must be calibrated for gain and radiation pattern with respect to a dipole around 1 m above the floor (ground). If the orientation is nonoptimal, it must be documented in the test report.
4. The antenna for pedestrian reception can be considered as nonlinear and randomly low gain or even no gain. If possible, the antenna should be mounted around 1 m above the floor and calibrated for gain and radiation pattern with respect to a dipole. If the orientation is nonoptimal, it must be documented in the test report.
5. The antenna for the mobile reception is generally considered to be omnidirectional and mounted at a fixed position on the vehicle to maximize the received signal level. If possible, the antenna should be calibrated for gain and radiation pattern with respect to a dipole. The orientation of the antenna for the mobile reception is not considered.
6. Similar to pedestrian reception, the antenna for the hand-held device can be considered nonlinear and randomly with low gain or even no gain. If possible, the antenna should be mounted about 1 m above the floor and calibrated for gain and radiation pattern with respect to a dipole. The gain and orientation of the receiving antenna of the hand-held device are generally considered nonoptimal.

9.4.5 Measurement Time

The measurement time is determined by the reception scenario and test purpose. It has a wide range of duration, including seasonal (months or even years) time, very long time (days or a month), long time (minutes or hours), short time (seconds to minutes), and very short time (seconds to less than a second).

9.4.5.1 Coverage Test Coverage measurements are normally conducted for short durations. Fixed-position coverage measurements over long periods of time (hours, days, months, and years) provide useful information about the effect of weather, seasons, and day–night variations.

9.4.5.2 Service Test For the service test, the measurement time is generally 5 min, and the test for both multipath and signal strength should be completed during that time period.

9.4.6 Channel Characteristic Recoding

Channel characteristic measurements are conducted for short durations (with minimum duration of no less than 20 s) due to the storage capacity.

9.4.7 Test Location

The test locations must be well determined before the open-air test. The document should be prepared in advance, which includes accurate coordination and locations of test points, surrounding buildings, vegetation, and climate conditions. This will allow the testing personnel to find the test points easily and accurately.

9.4.8 Test Calibration

Before and after the field test, the transmitting and receiving systems should be calibrated to make sure that they are functioning properly. Before the test, transmitter power must be calibrated, and if possible, the transmitter signal should be monitored continuously during the test. If the significant signal change is found in the test, stop the test immediately and check the transmitting system.

9.4.9 Records and Documents

Results should be recorded and documented in a way such that the data can be efficiently processed and analyzed later. When preparing the test record form, one should consider how these data should be organized, how these data would be used, and which should be analyzed later. For further comparison between current test results and other test results, one also needs to consider the way other tests were performed and how the data were analyzed. During a pass or fail test, a reception can be considered successful (pass) only when the normal reception lasts for a certain time duration, i.e., at least 5 min. No matter whether the system passes the test, all related information and recorded data should be well preserved for future analysis or comparison.

According to the requirements, all test record forms should be prepared and attached as an appendix to the final report.

9.4.10 Test: Instruments and Auxiliary Equipment

The purpose and contents of the tests are different, and the test procedure, test setup, and auxiliary equipment required for the open-air test are listed below:

1. A functional block diagram giving the flow chart of how the test is conducted must be provided with the test report.
2. The dynamic range of the signal level and NF of the test instrument should be recorded and described in the report.
3. The antenna used for the test must be checked and calibrated prior to use.
4. Components or devices such as the cable, amplifier, filter, attenuator, power splitter, switch, and other devices that could affect the accuracy of the signal measurement must be calibrated individually. Special care must be taken of the reflection coefficients of these devices and the impedance match between the amplifier and the attenuator.
5. As the receiver' performance will have significant impact on the measurement results, the receiver used for the service tests must be described in detail and the calibration information should be documented.
6. Auxiliary equipment or devices used in the test should be listed in the test report and the calibration information should be documented.

Figure 9-12 gives a very simplified, high-level block diagram of the field test (indoors or outdoors). All the test equipment in Figure 9-12 can be conveniently installed inside a testing vehicle, typically mounted in a chassis, and the vehicle has a retractable antenna on top that can be raised up to 10 m. Tests may also be carried out with an omnidirectional or low-gain antenna at 1.5 m height if it is to evaluate the system performance under the portable or pedestrian reception scenarios. Equipment

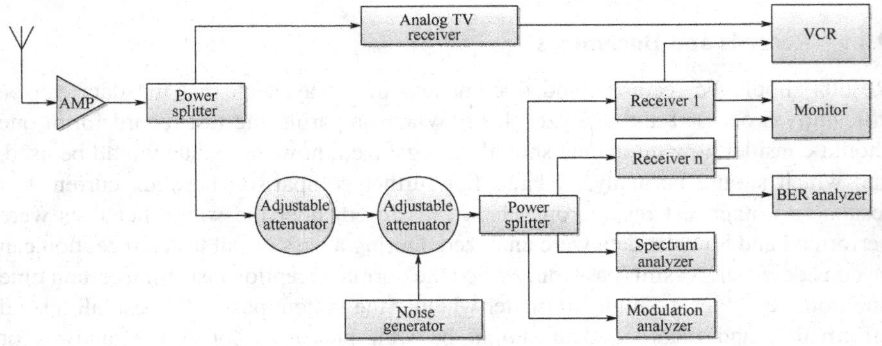

FIGURE 9-12 Block diagram for field test.

is selected and assembled according to the test plan and objectives. The low-power consumption and small-size equipment should be considered, and equipment that can be operated by a built-in battery is highly preferred.

9.4.11 Coverage Test Procedure

The methods and procedures for the coverage test are described below. The coverage test is based on field strength instead of BER measurements.

9.4.11.1 Measurement Methods The coverage test measures the field strength of the DTV signal. The common approach is to make the measurement on planned locations and, if necessary, conduct further measurements using an additional sector or so-called 30 M as defined in ITU-R BT 2035.

The *cluster* includes the initial test point and at least four additional test points a certain distance away from the initial test point, as shown in Figure 9-13. The initial test point in Figure 9-13 is taken as the central point.

In general, cluster measurement requires a minimum of five evenly distributed measurement points to capture the complete data within an area. If multiple frequencies are to be measured at one location, the cluster measurement area should be defined as 9 m^2 (3 m $\times$ 3 m). Suggested location patterns for the test are shown in Figure 9-13. This type of measurement should be applied when detailed investigation at the selected location is needed.

ITU-R BT 2035 introduces the 30 M run testing method if the cluster test is difficult to perform. According to the standard [9]: "If overhead obstacles preclude a cluster measurement, a 30 M run may be made in lieu of the mobile run. The run is characterized by positioning the antenna at height of 9.1 m (30 ft) above ground level

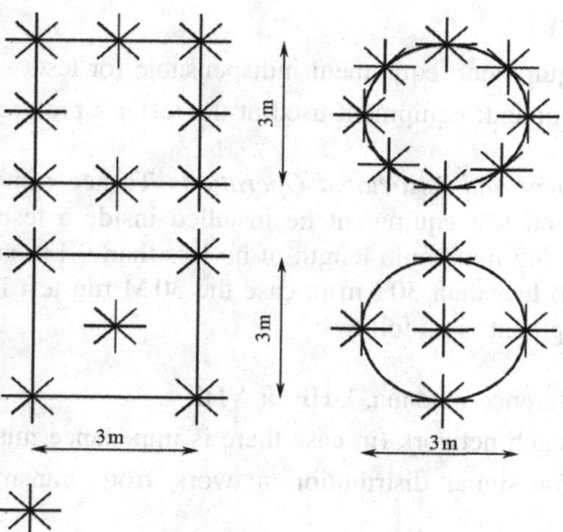

FIGURE 9-13 Illustration of cluster measurement method. (Figure 3 from ITU-R BT. 2035.)

(AGL) and displacing the vehicle back and forth on a straight line of 30.5 m for each side (total of 61 m). The average field strength as well as the field value of a minimum of five fixed points within 61 m of the centre point of the mobile run values shall be recorded. Continuous data collection over the length of the run is desirable."

An additional note in the same document states, "clusters and 30 M runs are of interest at measurement points where the orientation of the receiving antenna yielding the strongest signal differs from that of a direct bearing to the transmitter. Under this scenario, field strength shall be read and recorded with the antenna oriented toward the transmitter and with the antenna oriented toward the strongest signal."

Antenna Height The DTTB system coverage test should be conducted at an antenna height of 9.1 m above ground level.

Safety Since the measurement platform, antenna, mast, and cable feeder may represent potential safety hazards from electrical shock and/or falling objects, all measurement sites must be away from overhead power cables, steeply sloped terrain, wet surface, high wind, thunderstorms, and other natural/man-made obstructions or conditions that threaten the safety of testing personnel or property. The test plan should require that operators be trained in proper safety procedures to ensure worker safety.

Measurement Time When the coverage measurements are planned, the test should be taken when weather conditions match those of the anticipated service and environment. It is highly recommended that the coverage test not be done under the most severe weather conditions, such as thunderstorms and blizzards.

9.4.11.2 Field Test Equipment The field test equipment can be divided into two categories:

- Mandatory equipment: equipment indispensable for tests
- Optional equipment: equipment used at the tester's preference

Mandatory Equipment and Associated Operations To accommodate the test, it is recommended that all test equipment be installed inside a testing vehicle with a retractable mast having maximum length of no less than 9.1 m and the feeder cable length should be no less than 30.5 m in case the 30 M run test is needed.

Mandatory equipment is as follows:

- Calibrated reference antenna, UHF or VHF
- Impedance match network (in case there is impedance mismatch)
- Calibration for signal distribution network from transmit antenna to test terminal
- Calibration of measurement equipments
- GPS receiver

- Spectrum analyzer, which should have functions such as storage, display, signal power, and RMS power monitoring
- Digital TV receiver

Optional Equipment Optional equipment and associated operations are as follows:

- BER tester
- Noise generator
- Other analytical devices such as that for MER measurement
- Data acquisition and storage device—signal recorder
- Digital camera
- Vertical alignment adjustment equipment
- Antenna angle measurement device
- Barometer
- Measurement equipment for received signal margin

9.4.11.3 Measurement Data All the necessary information/data should be recorded during the testing process, as given below.

Mandatory Data The following data must be recorded in the test:

- Field strength (minimum, maximum, and average value) (dBμV/m)
- System margin—input signal attenuated in a controlled manner until the decision threshold (TOV) is reached
- Distance and orientation to transmitting antenna
- Ground elevation at test sites
- Time of day, date, weather, terrain, and traffic conditions
- Azimuth orientation of receiving antenna for best reception and for maximum field strength with vertical angle of mast/antenna support structure
- Detailed description of test-related equipment and system, including manufacturer, serial number, most recent calibration date, etc.
- Block diagram of coverage test
- Date, time, environment, and test results

Optional Data

- *C/N at TOV Threshold for Best Reception Angle and Maximum Field Strength.* In the field test, the TOV threshold is reached when the well-trained observer finds one image glitch in 2 min if video streams are used. This threshold can be reached when the BER is 3×10^{-6} in 1 min if the PN sequence is used as the test signal.

- *Spectrum Measured at Different Antenna Angles and Distances.* The spectrum records should include the narrowband display range (i.e., 7–9 MHz) as well as the wideband display range (i.e., 20 MHz) and the RBW (resolution bandwidth) is in the automatic mode.

Selection of Test Sites To accurately assess the DTTB system performance, there must be enough data sampling points. Generally, 30–100 sites are required for the coverage test. More points help obtain higher statistic confidence in the test results. Generally, the fixed outdoor measurement needs 100 sites while the test for other service scenarios requires a minimum of 20 sites. The test points can be selected by any of the following methods:

- Measurement points can be selected 3 km away from the transmitter location by following certain rules on distance and orientation. This means the measurement points will possibly be on each radiation path starting from the transmitter location. The angle between two adjacent paths should not be more than 20°.
- The measurement points can be selected based on population density and requirements for the service coverage.
- The measurement points can be selected based on existing analog TV reception coverage performance.

9.4.12 Service Test Procedures

Unlike the coverage test, measurements of C/N, receiving margin, multi-path, etc., are required for the service test. In general, the number of points of the service test is less than that of the coverage test.

9.4.12.1 Measurement Methodology The basic methodologies of the service test are given below.

Test Time The test time must be selected based on the environment and application scenario. The 5-min duration is usually selected as the test period in the field test.

Different testing duration may be selected depending on requirements. For example, 1 min is used to observe the effect of airplane flutter, 20 min will be appropriate to observe the trees (leaves) moving in the wind, and 10 min is good to observe the effect of traffic variations. If unordinary conditions are found during the test, the test duration should be extended and those phenomena recorded.

Antenna The antenna should be selected and used according to the receiving modes. Generally, the selected antenna should be able to represent users' typical receiving antennas, with the same polarization as the transmitting antenna.

Functionality Check It is preferable to perform the field test with the same receivers used in the laboratory tests. This provides an accurate assessment of the C/N value at the TOV threshold in both dynamic and static multipath channels.

Test Methodology Description The service tests are typically performed in a way that they can simulate the practical receiving situations. The tests for different receiving modes are described below:

Fixed Reception. Includes both indoor and outdoor tests. The outdoor test follows exactly the same procedure as coverage tests, but neither clusters nor the 30 M run is mandatory. In this test, the antenna orientation needed to successfully demodulate the received DTV signals must be recorded. It is appropriate to measure the full azimuth range of antenna orientations that result in successful operation of the receiver. A minimum of 100 sites is generally expected for the outdoor fixed reception tests.

Indoor fixed reception should be tested at a minimum of 20% of the receiving sites with high signal strength and good outdoor reception performance. During the measurement, the exact locations of existing analog reception with the antenna at 1.5 m high should be used. Tests should simulate the typical receiving conditions, including the controlled movement of persons nearby and the operation of home appliances such as a blender. Test site conditions must be accurately recorded.

Portable Reception. The same sites tested for fixed indoor reception should generally be used for portable reception. The site description and antenna-pointing direction should be recorded. Tests should mimic the typical receiving conditions, including the controlled movement of persons nearby and operation of home appliances such as a blender. Test site conditions must be accurately recorded.

Pedestrian Reception. The surrounding areas of the site used for indoor reception can generally be used for the pedestrian reception with a minimum of 20 sites. The receivers should be put at positions which can mimic real-world receiving situations.

Mobile Reception. The total path length for the mobile reception test should not be less than 10 km. For each segment of the route with a typical length of 1 km, the multipath channel, analog interference, traffic conditions, and other obstructions should be recorded. It is desirable to perform the channel characteristic measurement on each selected segment of the route. Testing should also include the signal reacquisition in the selected route when the receiver is moving.

Hand-held Reception. The same route used for the mobile reception test should generally be used for the hand-held test reception tests with the minimum route length of 10 km and the same antenna used for the pedestrian reception test.

9.4.12.2 Test Facility The equipment used for the service test is similar to that of the coverage test. However, the equipment for indoor and portable reception tests should be put into the user's house based on the test requirement.

9.4.12.3 Measurement Data There could be more than one set of measurement data obtained from the service test. Some of them can be considered as mandatory or minimum set(s) while the rest help enhance or describe the particular reception condition in detail and can be taken and kept as requested.

Mandatory Data The mandatory data include:

- Signal field strength
- Noise floor
- Noise added until TOV decision threshold is reached
- S/N threshold
- Receive margin
- BER
- Multipath characteristics
- Detailed location of the antenna
- Antenna characteristic description
- Antenna orientation
- Calibration of measurement system
- Detailed description of test points
- Time and date of test
- Description of building in which or around which measurements are made
- Nature of area immediately surrounding the antenna

Optional Data Other measurement sets include:

- Site location address
- Subjective audio and/or video impairments
- Log of the test activities

A spectrum analyzer can be used to store the spectrum received by the DTTB receiver for different antenna orientation when applicable. Spectrum records should include a narrowband range display (9 MHz) and a wideband range display (20 MHz).

9.4.12.4 Selection of Service Test Sites The test must collect enough data to accurately assess the service performance of the DTTB system. Generally, 30–100 sites are required and higher statistic confidence in the test results may need significantly more sites. The fixed outdoor measurements typically require 100 sites, while the other tests need a minimum of 20 sites. The sites can be selected by the following methods:

- Measurement points are selected to begin at a distance of 3 km from the transmitter location and are repeated at intervals of 3 km to the maximum distance at which measurements are performed, which is determined by a previous coverage prediction. The angle interval of each radiation path should not be more than 20°.

- The measurement points are selected based on coverage and population density.
- The measurement points are selected based on existing analog TV reception feedback.

If there are other particular factors such as multipath, aircraft flutter, buildings, and trees affecting the service test result, which means that the site selection is biased rather than random, it must be noted in the test report.

It is recommended to write down when and why measurements cannot be taken at a specific site. It is highly desirable to record the information on burst errors or continuous errors over the time during the test.

9.4.13 Measurement Guideline of Field Test for DTTB System

To make performance comparison tests among different DTTB systems, the following should be considered:

- Describe the methodology in detail together with the test results.
- Describe the field tests environment in detail.
- Describe the targeted service (indoor, outdoor, fixed, portable, mobile, etc.).
- Try to simulate the conditions of the targeted service as much as possible.
- Limit the number of variables as much as possible (antenna height, antenna orientation, seasonal effect, operation modes, etc.).
- Describe in detail the receiver type and provide the relevant laboratory test results in the AWGN and multipath channel.
- If the receiver is known using the internal IF, it helps to explain the unexpected results due to the adjacent or taboos channels used at the test locations.
- When unexpected results are found, further investigation is needed to find the possible causes. Using indoor sites as an example, where impulse noise from the household appliances, single-tone interference, and the dynamic multipath from the vehicular traffic or airplanes may come and go without any recognizable pattern.
- Find an appropriate calibration site and perform the daily check for the proper operation of both transmitter and receivers.
- Select testing sites that represent typical users' reception conditions as much as possible.
- Select the testing sites representing typical construction types and locations for the service reception.
- If possible, conduct the performance comparison tests simultaneously in time to minimize channel variations.
- If possible, use exactly the same site and test environment. It is highly preferable to take pictures of the site.
- List all possible limitations on the test results and the test methodology, such as what has been tested and why the testing is not done.
- Figure out the possible cause (s) of the reception failure.

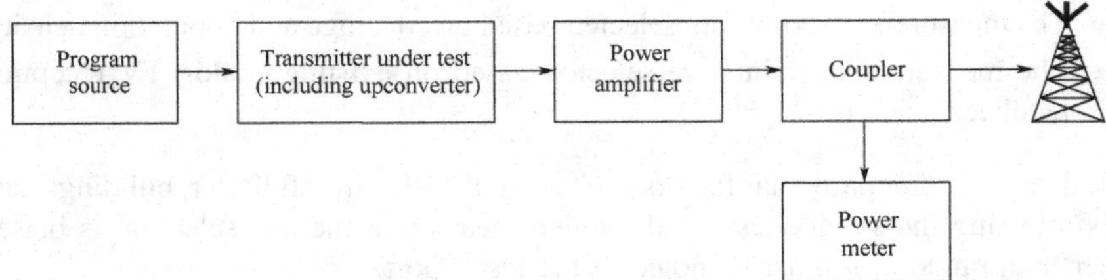

FIGURE 9-14 Transmitting site for field test.

Things that should not be done include:

- Eliminating one test site without explanation
- Changing the test procedure unless it is absolutely necessary
- Selecting the best site in the cluster or 30 M run to perform the service reliability test
- Selecting sites that are in favor of one particular system as compared to others
- Trying to test too many variables at once

9.4.14 Field Test Platform

The transmitting and receiving systems for the field test are shown in Figures 9-14 and 9-15.

9.4.15 Procedure for Fixed Reception Test

Test Purpose This test evaluates the reception performance at fixed locations.

Test Description

- Set up the receiver test platform as shown in Figure 9-15.
- Use the pseudorandom binary sequence as the input.

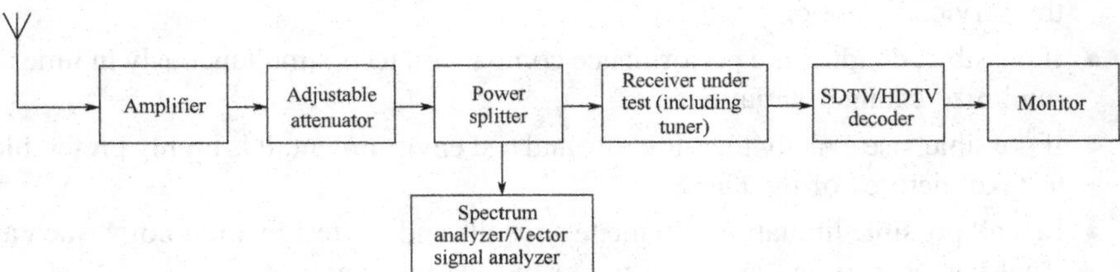

FIGURE 9-15 Constitution of receiving site for field test.

- Raise the outdoor receiving antenna to the height of 9.1 m at the selected site, adjust the antenna elevation angle, and change the antenna orientation toward the direction that maximizes the received signal level.

Test Procedure

1. Use the pseudorandom binary sequence as the input of the modulator.
2. Let the test vehicle stop at the test site, raise the receiving antenna to 9.1 m above the ground level, and try not to touch the tree branches, overhead power cable, and other obstructions.
3. At each site, obtain the latitude and longitude by the GPS receiver, and write down this information together with the altitude, linear distance, and orientation to the transmitting station. Record in detail the test date, time, location, weather, surroundings, etc. Take photos or a video of the testing vehicle and the surrounding environment if possible.
4. Change the orientation of the receiving antenna and fix it to the direction with the maximum receiving signal level. Record the antenna orientation.
5. Change the attenuation of the attenuator in the above figure so that the receiver input signal strength is roughly equal to the output level of the receiving antenna. Measure the signal level.
6. Measure and record the BER. If the BER is worse than the threshold, reduce the attenuation of the signal until it reaches the threshold and record the signal level at this time. If the successful reception still cannot be achieved even when the attenuation is reduced to 0, slowly rotate the antenna one round looking for the possible successful receiving direction. If the BER is better than the threshold, increase the attenuation until it reaches the threshold and record the signal level at this time.

9.4.16 Mobile Test

Test Purpose This test evaluates the mobile reception performance of the DTTB system.

Test Description

- Use a real-time program as the input of the modulator.
- Set up the receiver test platform as shown in Figure 9-15 with zero attenuation.
- In each section of the mobile reception test, continuously perform the subjective evaluation to the received signals to determine whether the reception is successful.
- Record the geographical position as well as the vehicle speed along the test path, subjective test results, received signal strength, and signal spectrum.
- Record the full path with a video camera.

9.5 SUMMARY

Each DTTB system is subject to specific objectives and has its own technical merits to accommodate all the design goals. From comprehensive performance tests using commonly accepted testing methodologies, the different parts of the ecosystem of the whole broadcasting industry will surely benefit: Broadcast service providers can use the information of the system performance under different working parameters to predict the coverage and quality of service while chip designers can use the same information as benchmarks to improve demodulator design.

REFERENCES

1. Advanced Television System Committee A/53, *ATSC Digital Television Standard*, Washington, DC: ATSC, 1995.
2. ETSI.300 744, *Digital Broadcasting Systems for Television, Sound and Data Services, Framing Structure, Channel Coding and Modulation for Digital Terrestrial Television*, Sophia-Antipolis, France: ETSI, 1999.
3. ITU-R WP 11A/59, *Channel Coding, Frame Structure and Modulation Scheme for Terrestrial Integrated Service Digital Broadcasting (ISDB-T)*, Tokyo: ARIB, 1999.
4. S. O'Leary, *Understanding Digital Terrestrial Broadcasting*, Boston and London: Artech House, 2000.
5. Y. Wu, "Performance comparison of ATSC 8-VSB and DVB-T COFDM transmission systems for digital television terrestrial broadcasting," *IEEE Transactions on Consumer Electronics*, vol. 45, no. 3, pp. 916–924, Aug. 1999.
6. N. Pickford,"Laboratory testing of DTTB modulation systems," Laboratory Report 98/01, Australia Department of Communications and Arts, June 1998, Available: http://happy.emu.id.au/lab/rep/rep/9801/9801_001.htm.
7. SET/ABERT, "Digital television systems Brazilian tests final report," 2000, Available: http://www.set.com.br/testing.pdf.
8. Ministerios de Transportes y Comunicaciones (MTC), *"Digital television systems Peru tests final report,"* MTC, Feb. 2009, Available: http://www.mtc.gob.pe/portal/tdt/tdt2.html.
9. ITU BT.2035-2, *Guidelines and Techniques for the Evaluation of DTTB Systems*, Geneva: ITU, Feb. 2003.
10. ITU BT.1368-11, *Planning criteria for digital terrestrial television services in the VHF/UHF bands*, Geneva: ITU, Feb. 2014.
11. Chinese National Standard GB 20600-2006, *Framing Structure, Channel Coding and Modulation for Digital Television Terrestrial Broadcasting System*, Aug. 2006, Available: http://gb123.sac.gov.cn/gb/gbInfo?id=1751.

10

DIGITAL MOBILE MULTIMEDIA BROADCASTING SYSTEMS

10.1 INTRODUCTION

As the switchover from analog to digital TV comes in full swing, the broadcasting industry enters into another golden era. With ever-increasing demand on mobile multimedia services, mobile multimedia broadcasting, also known as "mobile TV" or "mobile broadcasting," has been proposed and successfully deployed. The mobile multimedia services can be delivered in two ways. The first is by the telecommunication network, which can provide users with personalized services based on the existing mobile telecommunication network. This is essentially point-to-point transmission, combing both mobile telecommunication and streaming media technologies. The other is by the broadcasting network to deliver programs to all users' hand-held devices within the coverage area, which is by nature point-to-multipoint transmission. Obviously, mobile TV based on the mobile telecommunication network has the on-demand feature, supporting personalized services with the bidirectional link. The disadvantages of this scheme include higher service cost and possibly low data transmission rate due to network congestion. Mobile TV based on the broadcasting network has just the opposite features: Advantages include low service cost and high transmission data rate as there is no network congestion while it cannot support the interactivity as there is no bidirectional link and therefore has poor support for personalized service. The discussion of mobile multimedia broadcasting technologies and service implementation in this chapter is limited to the broadcasting network based scheme (satellite and terrestrial DTV broadcasting networks, the

Digital Terrestrial Television Broadcasting: Technology and System, First Edition. Edited by Jian Song, Zhixing Yang, and Jun Wang.

chapter focusing on the latter) and can be received by portable or hand-held devices with a small screen (which implies low-definition or at most standard-definition TV programs).

Similar to DTTB systems, digital mobile multimedia broadcasting systems are designed for high-data-throughput (to support multiple low-data-rate services) transmission through wireless channels (VHF/UHF bands in terrestrial, S- or L-band for satellite networks). It is therefore quite natural that it inherits some important features of the existing DTTB system, such as the frame structure, channel coding and modulation techniques, etc. This similarity is good because one can leverage the technologies in the DTTB system to build the mobile TV system or design one system which can support both terrestrial and mobile TV services simultaneously by choosing appropriate system parameters. On the other hand, even though the transmission rate per program has been reduced, the digital mobile multimedia broadcasting transmission system still faces stringent requirements on the network coverage, mobile reception performance, the flexibility of the service management, terminal power consumption, etc., and more cautions should be taken when the SFN is deployed. All these require that the conventional transmission schemes for DTTB systems must be modified (perhaps redesigned) and optimized accordingly.

It is quite easy to realize how important it is to provide mobile TV services to millions of users, how big the whole market will become, and how tough the competition from the telecommunication industry will be. The broadcasting industry, especially those organizations responsible for existing DTTB standards, is now actively exploring the possibility of either making the upgrade and evolution of existing DTTB systems or developing new systems to accommodate mobile TV services using the existing infrastructure. DVB-H (Digital Video Broadcasting Handheld) developed on the basis of DVB-T is considered the first big effort. It was officially approved as the European mobile TV standard by the European Telecommunication Standards Institute (ETSI) as ETSI standard EN 302 304 in November 2004 [6]. In March 2008, DVB-H was officially endorsed by the European Union as the "preferred technology for terrestrial mobile broadcasting." Compared to DVB-T, DVB-H terminals have much lower power consumption and superior mobile reception and anti-interference performance. The standard is applicable in mobile multimedia broadcasting services through the DVB-T network to portable or handheld devices such as mobile phones, PMP (Portable Multimedia Player), and notebooks. Although technically DVB-H and the China Multimedia Mobile Broadcasting (CMMB) are very solid systems, their commercialization has unfortunately not been successful. The lack of spectrum as well as the business model and the tough competition from other systems are widely believed to be the main reasons for their failure. Yet, we believe a brief introduction of these two systems will help readers have a better understanding of all the core technologies as well as major considerations in the mobile TV systems based on the broadcasting network.

The Advanced Television Systems Committee (ATSC) standard was originally designed to target the fixed reception of HDTV programs with the roof-top antenna. Therefore, the receiver power consumption is usually high and mobile reception performance is not good when facing Doppler frequency shift and multipath interference in wireless environments. This means that the ATSC standard needs to be modified to support mobile TV services [5]. The ATSC initiated its study on the mobile multimedia broadcasting system as early as 2007, and ATSC-M/H [7] (an extension to ATSC A/53 [8]) was officially announced as the free-to-air mobile DTV standard for the United States on October 15, 2009. Instead of using a new channel from the original ATSC network, ATSC-M/H provides new mobile TV services with the combination of existing terrestrial ATSC services in the same channel (which is just the opposite to the scenario of DVB-H), which has successfully stimulated great enthusiasm from TV broadcasters in the United States. By early 2010, the ATSC-M/H signal has been broadcasted over 20 large cities in the United States, covering about 30% of the area and about 70% of the population. Meanwhile, 12 television stations in Montreal and 1 station in Mexico began to broadcast the program. ATSC-M/H has been in full swing in North America. To date, almost 300 stations have ATSC M/H broadcasted and nearly 80% population is covered in the United States.

On October 24, 2006, the State Administration of Radio, Film and Television (SARFT) in China decided to choose the CMMB (China Mobile Multimedia Broadcasting) system as the broadcasting industry standard in China [2], which is based on the Satellite and Terrestrial Interactive Multiservice Infrastructure (STiMi). From August 2007 to May 2008, SARFT issued seven parts of its mobile multimedia broadcasting industrial standard, including channel frame structure, channel coding and modulation, multiplexing, e-business guide, emergency broadcasting, datacasting, specifications for conditional access system, and the technical requirement for the receivers. The field trials as well as deployment of the CMMB system were carried out from October 2007 in the six cities which hosted the 2008 Beijing Olympic Games, including Beijing, Shanghai, Guangzhou, and Shenzhen. The broadcasting was formally launched during the Beijing Olympics together with DTMB [1] in August 2008 and CMMB services have gradually expanded nationwide. Currently, the CMMB network has covered more than 300 large cities in China.

This chapter summarizes the features of DVB-H, ATSC-M/H, CMMB, and the new released DVB-NGH [9] standards with the focus on the technical merits of reliable transmission under wireless, mobile environment and the robust reception using the portable or hand-held devices [11].

10.2 DVB-H SYSTEM

The first generation of the DVB-T system was not originally designed for mobile reception. Although it is true that DVB-T can support the mobile reception of lower definition programs under the mobile environment and DVB-T demodulator chips can be built in to some cellular phones, it is not ideal for mobile terminals to receive these digital terrestrial television programs via the DVB-T network due to the relatively large receiving power consumption. The DVB organization has been studying the technical solutions of mobile reception for not only DTV programs but also a variety of other services. This effort was initiated in the hope of coming up with a solution to ensure the reliable and robust mobile reception of hand-held devices using the DVB-T-based standard with SFN. In 2002, the main requirements of the system were listed: broadcast services for portable and mobile devices with "acceptable quality"; a typical user environment, including geographical coverage and mobile radio; access to service in the vehicle at high speed (also the imperceptible handover when crossing cells); and maximal compatibility with the existing DVB-T standard for network and transmission equipment sharing.

In November 2004, the DVB-H standard (Digital Video Broadcasting-Handheld) was adopted as the ETSI standard EN 302 304, which gives the technical specifications for providing mobile broadcast services (other multimedia services in the future) to mobile handsets.

By including additional features to the DVB-T standard, the DVB-H standard ensures that portable devices such as mobile phones can stably receive broadcast television signals and other multimedia services. The DVB-H standard adopts new technologies such as time slicing to reduce the power consumption of hand-held devices, adding the 4 K transmission mode (i.e., the FFT size is roughly 4 K), using an in-depth symbol interleaver, MPE-FEC (multiprotocol encapsulation–forward error correction), etc. All these newly introduced techniques further improve the robustness of the system under mobile and noisy environments. To provide interactive mobile broadband services to users, DVB-H is also capable of optionally using the mobile telecommunication network as the return channel.

Unlike DVB-T, which is mainly used for fixed reception, the DVB-H standard supports terminals such as mobile phones with smaller antennas, and this provides considerable flexibility for mobility applications. Even though DVB-H is a standard based on DVB-T, it successfully meets special requirements as follows:

1. Since the receiving terminal is battery powered, it is able to turn off part of the receiving circuit periodically to save power and maximize battery life.
2. Roaming users are able to successfully receive the DVB-H services when they enter new service areas.
3. The transmission system is able to ensure smooth service reception at different moving speeds under different reception environments, such as indoors, outdoors, walking, and in moving vehicles.
4. The transmission system is able to effectively reduce the impact of various impulsive interferences from the transmission environment.
5. The transmission system can provide sufficient flexibility for different transmission data rates and channel bandwidths for various applications.

10.2.1 Block Diagram of DVB-H System

To be fully backward compatible with the DVB-T standard, DVB-H uses the same coded-OFDM modulation technology in the same VHF/UHF TV bands. This means that DVB-H can utilize all the existing multiplexers, modulators, transmitters, and other equipment for the DVB-T services.

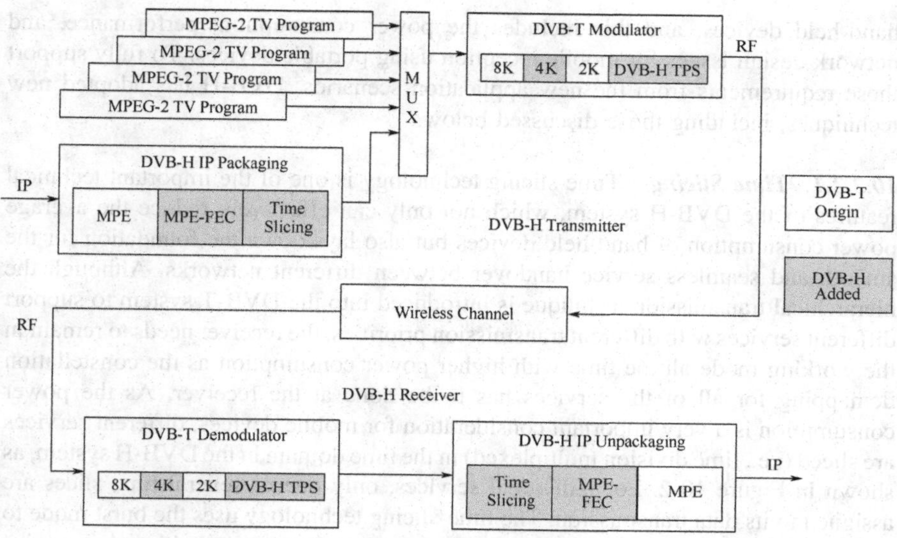

FIGURE 10-1 Block diagram of DVB-H system.

DVB-H also takes advantage of the Internet Protocol (IP) to reduce the deployment and operational cost of the services. With this, it is possible to be compatible with the IPTV and the 3G services platforms.

The block diagram of the DVB-H system architecture is shown in Figure 10-1. The DVB-H transmitter system consists of the DVB-H encapsulator and the DVB-H modulator. The DVB-H encapsulator is responsible for encapsulating IP data into the transport stream format, such as the MPEG-2 TS packet shown in Figure 10-1, and the DVB-H modulator performs the channel coding, modulation, and other baseband processing functionalities. Correspondingly, the receiver consists of the DVB-H demodulator and the DVB-H decapsulator, which performs the opposite operation from the DVB-H transmitter after the synchronization, channel estimation, and equalization are done first at the receiver.

The design philosophy of the DVB-H system is that the system can either operate as a stand-alone system with a dedicated spectrum or coexist with the DVB-T system by multiplexing the DVB-H service with the DVB-T service using the same spectrum to allow the reuse of the existing network infrastructure.

10.2.2 Technical Features of DVB-H System

To maintain maximal compatibility with the DVB-T standard, the DVB-H standard has adopted all the basic technologies of DVB-T, such as using concatenated codes of RS code and convolutional code and using C-OFDM as the basic modulation scheme. Although DVB-T has been proven it can successfully support fixed, mobile, and portable reception, further improvements are still required for mobile reception by hand-held devices, and this includes the power consumption, performance, and network design issues for mobile reception using portable devices. To fully support these requirements from the new application scenarios, DVB-H has adopted new techniques, including those discussed below.

10.2.2.1 Time Slicing Time-slicing technology is one of the important technical features of the DVB-H system, which not only can effectively reduce the average power consumption of hand-held devices but also lays down the foundation for the smooth and seamless service handover between different networks. Although the hierarchical transmission technique is introduced into the DVB-T system to support different services with different transmission priorities, the receiver needs to remain in the working mode all the time with higher power consumption as the constellation demapping for all of the services has to be done at the receiver. As the power consumption is a very important consideration for mobile devices, different services are sliced (i.e., time division multiplexed) in the time domain in the DVB-H system, as shown in Figure 10-2. For individual services, only one or several time slices are assigned to its data transmission. The time-slicing technology uses the burst mode to transmit the service data, and each time slice carries one service and lets that service occupy the entire system bandwidth. The information indicating the starting time of the next time slice for the same service will also be carried by the current time slice. This arrangement allows hand-held terminals to receive the selected service at the specified time slice only. Theoretically speaking, the receiver (the baseband processing module and RF circuitry) could be in the dormant (i.e., sleeping) mode during the time slices of other services, and the power consumption at the receiver can be effectively reduced.

When the time-slicing technique is used, one DTV program can always be sent by the DVB-H transmitter using many different time slices while other DTV programs or data services will be transmitted between any two time slices allocated for this

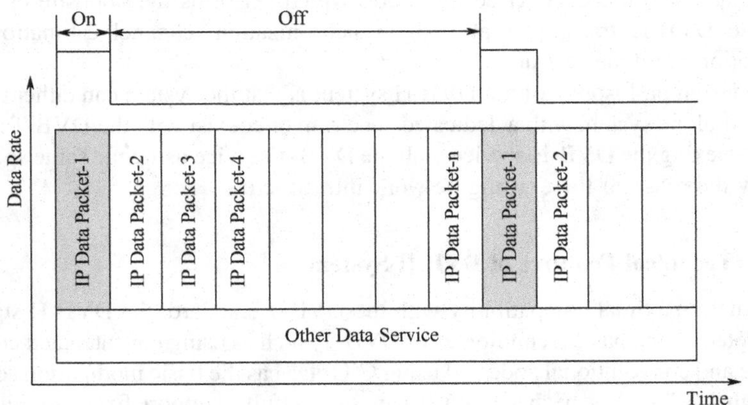

FIGURE 10-2 DVB-H time-slicing diagram.

program. From the receiving point of view, the DTV program is now received in the burst mode rather than in a continuous manner, which is different from the conventional DTTB receiver with the input of the constant data rate. This could be called burst transmission. If the decoding at the terminal requires a low yet constant data rate, the receiver can first buffer all the received burst data during the dedicated time slice and then generate the data stream with a constant data rate for the decoding. The burst data rate is usually much higher than the averaged constant data rate. Therefore, the power consumption can be lowered by this approach since the receiver can be back to the dormant mode and only wakes up when the slices containing the desired data arrive. Theoretically speaking, if the burst data rate is twice as much as the average data rate, 50% of the power might be saved. If this ratio is 10, 90% power consumption might be saved using this slicing technology. It should be noted that, in practice, the percentage value of the power saving will not be this high as a higher data rate always needs more processing power.

10.2.2.2 Switching The switching technique between DVB-H cells is similar to the soft handoff of the code division multiple-access (CDMA) mobile telecommunication networks. The widely used RAKE receiver of the mobile terminal in a CDMA system can simultaneously receive signals from more than one base station. The so-called soft handoff means that the mobile terminal does not disconnect from the original base station when it starts to communicate with a new base station, thereby increasing the probability of successful switching.

When the time-slicing technique is utilized, the DVB-H hand-held terminal can monitor the adjacent cells within its idle periods (i.e., time slices) by scanning other frequencies from adjacent transmitters and checking the signal strength at that frequency without interrupting its current reception. In this case, it is assumed that the MFN is used, that is, the same DTV program is carried by different frequencies within the DVB-H network. When the user enters the service area covered by another transmitter, the hand-held terminal can switch to a different transport stream of exactly the same program based on monitoring results, and suboptimal and seamless service switching can therefore be successfully achieved. Ideally, if the services have been arranged accurately and are synchronous well at the front end, it will enable the same service to appear at the different time slices from the adjacent transmitter in a timely manner, and the user will not notice this change. Note that in the case of an SFN the service switching is conducted only when the terminal is in a different network.

It should be noted that the monitoring operation will surely affect the low-power consumption requirement of the DVB-H system. In fact, this requirement can be satisfied when the power consumption of this effort is limited to an acceptable level (for example, when the time used to monitor the signal strength of one single time slice is less than 20 ms and the number of signals to be monitored can be reduced by intelligent prediction). In the DVB-T system, two tuners are required to achieve smooth handoff for one terminal. With the time-slicing technology, the adjacent cells can be monitored using the same tuner during the idle period to obtain suboptimal switching decisions.

10.2.2.3 Multiprotocol Encapsulation–Forward Error Correction DVB series standards can support four data broadcasting methods: data pipeline, data stream, multiprotocol encapsulation, and data carousel. In principle, they can all accommodate the IP data packet. Since both the data pipeline and the data stream are rarely used, they are not considered for the DVB-H system. The data carousel method is mainly used to transfer files and data (does not apply to the streaming service) and it is difficult for time slicing to be applied, and DVB-H only uses the MPE method. Unlike the other three methods, MPE technology not only supports both data and streaming services efficiently but also supports transport protocols other than IP well, and this provides great flexibility. Since it is easy to accommodate the time-slicing technique, it can be used to fully satisfy the technical requirements of the DVB-H system.

DVB-H uses the Reed–Solomon (RS) error-correcting code in the data link layer for the IP datagram as the MPE–FEC code, with the parity-check information being transferred in the specified FEC section. This is called multiprotocol encapsulation–forward error correction (MPE–FEC). MPE–FEC is designed to improve the C/N threshold, anti-Doppler frequency shift performance, and anti-impulse interference capability in the wireless channel.

As described above, the powerful FEC is introduced into the MPE layer. The parity-check information is obtained by encoding the data in the datagrams to be sent and will be transmitted in the separate MPE–FEC unit. The MPE–FEC frame is structured as a matrix with 255 columns and a flexible number of rows (up to 1024 rows), as shown in Figure 10-3, and each element in the matrix contains one byte. The left portion of the MPE–FEC frame in the figure consists of 191 columns, dedicated for the information of the IP datagram as well as the possible padding, and is called the application data table. The right portion of the MPE–FEC frame consists of the 64

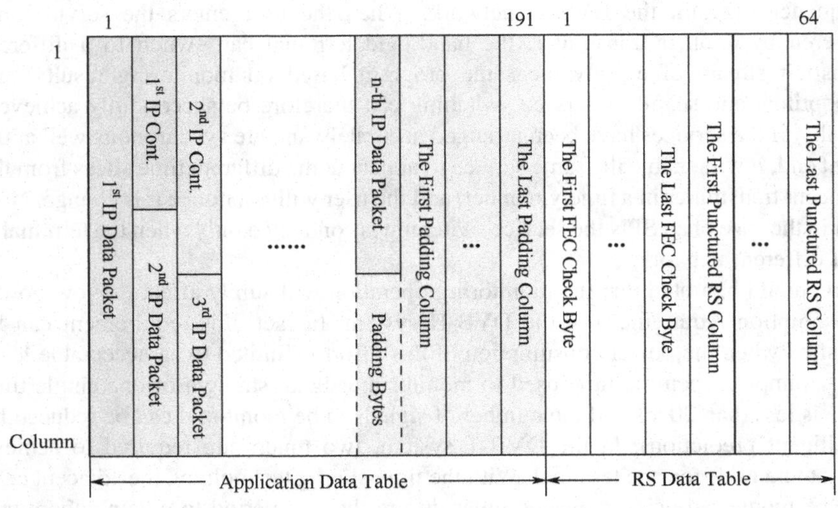

FIGURE 10-3 Structure of DVB-H MPE–FEC frame.

columns dedicated for the parity-check data of the FEC code and is called the RS data table. The IP data packets are carried in the MPE sections defined by the DVB standard regardless of whether MPE–FEC is used, and this ensures the full backward compatibility for those receivers without supporting the MPE–FEC technique.

At the receiver, the error correction is carried out based on the received parity check, which helps achieve error-free transmission in harsh conditions. The overhead from the parity-check information in MPE–FEC causes transmission efficiency loss, but this surely leads to better error correction performance: for example, if the overhead of MPE–FEC is 25% and the receiver can manage to achieve the C/N threshold improvement which is close to that of the dual-antenna transmit diversity without using MPE–FEC. In this case, the overhead of MPE–FEC can be compensated for by choosing a higher FEC code rate at the physical layer, and it can provide better performance than that of the DVB-T system operating at the same data throughput. It should be pointed out that the MPE–FEC scheme is also able to support high-speed DVB-T signal reception with a single antenna using 16QAM or even 64QAM constellation mapping, and the FFT size is 8 K.

10.2.2.4 4 K Mode and In-Depth Interleaver As described in previous chapters, the FFT size of the 2 K mode in the DVB-T system is known to be able to provide better mobile reception performance than the 8 K mode. However, since the OFDM data block duration (which is the reverse of the subcarrier spacing) of the 2 K mode is shorter than that of the 8 K mode, the associated guard interval duration (generally a percentage ratio of the OFDM symbol duration) is relatively shorter when the 2 K mode is used. This makes the 2 K mode more suitable for SFNs with small coverage area and difficult to support those SFNs with quite large coverage area. In addition to the 2 K and 8 K modes available in DVB-T, the 4 K mode (with 3409 modulated subcarriers) is added to DVB-H, giving more flexibility for network design with the compromise of mobile reception performance and SFN size. Meanwhile, the in-depth symbol interleaving technique is introduced for both 2 K and 4K modes in the DVB-H system to further enhance the anti-impulsive interference performance for the high-mobility scenario. The newly introduced 4 K mode has moderate OFDM symbol duration and guard interval length. Therefore, it is appropriate when medium-sized SFNs are built. This gives network designers more freedom to optimize the SFNs with respect to spectral efficiency. Although the spectral efficiency in the 4 K mode may not be as high as that of the 8 K mode, channel estimation can be performed more frequently due to its relatively short OFDM symbol duration. The 4 K mode thus can provide better mobile reception performance than the 8 K mode. In short, the overall performance of the 4 K mode is between that of the 2 K and 8 K modes (better spectrum efficiency than the 2 K mode and better mobile reception performance than the 8 K mode), which provides broadcasters an additional option when considering the trade-off among coverage, spectral efficiency, and mobile reception performance.

10.2.2.5 Transmission Parameter Signaling The objective of transmission parameter signaling (TPS) is to provide robust and easy-to-access signaling for DVB-H receivers, thus enhancing and speeding up the service recovery. TPS provides

TABLE 10-1 Parameters of DVB-H Physical Layer [6]

Parameter	2 K Mode	4 K Mode	8 K Mode
Number of modulated subcarriers *K*	1705	3409	6817
Number of data subcarriers	1512	3024	6048
Sample duration	7/64 us	7/64 us	7/64 us
OFDM symbol duration	224 us	448 us	896 us
Subcarrier spacing	*4464 Hz*	*2232 Hz*	*1116 Hz*
Effective bandwidth	*7.61 MHz*	*7.61 MHz*	*7.61 MHz*

Note: Numerical values in italics are approximate values.

a very robust signal, allowing fairly quick locking even under the very low C/N value at the receiver. The DVB-H system uses two additional TPS bits to indicate the presence of the time-slicing and optional MPE–FEC. In addition, the existing TPS bits defined in the DVB-T system are in the DVB-H system to indicate the 4 K mode, symbol interleaving depth, and cell identifier.

From the above introduction, it can be seen that DVB-H technology can be considered a superset of the DVB-T system for digital mobile TV broadcasting, with additional features to meet the specific requirements of hand-held, battery-powered receivers. Like any other DTV broadcasting system, the DVB-H system provides a downstream channel with high data throughput, and this can be stand alone or used interactively with mobile telecommunication networks for two-way services. Unlike DVB-T as the most successful DTTB standard, DVB-H is not commercially successful because of the spectrum issue and other challenges. Yet, technologically, it is very solid, and that is why it is introduced in this chapter.

10.3 ATSC-M/H SYSTEM

ATSC-M/H (ATSC-Mobile/Handheld) is the standard for the DTV mobile transmission developed by the ATSC in 2007, an extension to the available DTV broadcasting standard ATSC A/53, and this is similar to the relationship between DVB-T and DVB-H. ATSC uses 8VSB modulation as described in previous chapters, which was originally optimized for fixed reception in the typical North American reception environment and is not robust enough against Doppler frequency shift and multipath radio interference under mobile environments. To overcome these issues, additional channel coding mechanisms are introduced in ATSC-M/H to protect the signal and help broadcasters send DTV service to mobile/hand-held devices (including mobile phones) through the VHF/UHF channels. ATSC M/H was officially announced in October 2009.

ATSC allocates a portion of its 19.39-Mbps data throughput (8VSB modulated) for the mobile/hand-held broadcast services, while the rest can still be used for HDTV or multiple SDTV programs. By introducing extra training sequences and the FEC module, the ATSC M/H system can support mobile DTV reception successfully. The ATSC-M/H standard is fully backward compatible, allowing the operation of existing

ATSC services in the same RF channel. Broadcasters can use their granted license without additional restrictions, and mobile DTV service can be effectively delivered to users without a negative impact on legacy ATSC receivers.

10.3.1 Frame Structure of ATSC-M/H System

The ATSC M/H frame (or M/H frame for short) is used to carry both ATSC main data (for terrestrial service) and its M/H encapsulation (MHE) (for mobile service and to be introduced later) data block with the time duration of 968 ms (i.e., the time period of 20 VSB data frames). Each ATSC M/H frame is equally divided into 5 subintervals of equal length, called M/H subframes, and each M/H subframe is then divided into 16 ATSC M/H slots (to be more specific, each M/H subframe is first divided into 4 subdivisions of length 48.4 ms, which is the time it takes to transmit one VSB frame). These VSB frame time intervals are in turn divided into 4 M/H slots (altogether there will be 16 slots), as shown in Figure 10-4.

The ATSC M/H slot is the basic unit for multiplexing both M/H data and main service data, and each slot has the identical time needed to transmit 156 TS packets. The M/H slot consists of 156 main TS packets if there is no M/H packet or 118 TS M/H packets plus 38 packets of the main TS packets. The collection of 118 M/H packets transmitted within one slot is called an M/H group. The 118 M/H packets within an M/H group are encapsulated inside a special TS packet, known as an MHE packet.

10.3.2 ATSC-M/H System Block Diagram

The ATSC-M/H system shares the same RF channel of the traditional ATSC broadcast services (i.e., the "main service") while the M/H transmission system will receive two input streams: One consists of the MPEG TS packets of the main service data; the other consists of M/H service data. The function of the M/H

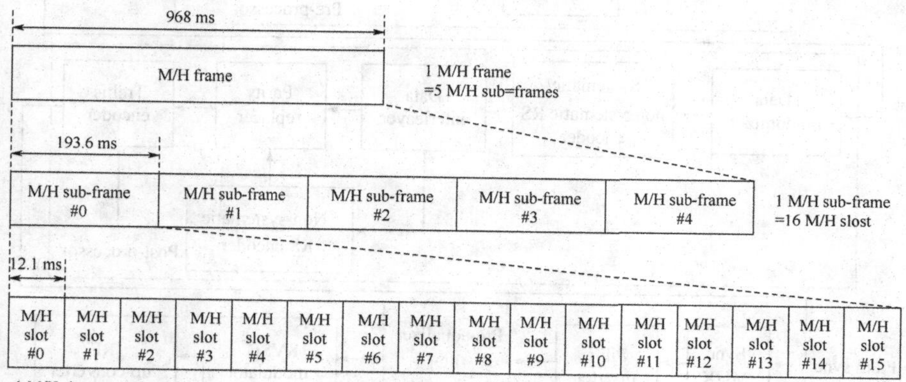

FIGURE 10-4 ATSC-M/H frame structure.

transmission system is to multiplex these two streams into one stream of MPEG TS packet, finish all the baseband processing, and eventually convert the input streams into one normal ATSC trellis-coded 8VSB signal, as shown in Figure 10-5.

To ensure full compatibility with legacy 8VSB receivers, the M/H service data are encapsulated in the special transport stream packets, named M/H encapsulation packets. The M/H transmission system is used to encapsulate the services data of various formats: the services carried by MPEG TS, including MPEG-2 video/audio, MPEG-4 video/audio, and other data of IP packets.

The operation of the M/H transmission system on M/H data can be divided into two blocks: preprocessor and postprocessor (shown in Figure 10-5).

The preprocessor includes the functional blocks of the M/H framing, FEC encoding, data block processing, group formatting, packet formatting, and M/H signaling encoding. The main purpose is to convert the M/H service data into the appropriate M/H data format and to enhance the robustness of the M/H service by introducing additional FEC as well as the training sequences and performing the service data encapsulation into the MHE transport stream packets.

The function of the postprocessor is to process the main service data by the normal 8VSB encoding and control the preprocessed M/H service data in the multiplexed stream for compatibility with ATSC legacy 8VSB receivers. The main service data in

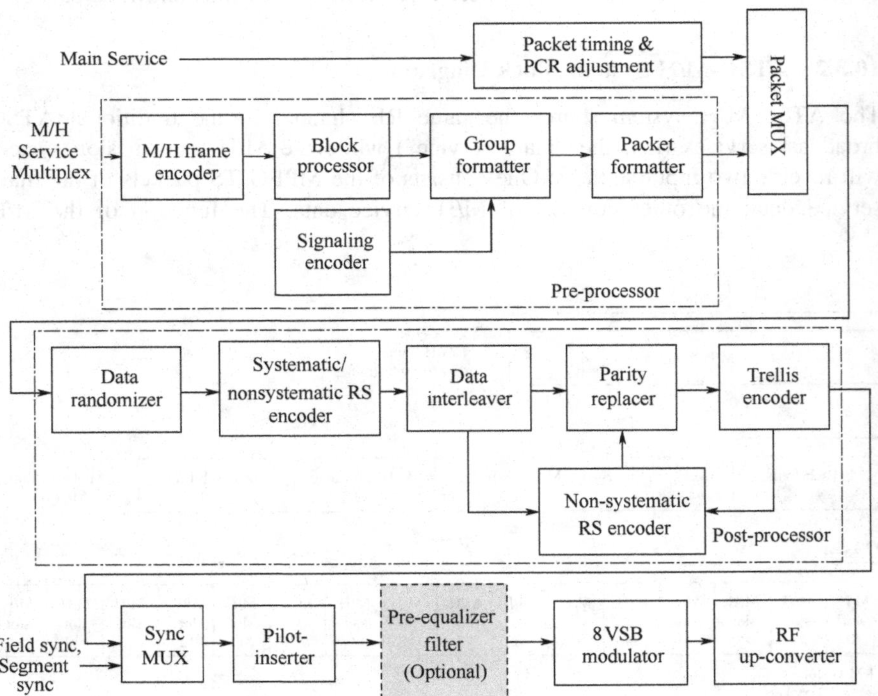

FIGURE 10-5 Structure of ATSC-M/H channel coding and modulation system.

the multiplexed stream are processed exactly the same as defined by ATSC A/53, which includes data scrambling, RS encoding, interleaving, and trellis coding.

The M/H service data in the multiplexed stream are processed differently from the main service data processing, and the preprocessed M/H service data do not pass through the data randomization block. The preprocessed M/H service data are processed by a nonsystematic RS encoder, and this nonsystematic RS coding allows insertion of regularly spaced long training sequences without disrupting reception by ATSC legacy receivers. Additional operations are taken on the preprocessed M/H service data for initialization of the trellis encoder when each training sequence starts, which has been included in the preprocessed M/H service data.

10.3.3 Frame Encoding of ATSC-M/H System

Figure 10-6 is a block diagram of the M/H frame encoder. The input demultiplexer separates the input M/H ensembles (defined as a collection of consecutive RS frames with the same FEC codes, wherein each RS frame encapsulates a collection of packetized data) and sends them to the corresponding RS frame encoders. Each RS frame encoder generates one or two RS frames and divides each RS frame into several portions with a fixed size of PL (i.e., RS frame portion length). Each segmented portion of the RS frame corresponds to the amount of data carried by an M/H group. The output multiplexer puts these RS frame portions from the RS encoders into the order required by the block processor.

The RS frame encoder in Figure 10-6 operates in either of two modes depending on the mode index assigned to this encoder. For the single-frame case with RS frame mode 00, the RS frame encoder generates one (primary) RS frame to be transmitted in regions A, B, C, and D of M/H groups. For the dual-frame case of RS frame mode 01, the RS frame encoder receives two M/H ensembles and will produce two RS frames accordingly: a primary RS frame to be transmitted in regions A and B of M/H groups and a secondary RS frame to be transmitted in regions C and D of M/H groups.

Data encapsulation: The RS frame is the basic unit for data delivery and has a two-dimensional frame structure encapsulating the IP datagram. The number of payload bytes of one RS frame should be a multiple of 187. The number of columns of one RS frame payload is N bytes, where N is determined by the transmission mode of the M/H physical layer subsystem and the parameters carried by the M/H ensemble. The

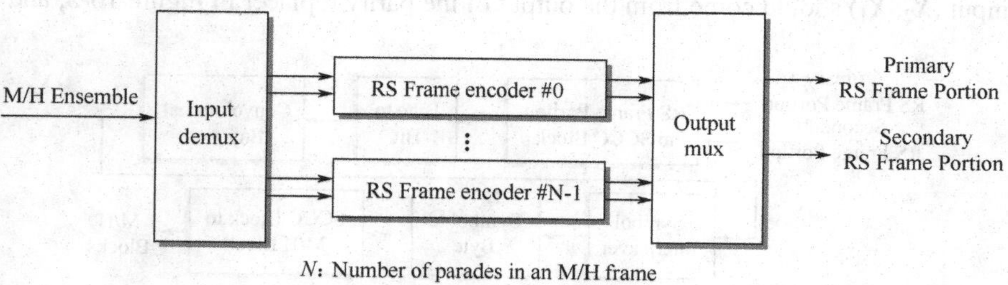

FIGURE 10-6 ATSC-M/H frame encoder.

number and size of the RS frame are also determined by the transmission mode of the M/H physical layer subsystem. Each RS row consists of a 2-byte pointer and the remaining bytes of this row are for the data. The data of the IP datagram will be sequentially written in each row of the RS frame with the order from left to right and top to bottom.

10.3.4 Block Processor of ATSC-M/H System

The main function of the block processor is to perform the outer encoding with the SCCC (Serial Concatenated Convolutional Code) for the output of the RS frame encoder. As shown in Figure 10-7, the operations include conversion from the RS frame structure to the SCCC block, byte-to-bit conversion, convolutional encoding, symbol interleaving, symbol-to-byte conversion, and SCCC-block-to-M/H-block conversion. Both the convolutional encoder and symbol interleaver can be in concatenation with trellis encoder in the postprocessor, helping reconstruct the SCCC effectively.

10.3.5 ATSC-M/H Trellis Encoder

The 8VSB transmission subsystem employs a 2/3-rate trellis coding, exactly the same as that in the ATSC system, and the data from the convolutional byte interleaver are demultiplexed into 12 independent streams (2 bits at a time). Those 12 streams of 2-bit symbols are individually and independently trellis coded to create 3-bit symbols, as shown in Figure 10-8. The modulation waveforms used for the trellis coding has eight levels with one-dimensional constellation, and the constellation mapping is also shown in the figure.

In the ATSC-M/H system, the initialization of the trellis encoder before the input of each training signal is required to guarantee the generation of the known training sequences for the M/H receiver. When the first 4 bits of the trellis initialization bytes (which have been inserted into the data by the group formatter) are input to one of these 12 modified trellis encoders, the trellis encoder should be initialized: When the first two 2-bit symbols converted from each trellis initialization byte are received, the input bits of the trellis encoder should be replaced by the fixed values in the memory of the trellis encoder.

One of the 12 modified trellis encoders is illustrated in Figure 10-9. Its 2-bit input (X_2, X_1) should come from the output of the parity replacer in Figure 10-5, and

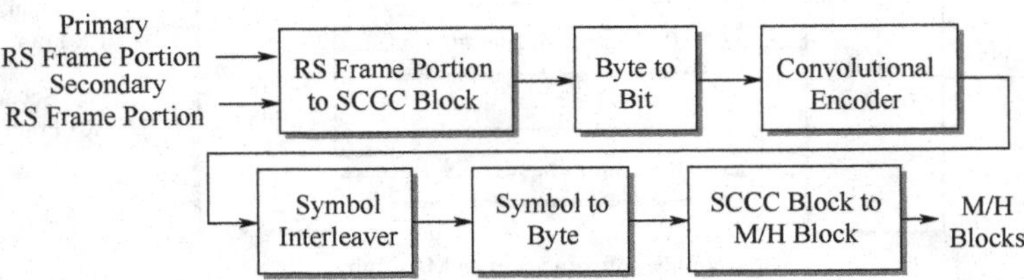

FIGURE 10-7 ATSC-M/H block processor.

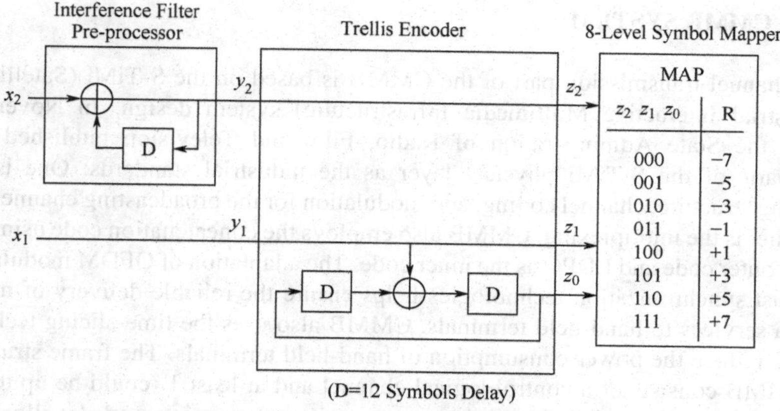

FIGURE 10-8 8-VSB TCM encoder and symbol mapper.

the 2-bit output (X_2', X_1') should be supplied to the input of the nonsystematic RS encoder. The 3-bit output (Z_2, Z_1, Z_0) should be mapped to the eight-level symbol (8VSB).

Each of the 12 modified trellis encoders has a multiplexer to switch between the normal input and an initialization input, which shall be fed back from the delay devices inside the trellis encoder. If initialization is required, the normal/initialization control input to the multiplexer selects (1) the initialization input path during the two-symbol interval of the trellis initialization byte immediately preceding each training sequence or (2) the normal path at all other times. The modified trellis encoder provides the input data to the non–system RS encoder for the trellis initialization and also the input data for the eight-level symbol mapper.

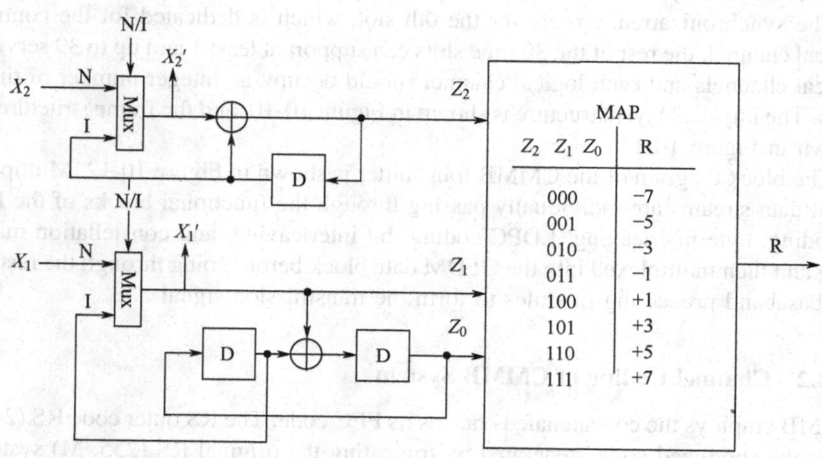

FIGURE 10-9 Modified trellis encoder.

10.4 CMMB SYSTEM

The channel transmission part of the CMMB is based on the S-TiMi (Satellite & Terrestrial Interactive Multimedia Infrastructure) system design. In November 2006, the State Administration of Radio, Film and Television published two key parts of the S-TiMi physical layer as the industrial standards: One is the framing structure, channel coding, and modulation for the broadcasting channel and the other is the multiplexing. CMMB also employs the concatenation code using RS as the outer code and LDPC as the inner code. The adaptation of OFDM modulation and fast synchronization technologies helps ensure the reliable delivery of multimedia services to hand-held terminals. CMMB also uses the time-slicing technology to reduce the power consumption of hand-held terminals. The frame structure of CMMB consists of a control logical channel and at least 1 (could be up to 39) service logical channel(s). The code rate, constellation mapping, and slot allocation of each logical channel can be independent. CMMB is designed to operate at either 470–798 MHz for terrestrial or 2635–2660 MHz for a satellite with system bandwidth of 8 or 2 MHz, and it supports SFN operation combining satellite and terrestrial networks.

10.4.1 Frame Structure of CMMB System

In the physical layer frame structure of CMMB, logical channel allocation is used: The physical layer is divided into a control logical channel (CLCH, carrying the system control information) and a service logical channel (SLCH, carrying the broadcasting services), respectively. There are a total of 40 time slots in the physical layer, and the duration of each time slot is 25 ms, consisting of one beacon and 53 OFDM symbols. Each beacon consists of a transmitter identifier (TxID) and two synchronization symbols that are used to identify the different transmitters as well as for the synchronization. Except for the 0th slot, which is dedicated for the control logical channel, the rest of the 39 time slots can support at least 1 and up to 39 service logical channels and each logical channel should occupy an integer number of time slots. The physical layer structure is shown in Figure 10-10, and the frame structure is shown in Figure 10-11.

The block diagram of the CMMB transmitter is shown in Figure 10-12. Multiple-input data streams are individually passing through the functional blocks of the RS encoding, byte interleaving, LDPC coding, bit interleaving, and constellation mapping and then multiplexed into the OFDM data block before going through the rest of the baseband processing modules to form the transmission signal.

10.4.2 Channel Coding of CMMB System

CMMB employs the concatenated code as its FEC code. The RS outer code RS (240, K) is the shortened code, generated by truncating the original RS (255, M) system code, and K has four choices: 240, 224, 192, and 176. With a length of 9216 bits, the LDPC inner code has two code rates of 1/2 (9216, 4608) and 3/4 (9216, 6912).

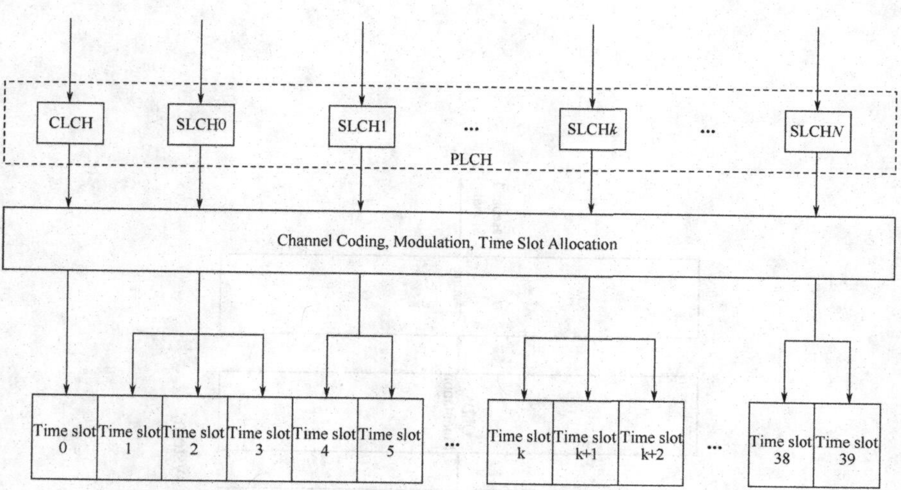

FIGURE 10-10 Logic channel allocation of CMMB system.

The 240-byte shortened RS code (240, K) is derived by adding 15 bytes of zeros in front of the K-byte information byte, letting the M = (15 + K) bytes pass through the RS encoder of (255, M) and then removing the added 15 bytes from the output of the RS encoder to make it 240 bytes long.

10.4.3 CMMB Byte Interleaving

In the CMMB system, the byte interleaving is performed between the RS encoder and the LDPC encoder, and this helps convert burst errors from the inner decoder into the random errors at the receiver. The output of the byte interleaver is then sent to the LDPC encoder in the order of from the least significant bit (LSB) to the most significant bit (MSB) within each byte.

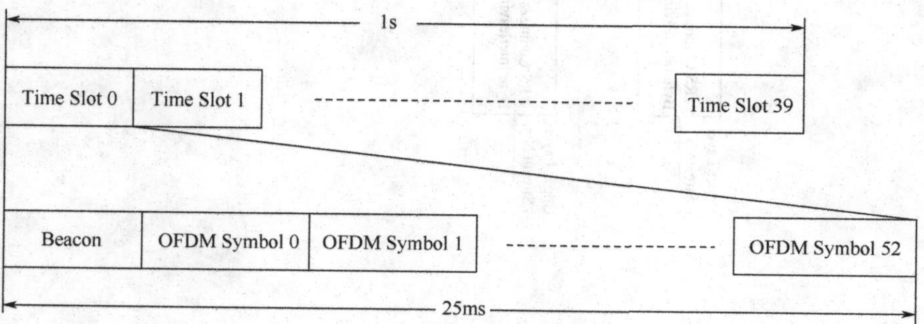

FIGURE 10-11 Time slot constitution of CMMB system.

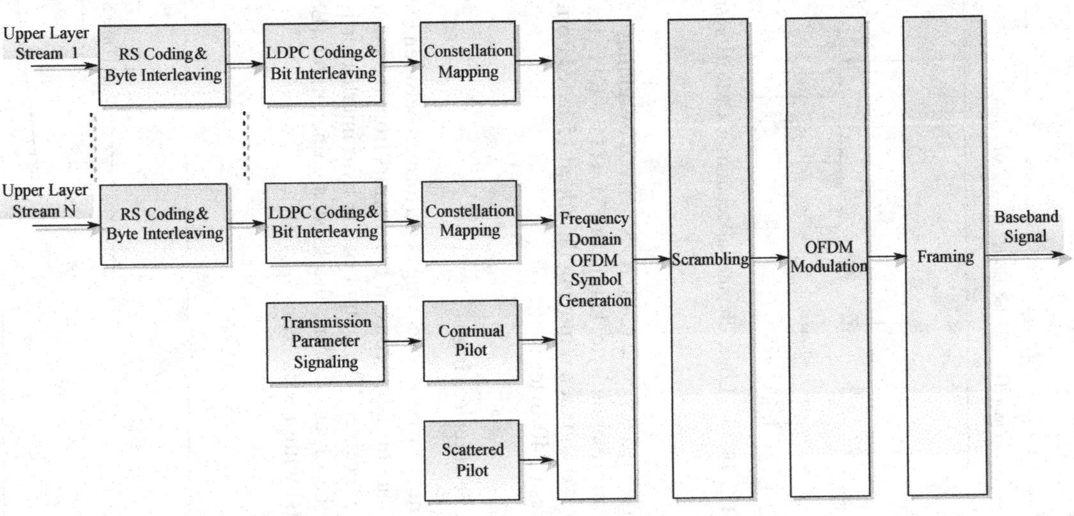

FIGURE 10-12 Block diagram of CMMB transmit system.

LDPC encoded bits must go through the bit interleaving before the constellation mapping. The block interleaving in CMMB is performed by write-in in rows and read-out by columns. Depending on the system bandwidth, there are two options for the interleaver size: 384×360 for 8 MHz bandwidth and 192×144 for 2 MHz bandwidth.

10.4.4 Modulation Scheme of CMMB System

The constellation mappings used for the CMMB system include BPSK, QPSK, and 16QAM, and OFDM (more specifically, cyclic-prefix OFDM) technology is adopted as the modulation scheme. The total number of subcarriers is 4096 and the effective number of subcarriers is 3076 with the system bandwidth of 8 MHz. Those corresponding numbers are 1024 and 628, respectively, with the system bandwidth of 2 MHz. The effective subcarriers include the data subcarriers, the scattered pilots, and the continuous pilots. The duration of the CP-OFDM symbol is 460.8 μs with the duration of OFDM data block, T_0, of 409.6 μs, and the duration of the cyclic prefix, T_1, of 51.2 μs.

To reduce the overhead of the guard interval (note that the GI mentioned in CMMB is different from that in the traditional CP-OFDM system), the GIs between adjacent OFDM symbols are overlapped, as shown in Figures 10-13 and 10-14, respectively. In Figure 10-14, each OFDM symbol is multiplied by a time-domain window function to ensure a constant signal power after overlapping. The window function is defined as

$$w(t) = \begin{cases} 0.5 + 0.5\cos(\pi + \pi t/T_G) & 0 \le t \le T_G \\ 1 & T_G \le t \le T_0 + T_1 + T_G \\ 0.5 + 0.5\cos(\pi + \pi(T_0 + T_1 - t)/T_G) & T_0 + T_1 + T_G \le t \le T_0 + T_1 + 2T_G \end{cases} \tag{10-1}$$

10.4.5 Payload Data Rate of CMMB System

The total payload data rate of the CMMB system with different system bandwidth, constellation mapping, and code rate are given in Table 10-2.

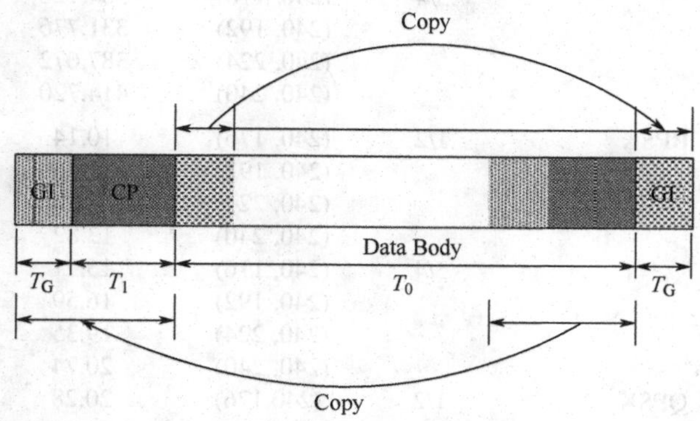

FIGURE 10-13 CP-OFDM frame structure in CMMB.

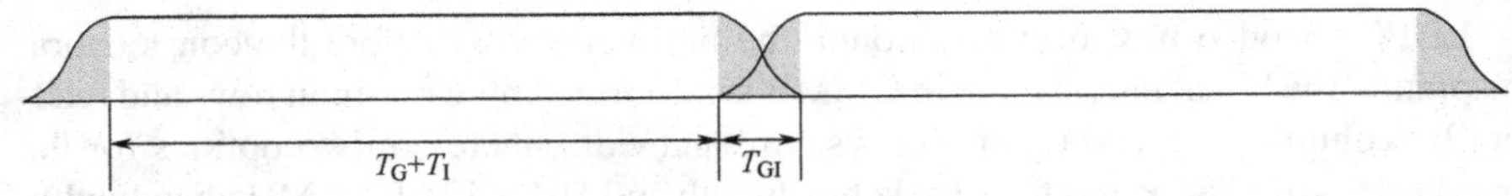

FIGURE 10-14 Overlapping between guard intervals in CMMB.

TABLE 10-2 Data Transfer Rate (Mbps) in CMMB

	Channel Configuration				
Bandwidth	Constellation Mapping	LDPC Code Rate	RS Code	Payload per Time Slot (kbps)	System Payload (Mbps)
$B_f = 8$ MHz	BPSK	1/2	(240, 176)	50.688	2.046
			(240, 192)	55.296	2.226
			(240, 224)	64.512	2.585
			(240, 240)	69.120	2.764
		3/4	(240, 176)	76.032	3.034
			(240, 192)	82.944	3.304
			(240, 224)	96.768	3.843
			(240, 240)	103.680	4.113
	QPSK	1/2	(240, 176)	101.376	4.023
			(240, 192)	110.592	4.382
			(240, 224)	129.024	5.101
			(240, 240)	138.240	5.460
		3/4	(240, 176)	152.064	6.000
			(240, 192)	165.888	6.539
			(240, 224)	193.536	7.617
			(240, 240)	207.360	8.156
	16QAM	1/2	(240, 176)	202.752	7.976
			(240, 192)	221.184	8.695
			(240, 224)	258.048	10.133
			(240, 240)	276.480	10.852
		3/4	(240, 176)	304.128	11.930
			(240, 192)	331.776	13.008
			(240, 224)	387.072	15.165
			(240, 240)	414.720	16.243
$B_f = 2$ MHz	BPSK	1/2	(240, 176)	10.14	0.409
			(240, 192)	11.06	0.445
			(240, 224)	12.90	0.517
			(240, 240)	13.82	0.553
		3/4	(240, 176)	15.21	0.607
			(240, 192)	16.59	0.661
			(240, 224)	19.35	0.768
			(240, 240)	20.74	0.823
	QPSK	1/2	(240,176)	20.28	0.805
			(240, 192)	22.12	0.877

Table 10-2 (*Continued*)

Channel Configuration					
Bandwidth	Constellation Mapping	LDPC Code Rate	RS Code	Payload per Time Slot (kbps)	System Payload (Mbps)
$B_f = 8$ MHz	BPSK	1/2	(240, 176)	50.688	2.046
			(240, 224)	25.80	1.020
			(240, 240)	27.65	1.092
		3/4	(240, 176)	30.41	1.200
			(240, 192)	33.18	1.308
			(240, 224)	38.71	1.524
			(240, 240)	41.47	1.631
	16QAM	1/2	(240, 176)	40.55	1.595
			(240, 192)	44.24	1.739
			(240, 224)	51.61	2.027
			(240, 240)	55.30	2.171
		3/4	(240, 176)	60.83	2.386
			(240, 192)	66.36	2.602
			(240, 224)	77.41	3.033
			(240, 240)	82.94	3.248

10.5 DVB-NGH

The DVB Next Generation Broadcasting System to Handheld (DVB-NGH) is the draft ETSI EN 303 105 standard, and DVB-NGH is the mobile evolution of the second-generation digital terrestrial TV broadcasting technology (DVB-T2 [4]). Its deployment is motivated by the continuous growth of mobile multimedia broadcasting services to hand-held devices such as tablet computers and smart phones. Its main objective is to provide superior performance, robustness, and increased indoor coverage compared with other existing DVB standards. DVB-NGH is based on the physical layer of DVB-T2 and has been designed so that it can be incorporated in DVB-T2 transmissions, allowing reuse of both the spectrum and infrastructure of the DVB-T2 network. The standardization started in March 2010 and was completed at the beginning of 2013. The main technical solutions of this new mobile TV broadcasting standard include:

1. Multiple input–multiple output (MIMO) for increased spatial diversity and transmission rate.
2. TFS (time-frequency slicing) for increased frequency diversity and more efficient statistical multiplexing.
3. Convolutional time interleaving for increased time diversity.
4. Improved LDPC codes and lower code rates.
5. Improved signaling robustness compared to DVB-T2.
6. SVC (scalable video coding) with multiple physical layer pipes (PLPs) for graceful service degradation.

TABLE 10-3 Constellation Mappings in DVB-NGH System

	Code Rate	
Modulation	1/3, 2/5, 7/15	8/15, 3/5, 2/3, 11/15
QPSK	2D/nonrotated	4D/nonrotated
16QAM		2D/nonrotated
64QAM or NU-64QAM		2D/nonrotated
256QAM or NU-256QAM		No rotation

7. Efficient transmission of local services within SFNs (single-frequency networks). Hierarchical local service insertion (H-LSI) is used to insert new services in an isolated transmitter or group of transmitters within a single-frequency network. In using this method, a local service PLP is hierarchically modulated over an appropriate regional service PLP, which already exists within the SFN. Receivers can either continue to receive the regional service PLP or switch to the hierarchically modulated local service PLP.
8. The use of nonuniform (NU) constellation and 2D and/or 4D rotated constellation; see Table 10-3

The standard consists of four parts each covering a different structure of the transmit network:

1. *Base Profile (Profile I).* Covers purely terrestrial transmission with single and multiaerial structures that require only a single aerial and tuner at the receiver.
2. *MIMO Profile (Profile II).* Covers purely terrestrial transmission with multi-aerial structures on both ends. Terminals suitable for this profile need to employ two tuners as well.
3. *Hybrid Profile (Profile III).* Covers a combination of terrestrial and satellite transmissions that requires only a single tuner at the receiver.
4. *Hybrid MIMO Profile (Profile IV).* Covers a combination of terrestrial and satellite transmission requiring a double aerial and tuner setup at the receiver. Once again, some of the configurations can be handled by profile II receivers, and other configurations require a special hybrid MIMO receiver. For the MIMO and hybrid profiles only, the differences between those and the base profile are additional functional blocks and that parameter settings are permitted in the MIMO or hybrid profile. The hybrid MIMO profile is not formulated solely as a list of differences to the other three profiles. Instead, it defines how previously described elements are combined to provide hybrid MIMO transmission and introduces profile-specific information. Functional blocks and settings that are the same as in the base profile are not described again but can be derived from the base profile.

Implementation of multiple antennas at the transmitter and receiver (MIMO) allows overcoming the Shannon limit of single-antenna communication without any additional bandwidth or increased transmission power. The first multiantenna transmissions for broadcasting systems were adopted for the digital terrestrial TV standard DVB-T2. However, in this case, a transmitter site distributed configuration was defined to exploit only diversity gain. Hence, DVB-NGH is the first broadcast system to employ pure MIMO as the key technology exploiting all the benefits of the MIMO channel.

The MIMO rate 1 codes exploit the spatial diversity of the MIMO channel without the need for multiple antennas at the receiver side (i.e., they can also be referred as MISO schemes). They can be applied across the transmitters of SFNs reusing the existing DTTB network infrastructure. DVB-NGH has adopted the distributed MISO scheme of DVB-T2 based on Alamouti coding as well as a novel transmit diversity scheme known as the enhanced single-frequency network (eSFN). MIMO rate 2 codes exploit the diversity and multiplexing capabilities of the MIMO channel. A new technique known as enhanced spatial multiplexing with phase hopping (eSM-PH) has been adopted to improve robustness and multiplexing capacity in the presence of spatial correlation.

DVB-NGH introduces new key features for terrestrial broadcast MIMO [10]. For the transmit diversity, DVB-NGH has specified a novel scheme known as the enhanced single-frequency network which is transparent to the receive terminals. The first type of techniques are known as MIMO rate 1 codes, which exploit the spatial diversity of the MIMO channel without the need for multiple antennas at the receive side. It can be applied across the transmit sides of SFNs to reuse the existing network infrastructure. The second type of technique is known as the MIMO rate 2 codes, which exploits the diversity and multiplexing capabilities of the MIMO channel. However, this requires additional investment at both sides of the transmission link.

10.5.1 Alamouti Scheme

One application scenario for multiple antennas transmitting identical signals is the SFN in Figure 10-15. Multiple transmitters sending identical signals synchronized in time and frequency are typically used to cover larger areas. Compared with the MFN

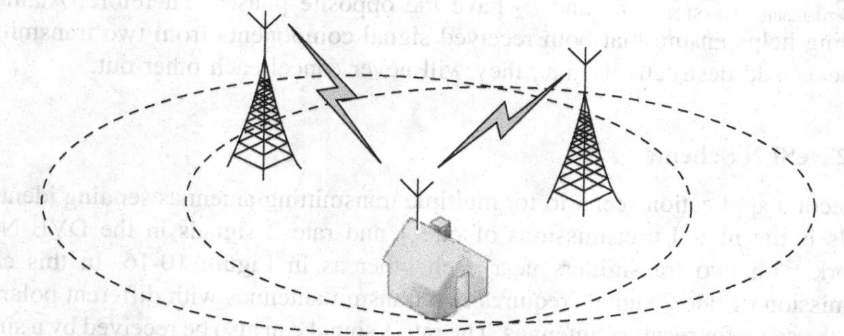

FIGURE 10-15 Example of SFN using multiple input and single output (MISO).

where each transmitter uses a specific frequency, the first benefit of the SFN is the high spectral efficiency, as only one single RF channel is needed to cover a larger area while multiple RF channels are required for the MFN. The second benefit is diversity gain as multiple transmitters sending the same signal can reduce the probability of deep fading.

MISO encoding based on the Alamouti scheme tries to minimize the impact of the frequency selectivity by sending space-coded signals by different transmitters and performing the simple decoding at the receiver. For this purpose and similarly to what has been discussed in Section 8.7.3, the Alamouti scheme divides the available transmitters into two groups (two transmitters in Figure 10-15 is a simple example): The signal is sent by the first group of transmitters without any modification while it needs to be modified before being sent by the second group of transmitters.

The Alamouti scheme helps increase the diversity since each transmitter-to-receiver path may suffer from severe fading, but the chance of having two independent paths suffer from the severe fading at the same time is relatively low.

One can use the theoretical channel capacity to estimate the potential gain when using Alamouti code. The achievable channel capacity for a classical SFN configuration is

$$C_{\mathrm{SFN}} = \log_2\left(1 + \frac{1}{2\sigma^2}\left(|h_1 + h_2|^2\right)\right) \tag{10-2}$$

where σ^2 is the noise variance of the channel and h_1 and h_2 are the channel states between transmitter 1 to the receiver and transmitter 2 to the receiver, respectively.

It is clear that the complex channel coefficients from both transmitters to the receiver add in amplitude and phase. The capacity equation for Alamouti code is

$$C_{\mathrm{Alamouti}} = \log_2\left(1 + \frac{1}{2\sigma^2}\left(|h_1|^2 + |h_2|^2\right)\right) \tag{10-3}$$

In contrast to the channel capacity given by 10-2 under the SFN configuration, the absolute squared value of both channel coefficients is added in (10-3). Since both transmitters always have the similar phases and add constructively, it is easy to show that $C_{\mathrm{Alamouti}} > C_{\mathrm{SFN}}$ if h_1 and h_2 have the opposite phases. Therefore, Alamouti encoding helps ensure that both received signal components from two transmitters will never add destructively, i.e., they will never cancel each other out.

10.5.2 eSFN Scheme

The second application scenario for multiple transmitting antennas sending identical signals is the mixed transmissions of rate 1 and rate 2 signals in the DVB-NGH network with two transmitters near each other as in Figure 10-16. In this case, transmission of rate 2 signals requires two transmit antennas with different polarizations, the same for receiver antennas. The rate 1 signal can also be received by a single receiving antenna.

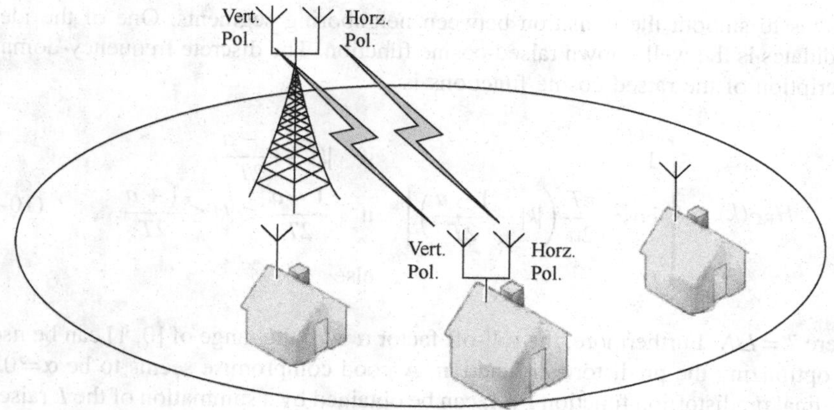

FIGURE 10-16 Example SFN with MIMO using different polarizations.

Simply switching one transmit antenna during the transmission of rate 1 signal may potentially disturb other receivers. Hence, it is preferred that both transmit antennas be used for the transmission of the rate 1 signal. The simple approach to transmit identical signals on both antennas will lead to frequency selectivity. In contrast to the distributed SFN, the delay between the signals of both antennas arriving at the receiver is almost zero. As a result, both signals can almost cancel each other out completely, making successful reception almost impossible. The impulse response recorded in the core SFN with two echoes arriving with similar power causes a deep fading over the entire signal bandwidth, and the system will not work well in this case.

To avoid the negative effects caused by the line-of-sight component, one can use a linear predistortion method at the transmitter. This predistortion should be unique for each transmitter and has to change over the different OFDM subcarriers. Hence, the transmitted signal can be expressed as

$$T_x(k) = S(k)P_x(k) \tag{10-4}$$

where k is the OFDM subcarrier number, $S(k)$ is the nondistorted complex frequency-domain representation of the OFDM symbol at the transmitter (which would be transmitted without predistortion), $P_x(k)$ is the complex predistortion function of the transmitter x on the OFDM subcarrier k, and $T_x(k)$ is the transmitted OFDM symbol for the transmitter x in the frequency-domain representation.

This approach divides the N OFDM subcarriers into L segments of equal size. Each of these L segments is then modulated by different phases (identical phases for the subcarriers within each segment). Additionally, the phases for the segments also differ among different transmitters. Consequently, the different phases decorrelate the signals, and the specific modulation of the phases for different transmitters allows for transmitter identification. Naturally, the sudden phase shift of two adjacent segments would significantly widen the equivalent impulse response. Therefore, one has to smooth the transition between neighboring segments. One of the ideal candidates is the well-known raised-cosine function. The discrete frequency-domain description of the raised cosine functions is

$$H_{\mathrm{RC}}(k) = \begin{cases} 1 & \text{if} \quad |k| \le \dfrac{1-a}{2T} \\ \cos^2\left[\dfrac{\pi T}{2a}\left(|k| - \dfrac{1-a}{2T}\right)\right] & \text{if} \quad \dfrac{1-a}{2T} < |k| \le \dfrac{1+a}{2T} \\ 0 & \text{else} \end{cases} \tag{10-5}$$

where $T = L/N$. Furthermore, the roll-off factor α with the range of [0, 1] can be used for optimizing the predistortion function. A good compromise seems to be $\alpha = 0.5$. The final predistortion function $P_x(k)$ can be obtained by a summation of the L raised-cosine functions that are frequency shifted against each other:

$$P_x(k) = \sum_{l=1}^{L-1} e^{j2\pi\Psi_x(l)} H_{\mathrm{RC}}\left(k - l\frac{N}{L}\right) \tag{10-6}$$

Actually, only $L-1$ raised-cosine functions are used because the edge carriers of the OFDM signal are virtual subcarriers which will not be modulated (as shown in Figure 10-17). To ensure that there is no correlation among different transmitters, the phase $\Psi_x(l)$ should be different for the different transmitters. Therefore, one can modulate the phase to allow for the unique transmitter identification. For example, one can use a differential encoding scheme with transmitter ID sequence $C_x = \{c_{0,x} c_{1,x}, \ldots, c_{L-1,x}\}$, which consists of L elements having the possible values of $c_{i,x} \in \{-1, 0, 1\}$. Finally, the phase term can be defined as

$$\psi_{(x)}(l) = \begin{cases} c_{0,x} & \text{if} \quad l = 0 \\ \psi_{(x)}(l) + c_{l,x}/L & \text{else} \end{cases} \tag{10-7}$$

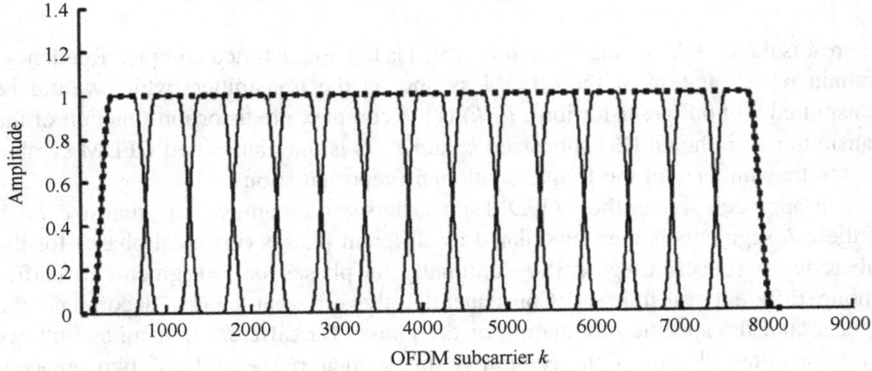

FIGURE 10-17 Division of 8 K OFDM signal into 16 raised-cosine functions.

Figure 10-17 illustrates the raised-cosine function for $N = 8192$, $L = 16$, and $\alpha = 0.5$. The dashed line shows the sum of all raised-cosine functions, which is equal to 1 if there is no phase change between adjacent segments. With the smooth function introduced by the raised-cosine function, the sum of all raised-cosine functions still roughly equals 1.

eSFN can offer performance gains by decorrelating the different paths and, hence, reduce the frequency selectivity. The advantage of eSFN is that no additional pilots are needed. One main application scenario for eSFN is the mixed SISO/MIMO scenario. In cross-polar MIMO transmission, one could have two transmit antennas on each transmitter site and two receive antennas at the receiver.

10.5.3 eSM-PH Scheme

Spatial multiplexing (SM) is specified in the DVB-NGH system where MIMO rate 2 code is used and the term "rate 2" stands for the transmission of two independent streams. MIMO rate 2 uses enhanced spatial multiplexing together with phase hopping (eSM-PH) to avoid the high correlation among different channels when strong LOS components appear. eSM exploits a precoding matrix to increase the spatial diversity compared to simple SM while PH can avoid the unfavorable channel conditions by periodic phase rotation on the modulated data symbols at one transmit antenna. Moreover, eSM-PH handles the transmitted power imbalance problem by transmission parameter optimization to further reduce the performance loss.

The SM precoding module processes the input symbols on each antenna stream or "layer" to increase the spatial diversity, known as eSM, and then the phase hopping will be applied. The block diagram is presented in Figure 10-18. The eSM scheme should be designed to satisfy the following three criteria:

1. The final MIMO encoded symbols must map onto all the different constellation points according to the group of bits to be delivered. For instance, if a pair of QPSK symbols is encoded by the MIMO precoder, the output should have a total of 16 unique projections on each antenna, as shown in Figure 10-19.
2. The resulting precoded SM will be uniform square QAM symbols to reduce SER (this is to maximize the minimum Euclidean distance among constellation points) if only this criterion is considered.
3. The bit to-symbol mapping needs to be carefully investigated to minimize the BER performance after symbol demapping. This can be achieved by

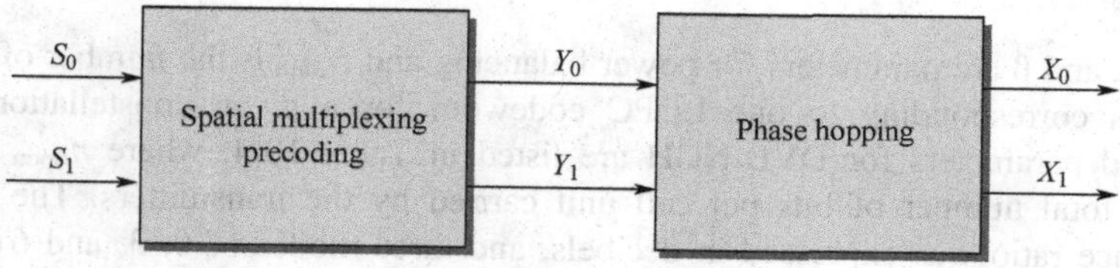

FIGURE 10-18 MIMO precoding block diagram.

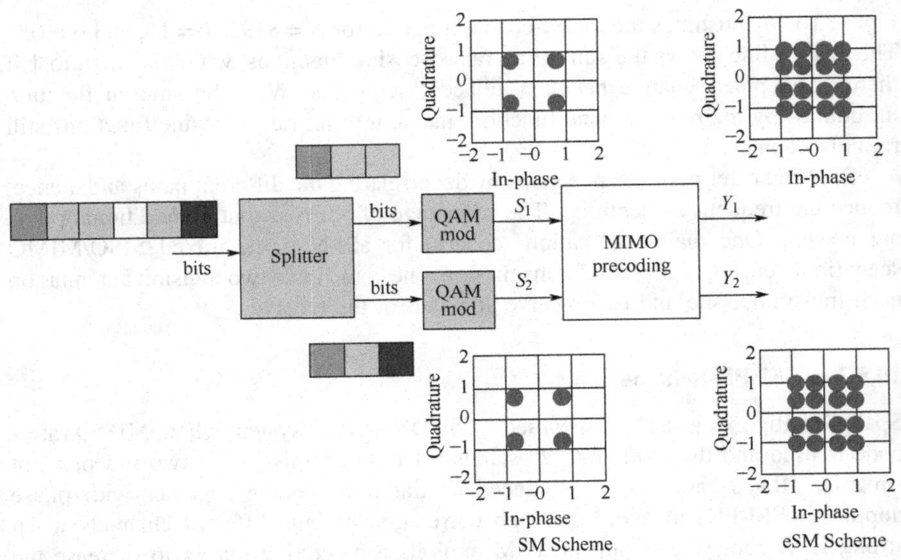

FIGURE 10-19 Block diagram of precoded SM with gray labeled 4QAM constellation.

maximizing both the minimum Euclidean distance and minimum Hamming distance. Gray bit labeling should also be an important factor in BER performance.

The eSM precoding matrix should be the best possible compromise among all three criteria above. High priority is the performance under high channel correlation while keeping the SM gain under the rich scattering condition. To satisfy all these conditions, the precoding matrix should be a unitary matrix and can be expressed as a rotation matrix with angle θ. The SM precoding processing is given as

$$\begin{bmatrix} Y_{2i} \\ Y_{2i+1} \end{bmatrix} = \sqrt{2}\begin{bmatrix} \sqrt{\beta} & 0 \\ 0 & \sqrt{1-\beta} \end{bmatrix}\begin{bmatrix} \cos\theta & \sin\theta \\ \sin\theta & -\cos\theta \end{bmatrix} \times \begin{bmatrix} \sqrt{\alpha} & 0 \\ 0 & \sqrt{1-\alpha} \end{bmatrix}\begin{bmatrix} S_{2i} \\ S_{2i+1} \end{bmatrix} \quad i = 0, \ldots, \frac{N_{\text{data}}}{2} - 1 \tag{10-8}$$

where α and β are parameters for power balancing and N_{data} is the number of QAM symbols corresponding to one LDPC codeword for a given constellation. The specified parameters for DVB-NGH are listed in Table 10-4, where n_{bpcu} stands for the total number of bits per cell unit carried by the transmitters. The power imbalance ratio was expressed in decibels, and three modes of 0, 3, and 6 dB are possible.

TABLE 10-4 Parameters for eSM in DVB-NGH System

Deliberate Power imbalance between two *T*x Antennas		0 dB			3 dB			6 dB		
n_{bpcu}	Modulation	β	θ	α	β	θ	α	β	θ	α
6	$S_q(T\times 1)$, QPSK; $S_q(T\times 2)$, 16-QAM	0.50	45.0°	0.44	1/3	0.0°	0.50	0.20	0.0°	0.50
8	$S_q(T\times 1)$, 16-QAM; $S_q(T\times 2)$, 16-QAM	0.50	57.8°	0.50	1/3	25.0°	0.50	0.20	0.0°	0.50
10	$S_q(T\times 1)$, 16-QAM; $S_q(T\times 2)$, 64-QAM	0.50	22.0°	0.50	1/3	15.0°	0.50	0.20	0.0°	0.50

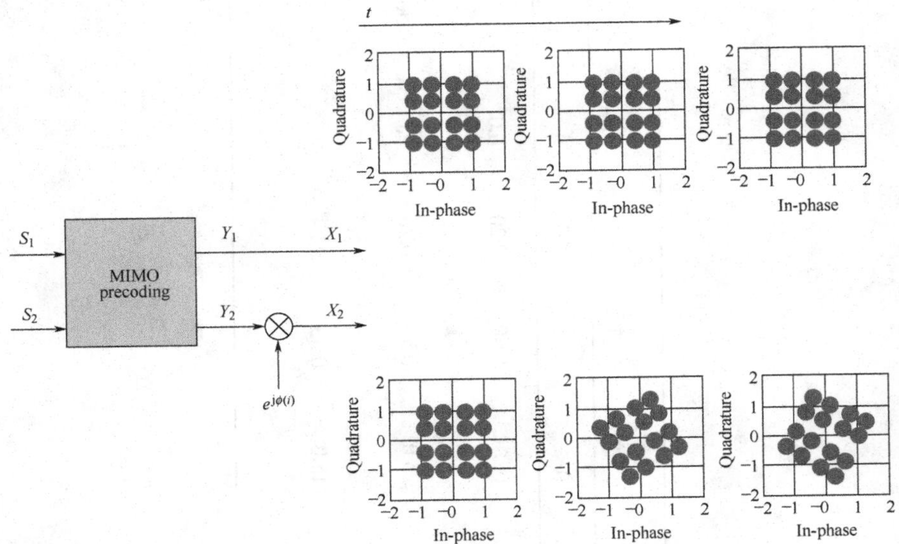

FIGURE 10-20 Block diagram of SM-PH with gray labeled 4QAM constellation.

The eSM performance can be further improved over various channel realizations in an average sense if randomization is applied to avoid the worst-case channel condition. In the DVB-NGH specification, the PH technique (shown in Figure 10-18) periodically changes the phase of symbols that can be sent by one transmitter:

$$\begin{bmatrix} X_{2i} \\ X_{2i+1} \end{bmatrix} = \begin{bmatrix} 1 & 0 \\ 0 & e^{j\phi(i)} \end{bmatrix} \begin{bmatrix} Y_{2i} \\ Y_{2i+1} \end{bmatrix} \quad \phi(i) = \frac{2\pi i}{N} \qquad N = 9 \tag{10-9}$$

Figure 10-20 illustrates a conceptual block diagram of eSM with PH on the second transmit antenna. The phase of the resulting constellation is periodically rotated. The amount of phase change is uniformly distributed in the range $[0, 2\pi)$ and the number of phase changes is $N = 9$. This value of N was determined by computer simulation and is sufficient for randomization purposes. Moreover, for every modulation, the number of cells corresponding to one LDPC codeword is a multiple of 9, providing an integer number of PH periods.

In the MIMO BICM system, performance can be further improved by employing iterative decoding at the receiver, where the MIMO demapper and the channel decoder exchange extrinsic information in an iterative way. However, iterative decoding significantly increases receiver implementation complexity, making it less suitable for mobile devices.

10.5.4 Hybrid System

The hybrid profile specifies the hybrid signal format, composed of a component coming from the terrestrial network and an additional component coming from the

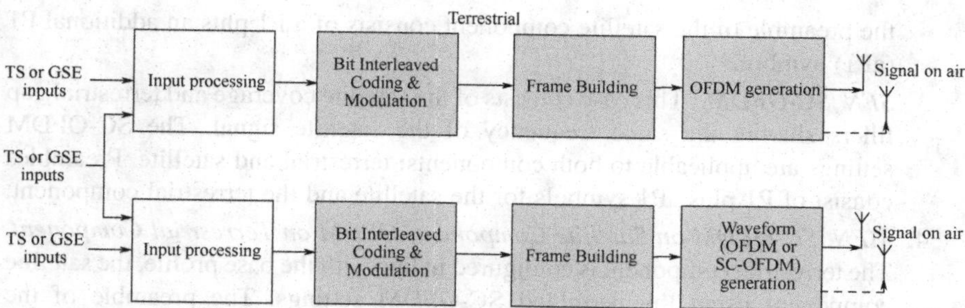

FIGURE 10-21 Block diagram of hybrid profile for NGH PHY layer.

satellite network. The satellite component of the hybrid profile is defined for channel bandwidths 1.7, 2.5, and 5 MHz (these three bandwidths are also covered by the base profile), respectively. Other than defining the hybrid signals, the hybrid profile defines the mechanisms to receive two signals simultaneously and to combine their outputs into a single stream.

Figure 10-21 presents the high-level NGH physical layer block diagram of the hybrid profile. Two chains are presented, one for the terrestrial component and the other for the satellite component. Compared to the base profile, the terrestrial and satellite chains of the hybrid profile present the potential functional differences in the BICM, OFDM frame building, and waveform generation. The system architecture of the satellite component (the lower branch) is identical to the terrestrial with the possibility of replacing the OFDM modulation block by the SC-OFDM modulation block. The system input(s) to the terrestrial and the satellite path may differ from each other in the MFN case. Time–frequency slicing can be applied to both of them.

Both SFN and MFN configurations are possible for the hybrid profile. In the SFN case, when the satellite and terrestrial components share the same frequency, the signal transmitted in the two components should be exactly the same. In the MFN case, the system architecture of the hybrid profile of DVB-NGH consists of two components: the terrestrial component and the satellite component. According to it, the following hybrid cases can be devised:

1. *SFN, OFDM.* The terrestrial network and the satellite share the same frequency and the same signal is transmitted on the two components. The signal waveform is OFDM and the preambles of both components consist of a P1 plus an aP1 symbol. The OFDM parameter set is applicable to both components: terrestrial and satellite. Alternatively, the base profile can be adopted for both components. In that case the P1 part of the preamble of both components consists of a P1 symbol only.
2. *MFN, OFDM.* The satellite signal is transmitted on different frequencies and OFDM is used on both components. The terrestrial component is transmitted according to the base profile, the satellite component according to the OFDM settings. The preamble of the terrestrial component consists of a P1 symbol and

the preamble of the satellite component consists of a P1 plus an additional P1 (aP1) symbol.

3. *SFN, SC-OFDM.* This case consists of the satellite coverage and terrestrial gap fillers sharing the same frequency of the satellite signal. The SC-OFDM settings are applicable to both components: terrestrial and satellite. Preambles consist of P1 plus aP1 symbols for the satellite and the terrestrial component.
4. *MFN, SC-OFDM on Satellite Component, OFDM on Terrestrial Component.* The terrestrial component is configured in line with the base profile, the satellite component using the permitted SC-OFDM settings. The preamble of the terrestrial component consists of a P1 symbol and the one of the satellite component of a P1 plus an aP1 symbol.

DVB-NGH includes multiantenna techniques as the key technology to cope with increasing demands for data rate and transmission reliability in digital mobile broadcasting systems. The system coverage area can be greatly extended with a high-robustness transmission through the MIMO rate 1 code by reusing the current transmitter infrastructure and having backward compatibility with single-antenna receivers. If delivery of high-data-rate services is the primary goal and no backward compatibility is needed, the MIMO rate 2 code can be used, which allows for even higher data rates through spatial multiplexing. The MIMO rate 2 scheme can best suit the outdoor medium- to high-signal-strength user cases.

10.6 SUMMARY

This chapter has described the technological features in detail of four multimedia mobile broadcasting standards: DVB-H, ATSC-M/H, CMMB, and DVB-NGH. A high-level comparison of the core technologies used in the physical layer of each standard is listed in Table 10-5.

As can be seen from Table 10-5, except for ATSC-M/H, which employs the VSB modulation technique of the single carrier to be compatible with the existing ATSC system, the other three standards adopt OFDM technology.

In the choice of channel coding scheme, DVB-H and ATSC-M/H inherit the encoding mode of the original system, i.e., concatenation of RS code and convolutional code, while CMMB and DVB-NGH use LDPC code as the inner code to enhance data transmission performance.

The time-slicing technology is used for all four standards to reduce the power consumption at hand-held terminals. The difference is that the time slicing is performed in the data link layer for the DVB-H system but in the physical layer for the other systems.

TABLE 10-5 Comparison of Parameters of Various Standards

Mobile Multimedia Broadcasting Standard Name	Main Physical Layer Technologies				
	Modulation	Constellation Mapping	Channel Coding	Multiservice Support Scheme	System Bandwidth (MHz)
DVB-H	OFDM	QPSK/16QAM	RS code + convolutional code	Time slicing	8
ATSC-M/H	VSB	8VSB	RS code + convolutional code	Time slicing	8
CMMB	OFDM	BPSK/QPSK/16QAM	LDPC + RS code	Time slicing	8 or 2
DVB-NGH	OFDM + MIMO	QAM/NU-QAM	LDPC + BCH code	Time–frequency slicing	1.7–20

REFERENCES

1. Chinese National Standard GB 20600-2006, *Framing Structure, Channel Coding and Modulation for Digital Television Terrestrial Broadcasting System*, Aug. 2006.
2. CMMB, *Mobile Multimedia Broadcasting Part 1: Broadcast Channel Frame Structure, Channel Coding and Modulation*, State Administration of Radio, Film and Television of China, Beijing: Oct. 2006.
3. DAB, *Specification of 30MHz–3000MHz Terrestrial Digital Audio Broadcasting System*, State Administration of Radio, Film and Television of China, Beijing: May 2006.
4. DVB Document A133, *Implementation Guidelines for a Second Generation Digital Terrestrial Television Broadcasting System (DVB-T2)*, Geneva: Digital Video Broadcasting (DVB), Feb. 2009.
5. C. Rhodes, "Some recent improvements in the design of DTV receivers for the ATSC standard," *IEEE Transactions on Consumer Electronics*, vol. 48, no. 4, pp. 938–945, Jan. 2002.
6. ETSI EN 302 304, *Transmission System for Handheld Terminals (DVB-H)*, Geneva: Digital Video Broadcasting (DVB), Geneva: Nov. 2004.
7. Advanced Television System Committee A/153. *ATSC Mobile/Handheld Digital Television Standard, Part 1—Mobile/Handheld Digital Television System*, Washington, DC: ATSC, 2009.
8. Advanced Television System Committee A/153, *ATSC Mobile/Handheld Digital Television Standard, Part 2—RF/Transmission System Characteristics*, Washington, DC: ATSC, 2009.
9. DVB-NGH, *Next Generation Broadcasting System to Handheld*, Geneva: Digital Video Broadcasting (DVB), Nov. 2012.
10. D. Vargas, D. Gozálvez, D. Gómez-Barquero, and N. Cardona, "MIMO for DVB-NGH, the next generation mobile TV broadcasting," *IEEE Communications Magazine*, vol. 51, no. 7, pp. 130–137, July 2013.
11. D. Gómez-Barquero, *Next Generation Mobile Broadcasting*, Boca Raton, FL: CRC Press, 2012.

INDEX

Digital Terrestrial Television Broadcasting: Technology and System, First Edition. Edited by Jian Song, Zhixing Yang, and Jun Wang.

D

THE COMSOC GUIDES TO COMMUNICATIONS TECHNOLOGIES

Nim K. Cheung, *Series Editor*
Thomas Banwell, *Associate Editor*
Richard Lau, *Associate Editor*

The ComSoc Guide to Next Generation Optical Transport: SDH/SONET/OTN
Huub van Helvoort

The ComSoc Guide to Managing Telecommunications Projects
Celia Desmond

WiMAX Technology and Network Evolution
Kamran Etemad and Ming-Yee Lai

An Introduction to Network Modeling and Simulation for the Practicing Engineer
Jack Burbank, William Kasch, and Jon Ward

The ComSoc Guide to Passive Optical Networks: Enhancing the Last Mile Access
Stephen Weinstein, Yuanqiu Luo, and Ting Wang

Printed in the United States
By Bookmasters